T0310278

Principles of Managerial Statistics and Data Science

# Principles of Managerial Statistics and Data Science

*Roberto Rivera*

*College of Business*
*University of Puerto Rico-Mayaguez*

This edition first published 2020
© 2020 John Wiley & Sons, Inc. All rights reserved.

All rights reserved. No part of this publication may be reproduced, stored in a retrieval system, or transmitted, in any form or by any means, electronic, mechanical, photocopying, recording or otherwise, except as permitted by law. Advice on how to obtain permission to reuse material from this title is available at http://www.wiley.com/go/permissions.

The right of Roberto Rivera to be identified as the author of this work has been asserted in accordance with law.

*Registered Office*
John Wiley & Sons, Inc., 111 River Street, Hoboken, NJ 07030, USA

*Editorial Office*
111 River Street, Hoboken, NJ 07030, USA

For details of our global editorial offices, customer services, and more information about Wiley products visit us at www.wiley.com.

Wiley also publishes its books in a variety of electronic formats and by print-on-demand. Some content that appears in standard print versions of this book may not be available in other formats.

*Limit of Liability/Disclaimer of Warranty*
While the publisher and authors have used their best efforts in preparing this work, they make no representations or warranties with respect to the accuracy or completeness of the contents of this work and specifically disclaim all warranties, including without limitation any implied warranties of merchantability or fitness for a particular purpose. No warranty may be created or extended by sales representatives, written sales materials or promotional statements for this work. The fact that an organization, website, or product is referred to in this work as a citation and/or potential source of further information does not mean that the publisher and authors endorse the information or services the organization, website, or product may provide or recommendations it may make. This work is sold with the understanding that the publisher is not engaged in rendering professional services. The advice and strategies contained herein may not be suitable for your situation. You should consult with a specialist where appropriate. Further, readers should be aware that websites listed in this work may have changed or disappeared between when this work was written and when it is read. Neither the publisher nor authors shall be liable for any loss of profit or any other commercial damages, including but not limited to special, incidental, consequential, or other damages.

**Library of Congress Cataloging-in-Publication Data:**
Names: Rivera, Roberto (Associate professor), author.
Title: Principles of managerial statistics and data science / Roberto
    Rivera.
Description: Hoboken, NJ : Wiley, 2020. | Includes bibliographical
    references and index.
Identifiers: LCCN 2019032263 (print) | LCCN 2019032264 (ebook) | ISBN
    9781119486411 (hardback) | ISBN 9781119486428 (adobe pdf) | ISBN
    9781119486497 (epub)
Subjects: LCSH: Management–Statistical methods. | Mathematical statistics.
    | Statistical decision. | Data mining. | Big data.
Classification: LCC HD30.215 .R58 2020 (print) | LCC HD30.215 (ebook) |
    DDC 519.5–dc23
LC record available at https://lccn.loc.gov/2019032263
LC ebook record available at https://lccn.loc.gov/2019032264

Cover Design: Wiley
Cover Images: © boutique/Shutterstock-Abstract Background, Courtesy of Roberto Rivera-graph

Printed in the United States of America.

V10016615_122719

*To Katka, Luca and Kai*

# Contents

# Preface

There is a wide array of resources these days that introduce Statistics and data science. Textbooks, videos, online courses, web-based interactive platform, and even mobile apps are available. This leads to the question of whether a new textbook is necessary. Yet, it is not difficult to argue that the answer is yes. Three main reasons should be emphasized. In order of importance, the recommendations on how to teach Statistics have not been adopted fully; the new fields of big data, data analytics and data science, have come into play; and the field of Statistics has changed.

The 2016 Guidelines for Assessment and Instruction in Statistics Education (GAISE) College Report emphasizes six recommendations to teach introductory courses in statistics. Among them, teach statistical thinking, focus on conceptual understanding, and use of real data with context and purpose. Multiple textbooks have succeeded in teaching statistical thinking. There has also been some progress on addressing conceptual understanding, although there is still some vagueness in the explanations of concepts. Furthermore, although some interesting examples and exercises have been made available to students, the "context and purpose" component of the data remains elusive. Most of the time, data sets (two words) are partially generic or heavily filtered and subsetted. One avenue that shows promise is the launching of open data portals by many cities around the world. By using local data, students become more engaged. Moreover, open data can be messy, may need preprocessing, and can sometimes be very large.

It has become less common to see in practice data sets with a few observations. In part thanks to digitalization, data sets can have hundreds, thousands of variables and millions of observations. For example, the City of Chicago has made available taxi ride data. The data set has over 113 million rides. We work with this data set in Section 14.9.1. The use of very large data sets brings new challenges and new opportunities for visualization. In Chapter 3, we present an animation to visualize over 14 years, 55 Baltimore housing observations on two variables. Population data is also included. Most examples and case studies based on open data used an open source software called R. Instructors and students are encouraged to reproduce results with the codes made available in the companion website. Alternatively, the codes may be adapted to applications with regional data of interest to the user. The fact that the field of Statistics is constantly changing is less important considering that our focus is introductory statistics. Still, some not-so-new changes in the field, like Bayesian statistics and resampling, deserve some discussion, even if superficial.

Chapter 1 provides an overview of applications of statistical methods. Chapters 2–5 introduce students to descriptive statistics, Chapters 6 and 7 present useful probability distributions, and Chapter 8 covers important concepts to understand statistical inference. The rest of the chapters explains basic procedures to perform statistical inference. Students will notice that from Chapter 11 onwards, there is a significant reduction in the use of mathematical formulation (and thus, less computation by hand) and more reliance on statistical software. Our opinion is that by presenting too much math, particularly in the more advanced topics, the objective of developing statistical literacy is hindered. As a remedy, emphasis is given to real data and interpretation of statistical software output.

# Acknowledgments

I would like to thank the University of Puerto Rico for their support on helping me finish this book. Special props must go to colleagues Maria Amador, Darik Cruz, Rosario Ortiz, and Mariel Nieves. Bodapati Gandhi and Herminio Romero gave great feedback on an early version of the manuscript. Graduate students Walter Quispe, Daniel Rocha, and Adriana Lebron helped immensely. Luenisse Rodriguez, Ruben Marin, Gabriel Ortiz, and Natalia Font also helped with early drafts of the book. The Wiley team contributed immensely during all stages of making the book happen: Kathleen Santoloci, Mindy Okura-Marszycki, Karthiga Mani, Elisha Benjamin, Grace Paulin, Gayathree Sekar and many others. I would also like to thank my mom Nora Santiago, who raised three kids all on her own. Last but not least, I must give thanks to my wife Katka for her love, patience, and delicious (yet healthy!) salads.

Aguadilla, Puerto Rico *Rob Rivera*
March 2019

## Acronyms

| | |
|---|---|
| ANOVA | Analysis of Variance |
| IQR | Interquartile Range |
| MAE | Mean Absolute Error |
| MAPE | Mean Absolute Percentage Error |
| pdf | Probability Density Function |
| RMSE | Root-Mean-Squared Error |
| VIF | Variance Inflation Factor |

# About the Companion Website

This book is accompanied by a companion website:

www.wiley.com/go/principlesmanagerialstatisticsdatascience

The website includes the following materials for students and instructors:

## Students

- PowerPoint Presentation Slides
- R Code to reproduce all data science case studies and some examples
- Portal Information
- Over 100 data files
- Motion charts to answer some problems.

## Instructors

- Instructors manual
- Undergraduate and Graduate Syllabus models

# Principles of Managerial Statistics and Data Science

This document serves as a guideline designed to help how to plan, conduct, analyze, and write a statistics project. The end goal is a cohesive, well written paper using statistical methods covered in class as a tool to convey insight garnered from data.

## Project Objective

The main objective of the project is to learn how to organize, analyze and write using statistical information. Data collection is not expected although students may potentially bring their own data. By attempting to do research, a student becomes aware of the problems that all researchers must face. Problems such as data wrangling, accurate reporting and recording, and depth of analysis are some examples. The student becomes a more critical consumer of data based information by completing a project.

Use of statistical software for the analysis is expected. No restrictions are given on the software to be used. Some well known software that allows statistical analysis includes, Excel, Minitab, SPSS, Stata, SAS and R/RStudio (the latter can be obtained for free over the internet) but there are many others.

### Project Topics

The project should focus on a data set NOT covered in class or a data set used in class **but applying a different analysis**. For example: assessing whether arrests (a variable available in the San Diego police stop data set) by San Diego police after a stop are associated to driver's race. Students should meet with the instructor or TA to discuss their term project. Here are some suggestions for various approaches to the project:

- Apply a method covered in class to a different open data set which may require a bit of adaptation or ingenuity.

- Investigate a method that is an extension of a topic we have covered. Potentially apply the method to data (examples: nonparametric regression, bootstrapping, data mining, simulation in business and Bayesian inference).
- Investigate the performance or sensitivity of a method across a class of problems and suggest/investigate adaptations for cases where performance is degraded.
- Use several methods to solve a challenging problem. Compare/contrast the results and the
- advantages/disadvantages of the methods.
- Develop the theory for a new method or extend an existing result. Apply the method to a small example.

## Format

- Recommended: at least 7 pages long of double spaced text (10 point arias font). This limit does not include figures, tables, appendices and other supporting documentation. Include the output (and final computer code if necessary) in an appendix. In the write-up, the student should describe the purpose of the project, discuss why the project is important, and provide at least one example (or data analysis) worked demonstrating the work.

One possible paper format would be: introduction, background, analysis or extension to method discussed in class, example(s), conclusions and references. At least 5 references are expected.

# 1

# Statistics Suck; So Why Do I Need to Learn About It?

## 1.1 Introduction

The question is a fair one, considering how many people feel about the subject. In this chapter, we go over some interesting aspects of the topic to convince the reader of the need to learn about statistics. Then we provide a formal definition.

*Statistics Are All Around Us*   Nowadays, we see and hear about statistics everywhere. Whether it is in statements about the median student loan amount of college students, the probability of rain tomorrow, association between two financial variables, or how an environmental factor may increase the risk of being diagnosed with a medical condition. These are all brief examples that rely on statistics. Information extracted from data is so ubiquitous these days that not understanding statistics limits our comprehension of the information coming to us. Therefore, understanding the main concepts from statistics has become imperative. In the world of business, making decisions based on evidence leads to a competitive advantage. For example, retailers can use data from customers to determine appropriate marketing campaigns, and manufacturing experiments can help establish the best settings to reduce costs. Investors can combine assets according to an adequate expected return on investment with an appropriate risk. We go over some specific applications of statistical methods in business in Section 1.2.

*The Importance of Understanding the Concepts*   Suppose you are invited to invest in a fund with an expected yearly return of 10%. Therefore, if you invest $100 at 10% yearly return, you expect to have $110 by the end of the year. So you decide to invest money; but at the end of the year, your portfolio is worth $90 (ignoring all fees). Specifically, the return on your money has been negative. That raises the question, based on the information you were given about the return you would receive, was your investment a mistake? Think about this question before reading ahead. The trick is in understanding the meaning of expected value. The information provided to you was that the fund had an "expected" yearly return

*Principles of Managerial Statistics and Data Science*, First Edition. Roberto Rivera.
© 2020 John Wiley & Sons, Inc. Published 2020 by John Wiley & Sons, Inc.
Companion website: www.wiley.com/go/principlesmanagerialstatisticsdatascience

of 10%. In some scenarios, such as a deposit account in a bank, the return given to you is guaranteed, and your principal (the money you start with) is protected. But in other scenarios, such as the stock market, the return you receive is not guaranteed, and, in fact, your principal may not be protected. Consequently, in any given year, you may have an outstanding return, much higher than the expected 10%, or you may get a return lower than the expected return of[1] 10%. To counter this risk, possible returns are higher in these investments than safer alternatives, such as a bank savings account. The expected return is what the company has determined you should receive over the long term. Because of this aspect, an investment such as a stock has to be looked at differently than a bank savings account, and if after a year your portfolio is worth $90, your decision to invest in the portfolio was not necessarily a bad decision. What matters is the return you obtain over the long term, say, 10 or 20 years, and how much risk you are willing to take as an investor.

Another relevant component in understanding statistics is being able to distinguish between good information and bad information. An example is a spurious relationship. A spurious relationship between two variables occurs when an association is not due to a direct relationship between them, but from their association with other variables or coincidence. Strange associations are not the only way we can wrongly use statistics. Many media outlets (and politicians) often misquote scientific studies, which often rely on statistical methods, to reach their conclusions.

**Example 1.1** *In the summer of 2014, several media outlets reported on a new scientific study; one of the titles was*[2] *"Study: Smelling farts may be good for your health." In reality the study was about a developed compound that delivered small amounts of hydrogen sulfide that helped protect cells.*[3] *Hydrogen sulfide is known to be a foul-smelling gas; hence the misunderstanding of the study results occurred. It is not enough to be able to correctly perform statistical analysis, but understanding the results is also key.*

The issues brought up are rather common and tend to occur for two reasons:

- Inappropriate use of statistics.
- Wrong interpretation of results (poor statistical literacy skills).

Either of these two issues will potentially lead to reaching a wrong conclusion, hence making the wrong managerial decision. In contrast, avoiding these issues offers valuable knowledge for a given context.

---

1 Actually, in the case of investments such as stocks, it is often possible to obtain a measure of the risk of that financial asset. Not all assets offering an expected 10% return are the same. We will go into this in more detail later.
2 Source: *The Week Magazine*, July 11, 2014. *The Guardian* and other media outlets had similar reports.
3 www.nbcnews.com (accessed July 26, 2019).

---

**Satiated Judge, More Lenient Ruling**

Did you know that judges used to give more lenient decisions after meals compared to before meals?[4] The finding comes from a study that looked at 1112 judicial rulings on parole. It was found that judge rulings were more likely to be favorable to the prisoner earlier in the day and that there was a jump in probability of parole ruling after a break. The study even considered the length of sentence, incarceration history of the prisoner, etc. Of course, the time of day of a parole hearing should not be a factor in determining a decision. The findings of this study helped establish a protocol to help eliminate the impact of time of day on parole ruling. Thus, an unintentional benefit of reading this book is that if you are facing the possibility of jail time, you should ask your lawyer to delay the sentencing until after lunch, just in case!

---

**Case Study 1**   A professor wants to answer the following question: Do quizzes help business undergraduate students get a better grade in their introductory statistics course? In academia, it is often simply assumed that quizzes will force students to study, leading to better course grades. The goal here is to answer the question based on evidence. There are two possible answers:

- They do not work: No difference between grades of students who took quizzes and students who did not take quizzes, or students who took quizzes perform worse.
- They do work: Grades of students who took quizzes are better than students who did not take quizzes.

Next is to determine how the comparison will be made. More specifically, what aspects associated with students will be taken into account? Hours of study per week, undergrad concentration, high school they went to, and the education level of parents are just some examples of "variables" that could be considered. Yet, the simplest way to make the comparison is based on two groups of students:

- One group that took quizzes during the course.
- The other group that did not take quizzes during the course.

Then, it is assumed that both groups are homogeneous in all their attributes, except that one group of students took the course with quizzes, while the other did not. This way, the average exam scores of both groups can be compared.

To compare average exam scores, two semesters were selected: one with quizzes and the other without. It was verified that the number of students that

---

4 For the statistical evidence see www.pnas.org (accessed July 26, 2019).

dropped the course or stopped attending class for both groups was similar (why do you think it is necessary to check this?).

- The group that took quizzes had an average test score of 79.95.
- The average of the group that had no quizzes was 75.54.

Now, these averages are just estimates, and different semesters will have different students in each group leading to new average values. This leads to uncertainty. Informally, what is known as statistical inference accounts for this uncertainty. When the inferential procedure was implemented, it was found that in reality there was no difference in average exam score among the two groups. Thus, there was no evidence that quizzes helped students improve their test scores.

Now, let's not get too excited about this. The interpretation is that quizzes do not help improve the performance of students who take the class with that professor. The conclusion does not necessarily extend to quizzes in other subjects, and it doesn't even necessarily extend to all introductory business statistics courses.

## Practice Problems

1.1 In Case Study 1, it was stated that "It was verified that the number of students that dropped the course or stopped attending class for both groups were similar." Why do you think it is necessary to check this condition?

1.2 In September 20, 2017, Puerto Rico was struck by Hurricane Maria, a powerful storm with 155 mph sustained winds. In May 2018, multiple media outlets reported the findings of a study[5] in which researchers used September to December 2017 data to estimate that 4645 people had died in Puerto Rico, directly or indirectly, due to Hurricane Maria. The news shocked many, since at the time the local government had stated a total of 64 people had died due to the storm. Indeed, the study was one of several indicating the official estimate of 64 was too low. Before the news broke, the Puerto Rican government had not publicly shared death certificate[6] data. Due to growing pressure from the study results, the government released data on the number of death certificates (Table 1.1). How does the released data suggests that the 4645 number may have been wrong?

    *Hint:* Compare death certificates after Hurricane Maria to the ones before.

---

5 Kishore, Nishant, Marqués, Domingo, Mahmud, Ayesha et al. (2018). Mortality in Puerto Rico after Hurricane Maria. *New England Journal of Medicine*, 379, pages 162–170.
6 A government must create a death certificate for every single death.

Table 1.1 Death certificates in Puerto Rico by month and year. Causes of death not provided. Numbers in bold include death certificates after Hurricane Maria.

| Month | Year | | | |
|---|---|---|---|---|
| | 2015 | 2016 | 2017 | 2018 |
| January | 2744 | 2742 | 2894 | **2821** |
| February | 2403 | 2592 | 2315 | **2448** |
| March | 2427 | 2458 | 2494 | **2643** |
| April | 2259 | 2241 | 2392 | **2218** |
| May | 2340 | 2312 | 2390 | – |
| June | 2145 | 2355 | 2369 | – |
| July | 2382 | 2456 | 2367 | – |
| August | 2272 | 2427 | 2321 | – |
| September | 2258 | 2367 | **2928** | – |
| October | 2393 | 2357 | **3040** | – |
| November | 2268 | 2484 | **2671** | – |
| December | 2516 | 2854 | **2820** | – |

*Source:* Demographic Registry Office, released in May 2018.

## 1.2 Data-Based Decision Making: Some Applications

Data-based decision making is a relatively new principle in management. Some people also use the term evidence-based management to mean data-based decision making, while others state that the latter is part of the former. We will use the term data-based decision making to avoid confusion. Researchers have found that the more data driven a firm is, the more productive it is, and the benefit can be substantial.

---

**Making Open Data Profitable**

In 2006, The Climate Corporation was founded. They obtained government data from weather stations, Doppler radar, US Geological Survey, and other freely accessible sources. Originally, the company provided weather insurance to clients. The company analyzed weather data, forecast future weather, and created an insurance policy from their analysis. A business would go to The Climate Corporation's website and buy insurance against bad weather nearby. Since 2010, The Climate Corporation focuses exclusively on agriculture. Monsanto acquired the company in 2013 for approximately $1 billion.

---

In marketing, statistical methods are applied to better understand the needs of customers, to create a profile of consumers, or to market products and services efficiently to potential consumers. Auditors sample company transactions to infer about its internal control system. In finance, statistics play a role in the **actuarial field**, where actuaries help control risk for banks, insurance companies, and other entities. **Financial engineering**, a field where investment strategies are developed, draws heavily from statistics.

---

**CityScore – Boston**

CityScore[7] is an initiative to inform city managers about the overall health of Boston. After its launch, the Boston Transportation Department noticed that the on-time percentage for sign installation from December 1, 2015 through January 14, 2016 was 73%, 7% below target. Looking into it, administrators discovered a backlog of 90 sign installation requests. Within seven days, the request backlog was reduced from 90 to 7 requests.

---

Making decisions based on data does come with challenges. Besides the obvious need to sometimes deal with massive amounts of data in a computationally effective and theoretically sound way, privacy issues have been raised, and failures have been pointed out.

Moreover, although nowadays there is access to data through open data portals, universities, government agencies, companies, and other entities, it is quite common to make data available without describing the exact source of the data and how it was gathered. These intricacies of the data set must not be taken for granted.

---

**Intricacies of Data: Google Trends**

Studies have exemplified the use of search query data extracted from Google Trends (www.google.com/trends) to forecast processes of interest. Search query data has been used to model tourism demand, auto sales, home sales, initial unemployment claims, influenza activity in the United States, consumer behavior, dengue fever, and more. However, the search query volume data given by Google Trends for a fixed past time period, in fact, varies periodically. That is, if one uses the tool today to gather search volume data for a given query in January 2006, one may find a volume of, say, 58. But when using the same tool under the same settings next week to gather search volume data in January 2006 for the same given query, one may find a volume of, say, 65, and the following week some other result. In their website Google only mentions that the search query volume data comes from a sample of the total search queries. Yet no information is given on sample size, sampling technique, and other important details of the search query volume data. Some researchers have found that even when

---

7 www.boston.gov (accessed July 26, 2019).

accounting for the search query volume data uncertainty, from a practical point of view, incorporating search query volume data is not universally better than simpler forecasting alternatives.[8]

Search query volume is an example of **proprietary data**, data shared by a private company. Companies have two strong reasons not to provide too much information about their proprietary data. First, it can threaten their competitiveness. Second, it would give the opportunity to others to "game the model."

*Leadership and Statistics*  We have presented a wealth of examples to convince the reader of the importance of statistics in making managerial decisions. In addition to the importance of evidence-based decisions being emphasized, we must acknowledge that to ensure the best gain from data, a company must have the proper environment. Designing a study, gathering data, and analyzing the results must be done efficiently. To do all these tasks adequately, good leadership is needed. Bad leadership leads to poor organization, and poor organization makes it harder to base decisions on evidence since it would be harder to filter out unwanted "noise."

**Example 1.2**  *Let's suppose that a company wants to determine the association between square feet of land and its value. There are two separate sources, source 1 and source 2, of land value and land size data from 75 purchase transactions. Each source collects land value differently so the entries vary. Both sources should imply the same type of association between these land metrics. But source 1 has been careful in their data collection procedure, while source 2 had poor record keeping, and the data was not preprocessed for errors. Figure 1.1 presents the association between land size (x-axis) and land value (y-axis) using data from source 1 and source 2, respectively. According to the data from source 1, there is approximately a linear association between land size and land value: as land size increases, on average, land value increases. On the other hand, due to the inefficiency-driven errors in the land value, source 2 has a lot of "noise," which has led to a wider "scatter" of the data. As a result, it is hard to tell if there is an association between land size and land value at all when using source 2 data.*

We cannot always control the error in measurements, but this example shows us that if we could, this would have a dramatic impact on the usefulness of our data to make decisions.

*Setting up the Proper Environment*  Once the leadership has recognized the benefits of evidence-based decisions, they must ensure the firm has the right structure to exploit data. Some points to consider are as follows:

---

8 Rivera, R. (2016). A Dynamic Linear Model to Forecast Hotel Registrations in Puerto Rico Using Google Trends Data. *Tourism Management*, 57, pages 12–20.

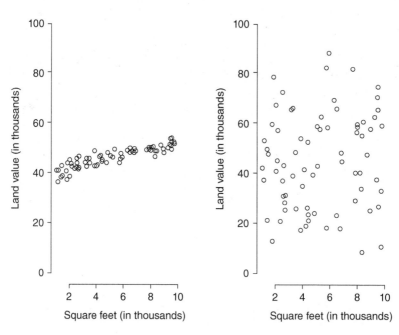

Figure 1.1 Land size versus land value from source 1, with little error in measuring land value (left), and from source 2, with large error in measuring land value (right).

- Does the entity have the right personnel? (To design a study, analyze data, etc.)
- Are hardware and software capabilities adequate?
- Is there good communication between offices that will need to work together?
- Has personnel been properly trained or oriented about any changes in procedures that may occur?
- Has a data management protocol been developed? (See Section 2.5.)

***Ethics and Data-Based Decision Making*** One cannot discuss making decisions based on data without talking about ethics. Several aspects must be considered to make decisions ethically, based on data. Some of the aspects apply in general, while others only apply in some situations. Specifically, the team must ensure the data is not misrepresented (intentionally or not), and they should avoid misinterpreting the results. For some analysis, researchers require the approval of boards that carefully evaluate the design of the process or the implementation of the work to make sure human subjects and animals are safe.

## 1.3    Statistics Defined

**Statistics** can be defined as the science and art of collecting, organizing, presenting, summarizing, and interpreting data. The inclusion of the word "art" in the definition is not an error. As we will see, there is strong theoretical evidence backing much of the statistical procedures covered in this book. However, in practice, the implementation of statistical methods requires decisions on building plots, organizing data, and relying on rules of thumb that make statistics also an art, not just a science.

The statistical tools at our disposal fall into two branches: descriptive statistics and inferential statistics. **Descriptive statistics** organize, present, and summarize data. For example, consider domestic violence data[9] from Edmonton, Canada. It provides number of reported criminal and noncriminal occurrences among intimate partners. The information is provided in a quarterly basis, from 2009 until the third quarter of 2018. Figure 1.2 displays a snapshot of the original data online.

Summarizing the data helps us make sense of it. Which type of data description to apply depends on the type of data, a topic that we discuss in Chapter 2. One useful description of this data is the mean criminal domestic violence cases reported per each quarter. Visual descriptive summaries are also extremely useful. One alternative here is to chart the number of criminal domestic violence cases through time, allowing us to explore whether the cases are changing in time.

Informally, **inferential statistics** are procedures that allows us to use a subset from a group of interest to reach conclusions about the group overall. This

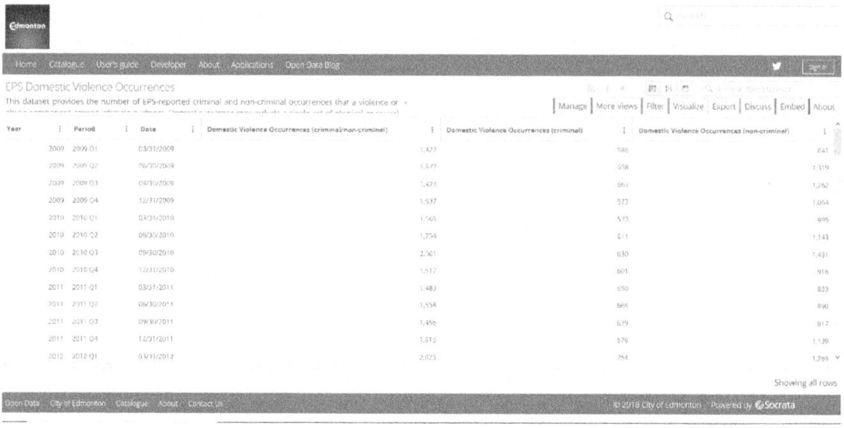

Figure 1.2    Quarterly domestic violence reported in Edmonton, Canada.

---

9 dashboard.edmonton.ca (accessed December 12, 2018).

subset is called a **sample**. Statistical inference is incredibly valuable to make decisions based on evidence, and we have seen many examples of how this is performed in the real world. In fact, even babies have been found to perform statistical inference.

---

**Babies Doing Statistics**

Laura Schulz and colleagues at MIT performed an experiment where 15-month-old babies were randomly assigned to two groups.[10] Randomization guaranteed that the characteristics of the babies were generally similar in both groups. Each group of toddlers saw how a research assistant:

- Removed three blue balls one at a time from a see-through box.
- Squeezed each blue ball for their squeak effect.
- And then handed a yellow ball from the box to the child.

The babies did not know that yellow balls did not squeak.

- In group 1, the baby could see that the see-through box had mostly blue balls.
- While in group 2, the baby could see that the see-through box had mostly yellow balls.

The study uses a sample of babies to infer about 15-month-old babies overall. It was concluded that, on average, babies from group 1 were more likely to try to squeeze the yellow ball handed to them than those from group 2. The researchers argue that this happens because, since the box from group 1 had mostly blue balls, babies saw the assistant's three blue ball sample as representative of the population. In contrast, since the box from group 2 had mostly yellow balls, babies did not see the assistant's three blue ball sample as representative of the population. In a way, babies in group 2 were able to determine that the assistant was "cherry-picking" the blue balls that could squeak.[11]

---

Descriptive statistics and inferential statistics are complementary. The former is always performed on data, while the latter tends to be performed when the data does not represent the entire group of interest.

***Statistics and Critical Thinking*** Critical thinking is clear, rational thinking that is informed by evidence. Evidence can be generated through

- observations
- experiments
- memory
- reflection (opinion)
- reason (theory)

---

10  See www.ted.com (accessed July 26, 2019).
11  Source: www.ted.com (accessed July 26, 2019).

Critical thinking pertains to the evaluation of the evidence to build – perhaps adjust – a belief or take action. Our emphasis is evidence gathered through data, and within this context, statistics is an essential part of critical thinking, allowing us to test an idea against empirical evidence. Also, as implied by some of the examples encountered earlier in this chapter, statistical literacy is required for proper interpretation of results, and critical thinking plays an important role in evaluating the assumptions made to conduct statistical inference. Given that critical thinking evaluates information, statistics allows us to do better critical thinking. At the same time, statistical inference requires making assumptions about the population of interest, and hence, critical thinking will allow us to do better statistics.

## 1.4 Use of Technology and the New Buzzwords: Data Science, Data Analytics, and Big Data

There is a wide selection of software available to perform statistical analysis. Computers have also become faster, allowing for new methodologies to be used and applications on bigger data sets. These technologies make the use of statistical procedures much easier, sometimes too easy. Specifically, just because a computer software creates a figure does not mean the figure is correct. It is up to the user to ensure that they choose the right type of chart for the data. Similarly, a computer software may calculate a statistic for you, but it will not necessarily determine if the statistic is adequate for your type of data or objective. Furthermore, computer software does not check all the assumptions that a statistical procedure entails. And the more complex the statistical procedure is, the greater the need for an expert to implement the procedure correctly. The misconception that with just the right computer software (or adequate programming skills), one can perform any type of statistical procedure without feedback from an expert in statistics is common in business, and it has become a topic of debate in the big data/data analytics world. Overall, advanced procedures should always be performed by a qualified data analyst.

### 1.4.1 A Quick Look at Data Science: Some Definitions

**Virus X Cure Found!**
Remarkably, recently a team of scientists was able to cure Virus X, a deadly antibiotic resistant disease in chickens. They reported that:

- Thirty-three percent of the chickens in the study showed improvement after treatment against the disease.

- One-third of the chickens showed no change.
- Unfortunately, the third chicken got away.

If you are confused with these results, do not fret. You should be confused. The results of the study, which did not really happen, help point out that evidence from very small data sets should not be trusted. Also, the joke is a way to introduce the concepts of data science and big data. A decade ago, **business intelligence** – methods to process data and influence decision making – was all the talk. But in the last few years, data science, big data, and data analytics have overtaken business intelligence in popularity.

**Big data** is defined as data that cannot be processed using traditional methods. Big data can hold petabytes[12] of information or may be sizable enough to not be processed through usual procedures. The field finds ways to query the data, store it, and analyze it.

**Example 1.3**   *Data from the computer-aided emergency management system (EMS) for New York City is available online.[13] The raw data includes almost five million records for incidents from January 1, 2013 until June 30, 2016.*

Although not massive in size (the original data file is a little over 1 gigabyte in size), this data is large enough to cause trouble. In fact, the raw data cannot be opened in most computers with software such as Excel or Minitab. It requires special software, and even then it is best to filter out unnecessary information from this data set at the beginning stages of the analysis.

**Data analytics** involve statistical methods to retrieve useful information from data. The main distinction it has from statistics is that in the latter, data is used to answer a very specific question, while in data analytics, data may be "mined" to gain insight. A data analytics branch called **predictive analytics** has become popular to forecast future observations.

**Example 1.4**   *In a 2018 working paper,[14] David Andrew Finer, a graduate student at University of Chicago's Booth School of Business, used open data on over one billion New York City taxi trips to assess if there was systematic information leakage from the Federal Reserve Bank. Specifically, he extracted cab trips starting at commercial banks and at the New York Federal Bank that converged on the same destination around lunchtime and those directly from banks to the New York Fed late in the evening.[15] He found evidence of those journeys rising sharply*

---

12  A petabyte consists of 1024 terabytes, which in turn consists of 1204 gigabytes.
13  data.cityofnewyork.us (accessed July 26, 2019).
14  What insights do taxi rides offer into federal reserve leakage?
15  Source: "Mining Data on Cab Rides to Show How Business Information Flows," *The Economist*, March 2018.

*around the dates of meetings when interest rates were determined by the Federal Reserve's monetary-policy committee in Washington. This is one way data mining works, using data collected by the New York City Taxi and Limousine Commission for official purposes, to gain insight on something else.*

**Data science** is an umbrella term that refers to the science that enables data management to convert data into knowledge while providing computational capabilities to carry out statistical methods. Data science merges aspects from statistics and computer science into one field. The fields of big data and data analytics fall within data science.

Figure 1.3 outlines the data science process. It starts with a problem statement, which may be rather ambiguous (Example 1.4). This is followed by getting the raw data and if necessary data wrangling, steps that sometimes demand big data methods. **Data wrangling** are methods to process the data for analysis. This includes filtering unwanted information out, sorting, aggregating information, or transforming variables into new ones. Preliminary data analysis serves as a first attempt to summarize the data, numerically and visually. The data must be screened for errors or any issues that may arise. For example, the data must be cleaned up when categories are not clearly defined, or numerical information is nonsensical, or missing values are present. If the data issue limits the usefulness of some information, more data wrangling is implemented. The exploratory data analysis will summarize the processed data, serving as a first look of its usefulness. Further statistical procedures are applied on the basis of the problem statement, followed by interpreting the outcome. Next, the results

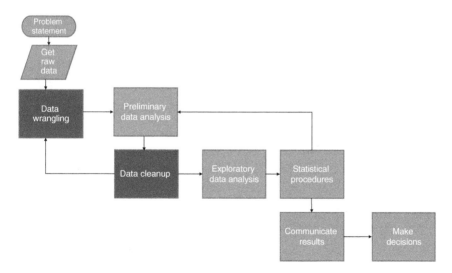

Figure 1.3 The data science process. Steps that are not always necessary (but most of the time are required) appear in purple.

are communicated to pundits. The importance of this step should not be underestimated, for it affects how pundits will make decisions from the results. This book highlights many parts of the exploratory data analysis and statistical procedure stages of data science, with parts of acquiring raw data, data wrangling, preliminary analysis, and data cleanup examined from time to time, especially in "Quick Look at Data Science" sections throughout the book.

## Chapter Problems

1.3 University administrators record data from their students. For example, using admission records, they determine that 1075 was the mean SAT score of the latest business program freshmen. Is the number provided part of descriptive statistics or inferential statistics?

1.4 A political candidate wants to know what are her chances of being elected in the coming elections. Her team takes a random sample of 2500 registered voters and ask them who they will vote for. Will the candidate's team need to perform statistical inference?

1.5 Provide examples of applying descriptive and inferential statistics.

1.6 In September 2018, Google launched a beta version of a search tool for data. Go to https://toolbox.google.com/datasetsearch, and enter domestic violence statistics. Download in CSV form one of the results.

1.7 Search online to see if cities near to your school have an open data portal. Download in CSV form data you consider interesting.

## Further Reading

Gurin, Joel. (2014). *Open Data Now*. McGraw-Hill.

Provost, Foster and Fawcett, Tom. (2013). *Data Science for Business: What You Need to Know About Data Mining and Data-Analytic Thinking*. O'Reilly.

Rivera, R., Marazzi, M., and Torres, P. (2019). Incorporating open data into introductory courses in statistics. *Journal of Statistics Education*. doi: 10.1080/10691898.2019.1669506

# 2

# Concepts in Statistics

## 2.1 Introduction

The first chapter served as an appetizer by discussing the importance of Statistics and the proper application of statistical methods. It also discussed the need to combine statistical procedures with critical thinking. But we need a deeper understanding of concepts about data, data collection, and statistics.

First things first, **data** is a collection of facts about multiple units. A person, a car, a company product are examples of what we call units. A particular value or fact of the data is known as an **observation**. The facts collected are typically described in words or numerical measurements. Each fact is termed informally as a variable. A **data set** collects several facts from multiple units. For the purpose of statistical analysis, it is best to organize the data set so that each row represents a series of facts on a unit and each column represents a variable.

**Example 2.1** *Chicago School's data for 2016–2017 is available online.*[1] *Below, we can see some of the observations from this data set. Some of the variables (e.g. School Type) are facts presented in words, while others (e.g. College Enrollment Rate) are numerical facts. Each row pertains to a school, while each column describes a fact or variable. The entire data set has over 660 rows and 90 columns.*

To extract useful information from this data set, we require statistics. Also worth noting is that some of the cells are empty, for example, the first few rows of college enrollment rate.[2] These are known as **missing values** and are one of the many issues one must deal with before extracting information from data (Figure 2.1).

---

1 Source of data: data.cityofchicago.org (accessed July 26, 2019).
2 For many of these schools, this data is missing because they are for children in lower grades.

*Principles of Managerial Statistics and Data Science*, First Edition. Roberto Rivera.
© 2020 John Wiley & Sons, Inc. Published 2020 by John Wiley & Sons, Inc.
Companion website: www.wiley.com/go/principlesmanagerialstatisticsdatascience

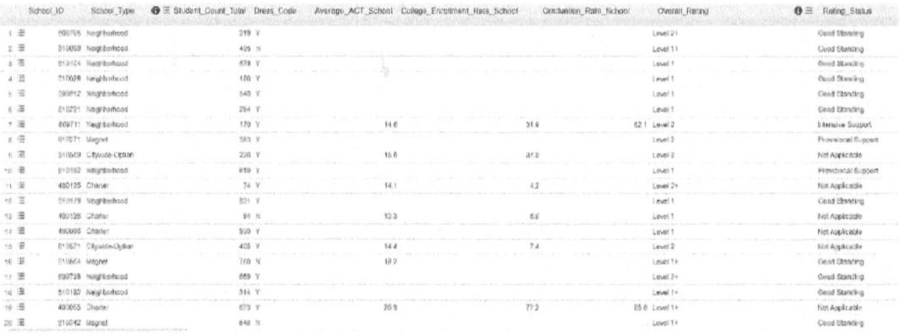

**Figure 2.1** A subset of the data available online for Chicago schools for the 2016–2017 school year.

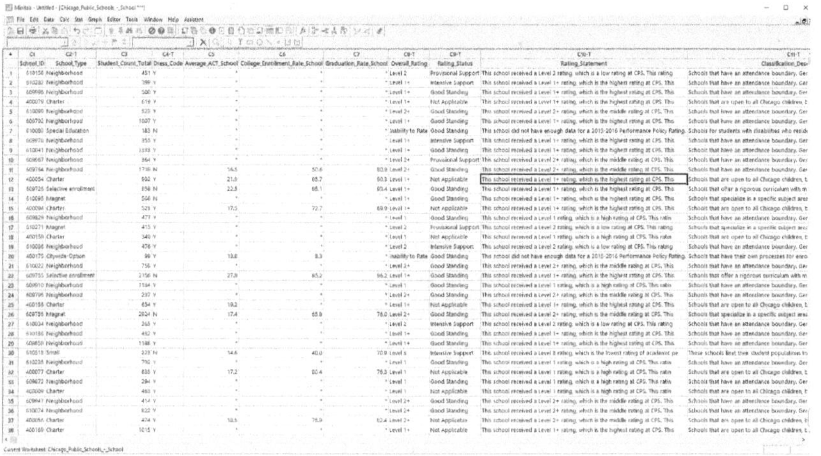

**Figure 2.2** Chicago school's data for the 2016–2017 school year as seen in Minitab.

This data set can be retrieved and opened with software such as Minitab (Figure 2.2). Besides the variable name, Minitab also has a column id, e.g. C1, which becomes useful when handling data sets with many variables.

---

**Unstructured Data**

The data definition provided in this section is more specifically referred to as **structured data**, collection of facts that can be set up in a table format. In contrast, **unstructured data** cannot be set up in a table format. Examples of unstructured data include photos, video feeds, and text from tweets or blogs. If data has no structure, it becomes challenging to gain insight from it. Thanks to digitalization, massive amounts (big data) of unstructured data are being generated; and the use of this type of data is becoming more commonplace. However, this book gives almost exclusive attention to structured data.

---

## Practice Problems

2.1 Data on daily units and legal sales (in $) of several marijuana products in Washington since July 2014 are available online.[3] A Minitab snapshot of the data is presented below.
  a) What variable does "C5" represent?
  b) How many "Marijuana Mix Packaged" units were sold May 16, 2016?
  c) How much was made in sales of "Marijuana Mix Packaged" on May 16, 2016?

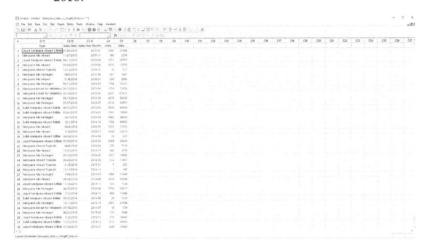

2.2 Below you will find 2015 data for several community statistical areas within Baltimore, as seen with Excel. The top row is the variable name, and the second row defines each variable. The rest of the rows give values of the variables.
  a) Identify the unit.
  b) What is the definition of "tanf15"?
  c) What is the "Percent of Babies Born with a Satisfactory Birth Weight" at Greater Govans?
  d) Which community has a higher value for "liquor15," Claremont/Armistead or Fells Point?

---

3 data.lcb.wa.gov (accessed July 26, 2019).

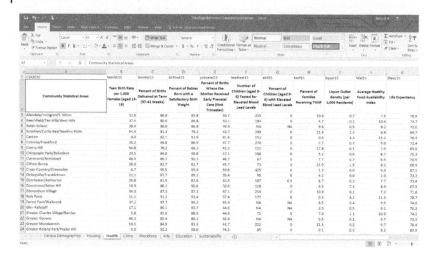

2.3 2012 mean total SAT scores per New York City school are available to the public.[4] The original data set provides location, number of test takers, and scores on Critical Reading, Math, and Writing.

   a) Open "NYC2012_SAT_Results.csv" using Minitab.

   b) If we wish to assess overall "average" math scores in New York City, why would you argue that schools with few students, say, less than 30 students, should not be used?

2.4 It is not always straightforward to open data files using statistical software.

   a) Open "Chicago_Public_Schools_-_School_Profile_Information_ SY1617.csv" using Excel.

   b) Open "Chicago_Public_Schools_-_School_Profile_Information_ SY1617.csv" using Minitab. How does the data set compare with part (a)? Anything strange?

   c) Save "Chicago_Public_Schools_-_School_Profile_Information_ SY1617.csv" using Excel as an xlsx file. Now open the xlsx file in Minitab, and compare the result with part (b). Can you tell what was the issue before?

2.5 Getting the Minitab screen grab in Figure 2.2 required filtering out the raw data. Open the original data file (see Problem 2.4), and eliminate unnecessary variables to reproduce Figure 2.2.

---

4 Source: data.cityofnewyork.us (accessed July 26, 2019).

2.6  Determine if the following are structured or unstructured data.
  a)  Driver's reaction time while using a cell phone.
  b)  Brand of tablet.
  c)  Blog content.

2.7  Determine if the following are structured or unstructured data.
  a)  Video feed
  b)  Tweets by a celebrity
  c)  Vehicle average speed

## 2.2  Type of Data

Recall that data is a collection of facts or variables. Generally, there are two types of variables: categorical variables and numerical variables. **Categorical variables**, also known as **qualitative variables**, group observations into categories. Town of residence, smartphone brand, and professor evaluation ("Excellent," "Good," etc.) are examples of categorical variables. **Numerical variables**, also known as **quantitative variables**, are obtained from counting, measuring, or applying a mathematical operation to something. Grams of sugar in a power bar, amount of debt of a loan applicant, and number of siblings are examples of numerical variables.

Numerical variables can be divided further. A **discrete variable** has a countable number of distinct values. Generally, the possible values are whole numbers: 0,1,.... The number of insurance claims received by a company and the number of passengers attended in a given hour at the Los Angeles airport are examples of discrete variables. When data can have any value within a given interval, we say it is **continuous**. Duration of an online video ad, miles until the engine of a car fails, and strength of a new construction material are examples of continuous variables. Since a discrete variable is countable, its values can be set up in order, something that cannot be done with a continuous variable, within which any given interval has an infinite amount of possible values. This distinction between discrete and continuous variables leads to the need of different mechanisms to summarize and visualize discrete and continuous variables. When a discrete variable has a large range of values, small differences become irrelevant, and the data may be treated as continuous.

*Scales of Measurement*   When conducting statistical analysis, it is valuable to be even more specific about the data. The scale of measurement will determine the kind of summaries and inference that can be done with data. There are numerical and categorical scales of measurement, but the distinction is more important for categorical variables.

There are two types of categorical scales: nominal scale and ordinal scale. **Nominal scale variables** are merely unstructured categories as observations.

Pasta brand, cable TV provider, and car color are some examples of nominal scales. The only possible mathematical operation that can be performed on a nominal scale is counting the different categories. This counting can also be done relative to the total amount of observations. An **ordinal scale variable** is a categorical variable that has some order defined into its categories. Since an ordinal variable is a categorical variable, a mathematical operation such as an average does not make any sense. A work-around is that scores are assigned to the ordinal scale and mathematical operations are then performed on the scores. Product evaluations and placing in a race are examples of ordinal scales.

Nominal variables may be recoded as numbers, but that does not turn the variable to a numerical variable. Hence, taking the average of a nominal variable coded into numbers is meaningless. However, we may count resulting categories. When doing this, essentially new numerical data is created based on categorical data. For example, if a group of people are asked if they have a college degree, they answer yes or no. It can be useful to define new data that counts the number of people who answered "yes." Not only are these counts used to summarize the data, but inference can be drawn from them as well.

Conversely, a numerical variable can be turned into a categorical variable. This is useful when the precision of the numerical variable is not needed. For example, body mass index is recorded numerically, but for analysis it may be enough to use body mass classifications: underweight ($<18.5$), normal ($18.5, 24.9$), overweight ($25, 29.9$), and obese ($\geq 30$).

## Practice Problems

2.8   Classify each variable as numerical or categorical.
   a) Number of hours of study a week.
   b) Favorite class this semester.
   c) Weight of your cell phone.
   d) Miles per gallon of a car.
   e) Car model.

2.9   Determine the level of measurement (nominal, ordinal) in the following problems.
   a) How much a person likes a pair of shoes (from not at all to a lot).
   b) Gaming app installed on phone (Yes, No) by citizens who are 50 years old or older.
   c) Customer's ranking of a service provided by an eye institute as either excellent, good, satisfactory, or poor.
   d) Marital status of guests completing a survey from a resort.

2.10 In the following, state if the numerical variable is discrete or continuous according to their possible values.
   a) $X = 0, 1, 2, 3$
   b) $Y = 10, 15, 42$
   c) $5 < Y < 73$
   d) $Z \geq 152$

2.11 In the following, state if the numerical variable is discrete or continuous.
   a) Number of times there was a power outage somewhere in the United States last year.
   b) Volume of chocolate by 25 kids sampled in October.
   c) Maximum winds of a hurricane in the last 24 hours.

2.12 As in Problem 2.11, state if the numerical variable is discrete or continuous.
   a) Distance traveled by an automobile until it makes a full stop.
   b) Number of faulty products that come from an assembly line.
   c) Fare paid for a cab drive.
      *Hint:* Think of it in cents.
   d) Time it takes to commute to work.

2.13 Placing at a marathon, do you think it counts as a categorical or numerical value?

2.14 A player's sport team number, is that considered a numerical or categorical variable?

2.15 There are also two levels of measurement for numerical variables. **Interval scale** applies to numerical variables with a zero value defined arbitrarily. Temperatures in Celsius are an example; the zero is conveniently defined as the reference temperature at which water freezes. Fahrenheit is also interval scale. Zero does not mean absence of the quantity being measured, and it does not necessarily mean the lowest possible number. Interval data can be counted, ranked, and averaged. However, since zero is arbitrary, ratios are meaningless. That is, a 90 $°F$ is not twice as hot as 45 $°F$. **Ratio scale variable** is a numerical variable where the zero is not defined arbitrarily. In a ratio scale, zero is the outcome of a count or measurement representing absence of the quantity. Zero does not have to be a necessary outcome in a given situation; what matters is its meaning in theory (i.e. a person of height zero, height is a ratio type of measurement). Ratio data can be counted, ranked, and averaged. Also ratios are meaningful. Of the following, which represents an interval scale and which represents a ratio scale?

a) Profit by Apple Inc. last quarter.
b) Year of birth of a person.
c) Longitude of a volcano.
d) Direction of wind in degrees.
e) Distance from your school to the nearest bus stop.

## 2.3   Four Important Notions in Statistics

Throughout the book, notions of population, parameter, sample, and statistic will constantly mentioned. We define each one in this section.

**Population** can be defined as the set of all the units of interest in an analysis. Notice that our definition of population is dependent on the analysis. Specifically, what the question of interest is.

**Example 2.2**   *If a manager wants to study the mean number of purchases customers made every time they came into a store last year, then the customers that came into the store last year are the population.*

On occasion, there are multiple populations of interest. For example, when a company wants to compare the performance of two computer models on a computing task, there are two populations: model 1 and model 2. It is convenient to represent our questions of interest about a population in terms of some numerical trait of that population, what we refer to as a **parameter**. The percent of businesses in Oceanside, San Diego, with at least five full-time employees, is an example of a parameter.

When data is collected from the entire population on the measurement of interest, data is called a **census**. Then the parameter can be explicitly determined. It is not always possible to obtain the numerical trait directly by counting or measuring from the population of interest. It might be too costly and take too long; the population may be too large or unfeasible for other reasons. If a car company is testing the safety of a car model during a collision, it cannot just smash all the models available. A **sample** is a subset of the population of interest. There are many ways to take samples from a population, yet it is key to make the sample as representative as possible of the population regarding the traits of interest. But there are other things to consider, such as convenience, cost, time, and availability of resources. The sample allows us to ESTIMATE the parameter of interest through what is known as a **statistic**, a numerical summary of the sample. For example, in a sample of 200 businesses taken in Oceanside, San Diego, the percent of them with at least five full-time employees is a statistic. Although the statistic estimates the parameter, there are some key differences between them. Figure 2.3 illustrates the relationship between the concepts of parameter, population, sample, and statistics.

Figure 2.3 A diagram on the concepts of
population, parameter, sample, and statistic.

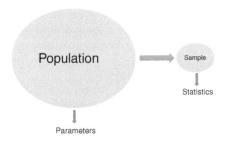

*Why Take a Sample?* Let us take another look at the distinction between a parameter and a statistic. A parameter uses the entire population to extract a numerical summary. For a given period of time, this parameter is constant. On the other hand, a statistic is a numerical summary of a sample. It is not constant but random, because it comes from a random process that allowed just a few units from the population to be part of the sample. If the sampling were to be performed again, the selection of different units will lead to a different value of the statistic.

---

**Statistics Nomenclature**
Due to this distinction between parameters and statistics, we will be using different notation for parameters and statistics. Generally, Greek symbols will denote parameters. For example,

- $\mu$ will represent the mean of a population.
- $\pi$ will represent a population proportion.
- $\sigma$ will represent a population standard deviation.

In contrast, we will use $\bar{x}$ to represent the mean from a sample and $\hat{p}$ to represent the sample proportion. Refer to Appendix A.6 for more about this nomenclature.

---

**Parameter in Practice**
The goal is to answer a question about a parameter. When this parameter can be measured in a precise matter through a census, then by all means it should be done. For example, if a local company wants to determine mean salary of its employees, all it has to do is go to its payroll files to find out (why?). For most of this book, we focus on scenarios where it is not feasible to determine precisely the parameter, and therefore, we must rely on the use of statistics.

---

As you will learn later, sample statistics can provide reliable estimates of parameters as long as the population is carefully specified relative to the problem at hand and the sample is representative of that population. It may sound trivial that the sample should be representative of the population, but it can sometimes be troublesome to achieve. Returning to the concept of inferential statistics, the goal is to use the information drawn from the sample, the statistic, to reach a conclusion about a parameter. The steps to perform a full statistical analysis are summarized within the context of data science in Figure 1.3.

**Example 2.3**   *Every year, administrators of the San Francisco airport conduct a survey to analyze guest satisfaction with facilities, services, and amenities. Airport administrators later compare results with previous surveys to look for areas of improvement and discover elements of the guest experience that are not satisfactory. For 2011, data was collected via customer interviews held at all airport terminals and boarding areas from May 11 to 26. During the survey period, interviews were conducted between 4:00 a.m. and 12:30 p.m. The questionnaires were available in English, Japanese, Chinese, and Spanish, and a total of 4445 guests were interviewed. One of the questions was: "Did you visit any airport restaurants today, but did not make a purchase?"*

The population of interest is all people traveling through the San Francisco airport in 2011. The parameter could be the proportion of all travelers who in 2011 visited an airport restaurant, but did not make a purchase. The sample consists of the 4445 travelers that answered the survey, and the statistic would be those respondents who answered they had visited an airport restaurant, but did not make a purchase.

## Practice Problems

2.16   Give an example of a parameter and a statistic for the following.
   a) Political elections
   b) Subaru 2017 Forester vehicles
   c) College students
   d) Adults in the United States

2.17   Insurance company $X$ would like to determine the proportion of all medical doctors who have been involved in at least one malpractice lawsuit during their career. The company selects 100 doctors at random from a professional directory and determines the number in the sample who have been involved in a malpractice lawsuit.
   a) Determine what the population is.
   b) What is the parameter?
   c) Define the sample.
   d) What is the statistic?

2.18 The 2016 General Social Survey included the following question: do you happen to have in your home (IF HOUSE: or garage) any guns or revolvers? Identify below the population, parameter, sample, and statistic.
   a) 2877 people participated in the survey.
   b) Proportion of US residents that own any guns or revolvers.
   c) People who reside in the United States.
   d) 593 people in the sample answered yes.

2.19 You are in charge of a local company who wants to obtain mean income of the company's engineers to compare with the mean income of engineers in the industry. You can either obtain the population mean (parameter) or a sample mean (statistic) to reach a managerial conclusion (both describing the same property, and assume both are feasibly possible to get). Which one would you rather use? Explain.

2.20 Choose between a sample or a census to measure each of the summaries below. Explain how you reached your decision.
   a) Mean time it takes you to get to school from your residence.
   b) Mean price paid for a laptop computer by three of your closest friends.
   c) Mean price paid for a laptop computer by students in your campus.
   d) Percentage of students absent to your statistics course last Tuesday.
   e) Percentage of students absent per day at any course at the university.

## 2.4 Sampling Methods

A sample allows for the characterization of a population under study. How reliable is the information obtained from the sample depends on a series of factors; among them is that the sampling method provides data that is representative of the population. As obvious as this sounds, the sampling method must be carefully chosen to ensure this property. However, there are other aspects to take into consideration when choosing a sampling method. In this section, the most common sampling methods are introduced, and their advantages and disadvantages are summarized.

### 2.4.1 Probability Sampling

In **probability sampling**, any unit of the population has a chance (greater than zero) of being chosen for the sample, and such a probability can be determined. In **nonprobability sampling**, either some units have no chance of becoming part of the sample, or the probability of any unit being part of the sample cannot be assessed. Nonprobability can lead to a sample that is not truly representative of the population. Let's start looking at some probability sampling methods.

*Simple Random Sampling*   In **simple random sampling** (often called random sampling for short), each unit has the same chance of being selected for the sample. When a unit is eligible to be selected again, the sampling is called **random sampling with replacement**. If a unit chosen to be part of the sample is not eligible to be selected again, the sampling is called **random sampling without replacement**. The main point to be aware of is that by definition, when sampling without replacement, a unit already sampled cannot be chosen again. Thus, at first all units have the same chance of being chosen, but once a first unit is chosen for the sample, the chances of the other units belonging to the sample change, altering the principle of simple random sampling. In practice, however, as long as the population is large enough (sample corresponds to only 5% or less of the population size), this concern is only academic, since then the chance of selecting any unit twice or more is small. For samples that correspond to a larger subset of the population, adjustments are needed when sampling without replacement.

To create a random sample, population units are first indexed from 1 to $N$. Then a computer is used to generate the random sample (of, say, size $n$ where $n < N$). Computers help avoid **judgment bias**. Specifically, if a large random sample with replacement from the numbers 1 to 100 is conducted using a computer, each number will show up in the sample about the same amount of times. However, if a large random sample with replacement from the numbers 1 to 100 is conducted by asking people to choose, each number will not show up in the sample the same amount of times, because people will tend to choose preferred numbers: 1, 25, 100, 50, and so forth. Many odd numbers 23, 37, 73, ... would be selected by people less often.

*Systematic Sampling*   Systematic sampling chooses every $k$ unit from a sequence to be part of the sample. The sampling method applies to finite and infinite populations. For example, suppose the population is finite of size $N$ and the goal is to obtain a sample of size $n$. Systematic sampling chooses every $k$ item in an ordered list as part of your sample, and $k$ is approximately equal to $N/n$. Either a random starting point is chosen from the first $k$ items, or in some cases (such as with simulations) every $k$ unit is taken to be part of the sample. Systematic sampling may also help to eliminate dependence of some samples from simulations. As long as there is no dependence of periodicity $k$, systematic sampling will result in a representative sample from the population. Otherwise, a pattern may influence the results. For example, suppose a bank decides to sample bank balances of customers every two Friday's to determine the average bank balance of customers. The sample may be influenced by customers on a biweekly pay program. The more bank customers are on a biweekly program, the greater the influence on bank balance results.

*Stratified Sampling*   Suppose a company would like to know what proportion of their products are defective. The company owns five factories, and they plan to take a sample of 1000 products. To ensure that each factory is represented, they can take random samples of products per factory. This is an example of **stratified sampling**, where the population is divided into non-overlapping groups (called **strata**) and random samples are taken from each group. To obtain the samples, either simple random sampling or systematic sampling is used. In the example introduced in this paragraph, the sample size for each factory may be the same, or it may be different. The decision will depend on ensuring the results are representative of the population and with the least uncertainty possible. Stratified sampling is a favorite technique when conducting political surveys to ensure participation from minority groups.

*Cluster Sampling* **Cluster sampling** divides the population into groups, just like stratified sampling. But then a random sample of groups is taken, different from stratified sampling, where ALL groups are selected. In cluster sampling, once the groups are randomly chosen, one either chooses each unit within the group to be part of the sample or takes a random sample of these units. Stratified sampling increases precision, while cluster sampling is implemented for the sake of cost effectiveness.

### 2.4.2 Nonprobability Sampling

By definition, probability sampling provides a certain confidence that the sample will be representative of the population under study. This is a key aspect of sampling. Unfortunately, probability sampling is not always possible. A list of units may not be available. Or sometimes attempting to sample people randomly for a study does not lead to sufficient participation (many people prefer not to participate in studies). Cost and resources must also be taken into consideration. Although results from a nonprobability sampling are not as trustworthy as those from a probability sampling, they are easier to implement and sometimes are the only alternative. Some nonprobability sampling methods are described in what follows.

*Convenience Sampling* As the name implies, this method is all about convenience. Although this can be interpreted in many different ways (convenience in terms of access, cost, or the results we want), the meaning here is in the sense of easily accessible samples. A common convenience sample uses volunteers to collect information. Volunteers are recruited through advertisements in different media outlets.

*Judgment Sampling* A sample based on the opinion of an expert is called **judgment sampling**. Since the sample is based on opinion, bias may occur. Just like stratified sampling, in **quota sampling** the population is divided into non-overlapping groups, but then the samples within these groups are not considered random.

Returning to the factory example, one possibility is to choose sample sizes based on the proportion of total products coming from each factory. If factory 1 and factory 5 produce 20 and 50%, respectively, of the total products, then a sample of 200 products can be obtained from factory 1 and 500 products from factory 2. The rest of the sampled products would come from the other three factories accordingly.

*Focus Groups* A **focus group** comprises a panel of people considered to be representative of the population. People are then asked about their opinion about an issue. Its nonrandom basis can lead to bias. Focus groups tend to offer qualitative information in addition to quantitative information, which is one of the reasons they are still used.

*Other Sources of Data* Samples are not the only way to gather data, and these days we have access to data through an extraordinary amount of sources. Generally, data may be primary or secondary. **Primary data** is collected firsthand by the team, while **secondary data** comes from somewhere else and perhaps is intended for something else. The advantage of primary data is that it is gathered for the specific task at hand. Moreover, the team has control over the quality of the data. However, cost, personnel, and other factors tend to limit the amount of primary data one can obtain. In contrast, secondary data is more likely to have many errors, and you may not even know how the data was gathered. But the digital age has made a lot of secondary data easily accessible, even at no cost. It is futile to attempt to list all sources of existing secondary data. Table 2.1 shows

Table 2.1 A small list of useful sources of data.

| Data sources | Some data available | Website |
| --- | --- | --- |
| Bureau of Labor Statistics | Unemployment rate, wages per profession | www.bls.gov |
| Census Bureau | Household income, population number | www.census.gov |
| National Center for Education Statistics | College enrollment, high school dropout rate | nces.ed.gov |
| Statistics Canada | Building permits, consumer price index | www.statcan.gc.ca |
| World Bank | Governance indicators, GDP ranking | www.worldbank.org |
| World Health Organization | Child malnutrition, immunization coverage | www.who.int/en |
| Google Dataset Search | Thousands of data repositories | https://toolbox.google.com/datasetsearch |

just a small number of existing sources of data. The organizations shown there provide reliable data on many important topics.

Some companies will provide data for anyone to use (e.g. Google) and for researchers (e.g. America Online), all for purchase (e.g. S&P 500).

---

**Open Data**

In the last few years, local city governments have also made their data available through open portals. **Open data** is defined[5] as:

- information freely available to everyone;
- without limits of reuse and redistribution; and
- are machine readable (in formats easily processed by computing devices).

Open portals have been growing quickly around the world, and these days, anyone can find data from relatively close to where they live. Crimes committed, police traffic stops, copper content in water, college enrollment, and air traffic are some examples of data available through open portals. In this book, we use a small fraction of these data sets while introducing statistical concepts. The list of these data sets, in alphabetical order by location or organization handling the data, is available in the Index of the book under "open data."

---

It is important to emphasize that not all data sets are created equal. For some data sets, great care is taken in its creation with the intention of making its use as efficient as possible. In contrast, some data sets are not even intended for statistical analysis and instead are created for administrative purposes (e.g. in case an audit is conducted). These type of data sets are still useful for statistical analysis, but they tend to come with many errors or may be hard to determine the meaning of the information provided.

---

**Experimental Study**

Primary data can be collected through experimental studies or observational studies. In an **experimental study**, there is a variable of interest known as a **response variable**. Other variables, called **treatments**, are controlled to determine if they have an effect on the response variable.

---

5 Manyika, J., Chui, M., Groves, P., Farrell, D., Van-Kuiken, S., and Doshi, E. A. (2013). *Open Data: Unlocking Innovation and Performance with Liquid Information*. McKinsey Global Institute, 21.

**Example 2.4**   *In an experiment, a group of computers will run a task with a company's processor, and another group of computers will run a task with a competitor's processor. Both computer processors are randomly assigned to each group. That way, the two groups are identical with the exception that one group has the company's processor installed and the other has the competitor's processor. The aim is to determine if there is a difference in mean running time of the task of the two groups.*

In this example, the response variable is the running time of the task, and the variable that may affect running time (i.e. the treatment) is the type of processor. If a difference is found between the mean running times, the team feels more comfortable in arguing that the difference is due to the type of processor since random assignment makes it likely that the two groups are homogeneous in terms of other traits.

---

**Observational Study**

In an observational study, there could still be a main variable of interest, but no attempt is made to control potential influential variables, and other variables are simply recorded from the units. For this reason, observational studies are also called nonexperimental studies. Surveys are a preferred tool when running an observational study.

---

Experimental studies are better than observational studies to argue **causal associations**, since randomization makes it unlikely to explain any difference by anything other than the treatments. However, costs, ethical considerations, time restrictions, reliability goals, and resources can make it difficult to perform an experimental study.

## Practice Problems

2.21  In the following scenarios, state whether primary or secondary data will be used.
   a)  For her research project, a student created a survey to collect data from 200 participants.
   b)  A company that specializes in lithium batteries takes a random sample of 10 new batteries from the assembly line to measure their quality.
   c)  A startup where phone apps are designed extracts restaurant quality data from an open data portal.

2.22  Considering the study objectives presented below, provide an example of an experimental and an observational study.

a) Measure drone battery life.
b) Determine if feeding chickens organic corn is associated to the number of eggs they yield.

2.23 From the provided study descriptions, state whether the study was experimental or observational.

a) A researcher is carrying out a study on the reliability of four different solar systems. She has sixteen 3 kW properties and plans to randomly assign each to one of the four treatment groups.
b) To examine first graders' hours of screen time a week, researchers prepared a questionnaire to be filled by parents.
c) Is the perspective of business owners in Chattanooga, Tennessee, as a place of business, independent of their perspective of its affordability? To answer this question, the survey answers from 2172 participants will be used.
d) To test new drone propellers, 20 unmanned aircrafts were randomly assigned to fly with either the new propellers or the traditional ones. Then their average speed was estimated.

## 2.5   Data Management

Data management is a collection of steps taken to benefit from extracting information from data most efficiently. It is a topic that is often completely omitted in statistics courses. Yet, when performing some type of statistical procedure, proper data management is of foremost importance. The aim of data management is to access the data as quickly as possible, while ensuring the best data quality. For projects requiring thousands of measurements, error is virtually guaranteed to occur. Poor quality data will not represent the population of interest well and hence lead to misleading conclusions. Moreover, effective data management reduces data errors, missing values, and reanalysis. This can translate into adequate conclusions and savings in time, resources, and costs. In this section we summarize the components of data management to help ensure the proper management and best data quality. For data management, establishing a data quality protocol is best. A data quality protocol can be set up according to the following stages of research:

- Study design.
- Data collection/measurement.
- Data entry/recording.
- Data wrangling and data screening.
- Data analysis.

Each stage affects data quality. A data quality protocol establishes a coherent data management procedure that ensures efficient creation and use of research data. Through the study design, the researcher intends to answer questions of interest as effectively as possible. When it comes to data management, the steps taken at this stage of the research have an impact on the quality of data. Also, it must be recognized that just because the study design was done well does not mean that there will not be errors nor that data will be of high quality. The study must be monitored to ensure the implementation proceeds as designed. Furthermore, efficient data collection and data entry are also key. We go over these topics below.

Transforming raw data into usable form is often called data wrangling. This step involves filtering, merging, formatting, and aggregating data. It also requires screening of the data for errors or issues in general.

The analysis of the data requires its entry into a computer in digital form. Sometimes, data collection and data entry occur simultaneously, but this is not always the case. Some ideas on how to do data entry efficiently are as follows:

- Use code when possible. For example, if age will be recorded multiple times, enter date of birth and date of measurement to determine age. Often the case, measurements on several subjects are performed in a day. Using code, less individual entries are needed, and as a result, human error is reduced.
- Use appropriate software for data entry. User-friendly software with the capability of minimizing errors is recommended.
- Keep a record of codification, also known as a **variable dictionary** or **code-book**. A variable dictionary prevents relying on memory when returning to the data and hence leads to less errors. It also makes it easier to share the data for given tasks and helps in building convenient, consistent codification, based on similar variables for future measurements. The variable dictionary should include the following:
  - Variable abbreviation used.
  - Definition of variable (continuous/discrete?); what does the abbreviation represent.
  - If a numerical variable, the unit of measurement.
  - For coded categorical variables what each coded value represents (e.g. M male and F female or 1- female and 0- male).
  - For some categorical variables an explanation of how categories were defined. For example, defining what income is considered low, moderate, or high.
- Use intuitive names for variables. For example, use gender or sex for gender abbreviation instead of something like $X4$ so it is easily recognized. Keep variable names consistent throughout if more data is collected later on. The

more variables one has, the more important it is to make variable names recognizable and consistent.

- Limit use of codification of categories of variables. For example, writing 1 for male and 0 for female is convenient for analysis, but not necessary for data entry, and it is best avoided.

Perhaps the most downplayed step of data wrangling is **data screening**, which consists of checking data for unusual observations. It is a procedure conducted before analysis, during exploratory analysis, and when conducting diagnostics on the inference methods used. Unusual observations may simply be accurate but may also be due to errors or to sampling issues (e.g. questions or answers were not understood by respondents). Sometimes, data issues are straightforward to detect. For example, one may find that a study participant's height was entered as 150 inches, and the issue would potentially be detected through summaries or model diagnostics. But data issues are not always easy to detect. A height may be erroneously entered as 65 inches and hence go undetected. Tips for data screening:

- Start screening data early in the project. By starting early, one is allowed to change patterns of ineffective data entry, data collection, or correct errors by contacting respondents.
- Conduct data screening "regularly." This includes data audits, where data management staff checks if IDs in the data set can be traced back to original documents and other tasks. For example, any missing value or skipping patterns by study participants in a survey may indicate privacy concerns. Any flaws or concerns should be properly documented.
- Check if observations are feasible. Are there any values that are simply impossible? This can be done through numerical and graphical summaries to detect "abnormalities." Maximums and minimums are very useful for quantitative variables.
- Error rates can be calculated over all variables or per variable.
- Refer to original data collection forms when in doubt.
- Plot the data: Plots help us explore features of the population of interest but also help us detect data issues. For example, suppose data includes an age variable and years of education variable (both numerical variables). You should not expect, say, a 15-year-old to have over 20 years of education. A scatterplot (Chapter 3) would be able to detect this type of issue.
- Search for unlikely combinations of values in variables. This method allows to screen categorical and quantitative data. For example, a male subject cannot be pregnant (an error the author once found while analyzing health insurance data). A search will easily detect these types of data issues.
- If multiple people are in charge of data entry, ensure the use of variable dictionary and communication between personnel.

---

**What to Do When Detecting a Data Issue?**

- Try to determine if there is an error from original data or from primary source (contacting personnel who collect/record data or study participants). Contact personnel within and outside the data management branch for necessary adjustments.
- If issue is resolved, make a record of the data issue, so that others using the data for analysis are aware. Also, if a source of error is detected, bring up the error source to the appropriate team to avoid further errors.
- If an issue is not resolved, record the issue in a log. The observation may be turned into a missing value during analysis depending on the extent of the error.

---

We will return to the topic of data screening several times throughout the book.

During the analysis stage, data is still evaluated for quality. For example, a study may require the construction of a statistical model. Yet, all statistical models require certain assumptions. Later on, we will see tools used to evaluate these assumptions. These tools may also lead to discovering data issues.

The statistical procedures discussed in this book help diminish the impact of any errors present in the data. By implementing these data quality tools, the statistical techniques are allowed to filter out noise more effectively and hence perform better inference. It is recommended that personnel from each branch of data management have periodical meetings. Empirical analysis is a dynamic process; hence changes in machinery, tools, personnel, and the environment occur. Regular meetings help ensure data quality throughout these dynamics. Lastly, when sharing data involving people, one must remember to protect the privacy and confidentiality of participants by only sharing the necessary part of the data.

### 2.5.1   A Quick Look at Data Science: Data Wrangling Baltimore Housing Variables

Baltimore officials compile housing data from several reliable sources, bolstering data-based decision making for the sake of the City's communities. Specifically, authorities have established 55 community statistical areas for Baltimore. In Chapter 3, we will visualize some of these housing variables. The variables that we wish to visually assess, across several years, are as follows:

- I1 – Percent of residential properties vacant.
- I2 – Median number of days on the market.
- I3 – Median price of homes sold.
- I4 – Percent of housing units that are owner occupied.
- I5 – Percent of properties under mortgage foreclosure.

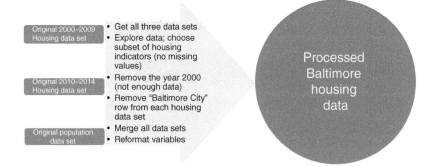

Figure 2.4 Steps taken to wrangle the Baltimore housing data set.

Considerable data wrangling is required to prepare these housing variables for visualization. Specifically, several separate collections of variables were retrieved.[6] Figure 2.4 sums up the data wrangling.

Note the following:

- Once all three collection of variables were retrieved, the features of the data were explored.
- Because housing data was spuriously available for the year 2000, this year was not included in the final data set. Population data was only available for 2010.
- The last row of the data set provides information for all of Baltimore City. Since our intention is to summarize information from the communities within Baltimore City, this last row must be removed.
- Next the housing indicators of interest are merged and reformatted.

Table 2.2 shows data for 2 of the 55 communities. Each column represents a variable, and each row an observation. So the first community shown had 16

Table 2.2 Yearly housing variables and 2010 population for two of the 55 community statistical areas in Baltimore.

| Year | Community | Population (2010) | I1 | I2 | I3 | I4 | I5 |
| --- | --- | --- | --- | --- | --- | --- | --- |
| 2001 | Allendale/Irvington/S. Hilton | 16217 | 2.27 | 75 | 55000 | 74.36 | 3.85 |
| 2001 | Beechfield/Ten Hills/West Hills | 12264 | 0.26 | 42 | 78500 | 84.76 | 2.26 |
| ⋮ | | | | | | | |

---

6 At data.baltimorecity.gov (accessed July 26, 2019), the following collections of variables were retrieved: "Housing and Community Development Vital Signs 2000–2009," "Housing and Community (2010–2014)," and "Census Demographics 2010–2012."

217 residents in 2010, while in 2001, 2.27% of the residential properties were vacant.

We return to this data set in Section 3.5.1.

## 2.6 Proposing a Statistical Study

Almost always, our textbook will proceed as follows:

- Present a hypothetical scenario.
- Introduce statistical concept(s).
- Provide examples of either the hypothetical scenario or similar scenarios, and apply the use of the statistical concept.
- Practice problems.

This is the main recipe for statistical education. Yet, by presenting the hypothetical scenario, we are sidestepping an important process: proposing a statistical study or analysis. Three of the main components of a proposal for a statistical study are as follows:

- Establish the objective.
- Present your methodology.
- Talk about your data.

Let's address each of these components in turn.

*Objective*  The objective can be expressed in terms of a **statistical question**, which is answered with data that exhibits variation. For example, the question "what is your height?" is not a statistical question, since there is no variation in your height. In contrast, "how long do college freshmen study for their statistics class?" is a statistical question, since some students study more than others.

---

**Statistical Question Cheat Sheet**
The main features of a statistical question are as follows:

- Question must be answered with data.
- Answer is not exact and has variation.

---

*Methodology*  In this component of a proposal for a statistical study, the researcher presents the statistical methods to be used[7] to answer the statistical question. The selection of the statistical method is influenced by the objective, available expertise, study budget, and time constraints.

---

7 And sometimes alternative statistical methods in case the primary one is deemed inappropriate.

*Data*  In conjunction with the methodology, the data helps you achieve the proposal objective. Thus, if the objective is to assess whether annual sales are increasing, using only 2019 sales data will not answer the statistical question.

## Chapter Problems

2.24  Suppose you ask all your friends whether they believe in the afterlife. Can you use this data to make conclusions on the percentage of all people who believe in the afterlife? Explain in one sentence.

2.25  What is the difference between a discrete and a continuous variable. Give an example of each one.

2.26  Summarize the advantages and disadvantages of primary and secondary data.

2.27  In the following scenarios, state whether primary or secondary data is being used.
   a) In an effort to understand their education system and how it can be improved, city managers download data from the federal US census website.
   b) The owner of a company that provides house cleaning services through online requests downloads their website and app data for analysis.

2.28  Eight years ago, company $X$ acquired new machines to can food. The machines were designed to produce a box of 24 cans in two minutes. In an effort to assess productivity, managers at company $X$ will look at the digital record for the time (in seconds) it took to produce 10 000 boxes of canned food. Explain how it would help to assess the minimum and maximum production time in the database.

2.29  Determine if the following is a statistical question.
   a) How many bathrooms does your house have?
   b) Did Jeff earn more in 2018 than in 2017?

2.30  Determine if the following is a statistical question.
   a) Does uncertainty on which candidate to vote for depend on political party affiliation?
   b) Is car tire diameter associated with miles per gallon?

2.31 In the following, state whether the data adequately addresses the (implied) statistical question.

a) To identify whether yearly cancer rates are increasing, 2018 cancer data will be used.

b) In an effort to determine how many sugary drinks first-year college students drink, 245 first-year students completed a survey during the first week of school. One of the questions was on average, how many sugary drinks (soda, sport drinks, energy drinks, flavored drinks) do you drink a day?

c) Is there a difference in speed between drones with propellers of differing propeller lengths? A company randomly selected 15 drones to use one of three different propeller lengths. Then their maximum speed in the lab was measured.

2.32 In the following, state whether the data adequately addresses the (implied) statistical question.

a) Have new bad weather protocols effectively reduced flight delay times at the John F. Kennedy (JFK) airport? JFK data from six bad weather events before the new protocols and from six bad weather events after the new protocol were used to compare the mean delay times (in minutes).

b) To determine if mean cell phone battery life has improved, the mean cost of five cell phone models was obtained.

c) For a research project, a student aimed to establish if there was a difference in the mean SAT scores of five different universities. The student requested last year admissions data from new students at his school.

2.33 With the topics below, (i) develop statistical questions, (ii) explain how to answer the questions with statistics, and (iii) try to find online data to conduct a statistical analysis (to answer your statistical question).

a) Determine if student tuition is becoming unsustainable.

b) Restaurant quality.

2.34 With the topics below, (i) develop statistical questions, (ii) explain how to answer the questions with statistics, and (iii) try to find online data to conduct a statistical analysis (to answer your statistical question).

a) Employee performance.

b) Suicides and natural disasters.

# Further Reading

Black, Ken. (2012). *Business Statistics*, 7th edition. Wiley.

Gurin, Joel. (2014). *Open Data Now: The Secret to Hot Startups, Smart Investing, Savvy Marketing, and Fast Innovation*. McGraw-Hill.

Scheaffer, Richard L., Mendenhall III, William, and Ott, Lyman R. (1995). *Elementary Survey Sampling*, 5th edition. Duxbury Press.

# 3

# Data Visualization

## 3.1 Introduction

Data collection generally consists of gathering information over many variables. It is imperative that we can summarize this data in a way that is quick and easy to understand. This will help us explore and understand the data and assist in making decisions through evidence. There are many ways to visualize data: through tables, graphs, and animations. Which data visualization should be used will depend on the type of data to be summarized. The ease of implementation is certainly an attractive attribute of data visualization techniques, but it can also lead to misuse. Visual techniques may be implemented in a way that it will not help summarize the data, mislead about what the data is showing, or flat out provide erroneous output. Just because a software program provides you with a graph, it does not mean that the graph is correct. The person preparing the visual display must provide the right type of data to the computer program. In this chapter, we will introduce the use of visualization tools and expand on these tools throughout the book.

## 3.2 Visualization Methods for Categorical Variables

As indicated in Chapter 2, data is classified as categorical (qualitative) or numerical (quantitative). A **numerical variable** either counts or measures something, while a **categorical variable** groups a trait of a data set based on classifications. Usually, data sets include both types of variables. With categorical variables, the least amount of mathematical operations are possible: counting and perhaps ordering the data, depending on the scale of measurement (Section 2.2). Therefore, there is only a handful of alternatives to summarize the categorical variable with tables and figures.

*Frequency Tables* One way to summarize a categorical variable is by presenting the categories with each of its counts. Each category must not overlap with other categories. The resulting table is also known as a **frequency distribution**.

*Principles of Managerial Statistics and Data Science*, First Edition. Roberto Rivera.
© 2020 John Wiley & Sons, Inc. Published 2020 by John Wiley & Sons, Inc.
Companion website: www.wiley.com/go/principlesmanagerialstatisticsdatascience

Table 3.1 Majors of new business students at a US university.

| Business major | | | | | | | | | | | | | | | |
|---|---|---|---|---|---|---|---|---|---|---|---|---|---|---|---|
| ACCT | ACCT | HUMR | ACCT | ACCT | ACCT | ACCT | FINA | ACCT | ACCT | FINA | MKTG | FINA | MKTG | FINA | MGMT |
| ACCT | ACCT | MKTG | FINA | ACCT | MGMT | MKTG | ACCT | INFS | ACCT | MKTG | ACCT | ACCT | HUMR | FINA | ACCT |
| INFS | FINA | FINA | FINA | ACCT | FINA | INFS | MKTG | INFS | FINA | MKTG | MKTG | ACCT | ACCT | ACCT | MKTG |
| ACCT | ACCT | FINA | ACCT | FINA | MGMT | ACCT | ACCT | FINA | MKTG | MGMT | FINA | MGMT | ACCT | ACCT | MKTG |
| INFS | FINA | INFS | FINA | ACCT | INFS | ACCT | ACCT | MKTG | MGMT | MKTG | ACCT | INFS | ACCT | ACCT | ACCT |
| ACCT | ACCT | ACCT | FINA | ACCT | ACCT | ACCT | ACCT | INFS | INFS | INFS | MGMT | MKTG | ACCT | ACCT | ACCT |
| FINA | ACCT | MGMT | MKTG | MKTG | FINA | ACCT | ACCT | ACCT | ACCT | ACCT | MGMT | MGMT | ACCT | ACCT | ACCT |
| INFS | INFS | ACCT | ACCT | ACCT | INFS | MGMT | ACCT | INFS | ACCT | ACCT | MGMT | ACCT | MGMT | ACCT | INFS |
| ACCT | MKTG | FINA | INFS | FINA | ACCT | ACCT | MGMT | ACCT | ACCT | ACCT | ACCT | MGMT | MGMT | INFS | ACCT |
| MKTG | ACCT | INFS | INFS | ACCT | ACCT | INFS | FINA | ACCT | MGMT | MKTG | INFS | FINA | ACCT | INFS | MGMT |
| MKTG | ACCT | MKTG | MGMT | MGMT | MGMT | INFS | FINA | MKTG | FINA | HUMR | FINA | ACCT | ACCT | ACCT | INFS |
| FINA | MKTG | MGMT | ACCT | ACCT | HUMR | INFS | ACCT | MKTG | ACCT | ACCT | FINA | ACCT | ACCT | MKTG | ACCT |
| INFS | ACCT | ACCT | INFS | ACCT | ACCT | ACCT | MKTG | INFS | ACCT | ACCT | ACCT | INFS | MKTG | ACCT | ACCT |
| ACCT | MKTG | MKTG | MKTG | ACCT | INFS | ACCT | ACCT | ACCT | ACCT | ACCT | MGMT | ACCT | ACCT | FINA | MGMT |
| ACCT | FINA | HUMR | INFS | ACCT | FINA | ACCT | MKTG | MGMT | FINA | ACCT | ACCT | INFS | MGMT | MGMT | INFS |
| ACCT | ACCT | INFS | ACCT | ACCT | INFS | ACCT | ACCT | FINA | INFS | ACCT | MGMT | ACCT | ACCT | MGMT | MKTG |
| MKTG | ACCT | MGMT | ACCT | MKTG | ACCT | FINA | ACCT | MGMT | HUMR | MKTG | INFS | ACCT | ACCT | MKTG | MGMT |

ACCT, FINA, HUMR, INFS, MGMT, and MKTG stand for accounting, finance, human resources, information systems, management, and marketing programs, respectively.

Besides counts in each category, another possibility called **relative frequency** presents counts in each category, relative to the number of observations, $n$. Since the counts of each category can never exceed $n$, a relative frequency distribution table would present numbers between 0 and 1, and total relative frequencies sum to 1. This often allows a better comparison of occurrence of each category. One can also take the relative frequencies and multiply them by 100 to get the **percent frequency**. Conceptually, it provides the same information as a relative frequency, but now scaled from 0 to 100.

**Example 3.1** *Let us illustrate the use of a frequency distribution. Table 3.1 shows the majors selected by new business students to a university a few years ago. With a total of 272 students admitted to the business school, Table 3.1 makes it hard to determine admissions by major.*

A frequency distribution would provide a clearer picture of the demand for business majors at the university. Table 3.2 presents all the scales of frequencies for this data. The counts show that the accounting major received three times more new students than any other business major, while human resources received the least amount of new students. The relative and percent frequencies are better in showing how the new admissions compare. The summary can help the school administration align the admissions with their resources and evaluate not only the job market demand, but also the student's perspective of the job market demand.

Table 3.2 New business students' majors at a US university.

| Business major | Frequency | Relative frequency | Percent frequency |
|---|---|---|---|
| ACCT | 125 | 0.46 | 46 |
| FINA | 35 | 0.13 | 13 |
| HUMR | 6 | 0.02 | 2 |
| INFS | 38 | 0.14 | 14 |
| MGMT | 32 | 0.12 | 12 |
| MKTG | 36 | 0.13 | 13 |
| Total | 272 | 1 | 100 |

ACCT, FINA, HUMR, INFS, MGMT, and MKTG stand for accounting, finance, human resources, information systems, management, and marketing programs, respectively.

*Bar Charts*   A **Bar chart** is a graphical representation of frequencies. Any frequency scale can be used to construct a bar chart. Normally, the $x$-axis is used for each category of the variable, while the $y$-axis will display the frequency scale. Each bar has a gap between them to point out that the categories do not overlap. Bars of constant width will display the frequency scales of each category, hence comparisons of the frequencies are based solely on the heights of each bar. Sometimes in the media, bars with different width are used, but this will distort the comparison of frequencies since the observer will base comparisons in two dimensions (height and width of bars) instead of one dimension (height of bar).

Let us return to the university admission data introduced in Example 3.1. Figure 3.1 presents the bar charts at different frequency scales. Although the height relationship among bars is the same for all frequency scales, the relative frequency and percent frequency are more useful in comparing interest in programs, while the frequency scale gives us the actual number of students per program.

On occasion, the graphical display will present categories ordered from highest frequency to lowest frequency, a type of bar chart called a **Pareto chart**. The Pareto chart may also include a line with dots indicating the cumulative relative or percentage frequencies. In the case of analyzing quality of a service or a product, this type of chart helps the company determine the top issues that require attention. Figure 3.2 shows a Pareto chart summarizing 900 main complaints from customers. Over 93.22% of the main complaints involved packaging, shipping, or customer service. Management can use this information to revamp these services.

*Bar Charts (and Other Figures) in Minitab*   To construct a bar chart in Minitab, go to Graph > Bar Chart. Figure 3.3 illustrates how to construct a simple bar chart

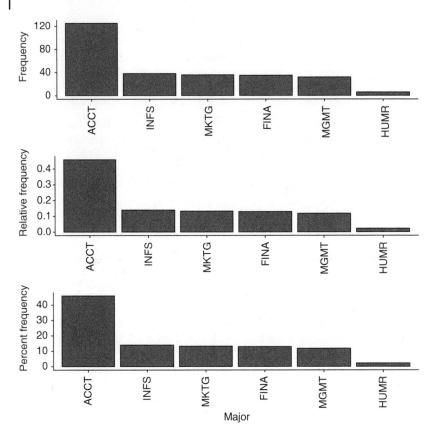

Figure 3.1 Bar chart displaying majors of new business students for a US university a few years ago.

for the main complaints example above (since the complaints are summarized by category, the Bar Represents box in the bar chart window must be changed to Values from a table). It is easy to create other types of summaries through the Graph window.

*Pie Charts* **Pie charts** are a commonplace graphical display for qualitative data. They are expected to provide the same type of summary as a bar chart, but do so differently. Pie charts take a circle and divide it into slices, with each slice corresponding to a category and the area of the size depending on the frequency of occurrence of the category. The slices add up to the entire area of the circle. A pie chart (Figure 3.4) is less effective summarizing the data than a bar chart. The main problem is that it is hard to compare frequencies of all categories because the visual comparison is now based on area, not just length like in the

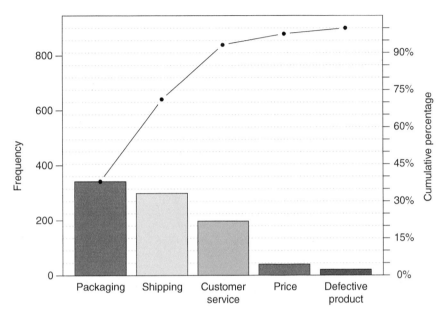

Figure 3.2 Pareto chart displaying 900 main complaints from customers. Over 90% of main complaints involve packaging, shipping, or customer service.

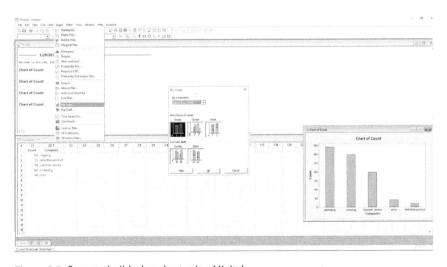

Figure 3.3 Steps to build a bar chart using Minitab.

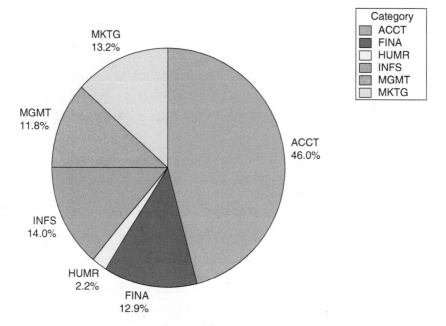

Figure 3.4 Pie chart displaying percentage of new students for Business majors in a US university a few years ago.

case of a bar chart. Returning to Example 3.1, it is easy to see how accounting admissions are most common, but it is hard to determine how admissions in marketing, management, information systems, and finance compare from the pie chart. The percentage labels are required in the pie chart for the comparison, but not in the bar chart.

The more categories in a qualitative variable, the harder it is to interpret a pie chart. In general, this visualization device should be avoided when there are too many categories.

## Practice Problems

3.1 Answer the following questions using the relative frequency distribution below:

| Category | Relative frequency |
|----------|--------------------|
| A | 0.46 |
| B | |
| C | 0.32 |
| Total | 1 |

a) What is the relative frequency of Category B?
b) If the data set consisted of 400 observations, how many times did category C occur?

3.2 The Canadian Dairy Information Centre compiles from several companies' information on the cheeses made from cow, goat, sheep, and buffalo milk.[1] The chart below summarizes the cheese type.
a) Which are the three most common types combined?
b) Which type is the least common?
c) The data consisted of 1296 companies. How many companies stated that their Cheese type is Firm if the relative frequency of that category was 0.344 135 8?

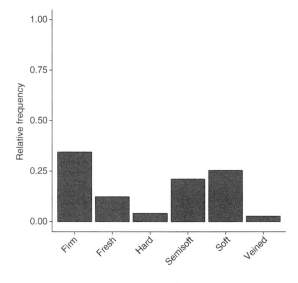

3.3 In Problem 3.2, some Cheeses are organic. The chart below summarizes the cheese type based on whether they are Organic (Yes, No):

1 Source: open.canada.ca (accessed September 1, 2017).

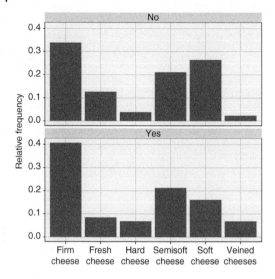

a) Does it seem that whether Cheese is organic or not has an impact on type of cheese?

b) Why can't we be sure that whether Cheese is organic or not has an impact on type of cheese?

3.4 The gender of employees according to Charlotte-Mecklenburg Police Department demographic data[2] is summarized below:

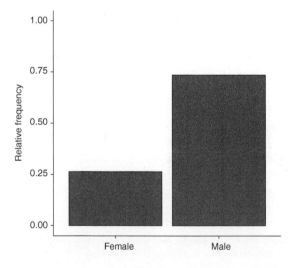

_____
2 Source: clt-charlotte.opendata.arcgis.com (accessed August 15, 2017).

a) Which gender is more common in Charlotte-Mecklenburg Police Department?
b) What is the approximate relative frequency of women in the police force there?

3.5 The following table summarizes the medical and mental healthcare services provided to people who may still be sick from[3] 9/11 attacks from 2008 to 2011. Construct a barplot of the counts.

| Category | Count |
| --- | --- |
| Resident | 154 |
| Local worker | 378 |
| Resident+ local worker | 77 |
| Cleanup worker | 51 |
| Cleanup+ local worker | 20 |
| Rescue/recovery | 4 |
| Passerby, commuter | 32 |
| Student | 25 |
| Other | 13 |

3.6 The file "2015_APD_Traffic_Fatalities_Austin" has data on fatal traffic accidents at Austin, Texas, during[4] 2015.
a) Construct a bar chart for the number of fatal accidents by type of accident.
b) Construct a pie chart for the number of fatal accidents by type of accident.
c) Construct a bar chart for the relative frequency of fatal accidents by the variable "Day" (day of the week). Is there any pattern seen in this chart?

3.7 Administrators for the City of Austin, Texas, randomly send surveys to residents with the aim to gauge their opinion about services. The Pareto chart below summarizes the answers to the evaluation of the city as a place to raise children.
a) Approximately what percentage of responses were "Very Satisfied" or "Satisfied"?
b) Approximately what percentage of responses were "Dissatisfied" or "Very Dissatisfied"?

---

3 Data: data.cityofnewyork.us (accessed July 26, 2019).
4 Source: data.austintexas.gov (accessed June 26, 2018).

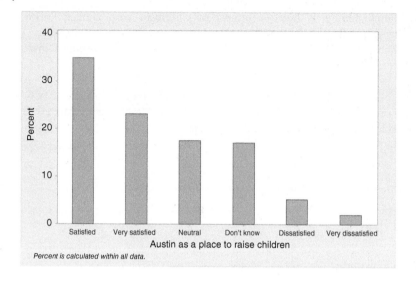

Percent is calculated within all data.

3.8 Answers to the 2016 community survey from respondents in Chattanooga are available[5] (filename 2016_Community_Survey_Results_Chattanooga). Use a Pareto chart to summarize the respondents' answers to survey item "Q22b: Overall direction the City is taking." 
*Hint:* If performed with Minitab, use the Chart Options button in the secondary bar chart window to properly order the categories.

3.9 Administrators of the San Francisco airport conduct a survey every year to analyze guest satisfaction.
   a) Download the 2011 data[6] "2011_SFO_Customer_Survey."
   b) Visualize the responses of guests for "On-time status of flight" and interpret the results. Note that to interpret the results, you will have to download the variable dictionary "AIR_DataDictionary_2011-SFO-Customer-Survey." Also, be aware that the dictionary refers to the variable as "LATECODE" instead of "On-time status of flight."

## 3.3 Visualization Methods for Numerical Variables

Quantitative variables are summarized differently than categorical variables. Depending on the type of quantitative variable (discrete or continuous) and the objectives of the analysis, we choose what type of property of the data we want

---

5 data.chattlibrary.org (accessed July 26, 2019).
6 Source: data.sfgov.org (accessed July 26, 2019).

to summarize. These properties tend to be related to the distribution of the data, which indicate how frequent different values of the variable are. Typical traits of interest are as follows:

- **Central tendency**: tells us what values of the variable are considered typical. Some figures help us have an idea of what the central tendency of a variable is without using any math.
- **Dispersion**: this property of a variable tells us how spread out values of the variable are or how far away from the central tendency we may expect any individual value of the variable to be.
- **Shape**: if we were to visualize the distribution of values, what kind of shape would these frequencies have? Informally, if the small values of the variable are much more likely than the large values of the variable, or if the large values of the variable are much more likely than the small values, we say that the distribution is **skewed** or asymmetric. Otherwise, we say the distribution is **symmetric**.
- **Unusual observations**: at times, observations that do not occur often are of interest. This may be the maximum or the minimum or a group of values that are far from central tendency. Observations that are too far from the values we generally get for a variable are known as **outliers**.
- **Trend, seasonality, and temporal dependence**: there are occasions when we have data that varies in time. Weekly sales and monthly gas prices are some examples. Over time, the data may exhibit increasing or decreasing trend. Seasonality may also be present.

Most of the time, interest will be on more than one of these data traits. Some type of figures allow us to look at several of these traits simultaneously, and how these traits compare among two or more variables, whether categorical or numerical.

*Dot Plots*    The simplest visualization tool for a numerical variable is known as a **dot plot**. They can be used for discrete or continuous data. For each datum, the dot plot displays a dot, typically on a horizontal line. Dot plots are useful to explore central tendency, dispersion, and outliers. Dot plots may also give insight on the shape of the distribution although we will see better graphical displays for this later. Figure 3.5 illustrates the use of a dot plot to visualize Levels of Copper (µg/l) in water from samples from residential customers taken in 2016 in[7] Bloomington, Indiana. The authorities define a threshold of 1300 µg/l to take action. We see, that levels of copper fall well below the established limit with a range of 0.001–0.03 µg/l. Overall, a typical level of copper according to the sample is about 0.01 µg/l.

The more unique values the numerical variable has, the harder it becomes to see centrality, dispersion, and shape traits of the data through dot plots.

---

7  Source of data: data.bloomington.in.gov (accessed July 26, 2019).

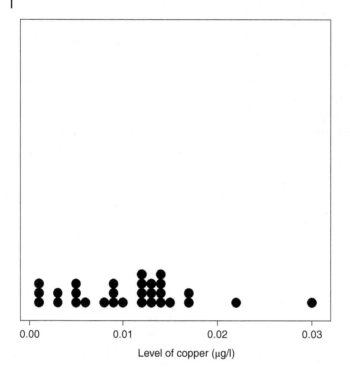

0.00          0.01          0.02          0.03

Level of copper (µg/l)

Figure 3.5 Levels of Copper in water from 30 samples from residential customers taken in 2016 in Bloomington, Indiana.

**Frequency Distributions** It is also possible to use frequency distributions to summarize numerical variables. Since a numerical variable tends to have many possible values, the data is ordered and categorized before tabulation. One must ensure that the categories do not overlap. The choice of lengths of nonoverlapping intervals of values for the numerical variable must summarize the data effectively, a balance of reducing "noise" in the data without loosing important information. Interval lengths are held constant. Each interval represents classes or bins. Computer software will construct a frequency distribution for numerical variables by first sorting it in ascending order, choosing the number of bins, and setting the bin limits. The last step is to set each data value in the corresponding bin to determine frequency. The number of bins is what determines how effective the frequency distribution is in summarizing the data. Traditionally, Sturge's rule has been the default choice to determine number of bins in the frequency distribution. With data sets becoming larger, some have recommended the use of the more robust Freedman–Diaconis rule instead. Just as with categorical variables, counts in each category, relative frequency, or percent frequency scales can be used.

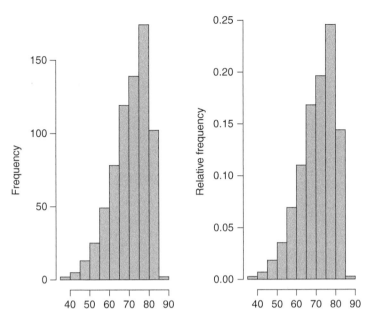

Figure 3.6 Frequency histogram (left) and relative frequency histogram (right) for recycling diversion rate data in New York City for 2018 fiscal year.

*Histograms*   The visual representation of a frequency distribution for a numerical variable is known as a **histogram**. The height of the bins expresses the frequency of each class. The bins are now displayed next to each other because each one is connected to the other through the limits of the intervals. This way, the histogram can also present other attributes of the data such as centrality, dispersion, and shape of the distribution.

**Example 3.2**   *Figure 3.6 presents a histogram of 2018 recycling diversion rates (total recycling / total waste × 100) for several New York City zones[8] in terms of frequencies (left) and relative frequencies (right).*

For this data, Freedman–Diaconis and Sturge's rules both resulted in the same number of bins, 11. The height of the bin represents the frequency measurement of each category. Focusing on the frequency version, we see that typically, zones have a diversion rate roaming around 65–80%. Some zones have recycling rates outside that range. Based on the discussion of distribution shape in Section 3.3, the diversion rates have an asymmetric distribution, skewed left.

_____

8  Source: data.cityofnewyork.us (accessed November 28, 2018).

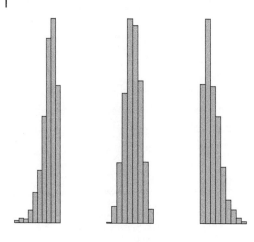

Figure 3.7 From left to right: a left-skewed distribution, a symmetric distribution, and a right-skewed distribution.

---

**Type of Distribution Skewness**
When the lower values of the variable are much more likely than higher values, we say that the distribution is **right skewed**. When the higher values of the variable are much more likely than lower values, we say that the distribution is **left skewed**. In a histogram, the aim is to compare the height of the rectangles of the high values of the variable with the height of the rectangles of the low values of the variable. The rectangles of the low and high values are said to approximate the tail of the distribution of the values of the variable.

---

Figure 3.7 shows examples of a left-skewed distribution (the left "tail" of the distribution is longer than the right "tail"), a symmetric distribution (left and right "tail" have about equal length), and a right-skewed distribution (the right "tail" of the distribution is longer than the left "tail").

Small sample sizes, say less than 30 observations, may not give enough information about the frequency of values. So the histogram can give an unreliable picture of the data distribution when few observations are available. The larger the sample size, often the better a histogram will represent the distribution of the data. However, it has been found that for large data sets Sturge's rule may result in not enough bins. Also some data, even when large, may exhibit very large fluctuations in measurements leading to gaps (empty bins) in the histogram. Another possible outcome is to encounter data that produce a histogram with multiple peaks.

*Line Charts* **Line charts** are used to display a numerical variable that has an index of time points at constant intervals, time series data. Monthly company sales, yearly exports, and quarterly earnings are examples. One axis presents the time index, and the other axis presents the numerical variable. The idea

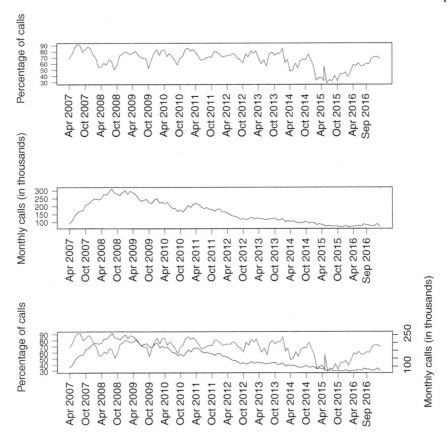

**Figure 3.8** Upper panel shows a line chart for the monthly percentage of 311 calls in San Francisco attended within 60 seconds. Mid panel displays a line chart for the monthly number of 311 calls attended in SF. The lower panel presents both time series in one panel: percentage of calls attended within 60 seconds in red and monthly calls in blue.

is to explore the temporal attributes of the data: mainly if there is a seasonal component or a global trend component.

**Example 3.3** *The upper panel of Figure 3.8 presents the monthly percentage of 311 calls made in San Francisco and attended within 60 seconds, from April 2007 to November 2016.[9] We see that a high percentage of calls were attended within 60 seconds from 2007 until 2013. Then in 2015, a drastic drop in speed of attending calls is implied with an apparent recovery in speed of service by 2016. The middle panel of Figure 3.8 shows the total number of 311 calls attended. The bottom panel of Figure 3.8 charts both time series together. It shows that*

---

9 Source: data.sfgov.org (accessed July 26, 2019).

the drop in performance (red) occurred around 2015 although the total number of monthly calls attended (blue) in recent years are about a third of the calls attended in 2008. The percentage of calls attended within 60 seconds improved in 2016.

---

**Logarithmic Scale**

Some time series data have a really wide range of values. When data changes at a compound rate over time, the line chart of the logarithm of the data can help assess the percentage change in the value of the data over time. The log scale is commonly used with economical and financial data.

---

## Practice Problems

3.10 The following figure presents a histogram of the number of monthly catch basins cleaned by Los Angeles Sanitation, during the Fiscal Year 2014 until mid Fiscal Year 2017.

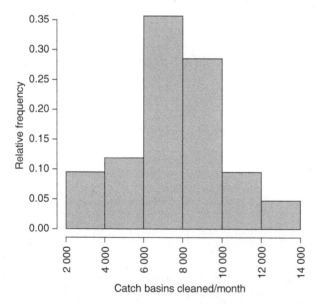

a) What is roughly the typical number of monthly catch basins cleaned by sanitation?

b) What approximate shape does the number of cleaned catch basins have?

c) According to the histogram, how likely is it that in a month the number of basins cleaned is between 6000 and 10 000?

3.11 Download the file "Air_Traffic_Landings_Statistics" with data[10] on Air
traffic landings for San Francisco:
a) Check the minimum and maximum "Landing Count."
b) Draw a histogram of Landing Count.
c) Roughly, what is the typical number of landings?
d) Describe the approximate shape of the distribution of number of landings.

3.12 If the intention is to get a measure of monthly United Airlines landings at
the San Francisco airport, the data used in Problem 3.11 is too detailed.
Specifically, at least one category of the qualitative variables is different for
every row in the data set. Thus, United Airlines data is separated by type of
"Aircraft Version" and categories of other qualitative variables. As a result,
the population is not specific enough. A better[11] option is to summarize
the monthly Landing Count of United Airlines. First, the original data set
needs to be filtered by United Airlines[12] and then aggregation of Landing
Count by Activity Period is necessary. Based on data from July 2005 to
June 2017, the histogram for the 144 monthly landing counts of United
Airlines arrivals is shown below:

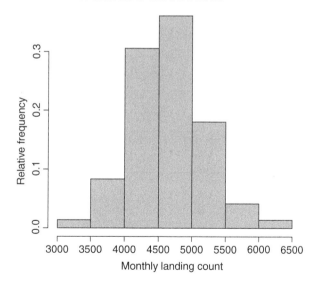

_____
10  Source: data.sfgov.org (accessed August 29, 2017).
11  However, this new alternative is not best either because it does not consider the seasonal
nature of flights.
12  By setting "Operating Airline IATA Code"= "UA".

a) Roughly, what is the typical number of monthly United Airlines landings?

b) Describe the approximate shape of the distribution of number of United Airlines landings.

3.13 Download data[13] on London animal rescue incidents, "Animal Rescue London":

a) Draw a histogram of Incident nominal cost for 2017. Interpret the results.

b) Are incidents more common with dogs or cats? To find out, build a bar chart of the variable "AnimalGroupParent."

*Hint:* You may want to subset the data first.

3.14 The chart below presents quarterly criminal domestic violence data[14] from Edmonton, Canada. Q1 stands for the first quarter of every year.

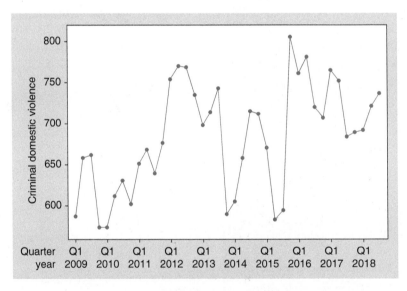

a) Approximately how many cases of criminal domestic violence were reported for the first quarter of 2013?

b) Is a seasonal behavior evident?

3.15 Use the data set "EPS_Domestic_Violence" to reproduce the chart from Problem 3.14.

---

13 Source: data.london.gov.uk (accessed November 14, 2018).
14 dashboard.edmonton.ca (accessed December 12, 2018).

3.16 New York City chronic lower respiratory disease mortality rates for White Non-Hispanic men and women are available below[15]:
  a) Build a line chart for the mortality rates of men. Does there appear to be a trend over time?
  b) Build a line chart for the mortality rates of women. Does there appear to be a trend over time?

| | | | | Year | | | | |
|---|---|---|---|---|---|---|---|---|
| Gender | 2007 | 2008 | 2009 | 2010 | 2011 | 2012 | 2013 | 2014 |
| Male | 20.0 | 18.3 | 16.9 | 20.6 | 21.7 | 21.3 | 18.5 | 20.6 |
| Female | 25.9 | 26.4 | 27.1 | 26.7 | 32.9 | 29.3 | 30.0 | 29.5 |

3.17 Yearly water use per capita in New York City is available[16] ("NYC_Water_ Consumption"). Draw a line chart for water use per capita. Interpret chart.

## 3.4 Visualizing Summaries of More than Two Variables Simultaneously

It is also possible to create tables and graphical displays summarizing two variables simultaneously. The bottom panel of Figure 3.8 already presented an example of visualizing two variables simultaneously.

*Frequency Table Summarizing Multiple Categorical Variables* When there is an interest in the relationship between two categorical variables, it is possible to build a cross tabulation of the occurrence of each category of variable $A$ versus the occurrence of each category of variable $B$, a table known as a **contingency table**. Each row of the contingency table will represent the category of one variable, say variable $A$, while each column of the table represents the category of the other variable, say variable $B$. If variable $A$ has $r$ categories and variable $B$ has $c$ categories, then the contingency table will have $r \times c$ cells with numerical entries. Row totals will present the total number of times each category of variable $A$ occurs. Similarly, column totals will present the total number of times each category of variable $B$ occurs. The Table below summarizes alcohol related deaths in England in 2014, considering death classification and

---

15 Source: data.cityofnewyork.us (accessed July 26, 2019).
16 Source: data.cityofnewyork.us (accessed August 17, 2018).

gender.[17] Below are just a few statements concerning alcohol-related deaths in England during that time period:

| Category | Gender | | |
|---|---|---|---|
| | Male | Female | Total |
| Mental/behavioral disorders | 339 | 150 | 489 |
| Alcoholic cardiomyopathy | 68 | 16 | 84 |
| Alcoholic liver disease | 2845 | 1488 | 4333 |
| Fibrosis and cirrhosis of liver | 911 | 609 | 1520 |
| Accidental poisoning | 242 | 127 | 369 |
| Other | 28 | 8 | 36 |
| Total | 4433 | 2398 | 6831 |

- There were 6831 alcohol-related deaths.
- Most records of alcohol-related deaths are of men.
- 1488 women died of alcoholic liver disease.

***Bar Charts Summarizing Two or More Categorical Variables*** It is straightforward to add more dimensions (other categorical variables) to bar charts using statistical software. There are many ways to do this. One way to do this is using **stacked bar charts** where bars now divide a category by categories of another categorical variable. Then, bar heights can be compared by how each subcategory height compares to the overall rectangle height it is within. Suppose you have two categorical variables. Variable 1 has multiple categories while variable 2 has just two categories. A stacked bar chart builds first a traditional bar chart for variable 1, and then color splits each bar per the categories of variable 2. The aim is to assess if the frequencies of categories of variable 2 change drastically by category of variable 1.

***Dot Plots and Histograms*** Another dimension may also be added to dot plots, the idea being that the centrality and dispersion is presented for some numerical variable accounting for a categorical variable. In Figure 3.9, dot plots are presented for all merchandise imports from around the world for every state[18] of the United States.

Multiple histograms allows us to explore the distribution of a numerical variable according to a categorical variable. For data on vehicle stops in San Diego,

---

17 Source of data: data.gov.uk (accessed July 26, 2019).
18 Source of data: tse.export.gov (accessed July 26, 2019).

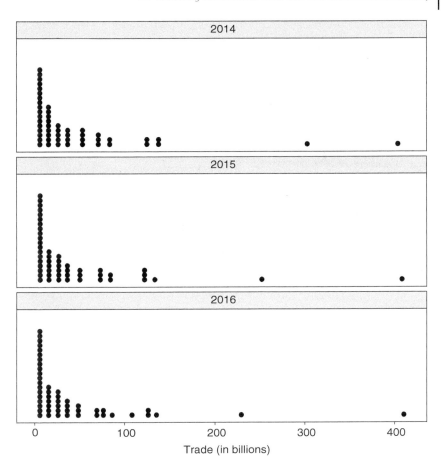

Figure 3.9 All merchandise imports from world (in billions) for every US state.

it may be informative to explore the distribution of ages by race of drivers who were searched. Figure 3.10 does that through the use of multiple histograms.[19] Searches are uncommon for Asian and "Other" drivers of any age. The typical age of a driver who was searched is about the same for White, Black, and Hispanic drivers. On the other hand, the ages of White drivers that were searched are more dispersed than that of Hispanic drivers. It appears that searches for Hispanic and Black drivers under 25 occur more frequently than for White drivers.

---

19 Only data where a moving violation was the cause of the vehicle stop was used.

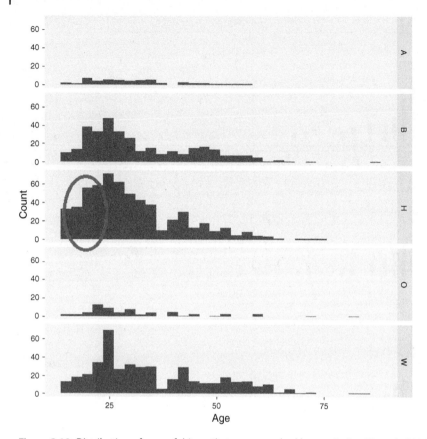

Figure 3.10 Distribution of ages of drivers that were searched by race in San Diego in 2016. Race categories are A, Asian; B, Black; H, Hispanic; O, Other; and W, White. The red oval indicates that searches occur more frequently for Hispanics under 25 than for other young drivers.

*Scatterplots* A **scatterplot** is a graphical representation of $n$ data pairs of two numerical variables, $X, Y$. Think of a scatterplot as two-dimensional version of a dot plot: it represents the features of central tendency and dispersion of the data while also providing an opportunity to screen the observations. But an important contribution of the scatterplot is that a potential association among $X$ and $Y$ is explored. Aspects of the association that are important are as follows:

- Type of association: what approximately happens to one variable as the other increases (whether the association is positive, negative, or perhaps a combination)?
- Form of association: is the association approximately linear, nonlinear, or nonexistent?

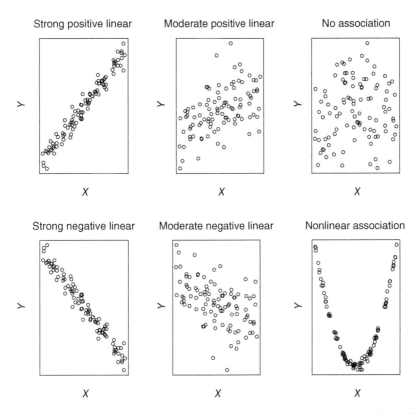

Figure 3.11 Scatterplots showing different associations between two numerical variables $X, Y$.

- Strength of the association: how close to a smooth curve are the points in the scatterplot?

It is key to understand that the association between the two variables are approximate, not exact. Figure 3.11 presents a series of scatterplots.

- On the upper left panel, it is seen that as $X$ increases, on average, $Y$ also increases. The pairs appear to fall on a hypothetical line, supporting the argument that there is a strong positive linear association between the two variables.
- The lower left panel shows an almost identical pattern except that as $X$ increases, on average, $Y$ decreases. Thus, it is said that the variables have a strong negative linear association.
- On the middle upper panel, as $X$ increases, on average, $Y$ increases but this is not as obvious as in the upper left panel. In this case, we say there is moderate linear association between $X$ and $Y$.

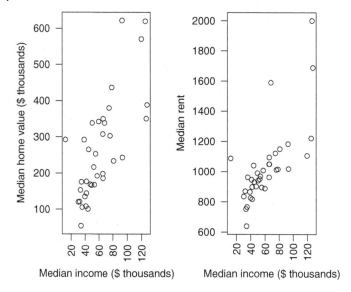

Median income ($ thousands)    Median income ($ thousands)

Figure 3.12 Scatterplots showing different associations of housing market variables from 37 Austin, Texas, ZIP codes.

- The lower mid panel shows the analogous scatterplot for a moderate negative linear association.
- In the upper right panel, there is no identifiable pattern in what approximately happens to $Y$ as $X$ increases: there does not appear to be an association between the two variables.
- The lower right panel presents a situation where at first as $X$ increases, on average, $Y$ decreases. But after a certain point, as $X$ continues to increase, on average, $Y$ now increases as well. This last pattern is an example of nonlinear association.

Scatterplots are also useful in detecting potential aberrant observations. That is, observations with values that appear to deviate considerably from the range of values of most of the observations.

**Example 3.4** *data.austintexas.gov provides 2014 housing market data from 37 ZIP Codes in Austin, Texas, including median household income, median home value, and median rent. In the left panel of Figure 3.12, we see the scatterplot of ZIP Code median household income and median home value.*

- *As median household income increases, on average, median house value also increases.*
- *The association does not appear strong, though one ZIP Code has a median household income of $11 917 and a median house value of $292 500. On*

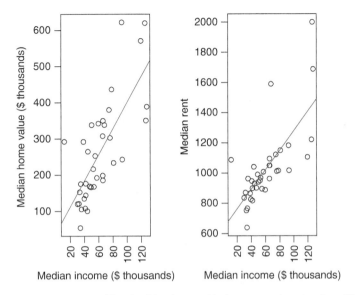

Figure 3.13 A trend line (red) has been added to scatterplots showing different associations of housing market variables from 37 Austin Texas ZIP codes.

*the other side of the spectrum, some ZIP Codes with relatively large median household income have median house values below $400 000.*

*The right panel presents ZIP Code median household income against median rent.*

- *An association between the two variables is evident, but it may not be linear.*
- *Note that for median household incomes below $40000, there is a wide range of median rents. After a median household income of $40000, as median household income increases, median rent increases in an approximately linear fashion until a median household income of about $120000, when again a wide range of median rents occur.*

*Scatterplots may be constructed with a trend line to help envision if a linear association is truly present (the trend line has a nonzero slope when it is). The trend line for the median household income versus median home value scatterplot does have a nonzero slope (left panel, Figure 3.13). Meanwhile, the trend line for the median household income versus median rent scatterplot does support a possible nonlinear relationship (right panel, Figure 3.13).*

When exploring the association between two variables with scatterplots, an implicit assumption has been made that both variables are approximately continuous. If one of the variables is discrete and with a small range of values, then gaps will be visible between points in the figure and geometrically the visualization of an association is slightly different than when both variables are at least

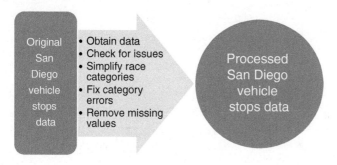

Figure 3.14 Procedure to preprocess San Diego vehicle stops data.

approximately continuous. Nevertheless, scatterplots also come handy in this situation. Let $X$ be the discrete variable with a small range of values. The idea is to determine from the graph if on average, the continuous variable $Y$ shows a trend as the discrete variable $X$ changes value.

### 3.4.1 A Quick Look at Data Science: Does Race Affect the Chances of a Driver Being Searched During a Vehicle Stop in San Diego?

To answer this question, San Diego vehicle stops data was downloaded from an open portal.[20] The original data consisted of over 100 000 incidents of vehicle stops but required some wrangling (Figure 3.14). After exploring the data, one aspect of notice is that the race variable is very specific (Korean, Japanese, Indian, etc.). To simplify our task, we reclassified race into five categories: Blacks, Hispanics, Whites, Asians, and Others.

Figure 3.15 summarizes vehicle stops in San Diego in 2016 by race and whether the driver was searched.[21] Races were ordered by frequency of vehicle stops and search incidences were stacked per driver's race. Overall, White drivers were stopped the most.[22] Stacked bar charts also help visualize the relative frequency of an event based on some category. In this example, we can visualize the frequency of searches per race by evaluating the ratio of each search bar to its corresponding race bar. Drivers who are Black or Hispanic appear to be searched more often than drivers from other races. Specifically, for Black drivers, the red bar covers more of the entire race bar than for White drivers. The same can be said for Hispanic drivers.

What if we take gender into account? Figure 3.16 summarizes vehicle stops in San Diego in 2016 by race, gender, and whether the driver was searched.

---

20 data.sandiego.gov (accessed July 26, 2019).
21 Missing values in the variables were first removed.
22 This does not mean that White drivers are more likely to be stopped by police than drivers from other races. Assessing likeliness of being stopped by race could benefit from using other data such as demographics and time of day the vehicle stop occurred.

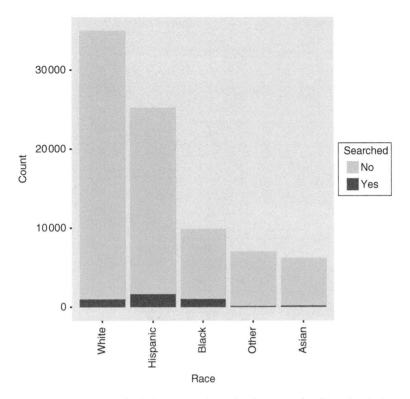

Figure 3.15 Number of vehicle stops and searches by race in San Diego in 2016.

Search counts were stacked per driver's race, while the stacked bar charts were separated by gender. Overall

- men are generally stopped more frequently than women, White men being stopped the most.
- men are also searched more frequently than women, regardless of race.
- for male drivers, the search bar covers more of the entire bar for Black drivers than for White drivers. The same can be said for male Hispanic drivers.
- it is hard to tell with female drivers, but it also appears that race plays a role in chances of a search.

To have a better idea of what is suggested by these bar charts, numbers are needed. The percent of searched male drivers according to whether they were Black, Hispanic, or White were 12.41, 7.84, 3.57%, respectively. Unfortunately, this summary does not answer why there is such a discrepancy in searches of drivers by race. For one, the summary is mostly visual and more numerical statistics are needed to better assess the scale of the differences by race.

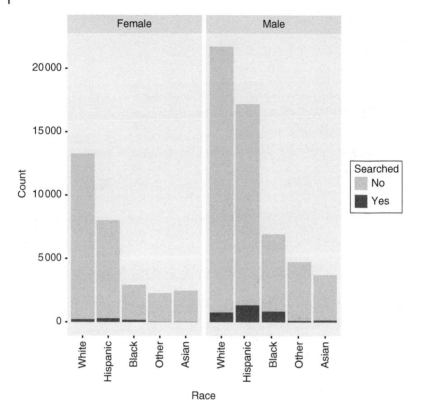

Figure 3.16 Number of vehicle stops and searches by race and gender in San Diego in 2016. Female and male correspond to the driver's gender.

Also the summary does not consider the cause for the vehicle stop[23] or location in San Diego. Furthermore, the data does not include the race of the officer, which may (or may not) be associated to the chance that a vehicle stop involves a search. Finally, statistical inference would be needed to draw conclusions from the current data accounting for several variables, like location of stops, the age of drivers, and the time of day. See Problems 5.44 and 5.45 in Chapter 5 for more.

> **Want to Know More?**
> If you want to know more about this case study, in the companion website you will find the R code that reproduces the results presented here. This data set is rich, and you can modify the code to serve your own needs.

---

23 For visualization purposes, the cause of the vehicle stop was not taken into account. This, turns out, will not affect the overall interpretation.

# Practice Problems

3.18 Returning to 2014 alcohol-related deaths in England, when considering death classification and gender[24] (table below), determine if each statement is appropriate.

| Category | Gender | | |
| --- | --- | --- | --- |
| | Male | Female | Total |
| Mental/behavioral disorders | 339 | 150 | 489 |
| Alcoholic cardiomyopathy | 68 | 16 | 84 |
| Alcoholic liver disease | 2845 | 1488 | 4333 |
| Fibrosis and cirrhosis of liver | 911 | 609 | 1520 |
| Accidental poisoning | 242 | 127 | 369 |
| Other | 28 | 8 | 36 |
| Total | 4433 | 2398 | 6831 |

a) According to the data, 369 people in England died of accidental poisoning in 2014.

b) In England, according to the data, more men than women died of fibrosis and cirrhoses of the liver in 2014.

c) In England, alcohol was one of the most common causes of death in 2014.

d) In 2014, alcohol-related deaths were just as frequent for women as for men in England.

3.19 Dashboards: visual displays of multiple data set variables have become very popular because they provide decision makers a quick presentation of the overall status of a situation. The figure below[25] is a dashboard of medical and retail gross sales of marijuana in Denver Colorado. State whether each statement is true or false.

a) From 2010 to 2018, there has been $1.2 billion in retail gross sales.

b) In 2015, combined medical and retail sales exceeded $500 million.

---

24 Source of data: data.gov.uk (accessed July 26, 2019).
25 Source: app.powerbigov.us (accessed August 27, 2018).

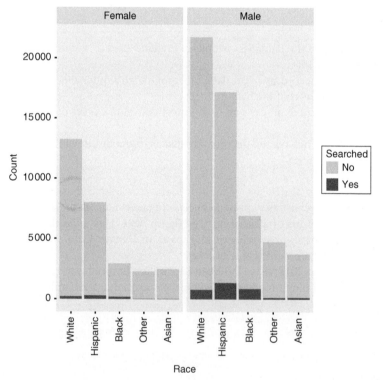

Figure 3.17 Race, gender, and whether a driver was searched during vehicle stops in San Diego in 2016.

3.20 Using the dashboard above, state whether each statement is true or false.
  a) Yearly retail gross sales are usually lower than medical gross sales.
  b) August is usually among the months with the highest retail and medical gross sales.

3.21 In the 2016 San Diego vehicle stops case study, the indication was that vehicle stops were more frequent for White drivers followed by Hispanics (see Figure 3.17). Explain why this makes sense.

3.22 Figure 3.10 shows age histograms by driver's race for moving violation stops in San Diego involving a search. The figure below displays the age histograms by driver's race for moving violation stops in San Diego that do not involve a search. How do the distributions per driver's race compare? Any differences from Figure 3.10?

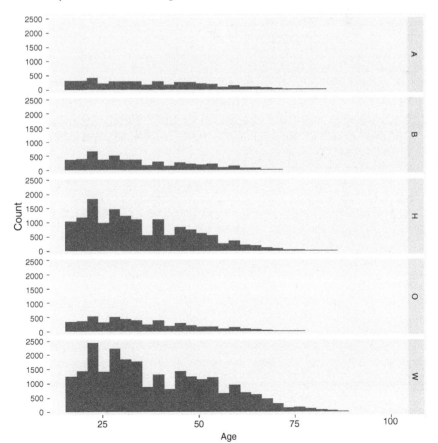

**3.23** The dot plots below show the use of energy over several ZIP Codes in the area surrounding Nashville, Tennessee, in February 2017.[26] Interpret.

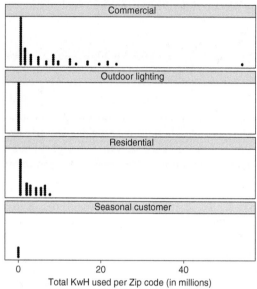

**3.24** According to 2016–2017 Chicago school data,[27] the following scatterplot shows school college enrollment rate versus average American College Testing (ACT) test scores.

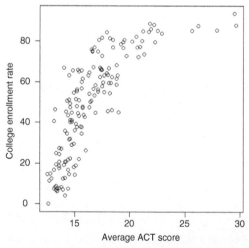

---

26 Source of data: data.nashville.gov (accessed July 26, 2019).
27 Source of data: data.cityofchicago.org (accessed July 26, 2019).

a) What happens to college enrollment rate as average ACT score increases?

b) Is there a linear association between the two variables?

3.25 The following scatterplot displays the number of people exposed to 65 dB or more of railway road noise at night versus number of people exposed to 65 dB or more of industrial road noise at night for 65 agglomerations in England[28] in 2011.

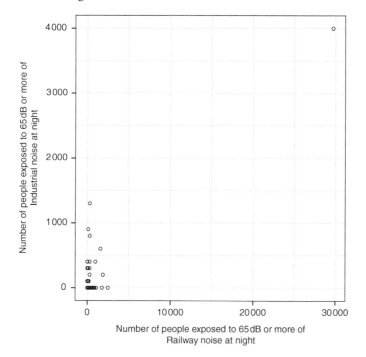

a) What happens to the number of people exposed to industrial road noise as the number of people exposed to railway road noise increases?

b) Is there a linear association between the two variables?

c) Do you think industrial and railway noise have symmetrical distributions?

3.26 Using the data file "2014_Housing_Market_Analysis_Data_by_Zip_Code," reproduce the scatterplots of Austin, Texas, housing market data presented in Figure 3.12.

---

28 Source of data: data.gov.uk (accessed July 26, 2019).

3.27 Using the data "2014_Housing_Market_Analysis_Data_by_Zip_Code," produce a scatterplot of unemployment and median rent at Austin, Texas. Interpret.

3.28 Scatterplots are useful tools to visualize the association between two numerical variables. When the data is highly discretized – there are large gaps among variable values – the scatterplot can have an awkward look. Download the file "Air_Traffic_Landings_Statistics" with data[29] on air traffic landings for San Francisco. Create a scatterplot of Total Landed Weight versus Landing Count. How do you explain the result?

3.29 Which countries provide their citizens with the best higher education? The figure below presents one summary the World Economic Forum used to address this question.[30] On the left, it shows the number of Top 200 universities by country. On the right, it shows the percentage of universities of each country that are part of the Top 200. Mention some ways to better illustrate the point they are trying to make.

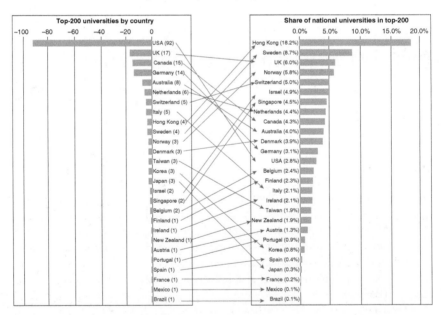

---

29 Source: data.sfgov.org (accessed July 26, 2019).
30 Source: Which countries provide their citizens with the best higher education? World Economic Forum, Switzerland, 2017.

## 3.5   Novel Data Visualization

The techniques examined in this chapter only scratch the surface of data visualization, which with the proliferation of data has become more important. Many other visualization methods are available, some are simply extensions of the tools seen in this chapter. For example, scatterplots can be constructed in three dimensions, or circles representing each $x, y$ pair may change area based on a third numerical variable, a chart sometimes called a **bubble chart**. Moreover, $x, y$ pair circles may have different colors based on some qualitative variable. If the data is available over time, the scatterplot can be made dynamic, often called a **motion chart**, a visualization method popularized by Hans Rosling. The motion chart allows to see how the association between the variables evolves through time. Other popular visualization methods include dashboards for multiple variables (see Problem 3.19), stacked area charts for frequencies that vary in time (see Problem 3.62), word clouds (see Problem 3.64) for unstructured text data, and many others.

Generally, the more variables there are in the data set, the more optional visualization techniques are at our disposal. An analyst will attempt several of them and decide which one better summarizes the data. Figure 3.18 displays the association between cheese fat percentage and moisture percentage, using data from Canada while also considering the type of cheese, color coding the cheese types in the chart. The objective is to explore the possibility that the association between fat percentage and moisture percentage may depend on the type of cheese. It is hard to tell if the cheese type has any impact in the type of association between fat percentage and moisture percentage. Figure 3.19 on the other hand draws separate panels of scatterplots for each cheese type, while also adding a trend line. In this display, it is easier to see that there is some evidence of the cheese type being influential on the type of association between fat percentage and moisture percentage. Specifically, the slopes of the trend line for the firm and soft cheese are very similar. However, the slope of the hard cheese is very different to the firm and soft types.

***Tips on Building and Using Effective Visual Displays***   Apps, websites, and computer software are making interactive data visualization more commonplace. With a movement of a finger or a click of a computer mouse, the end user can now manipulate visualization settings to fit their needs.

We have shown how visual displays help us understand certain features from data. Yet by focusing on the numeric displays, other important considerations have been left out. Some recommendations to effectively use graphical representations of data are as follows:

- Make sure you are using the right visual tool. Statistical software will not necessarily give you an error when applying the wrong visualization tool, it is up to the end user to make the right choice.

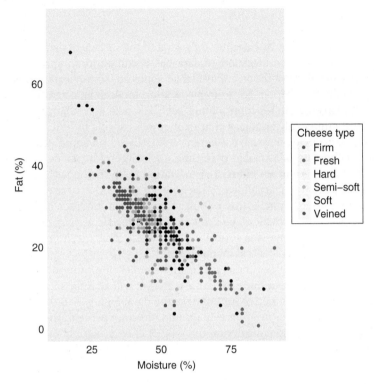

Figure 3.18 Scatterplot of cheese fat percentage and moisture percentage color coded by cheese type.

- Use axis labels. For charts to function properly as aids to understand the data, axis labels must be included specifying what is being presented. For example, in the case of a bar chart, are frequencies being presented, or percentages? It may not be immediately clear without labels. In the case of a numerical variable, include the unit of measurement in parenthesis.
- Keep acronyms or abbreviations to a minimum. Categorical data frequently use abbreviations to represent categories. Too many abbreviations in a chart hinder its interpretation.
- Use captions: a short summary of what the visual displays shows and a description of symbols used if needed.
- When using visual displays in an article or report, make reference to the chart or table and explain what is summarized.
- Be aware that each visual tool has its own attributes and these attributes affect the effectiveness of the displays. For example, statistical software use different aspect ratios[31] for scatterplots or use different algorithms to

---

31 Width/height of the chart.

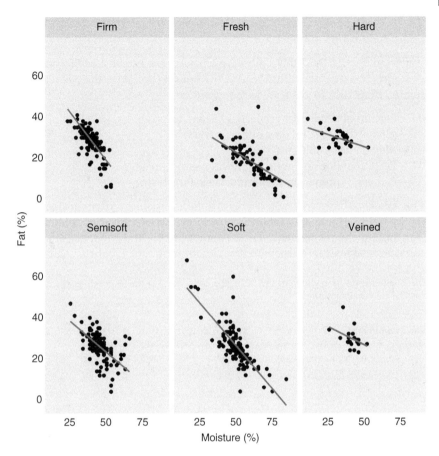

Figure 3.19 Scatterplots of cheese fat percentage and moisture percentage per cheese type.

determine the number of bins in a histogram. It is possible to modify these attributes but this should be done with care.

- Attempt several alternatives. Many times more than one way to visualize data is possible. When unsure of which option is best, a few can be done to see which method best summarizes the data. Two features are key: it should be as simple as possible and easy to understand.
- Understand that charts or tables do not prove beyond any reasonable doubt a point. They may help support a point, yet there is a level of subjectivity in not only building some visual representations, but also in understanding the summary. There are statistical inference procedures that can provide more support to arguments.

### 3.5.1 A Quick Look at Data Science: Visualizing Association Between Baltimore Housing Variables Over 14 Years

Recall from Section 2.5.1 that Baltimore officials have established 55 community statistical areas for Baltimore, and the housing variables that we wish to visualize, from 2001 to 2014, are as follows:

- I1 – Percent of residential properties vacant.
- I2 – Median numbers of days on the market.
- I3 – Median price of homes sold.
- I4 – Percent of housing units that are owner occupied.
- I5 – Percent of properties under mortgage foreclosure.

Our aim is to build a motion chart with these variables. After considerable data wrangling (Figure 2.4), Table 3.3 shows data for 2 of the 55 communities.

The motion chart is found in the companion site for the book, and students are encouraged to experiment with that visualization tool. Figure 3.20 presents

Table 3.3 Yearly housing variables and 2010 population for 2 of the 55 community statistical areas in Baltimore.

| Year | Community | Population (2010) | I1 | I2 | I3 | I4 | I5 |
| --- | --- | --- | --- | --- | --- | --- | --- |
| 2001 | Allendale/Irvington/S. Hilton | 16 217 | 2.27 | 75 | 55 000 | 74.36 | 3.85 |
| 2001 | Beechfield/Ten Hills/West Hills | 12 264 | 0.26 | 42 | 78 500 | 84.76 | 2.26 |
| ⋮ | | | | | | | |

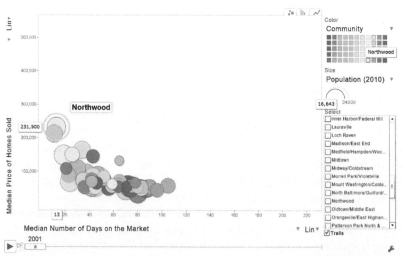

Figure 3.20 A screenshot of the Baltimore housing variables motion chart. Northwood has been selected to stick out from other communities.

one screenshot of the motion chart. It shows the median number of days residential houses are in the market versus the median prices of homes sold for each of the Baltimore communities in 2001. Capitalizing on the interactivity on the motion chart, we can state the following:

- In 2001, there appeared to be a nonlinear negative association between the median number of days residential houses were in the market and the median prices of homes sold.
- In that year, the median number of days residential houses were in the market in Northwood was 13 days, while the median prices of homes sold there was \$231 500. Also, the 2010 population of that community was[32] 16 643.

Figure 3.21 presents the relationship of the same two housing variables, but in 2010. For that year:

- There is no strong evidence of an association between the median number of days residential houses were in the market and the median prices of homes sold.
- The median number of days residential houses were in the market in Northwood was 96 days, while the median prices of homes sold dropped to \$138 500.

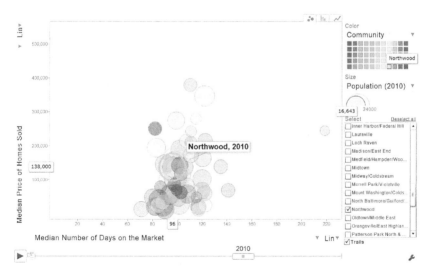

Figure 3.21 Another screenshot of the Baltimore housing variables motion chart. The "play" button in the lower left corner allows to see the scatterplots through all years. Specific years are selected with the "slider," also down at the bottom.

---

32 It was not possible to access community populations for each year between 2001 and 2014. Thus, we only used the 2010 estimate.

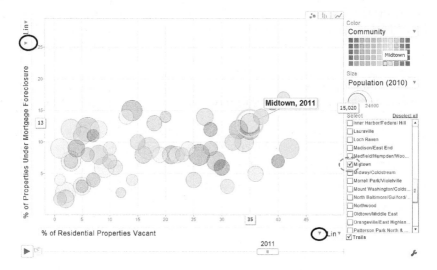

Figure 3.22 Screen shot of the Baltimore housing variables motion chart, showing percent of residential properties vacant versus percent of properties under mortgage foreclosure. Variables to show can be selected from pull-down menus that appear when the arrows are pressed (black circles). Selecting a community (green circle) and then running the animation will track the chosen community over time.

Additional features of the motion chart include the following:

- Tracking specific communities by selecting them (Figure 3.22, box circled in green).
- Selecting different variables for the animated scatterplot (Figure 3.22, arrows circled in black).
- Pressing on the tabs on the upper right corner to draw bar charts or line charts (Figure 3.23).

---

**Get Your Hands Dirty with Motion Charts**

The motion chart summarizes the association between housing variables for each of the 55 communities across 14 years, all in one place. Make sure you try the animated version of the motion chart discussed in this section (see the textbook companion site). If you want to create the motion chart yourself, access the code in the supplementary website. It permits reproducing the application presented in this section. Better yet, you can modify the code to create a motion chart using a different data set.

---

One disadvantage of motion charts is that they sacrifice some data details. Specifically, to be able to animate the scatterplots, the scale of the $x$ and $y$ axes is

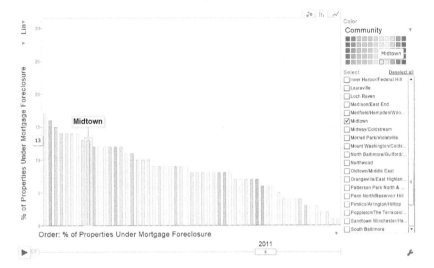

Figure 3.23 The motion chart can also build bar charts by ordering values of variables. Line charts (not shown) are also possible.

set to fit the largest values of the variables across all years covering the data set. A consequence of this is that for some years it may be difficult to determine the association between two variables from the motion chart. Furthermore, with the opportunity to display so many features from data, one can easily lose sight of the goal, to summarize data in a useful way.

## Chapter Problems

3.30 You want to make a graph to display the proportion of faculty members in each of the academic ranks at your institution. The best choice of graph is
  a) scatterplot.
  b) bar chart.
  c) histogram.
  d) pie chart.

3.31 The bar chart below summarizes the 2014 percentage of total electricity output that comes from renewable sources for several countries.[33] According to the chart,
  a) for Brazil, what was the approximate percentage of total electricity output coming from renewable sources?

---

33 Source: data.worldbank.org (accessed July 26, 2019).

b) for India, what was the approximate percentage of total electricity output coming from renewable sources?

c) which country had the lowest electricity output?

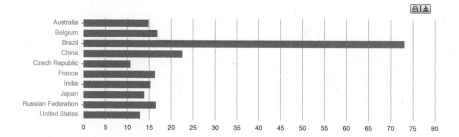

**3.32** Use the complaint data below to reproduce Figure 3.2, a Pareto Chart.

| Category | Frequency |
| --- | --- |
| Shipping | 299 |
| Defective product | 21 |
| Customer service | 198 |
| Packaging | 342 |
| Price | 40 |
| **Total** | 900 |

*Hint:* If performed with Minitab, use the Chart Options button in the secondary bar chart window to properly order the categories.

**3.33** Many places establish rebate programs to promote energy savings. For example, Delaware has a rebate program for the purchase or lease of alternative fuel vehicles. Data is available to the public[34] including the amount of the rebate. Before seeing the data, what type of visual summary do you think is adequate for the amount of the rebate? Download "Delaware_State_Rebates_for_Alternative-Fuel_Vehicles." Construct a

a) frequency table of the amount of rebate. Interpret.

b) bar chart of the amount of rebate. Interpret.

**3.34** Using the data set from Problem 3.33, construct a

a) frequency table of the vehicle type. Interpret.

b) contingency table of gender and vehicle type.

---

34 Source: data.delaware.gov (accessed July 26, 2019).

3.35 A survey of members of the LGBT community in the European Union included questions about whether the person experienced violence or harassment in the last five years, which happened at least partly because they were perceived to be LGBT. The chart below[35] summarizes the responses.

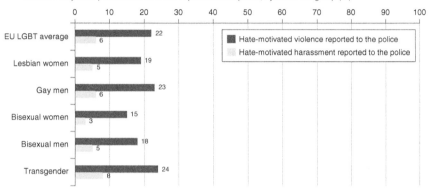

Most serious incident of violence or harassment in the last five years which happened partly or completely because they were perceived tobe LGBT reported to the police, by LGBT subgroup (%)

a) Which LGBT group had the highest proportion of hate-motivated violence?
b) Which LGBT group had the lowest proportion of hate-motivated harassment?
c) What is the implied proportion of EU LGBT that did not report to the police hate-motivated violence?

3.36 In an article published in 2017 in *The Economist* in collaboration with Igarapé Institute, homicide data from many cities around the world were analyzed. The figure below[36] summarizes homicides for the most violent cities. According to the figure,
a) Which city has the highest total number of homicides?
b) Which city has the highest homicide rate?
c) Which city in the United States has the highest homicide rate?
d) Of the countries shown, which country has the lowest national homicide rate?

35 Source: FRA – European Union Agency for Fundamental Rights. (2013). *EU LGBT Survey, Results at a Glance*. Luxembourg: Publications Office of the European Union (figure 11).
36 https://www.economist.com/blogs/graphicdetail/2017/03/daily-chart-23 (accessed July 26, 2019).

e) Can you state which country in the world has the lowest national homicide rate, from this chart?

f) Which country shows the most obvious increase in national homicide rate over the last few years?

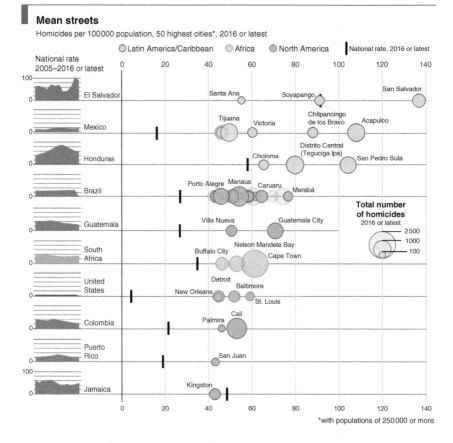

**Mean streets**

Homicides per 100000 population, 50 highest cities*, 2016 or latest

*with populations of 250000 or more

3.37 From Table 3.4,

a) construct an appropriate visual summary for the total poverty ratio data presented in the table.

b) construct an appropriate visual summary for the poverty ratio data by age.

3.38 Using Buffalo New York curb recycling data from January 2011 until June 2018, the line chart below was constructed. Discuss the global trend behavior.

Table 3.4  2014 Poverty ratio for several countries by age category.

| Country | Total | ≤17 year olds | 18–65 year olds | 66+ year olds |
|---|---|---|---|---|
| Australia | 0.1 | 0.1 | 0.1 | 0.3 |
| Finland | 0.1 | 0.0 | 0.1 | 0.1 |
| Hungary | 0.1 | 0.1 | 0.1 | 0.1 |
| Israel | 0.2 | 0.2 | 0.1 | 0.2 |
| Korea | 0.1 | 0.1 | 0.1 | 0.5 |
| Mexico | 0.2 | 0.2 | 0.1 | 0.3 |
| The Netherlands | 0.1 | 0.1 | 0.1 | 0.0 |
| United States | 0.2 | 0.2 | 0.2 | 0.2 |

*Source:* OECD (2017), Poverty rate (indicator). doi: 10.1787/0fe1315d-en (accessed March 10, 2017).

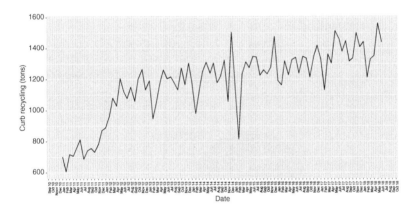

3.39  The file "Buffalo_Monthly_Recycling" has the data used to construct the line chart above, and more. Download the file and construct a line chart for type "Curb Garbage." Interpret.
*Hint:* You will need to subset the data first.

3.40  Every year, administrators of the San Francisco airport conduct a survey to analyze guest satisfaction. Download the 2011 data[37] "2011_SFO_Customer_Survey."
   a) Visualize the responses of guests for "On-time status of flight" by STRATA (time of day). Note that to interpret the results, you will have to download the variable dictionary "AIR_DataDictionary_2011-SFO-

---

37  Source: data.sfgov.org (accessed July 26, 2019).

Customer-Survey." Also, be aware that the dictionary refers to the variable as "LATECODE" instead of "On-time status of flight."

b) Visualize the responses of guests for "On-time status of flight" by DESTGEO. Note that to interpret the results, you will have to download the variable dictionary "AIR_DataDictionary_2011-SFO-Customer-Survey." Also, be aware that the dictionary refers to the variable as "LATECODE" instead of "On-time status of flight."

3.41 Technology has made data visualization accessible to everyone. Indeed, this is a good thing. Yet it is up to the user to properly use visualization tools. Below is a pie chart built a few years back, using an open portal platform. The intention was to summarize school performance data.[38] Specifically, the data included two variables:

- One indicating school subject and school year (e.g. M10-11 for Math in the 2010–2011 academic year).
- The second, establishing the target: minimum percentage of students that pass standardized tests.

These two variables were used to create the chart. Mention at least three things that are wrong with this chart.

*Hint:* The first two columns of the data set "Acad_Perf_PR_2006-2011" are the variables in question.

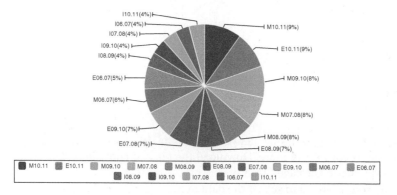

3.42 In practice, bar charts are used occasionally to summarize a numerical variable over time, which can be handy. The bar chart below[39] shows the inflation rate based on the Consumer Price Index in India. For each month, inflation rate is grouped by product class, and class rectangle

---

38  Source: data.pr.gov (accessed July 26, 2019).
39  Source: Ministry of Statistics and Programme Implementation, mospi.nic.in (accessed July 26, 2019).

heights are based on the percentage the product group contributed to inflation rate.

a) What is the inflation rate for September 2014?

b) Which product group consistently has the highest impact on inflation rate?

c) Which product group consistently has the least impact on inflation rate?

d) During what month and year does it seem that the Miscellaneous product group accounted for the biggest percentage of the inflation rate?

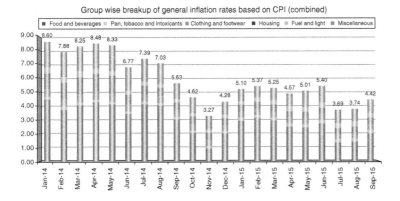

Group wise breakup of general inflation rates based on CPI (combined)

3.43 The figure below presents the association between cheese fat percentage and moisture percentage using data from Canada. Points are color coded by whether the cheese is organic. Interpret.

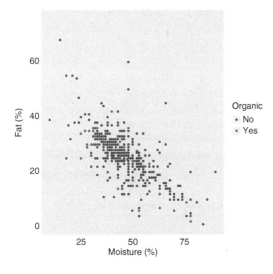

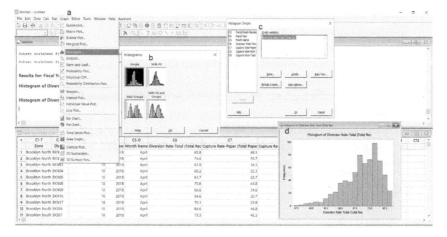

Figure 3.24 Steps, a–d, to obtain a histogram using Minitab.

**3.44** When constructing histograms, the statistical software being used will commonly determine how the graph looks. In Example 3.2, histograms were developed to summarize diversion rates in New York City (Figure 3.6). The required steps and output to get a histogram summarizing the same data using Minitab can be found in Figure 3.24. Why does the resulting histogram have a different shape than the ones seen in Figure 3.6?

**3.45** Continuing from Problem 3.44, using Minitab and the file "Recycling_Diversion_and_Capture_Rates":
a) Construct a histogram for diversion rates, fiscal year 2018, in New York City.
   *Hint:* Subset the data by Fiscal Year = 2018.
b) Modify the histogram to look like Figure 3.6.
   *Hint:* Double-click a bin from the histogram produced in (a) and then fiddle with the options available in the Binning tab.

**3.46** Answers to the 2016 community survey from respondents in Chattanooga are available[40] (filename "2016_Community_Survey_Results_Chattanooga"). Use an appropriate chart to summarize the respondents' answers to survey item "Q1c: Chattanooga as a place to work."

---

40 data.chattlibrary.org (accessed July 26, 2019).

3.47 Using the data from Problem 3.46:
a) Cross tabulate the answers to questions "Q1a" and "Q1c."
b) Build a bar chart of "Q1c" stacked by "Q1a."
c) Based on the stacked bar chart, do you think the answers to the two questions are related?

3.48 Data on leading cause of death between 2007 and 2014 for New York City are available online.[41] Build a tabular summary by "Year" and "Race Ethnicity" variables of the deaths for which the "Leading Cause" was diabetes mellitus.
*Hint:* If you use Minitab, first subset the data. Then sort new data by year and race. Next get the sums of "Deaths" by "Year" and "Race Ethnicity" (ignoring "Sex").

3.49 The figure below summarizes domestic violence data from Bloomington, Indiana, from October 2011 to April 16, 2017. It focuses on cases where the authorities determined there was no alcohol, mental health, or drugs involved.[42] What is the main problem with this Figure?

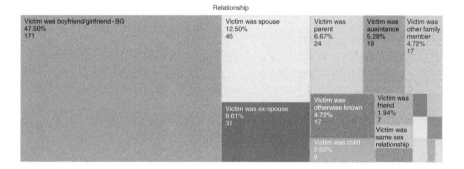

3.50 In 2005, Florida lawmakers passed the controversial "Stand Your Ground" law, which broadened the situations in which citizens can use lethal force to protect themselves against perceived threats. The figure below summarizing murders committed using firearms was released by *Reuters* in 2014. Interpret what the figure suggests. Would you recommend a way to improve the design of the chart?

---

41 Source: data.cityofnewyork.us (accessed July 26, 2019).
42 Source: www.burlingtonvt.gov (accessed July 26, 2019).

**Gun deaths in Florida**
Number of murders committed using firearms

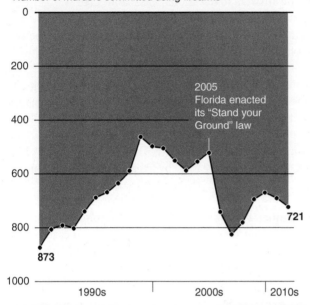

3.51 Download "Minneapolis_Property_Sales_20102013" a data set[43] of property sales in Minneapolis.
   a) Using statistical software, extract data for transactions such that "*PropType = Residential.*"
   b) Construct a scatterplot of Gross Sale Price versus Adjusted Sale Price.
   c) Identify unusual observations from the scatterplot.

3.52 Using the data "2014_Housing_Market_Analysis_Data_by_Zip_Code," produce a scatterplot of "rental units affordable to average artist" and "owner units affordable to average artist" at Austin, Texas. Interpret.

3.53 Using the data "2014_Housing_Market_Analysis_Data_by_Zip_Code," produce a scatterplot of "rental units affordable to average teacher" and "owner units affordable to average teacher" at Austin, Texas. Interpret.

3.54 In the textbook companion site, access the motion chart for the Baltimore housing data. Set the $x$ axis to "% of Residential Properties Vacant" and the $y$ axis to "% of Properties Under Mortgage Foreclosure."

43  Source: opendata.minneapolismn.gov (accessed July 26, 2019).

a) What was the 2001 "% of Residential Properties Vacant" in the Madison/East End community?
b) What was the 2001 "% of Properties Under Mortgage Foreclosure" in the Madison/East End community?
c) What was the 2010 population in the Madison/East End community?
d) What was the 2010 "% of Properties Under Mortgage Foreclosure" in the Madison/East End community?

3.55 In the textbook companion site, access the motion chart for the Baltimore housing data. Set the $x$ axis to "% of Housing Units that are Owner-Occupied" and the $y$ axis to "% of Residential Properties Vacant."
a) What was the 2001 "% of Housing Units that are Owner-Occupied" in the Lauraville community?
b) What was the 2001 "% of Residential Properties Vacant" in the Lauraville community?
c) What was the 2010 population in the Lauraville community?
d) What was the 2014 "% of Residential Properties Vacant" in the Lauraville community?

3.56 In the textbook companion site, access the motion chart for the Baltimore housing data. Set the $x$ axis to "% of Housing Units that are Owner-Occupied" and the $y$ axis to "% of Residential Properties Vacant".
a) For 2001, does there appear to be a linear association between "% of Housing Units that are Owner-Occupied" and "% of Residential Properties Vacant"? If so, is it negative or positive?
b) For 2014, does there appear to be a linear association between "% of Housing Units that are Owner-Occupied" and "% of Residential Properties Vacant"? If so, is it negative or positive?
c) How is the 2014 association compared to the 2011 one?

3.57 In the textbook companion site, access the motion chart for the Baltimore housing data. Set the $x$ axis to "% of Residential Properties Vacant" and the $y$ axis to "% of Properties Under Mortgage Foreclosure."
a) For 2010, does there appear to be a linear association between "% of Residential Properties Vacant" and "% of Properties Under Mortgage Foreclosure"? If so, is it negative or positive?
b) For 2014, does there appear to be a linear association between "% of Housing Units that are Owner-Occupied" and "% of Residential Properties Vacant"? If so, is it negative or positive?
c) How is the 2014 association compared to the 2011 one?

**3.58** In the Baltimore housing data motion chart, go to the bar chart tab and set the y axis to "% of Properties Under Mortgage Foreclosure" (leave the x axis on "Order: % of Properties Under Mortgage Foreclosure").

a) Which community had the highest "% of Properties Under Mortgage Foreclosure" in 2005?

b) Which community had the highest "% of Properties Under Mortgage Foreclosure" in 2013?

**3.59** In the Baltimore housing data motion chart, go to the bar chart tab and set the y axis to "Median Price of Homes Sold" (leave the x axis on "Order: Median Price of Homes Sold").

a) In 2005, what was the "Median Price of Homes Sold" for Greater Roland Park/Poplar Hill?

b) In 2014, what was the "Median Price of Homes Sold" for Greater Roland Park/Poplar Hill?

**3.60** Returning to the Baltimore housing data motion chart. Set the x axis to "% of Residential Properties Vacant" and the y axis to "% of Properties Under Mortgage Foreclosure." What part of the design of the motion chart makes it difficult to use the chart to visualize the association between the two housing variables?

**3.61** In the textbook companion site, access the motion chart for Puerto Rico workforce data. Set the x axis to "Time" and the y axis to "Population."

a) Place a check mark on San Juan in the bottom left box. What is the general trend of the San Juan Population?

b) Place a check mark on Dorado in the bottom left box (and make sure San Juan or any other location has no check mark). What is the general trend of the Dorado Population?

**3.62** When frequencies of several categories of a variable are available over time, a stacked area chart can be constructed. The figure below illustrates the monthly behavior of different types of bankruptcies in Puerto Rico, an island in the Caribbean, from 2000 to mid 2018. The y axis shows relative frequencies and the x axis displays time.

a) Overall, what type (chapter) of bankruptcy is most common?

b) Overall, what types (chapter) of bankruptcy are the least common?

c) A new bankruptcy law to reduce the frequency of Chapter 7 bankruptcies was created in October 2005. Did the law work?

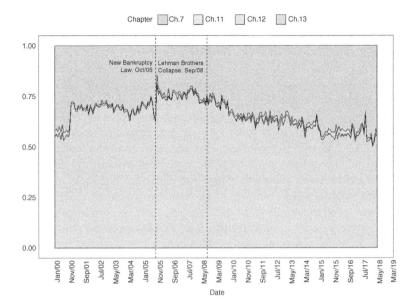

3.63 NOAA has records for many weather stations around the world. Temperature, cloud coverage, and rainfall are just some of measurements that can be extracted from these stations. The rainfall data is at a daily scale, but for water management purposes, it is useful to aggregate rainfall into monthly totals. The challenge is that weather station data is commonly riddled with missing values, which hinder the characterization of events such as droughts. The figure below summarizes the number of days with rainfall data per month for two weather stations in San Juan, Puerto Rico. Which weather station will allow to better characterize monthly rainfall behavior?

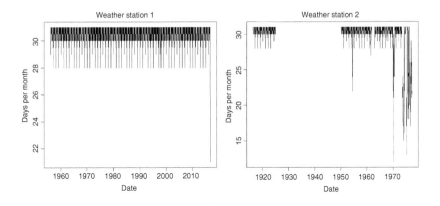

**3.64** Compared to structured data, it is harder to determine patterns when data strictly consists of unstructured text. In this case, visualization can be very useful. A word cloud is a visual representation of text data. One way to construct them is to present the words such that the most common words appear the biggest and closest to the middle of the cloud. As we depart the center of the cloud, the less common words are seen in smaller font. The Metro Codes Department in Nashville, Tennessee, handles property standards violations. The problems are recorded and are available online.[44] Determining what the most common violation is or frequency of violations can help the authorities establish policies to reduce violations and ensure they have the necessary resources to handle some violations (e.g. removal of an abandoned vehicle). With over 35 000 recorded violations and no categories, there are limited alternatives to extract information from the data. The figure below presents a word cloud of the reported problems.[45] What are the two most common complaints?

44  data.nashville.gov (accessed July 26, 2019).
45  Word Cloud is based on data downloaded on April 29, 2017.

## Further Reading

Baumer, Benjamin S., Kaplan, Daniel T., and Horton, Nicholas J. (2017). *Modern Data Science with R*. CRC Press.

Rivera, R., Marazzi, M., and Torres, P. (2019). Incorporating open data into introductory courses in statistics. *Journal of Statistics Education*. doi: 10.1080/10691898.2019.1669506.

Wickam, Hadley. (2009). *ggplot2: Elegant Graphics for Data Analysis*. Springer.

Wickam, Hadley and Grolemund, Garrett. (2017). *R for Data Science*. O'Reilly.

# 4

# Descriptive Statistics

## 4.1 Introduction

We have seen that visual descriptions of data can be very useful in giving us an overall idea of many traits of a sample: frequencies, centrality, dispersion, shape of the distribution, association between variables or groups, and trends were some of the traits we were able to assess from visual descriptions, depending on the type of variable at hand. These type of summaries also allowed detection of data errors. However, visual descriptions do not tend to be enough to describe data. There are two main reasons for this:

- By choice of scale, bin size, and other attributes, figures are subjective.
- Numerical summaries usually cannot be determined precisely from charts.

Take Figure 4.1, which shows the line chart of monthly hotel registrations at Puerto Rico, an island destination in the Caribbean. There is a seasonal aspect to the hotel registrations. But, it is hard to tell from the chart if the hotel registrations have decreased or increased overall. Perhaps there was a drop around 2009 and an increase since then, but it is unclear.

The situation seen above is not an exception. Scatterplots may not clearly indicate if there is a linear association between variables, and histograms may not be enough to visualize common values of a measurement or the shape of the data. We often need more specific summaries of the data to make decisions. For example, suppose we have the SAT scores of 100 randomly sampled freshman business students. A histogram would display the shape of the SAT scores in the sample and would give us an idea of centrality of scores and how they are spread out around the measure of centrality. But, instead of just having an idea of these traits, we may need more specific descriptions of them. We may want a number that indicates the "expected" SAT score in the sample, or a number that indicates how SAT scores are spread out around the "expected" SAT score. These numbers are known as numerical descriptive statistics. This chapter will cover many common type of numerical descriptive statistics.

*Principles of Managerial Statistics and Data Science*, First Edition. Roberto Rivera.
© 2020 John Wiley & Sons, Inc. Published 2020 by John Wiley & Sons, Inc.
Companion website: www.wiley.com/go/principlesmanagerialstatisticsdatascience

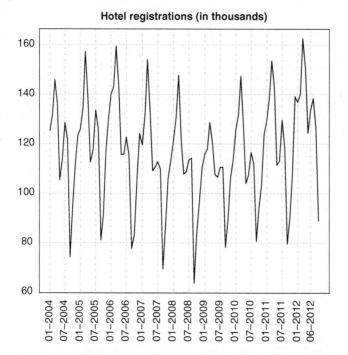

**Figure 4.1** Line chart of monthly hotel registrations (in thousands) at Puerto Rico, an island in the Caribbean from January 2004 to September 2012.

The first numerical summaries obtained from a quantitative variable should be the minimum and the maximum. Easily computed with the help of a computer, the minimum and maximum grant the opportunity to quickly check if there are any egregious errors in the data. For example, in a sample of baby car seat weights, an outcome of −25 or 300 would immediately flag the observation as a mistake. With sample sizes often being in the hundreds or more, the minimum and maximum become important error detection tools. In conjunction with visual summaries, the minimum and maximum help detect obvious errors and how many are present in the data. If an outcome is determined with certainty to be a mistake, it should be removed. Otherwise, numerical descriptions of the data could be grossly misleading. In what follows, we introduce some numerical summaries to describe a quantitative variable. Afterwards, methods to summarize the association between two variables are presented.

---

**Need a Math Refresher?**
From this point until Chapter 11, mathematical equations and calculations will be routine. Head to Appendix A if you wish to review mathematical concepts and notation.

## 4.2 Measures of Centrality

We already know that central tendency tells us what a typical value of a variable is and how to visualize the central tendency of a variable. Most of the time, we also need to quantify it. In the introduction, we gave an example about SAT scores from a sample of 100 randomly sampled freshman business students. Let us suppose the sample was taken by the school administration to answer the following question: What is the expected SAT score of a freshman business student? There are a lot of factors that affect a student's SAT score: the amount of time they study, type of high school, parental support, parents' education, etc. But the administration may not be interested in these other factors or may not have the resources to gather all this information. A measure of central tendency allows us to disregard all these factors to easily determine what is the typical SAT score of entering freshman students. Say the expected SAT score is 625. The SAT score of any entering student could be far away from 625. But among many students, their SAT score will typically be around 625. There are many ways to measure central tendency. The most typical ones are the mean and the median. We will learn that the one that is most adequate to use will depend on our interest and other properties of the variable.

### Mean

**Example 4.1** *Suppose there is a game where we can either get $2, $4, or $6. All possibilities are equally likely. We would like to know what we should expect to win if we were to play this game.*

One way of doing this is to use the fact that all possibilities are equally likely to assign weights for each possible amount. Our expected winnings will be the sum of the product of each possible amount and its weight. That is, $2 is just one of three possible amounts that we can win by playing this game, therefore we assign $2 a weight of 1/3. The other possibilities are assigned the same weight. We can then obtain the expected amount:

$$\frac{1}{3}(2) + \frac{1}{3}(4) + \frac{1}{3}(6) = \frac{1}{3}(2 + 4 + 6) = 4$$

So we should expect to win $4 if we play this game. Specifically, each time we play this game we will get one of the three possibilities, but after playing the game many, many times, on average we would have won $4 a game.

This concept can be generalized to any situation with $N$ equally likely possibilities. For a population with $N$ equally likely items, the **population mean** ($\mu$), also known as the expected value, is

$$\mu = \frac{\sum_{i=1}^{N} x_i}{N} = \frac{x_1 + x_2 + \cdots + x_N}{N}$$

In Example 4.1, we see $\mu$ gives us the value "in the center" of the ordered possible values, and that is why the mean is a measure of centrality. Often, it is not possible to measure $\mu$ directly. It might need too many people, a lot of money, or the population may be too big. An alternative is to estimate it based on a (representative) random sample of $n$ observations ($n$ is smaller, often much smaller than $N$):

$$\overline{X} = \frac{\sum_{i=1}^{n} x_i}{n} = \frac{x_1 + x_2 + \cdots + x_n}{n} \tag{4.1}$$

this is known as the **sample mean**.

**Example 4.2**  *To assess the typical commercial power consumption in February in the surrounding area of Nashville, Tennessee, we will use 2017 kilowatt-hour (kWh) data[1] from 86 ZIP Codes. The ordered data is*

| | | | | | | | |
|---|---|---|---|---|---|---|---|
| 40 | 471 | 668 | 1064 | 4605 | 5222 | 9755 | 10118 |
| 12713 | 16300 | 19126 | 23291 | 28025 | 44520 | 46800 | 49739 |
| 53640 | 63720 | 99794 | 121500 | 130944 | 133812 | 153037 | 153499 |
| 173118 | 176418 | 331200 | 362460 | 404021 | 419895 | 447340 | 491964 |
| 516000 | 535234 | 559975 | 581073 | 615747 | 673198 | 696971 | 754214 |
| 761900 | 764775 | 829661 | 831616 | 836826 | 907158 | 915635 | 970928 |
| 972945 | 1062664 | 1235538 | 1276576 | 1309065 | 1502947 | 1630032 | 1798960 |
| 1937568 | 2540797 | 2814565 | 2902660 | 2949790 | 3206050 | 4315417 | 4490491 |
| 5009918 | 5060494 | 6454158 | 6882965 | 8158407 | 8292035 | 8325783 | 8335953 |
| 8574076 | 9235521 | 9859196 | 12209694 | 12450797 | 12568467 | 13864886 | 16382338 |
| 16897434 | 19503463 | 21193679 | 21919235 | 23689663 | 54015896 | | |

The sample mean is

$$\overline{X} = \frac{40 + 471 + \cdots + 54\,015\,896}{86} = 4\,192\,673$$

According to data for February 2017, the typical commercial power consumption for the month is 4 192 673 kWh. Figure 4.2 shows the dot plot of the power consumption scaled in millions of kWh. It indicates that the consumption for most ZIP Codes tend to be below 4 000 000 kWh. So why is our average, a measure of "typical" power consumption, above that value? Because some ZIP Codes have a power consumption that exceed by far 4 000 000 kWh. As a matter of fact, one ZIP Code had a consumption of 54 015 896 kWh.

As can be seen from Eq. (4.1), what the sample average does is assigning an equal weight of $\frac{1}{n}$ to the sum of possibilities. That is, each observation value is given the same importance in computing the average. On occasion, not all possible values in a sample are equally likely. Sometimes, very high values are more likely than very low values or vice versa. When this happens, computation

---

1  Source of data: data.nashville.gov. (accessed July 26, 2019)

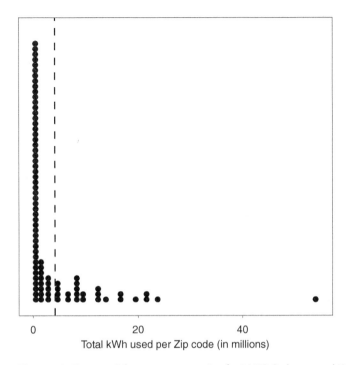

**Figure 4.2** Commercial power consumption for 86 ZIP Codes around Nashville, Tennessee. Consumption has been scaled to millions of kWh. The dashed line represents the sample mean.

of $\overline{X}$ is still possible but it may become unreliable as a measure of centrality by giving too much weight to unlikely outcomes.

---

**When to Use and When Not to Use the Sample Average**

English comedian Jasper Carrott once said "Did you know that the average human being has approximately one testicle and approximately one breast? – which just goes to show how useless averages are!"

There is a point to his joke, even if not exactly the one he meant. The use of $\overline{X}$ should be questioned:

- When observations are not homogeneous. For example, observations are taken over drastically different times on a trait that drastically changes over time.
- When sampling of observations are not equally likely. Since the mean assigns each possibility equal weight, when high values are more likely than very low values, this will result in a mean that is too low. On the other hand, when low

values are more likely than very high values, by the mean giving equal importance to all values, the central tendency estimate will be too large.
- When observations are not numerical. This sounds obvious enough, but data
  can be misleading sometimes (e.g. when categorical answers on favorite color
  are coded into numbers).

*Median*   The **median** is a value such that half of the ordered numerical observations are to the left of it, and the other half to its right. Think of the median
as the middle point of the data. One can show that when the sample size is
uneven, the median will be one of the observations, while when the sample size
is even, the median will not be one of the observations. Although one can easily
determine what the median is by sorting the data and counting the number of
observations, we use a more systematic approach to find its value. The reason
we implement the procedure this way will become apparent in Section 4.4. To
find the median, we proceed the following way:

- Check if the data is sorted (in ascending order). If not, sort the data.
- Calculate the index value $i$,

$$i = \left( \frac{50}{100} \right) n$$

  where $n$ is the number of observations.
- If $i$ is not an integer, round up to find the position in the ordered data where
  you will find the value of the median. If $i$ is an integer, then take the $i$th observation and average it with the $i$th $+ 1$ observation to find the value of the
  median.

Returning to Example 4.2, to get the median, the data must be sorted first.
Figure 4.3 presents the ordered data. The first number between square brackets in each row is not an observation; instead, it indicates the order of the first
observation in that line. Therefore the 41st observations was 761 900. According to the second step to find the median,

$$i = (50/100) \times 86 = 43$$

Since this number is an integer, the median is the average of the 43rd and 44th
observations, $m = 830\ 638.5$.

The median gives us a value that is substantially different than the average
$-4\ 192\ 673$ – although they are both measuring centrality. Which one is more
reliable? Figure 4.4 illustrates the dot plot of power consumption but now it
also includes the median (red dashed line). It is seen that the median is a better
indicator of the most typical commercial power consumption in comparison to
the mean.

| [1] | 40 | 471 | 668 | 1064 | 4605 | 5222 | 9755 | 10118 |
|---|---|---|---|---|---|---|---|---|
| [9] | 12713 | 16300 | 19126 | 23291 | 28025 | 44520 | 46800 | 49739 |
| [17] | 53640 | 63720 | 99794 | 121500 | 130944 | 133812 | 153037 | 153499 |
| [25] | 173118 | 176418 | 331200 | 362460 | 404021 | 419895 | 447340 | 491964 |
| [33] | 516000 | 535234 | 559975 | 581073 | 615747 | 673198 | 696917 | 754214 |
| [41] | 761900 | 764775 | 829661 | 831616 | 836826 | 907158 | 915635 | 970928 |
| [49] | 972945 | 1062664 | 1235538 | 1276576 | 1309065 | 1502947 | 1630032 | 1798960 |
| [57] | 1937568 | 2540797 | 2814565 | 2902660 | 2949790 | 3206050 | 4315417 | 4490491 |
| [65] | 5009918 | 5060494 | 6454158 | 6882965 | 8158407 | 8292035 | 8325783 | 8335953 |
| [73] | 8574076 | 9235521 | 9859196 | 12209694 | 12450797 | 12568467 | 13864886 | 16382338 |
| [81] | 16897434 | 19503463 | 21193679 | 21919235 | 23689663 | 54015896 | | |

**Figure 4.3** Ordered commercial power consumption for 86 ZIP Codes around Nashville, Tennessee. Consumption has been scaled to millions of kWh.

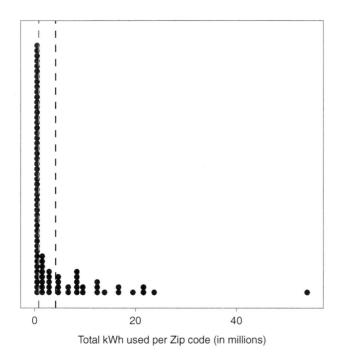

Total kWh used per Zip code (in millions)

**Figure 4.4** Commercial power consumption dot plot. The black dashed line represents the sample mean and the red dashed line represents the sample median.

***Comparison of the Mean and the Median*** These two measures of centrality will not always be equal. So which one should we choose? To answer this question, the shape of the distribution will play an important role. To picture the shape of the distribution of values of the data, it is convenient to use smooth curves to "model" the shape of the histogram (Figure 4.5). The $x$ axis represents the possible values of the variable while the $y$ axis indicates how often they occur.

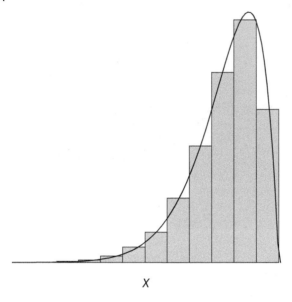

Figure 4.5 A smooth curve alternative to the histogram to visualize the shape of the distribution. When the left tail is longer, the distribution is said to be left skewed.

X

Recall from Chapter 3 that the tails of the distribution of the values of the variable pertain to the higher values and smaller values of the variable. When the distribution is left skewed, the left tail is longer than the right tail (Figure 4.5).

Let us consider the behavior of the mean and median in situations where we have a fairly large amount of observations, say $n \geq 30$. Figure 4.6 illustrates the association between the mean and median of data for distributions with left-skewed, symmetric, and right-skewed shapes.

- It is shown that when the data is left skewed, the mean is smaller than the median. This occurs because the mean is a function of the value of each observation. Thus, although low values of the variable are unusual, they are given equal importance when calculating the mean than higher, more typical values. For left-skewed data, Figure 4.6 shows that the median in this case is better at characterizing the typical values of the variable because it uses the order of the data and only the values of the observations close to the middle of all values.
- In contrast, when the data is right skewed, the mean is greater than the median. This occurs because although high values of the variable are unusual, they are given equal importance when calculating the mean than lower, more typical values. For right-skewed data, Figure 4.6 shows that the median is better at characterizing the typical values of the variable.
- In the case of a symmetric distribution, then the mean and median are the same, so both measures are equally effective at characterizing the typical values of the variable.

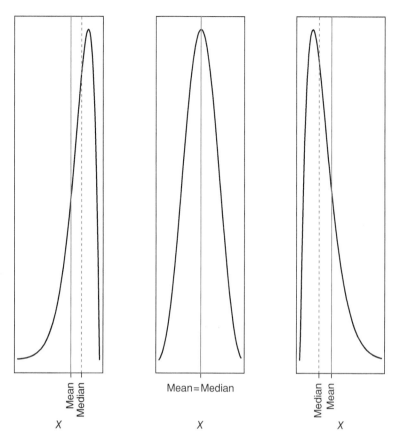

Figure 4.6 How the mean and median of the data compare when the distribution of the data is left skewed (left panel), symmetric (mid panel), and right skewed (right panel).

By knowing the shape of the data, it is possible to determine the relationship between the mean and the median before calculating them. Furthermore, by knowing the relationship between the mean and the median, it is possible to determine the shape of the distribution. Admittedly, the shape of the distribution is not always perfectly left skewed, right skewed, or symmetric. It is fairly common for the data to be approximately symmetric, and then the mean and median will be approximately equal.

***When to Use and When Not to Use Sample Median*** The comparison above appears to imply that for skewed distributions the median should be preferred over the mean in general. However, the context of the situation is important. In a nutshell, the median may leave some information out when the variable has only a few possible numerical values.

Table 4.1 Number of live birth children for teenagers who reside in community *X*.

| | Frequency | |
| --- | --- | --- |
| No. of Children | Male | Female |
| 0 | 48 | 43 |
| 1 | 10 | 13 |
| 2 | 2 | 4 |
| **Total** | **60** | **60** |

**Example 4.3** *Suppose there is a community that is interested in evaluating sex education among teenagers. 120 teenagers answered a survey, and one of the questions was "how many live birth children do you have?" Table 4.1 summarizes the answers.*

For male and female, the median number of children is 0. In contrast, the mean for males is 0.23, and for females it is 0.35. By using the numerical value of the observations and not just the ordering, the mean indicates there is a difference in "typical number of children" among teenage females and males. On average, teenage females have more children than males in this community.

Also, as we will discuss in Chapter 8, the sample mean has good mathematical properties, making it a better measure of centrality in some situations.

*Other Measures of Centrality* The **mode**, which is the most frequently occurring value of the data, is another measure of centrality. For a value to be considered the mode, it must occur at least twice in the data. Based on its definition, the mode applies to numerical and categorical data. But the mode as a measure of centrality only makes sense when applied to numerical or ordinal data.

**Example 4.4** *A graduate student is working on a project to determine the amount of hours that second-year business students dedicate to overall studying every week. Eight of the answers from the random sample were*

$$2 \quad 1 \quad 4 \quad 2 \quad 3 \quad 8 \quad 5 \quad 2$$

The most frequent response was 2 hours. In contrast, the mean is 3.38 hours and the median is 2.5 hours.

The mode has a series of disadvantages that hinder its usefulness. First, for continuous data there may be no mode, which occurs when no observation was repeated. Second, samples may contain multiple modes, multiple values that occurred most frequently.

Returning to Example 4.4, assume that the eight observations were instead:

  2  1  4  2.25  3  8.25  5  2.5

This new version of the data set has no mode, and the mean and median are now 3.5 and 2.75 hours, respectively. On the other hand, had the three observations been

  2  3  5  1  3  8  5  2

the modes are 2, 3, and 5, while the mean and median are now 3.62 and 3 hours, respectively.

The are other alternative measures that quantify centrality. For example, on some occasions it is helpful to assign more weight to some observations than to others when determining the mean.[2] The **weighted mean** assigns different weights to each observation to estimate the measure of centrality. Moreover, the **trimmed mean** trims away from the data a certain percentage of the lowest and highest values of the observations. The idea is to minimize the impact that extreme values have in the calculation. As a last example, the **geometric mean** is a mean based on the $n$th root of the product of $n$ observations, and it is useful with data that has a certain behavior.

The usefulness of measuring the centrality of a numerical variable depends on the context of the intended use. Suppose you need to cross a river but do not know how to swim. If you are told that the average depth is 2 feet, should you cross the river? Hopefully you answered no. Even assuming that the average depth of the river is 2 feet, this information is not useful enough to a person who does not swim. In some parts, the river may be 9 feet, while in others it may be 0.5 feet, but overall the average is 2 feet. However, a swimmer may find this information useful, to determine the swimming approach they must take in the river.

*Descriptive Statistics in Minitab*   To obtain mean, median and other descriptive statistics in Minitab, go to `Stat > Basic Statistics > Display Descriptive Statistics` .... Figure 4.7 illustrates how to obtain the mean, median, and other statistics for Example 4.2. Do not forget that through descriptive statistics, we also want to preprocess the data. Thus, the minimum and maximum observations are useful summaries.

---

2  The arithmetic mean, $\bar{x}$ assigns the same weight, $1/n$, to every observation.

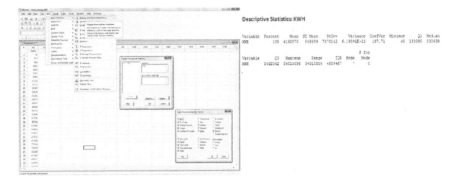

Figure 4.7 Steps to obtain descriptive statistics using Minitab.

## Practice Problems

4.1 Using the following data,

9.71   4.25   3.75   9.42   8.33   7.21   2.55

show that
a) $\overline{X}$ = 6.46.
b) median = 7.21.

4.2 For a sample: 4.5, 6.2, 7.4, 3.2, 4.5, 4.7, 5.1
a) Calculate $\overline{X}$.
b) Find the median.
c) Find the mode.

4.3 Here is a random sample of five "hours per week playing video games" answers.
2, 8, 4, 2, 7, 10
a) Calculate the average (mean) of this random sample.
b) Find the median.
c) Find the mode.

4.4 Data on daily units and legal sales (in $) of several marijuana products in Washington since July 2014 are available.[3] Use the Minitab snapshot of the data presented below:
a) To find the "Marijuana Mix Packaged" sales per unit for May 16, 2016.
b) To find the "Marijuana Extract for Inhalation" sales per unit for April 13, 2015.

---

3 data.lcb.wa.gov (accessed July 26, 2019).

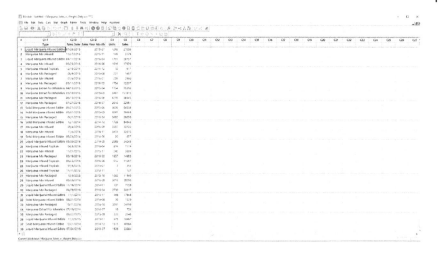

4.5 You are told that a sample of 170 cans of tuna was taken to test if too much tuna (more weight) occurs too often. The personnel tells you that although they do not know much about the stated study, they do know the weight distribution of tuna per can is highly skewed. If asked for a measure of central tendency, which one would you give preference to? Why?

4.6 You are asked about how 25 observations of light bulb power (in watts) are grouped (i.e. centralized), but only told that the sample median is 125 watts and the sample mean is 62 watts.
a) What is the shape of the data distribution?
b) Which measure of centrality would you prefer and why?
c) What if the sample median was 100 and the sample mean was 101. Which measure of centrality would you prefer?

4.7 According to the 2013 Canadian General Social Survey, the average hours a year, contributed by volunteers, is[4] 154, while the median hours a year is 53.
a) What is the shape of the data's distribution?
b) Which measure of centrality would you prefer to summarize typical hours dedicated by volunteers and why?

4.8 Madison Water Utility offers bill credits (rebates) for customers who replace high water using toilets with EPA WaterSense-rated High Efficiency Toilet models. Monthly average rebates can be determined from 2009 to 2013 data. From the figure, which is higher, the mean or the median?

---

4 Source: www.statcan.gc.ca (accessed July 26, 2019).

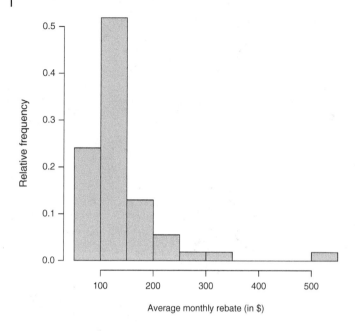

4.9 Refer to Example 4.3.
  a) Show that the mean number of children for teenage females is 0.35.
  b) Show that the median number of children for teenage males is 0.

4.10 Suppose that after three quizzes in a class, you have a mean of 75. If you get 92 in the next quiz, what would your new average be?

4.11 Say you work for a government agency that has sample median income of residents in Houston for every year from 1920 to 2018, with the exception of the year 1980 which is missing. A supervisor suggests taking the data of all those years and then taking the median as the median income of 1980. What do you think of this recommendation?

4.12 Using daily inmate data from the Orleans Parish Prison in New Orleans, from 2011 until mid 2017, it can be found that the average daily inmates was 2 181.69.
  a) Is the number provided a parameter or a statistic, when assessing inmates from 2011 until mid 2017?
  b) ResultsNOLA is a system implemented in New Orleans to track progress towards citywide and strategic goals. The figure below[5]

---

5 Source: datadriven.nola.gov (accessed July 26, 2019).

summarizes the quarterly number of inmates from 2011 until mid 2017. What can you say about the mean daily inmates given above?

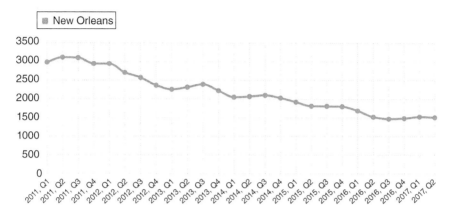

4.13 The number of tourist (in millions) visiting New Orleans from 2006 to 2014 was[6] 3.7, 7.1, 7.6, 7.5, 8.3, 8.8, 9, 9.3, 9.5, respectively. Does it make sense to take the average number of tourists here?

4.14 A professor of a business school course has been struggling to motivate students to put some effort into the course. Frustrated, he offers them a 5 point bonus in the Final exam if they all attend a Marketing presentation to be given in school. All students attended. When entering the Final exam grades, the professor did not use separate columns for the exam result and bonus for each student. The final exam mean he gets, as a measure of average student knowledge, is inflated because of the bonus. Now, if the professor really wants to assess how much the students knew in the final exam through the average score and the actual exam average was 78, how must the actual final exam average score be adjusted to account for the bonus given to all students? (*Hint:* If $X_i$ is what each student got in the exam without the bonus then, accounting for the bonus, the mean Final score was $\overline{Y} = \frac{\sum_{i=1}^{n}(X_i+5)}{n}$.)

## 4.3 Measures of Dispersion

Centrality is just one trait from a population or sample that we may need to determine. We already know that dispersion measures how spread out values of a variable are and how much the variable varies. In finance for example,

---

6 Source: data.nola.gov (accessed July 26, 2019).

measures of dispersion are used to determine risks of investments. Specifically, a fund may have an expected yearly return of 0.08, but each year, the returns may be markedly above or below its expected return of 0.08. That is, the dispersion of the yearly returns is high. As you can imagine, the possibility of having a yearly return markedly different than what is expected can make investors nervous, and in finance dispersion, is one way to measure the risk of an investment.

The question is how can we measure dispersion? Based on the definition, perhaps the obvious option is the difference between the largest observation and the smallest observation, which in statistics is known as the **range**. The range is easy to calculate, but comes with drawbacks. Since it depends solely on the largest and smallest observation, it is susceptible to extreme values and hence, it is quite likely that it will be an inflated estimate of dispersion. More importantly, the range only depends on two values from the sample. Philosophically, statistics perform better when they use the right amount of information available from the data to summarize a trait from the sample. We should expect that using just two observations does not provide enough information about how a variable is dispersed. However, the range is on the right track in the sense that whatever metric we use for spread must be defined based on distances, obtained through differences.

**Variance and Standard Deviation**   We still need to figure out what kind of differences the measure of dispersion must be based on. Since centrality measures how data is grouped, a good candidate is some function of the difference between the observations and a measure of centrality. A favorite measure of centrality is the population mean, $\mu$. Let's represent the difference between a possible value of a variable, $x_i$, and the population mean as $(x_i - \mu)$. Most likely though, $\mu$ will be unknown, so let's represent the difference between a possible value and the sample mean as $(x_i - \bar{x})$. An intuitive way to measure dispersion is through the average distance of each possible value from its mean. However, it can be shown that $\sum_{i=1}^{n}(x_i - \bar{x}) = 0$ for any combination of observations. An alternative is to measure dispersion in a sample using squared differences,

$$s^2 = \frac{\sum_{i=1}^{n}(x_i - \bar{x})^2}{n-1} = \frac{n\sum_{i=1}^{n}x_i^2 - \left(\sum_{i=1}^{n}x_i\right)^2}{n(n-1)} \tag{4.2}$$

this is known as the **sample variance**. $\sum_{i=1}^{n}(x_i - \bar{x})^2$ is known as the sum of squares and it will play a vital role in some inferential procedures later on. The two equations are equivalent with the first coming from the definition of variance, and the second equation being more convenient for computations by hand. Variance is wildly used in practice. The larger the sample variance is, the more spread out the data are. One drawback of the variance is that it is in squared units of the actual measured values. Hence, if we are summarizing heights of subjects in inches, variance is in $in^2$. We can deal with this by taking the square root of the sample variance,

$$s = \sqrt{\frac{\sum_{i=1}^{n}(x_i - \bar{x})^2}{n-1}}$$

a dispersion measure known as the **sample standard deviation**. Note that the variance and standard deviation must always be greater or equal to zero.[7] $s$ can be interpreted as the typical distance observations are from the mean.

**Example 4.5**  *Below are some of the responses by women when asked how many text messages they have sent in the last hour.*
  *3, 27, 3, 5, 2*
  *What is the standard deviation of the number of text messages?*

Table 4.2 breaks down the computations to find the variance using either of the two possible equations, Eq. (4.2). With the first equation,

$$s^2 = \frac{456}{4} = 114$$

while through the second equation,

$$s^2 = \frac{5 * 776 - 40^2}{5 * 4} = 114$$

Therefore, $s = \sqrt{114} = 10.68$. Typically, the number of texts a respondent answers is 10.68 texts messages from the mean. Since measures of dispersion

Table 4.2  Standard deviation computations using both formulas.

| i | $x_i$ | $x_i - \bar{x}$ | $(x_i - \bar{x})^2$ | $x_i^2$ |
|---|---|---|---|---|
| 1 | 3 | $(3 - 8) = -5$ | $(-5)^2 = 25$ | $(3)^2 = 9$ |
| 2 | 27 | $(27 - 8) = 19$ | $(19)^2 = 361$ | $(27)^2 = 729$ |
| 3 | 3 | $(3 - 8) = -5$ | $(-5)^2 = 25$ | $(3)^2 = 9$ |
| 4 | 5 | $(5 - 8) = -3$ | $(-3)^2 = 9$ | $(5)^2 = 25$ |
| 5 | 2 | $(2 - 8) = -6$ | $(-6)^2 = 36$ | $(2)^2 = 4$ |
| Sum | 40 | 0 | 456 | 776 |
| $\bar{x}$ | 8 | | | |

---

7  Under what circumstance is the variance equal zero?

such as the standard deviation determine how far observations are expected to be from the mean, it can be used to define a threshold that indicates which observations are rather far above or below the typical values. Observations that are beyond the threshold are considered to be **unusual values** or **outliers**. Outliers tend to be associated with data errors, but they may not be errors. When outliers are not due to error, the selection of which measure of centrality or dispersion becomes important. Now, variance and standard deviation are functions of $\bar{x}$. In situations of highly skewed data, $\bar{x}$ is not the best alternative to measure centrality and therefore, $s$ or $s^2$ are not the best alternatives to measure dispersion. We discuss an alternative to $s$ in Section 4.4.

---

**Average Squared Deviations from the Mean?**
Perceptive readers may have noticed the reference to $s^2$ as an average squared distance, yet it divides the squared difference by $n - 1$, not by $n$. We defer the details of this for later, but for the moment, we will just say that this helps ensure that $s^2$ is less likely to be strictly below $\sigma^2$.

---

Another measure of dispersion is known as the **coefficient of variation**. It is used to compare data with different means or different units of measurement

$$CV = \frac{s}{\bar{x}}$$

Applications include dispersion assessment of tomato sauce sold in different can sizes, or comparing the dispersion of investment funds with different returns.

*Central Tendency and Dispersion*   Organizations will frequently evaluate the central tendency and dispersion of data to ensure proper execution of work. For example, a plant that packs cans of food will evaluate if the mean weight of a sample of cans is close to the planned mean. Also, they may evaluate if the standard deviation of the cans' weight is not too large. Under certain circumstances, it will be possible to compare the distribution of two groups by simultaneously considering sample mean and standard deviation.

**Example 4.6**   *A while back, right at the beginning of a class, students were asked how many text messages they have sent (not received) during the last hour. Below are some of the responses by gender.*
  *Female: 3, 27, 3, 5, 2*
  *Male: 2, 1, 5, 8, 3*

In Example 4.5, the mean and standard deviation of text messages from women were found to be 8 and 10.68, respectively. It can be shown that for men, the mean number of texts is 3.8, while the standard deviation is 2.77.

According to this data, women texted more frequently. Moreover, the standard deviation of number of texts that women sent is far higher than that of men. Observe that one female sent a lot more texts than the other females, an example of an unusual observation. This observation is causing the mean and standard deviation of texts sent by women to be much higher than the analogous statistics for men.

## Practice Problems

4.15 Using the following data,

9.71   4.25   3.75   9.42   8.33   7.21   2.55

show that
a) $s^2 = 8.48$.
b) $s = 2.91$.

4.16 Find the variance and standard deviation for the following sample:

123, 101, 171, 154, 168, 188, 203

4.17 Find the variance and standard deviation for the following sample:

4.23, 1.75, 13.92, 9.65, 7.38

4.18 Show that in Example 4.6, the mean number of texts from men is 3.8, while the standard deviation is 2.77.

4.19 Have you ever wondered what is the typical weight of a meteorite landing on Earth? Physicists will tell you that it is best to consider the mass of the meteorite, not the weight, because the mass remains the same no matter where the meteorite is in space. NASA has provided the mass (in grams), class, and other details of all known meteorite landings[8]: "Meteorite_Landings."
a) With the help of a computer, construct a histogram of the mass of the meteorites classified as Acapulcoite.
b) Based on the shape of the histogram in part (a), do you expect the mean to be greater to the median, less, or equal? Explain.
c) Based on the shape of the histogram in part (a), which measure of centrality do you prefer?

---

8 Source: data.nasa.gov (accessed June 8, 2018).

d) With the help of a computer, find and compare the mean and median of the mass of the meteorites classified as Acapulcoite.

e) Find the standard deviation of the Acapulcoite meteorite mass with the help of a computer.

4.20 Returning to the meteorite mass Problem 4.19, use a computer:

a) To construct a histogram of the mass of the meteorites classified as Pallasite.

b) Based on the shape of the histogram in part (a), do you expect the mean to be greater than the median, less than, or equal to? Explain.

c) Based on the shape of the histogram in part (a), which measure of centrality do you prefer?

d) To find and compare the mean and median of the mass of the meteorites classified as Pallasite.

e) To find the standard deviation of the Pallasite meteorite mass.

4.21 Based on samples of residents in Barnstable County, Massachussetts from 2009 to 2014, the mean travel times to work in 15 towns are[9]

23.1, 27.2, 25.1, 23.7, 27, 23.4, 24.6, 21.1, 23.9, 21.3, 20.1, 29.9, 20.4, 23.5, 24.1

a) What is the average travel time to work for residents of Barnstable County?

b) Find the median travel time to work for residents of Barnstable County.

c) What is the variance of travel time to work for Barnstable County?

d) Find the standard deviation of travel time to work for residents of Barnstable County.

## 4.4 Percentiles

On occasion, it is handy to summarize data in terms of percentiles. **Percentiles** basically divide your data into 100 groups. The $p$th percentile is the value such that $p\%$ of the data has a smaller value. We have already seen one type of percentile; the median is the 50th percentile. The classic example application is SAT scores, which are reported in terms of percentiles. If you are told you are in the 92nd percentile, then 92% of people who took the test had a score lower than yours. Another way of putting it is that your score is in the top 8% of those who took the test. Other applications are to compare salaries of professionals in a given field. Moreover, in terms of development of children, doctors provide charts on weight and height percentiles to look at the progress of a child.

---

9 Data: ckansandbox.civicdashboards.com (accessed July 26, 2019).

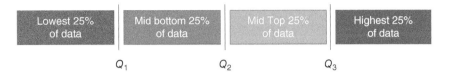

| Lowest 25% of data | Mid bottom 25% of data | Mid Top 25% of data | Highest 25% of data |

$$Q_1 \qquad Q_2 \qquad Q_3$$

Figure 4.8 The quartiles separate the data into four groups.

Computer software will often calculate percentiles through interpolation. Percentile algorithms used in computer software vary slightly, leading to slightly different percentile estimates. This difference does not tend to affect the overall interpretation of the result.

### 4.4.1   Quartiles

Several special cases of percentiles exist. For example, summaries can be based on dividing the data into 10, 5, or 4 groups. **Quartiles** are the 25‰, 50‰ and 75‰ and they divide the sorted data into four groups.

- The 25‰ is known as the first quartile, $Q_1$.
- The 50‰ is known as the second quartile, $Q_2$, or the median.
- The 75‰ is known as the third quartile, $Q_3$.

Figure 4.8 is a simple representation of how quartiles work.

There are several ways to obtain the quartiles from a sample.[10] The steps to find a quartile are[11] as follows:

- Check if the data is sorted (in ascending order). If not, sort the data.
- Calculate the index value $i$,

$$i = \left( \frac{p}{100} \right) n$$

where $n$ is the number of observations, and $p$ is the quartile of interest: 25, 50, or 75.
- If $i$ is not an integer, round up to find the position in the ordered data where you will find the value of the median. If $i$ is an integer, then take the $i$th observation and average it with the $i$th + 1 observation to find the value of the median.

**Example 4.7**   *Recall from Example 4.2 that the sorted February 2017 energy use data in Nashville, Tennessee, is*

---

10  This section will not be addressing the calculation of population quartiles. Instead, these will be estimated through sample quartiles.

11  It is possible to find the median according to whether $n$ is odd or even. However, this is not true for quartiles in general, making the method presented here more general.

| [1] | 40 | 471 | 668 | 1064 | 4605 | 5222 | 9755 | 10118 |
| [9] | 12713 | 16300 | 19126 | 23291 | 28025 | 44520 | 46800 | 49739 |
| [17] | 53640 | 63720 | 99794 | 121500 | 130944 | 133812 | 153037 | 153499 |
| [25] | 173118 | 176418 | 331200 | 362460 | 404021 | 419895 | 447340 | 491964 |
| [33] | 516000 | 535234 | 559975 | 581073 | 615747 | 673198 | 696971 | 754214 |
| [41] | 761900 | 764775 | 829661 | 831616 | 836826 | 907158 | 915635 | 970928 |
| [49] | 972945 | 1062664 | 1235538 | 1276576 | 1309065 | 1502947 | 1630032 | 1798960 |
| [57] | 1937568 | 2540797 | 2814565 | 2902660 | 2949790 | 3206050 | 4315417 | 4490491 |
| [65] | 5009918 | 5060494 | 6454158 | 6882965 | 8158407 | 8292035 | 8325783 | 8335953 |
| [73] | 8574076 | 9235521 | 9859196 | 12209694 | 12450797 | 12568467 | 13864886 | 16382338 |
| [81] | 16897434 | 19503463 | 21193679 | 21919235 | 23689663 | 54015896 | | |

*where the first number between square brackets in each row is not an observation but indicates the order of the first observation in that line. Let's determine what $Q_1$ and $Q_3$ are.*

According to the second step to find $Q_1$,

$$i = (25/100) \times 86 = 21.5$$

Since this number is not an integer, $Q_1$ is the 22nd observation: $Q_1 = 133\ 812$. Therefore, it is estimated that 25% of sampled observations are below 133 812 kWh. For $Q_3$,

$$i = (75/100) \times 86 = 64.5$$

Since this number is not an integer, $Q_3$ is the 65th observation: $Q_3 = 5\ 009\ 918$. Therefore, it is estimated that 75% of sampled observations are below 5 009 918 kWh.

Often in textbooks, quartiles or percentiles are calculated for small data sets. But that is done mainly so the calculations can be performed by hand. When the data set is small, it is questionable to use any percentile, especially if we have repeating values. For example, if you have only five observations, it is unreasonable to divide the data into 100 groups as percentiles do. Also, it should be noted that we will be calculating sample quartiles, so the 75% quartile may not have exactly 75% of the data below it, only approximately.

**Interquartile Range**    Another measure of dispersion is called the **interquartile range**, also referred to as the *IQR*:

$$IQR = Q_3 - Q_1$$

The *IQR* is the range associated with the middle 50% of the data (Figure 4.9). It is a useful alternative in situations when the distribution is highly skewed, or large outliers are present since $s$ or $s^2$ may be inappropriate to calculate dispersion in these cases. When in doubt, it is best to obtain the standard deviation and the *IQR* and see if there is a big difference between them.

**Example 4.8**    *Example 2.3 introduced us to a 2011 survey conducted by San Francisco airport (SFO) administrators to analyze guest satisfaction. One of the*

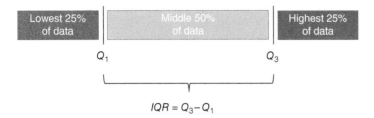

Figure 4.9 The *IQR* uses the first and third quartiles. This reduces the impact of outliers when measuring dispersion.

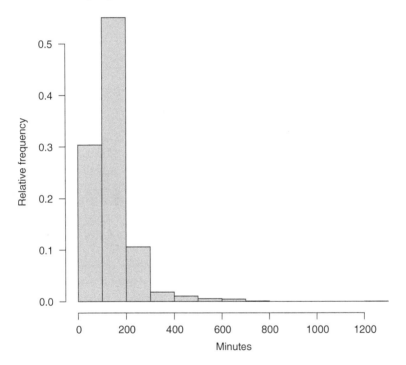

Figure 4.10 Histogram of amount of time from passenger arrival at SFO to flight departure (in minutes) according to a 2011 survey.

*variables collected was the amount of time from passenger arrival at SFO to flight departure (in minutes).*

Figure 4.10 examines the distribution of minutes from arrival at SFO to flight departure. Most passengers must wait between a few minutes to about 200 minutes. The data is clearly skewed to the right, with some passengers having to wait much longer than 200 minutes. In fact, one respondent had a substantially longer waiting time, 1300 minutes. Through computer software, it can be found that $Q_1 = 95$ and $Q_3 = 170$. Therefore,

$$IQR = 170 - 95 = 75$$

In comparison, $s = 84.97$, likely influenced by the outliers in the data.

***Five-Number Summary and the Boxplot*** The **five-number summary** consists of the following statistics:

$$X_{min}, Q_1, Q_2, Q_3, X_{max}$$

Most of the time, a graphical representation of the five-number summary called a **boxplot** or **box and whisker plot** is presented. The visual display presents a box with each edge representing the first and third quartile. Inside the box, a line indicates what the median of the variable is. Whiskers extend out from the box toward the minimum and maximum, yet modifications exist to point out unusual or extreme values of the variable.

**Example 4.9** *Sustainability data for several communities within Baltimore is available[12] from 2010–2012. One of the variables is the 2011 walking score, calculated by mapping out the distance to amenities in nine different categories (grocery stores, restaurants, shopping, coffee shops, banks, parks, schools, book stores/libraries, and entertainment) which are weighted according to importance. The distance to a location, counts, and weights determine a base score of an address, which is then normalized to a score from 0 to 100 (with 100 being best). The boxplot of the walking scores of 55 communities is displayed in Figure 4.11. It suggests the walking scores of the communities are approximately symmetric, with a median of 60. 50% of the walking scores are between 48.42 (the first quartile) and 76.80 (the third quartile).*

The boxplot is one of the most useful visual representations in statistics. Its summaries help us explore central tendency, dispersion, shape, outliers, and even distributions of different populations. A function of the $IQR$ is used to determine which observations are far enough below $Q_1$ or far enough above $Q_3$ to be treated as unusual observations or outliers. Some statistical software use different symbols for these potentially atypical values.

Boxplots are often used to explore whether there is a difference in centrality of different populations. Informally, if the boxes of the boxplots do not overlap, then there is some indication that the populations have different measures of centrality. However, if the boxplots overlap, then the visual comparison gives us no indication of their being a difference between their measures of centrality.

---

12 data.baltimorecity.gov (accessed July 26, 2019).

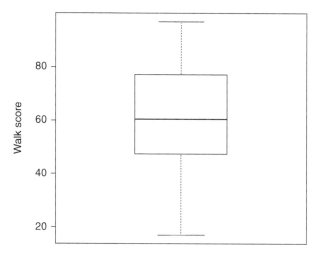

Figure 4.11 2011 Walking scores for 55 communities within Baltimore.

Figure 4.12 2016–2017 average American College Testing (ACT) scores by Bilingual School.

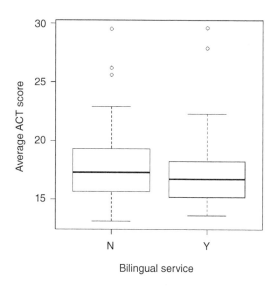

Figure 4.12 shows boxplots of the 2016–2017 American College Testing (ACT) scores for high schools in Chicago according to whether they have bilingual services or not. They do not show any strong evidence that bilingual schools have higher average ACT scores. The circles along each boxplot indicate that for non-bilingual and bilingual schools, some average ACT scores fall far above the typical values.

## Practice Problems

4.22 For a specific data set, you know the median is 22. You use a computer to find the 30th percentile for the same data and find it to be 47. Explain how you know there must be a mistake in the computation.

4.23 On a statistics test, the third quartile for scores was 82. What is the interpretation of the third quartile in the context of this situation?

4.24 The starting monthly salary for a marketing job from 10 randomly sampled people straight out of school was (in $)

   3333   3553   3961   3480   3715   3520   3490   3500   3425   3375

   Find $Q_1, Q_2, Q_3$.

4.25 You are given the following information on miles per gallon (mpg), based on samples collected from two car models: Model A and Model B
   Model A: Five-number summary = (30, 34.125, 34.75, 35.87, 38)
   Mean = 34.75
   Standard deviation = 1.98
   Model B: Five-number summary = (30, 31.5, 32, 32.5, 34.75)
   Mean = 32.20
   Standard deviation = 0.975
   a) What is the best mpg of Model A cars tested?
   b) What is the car with the highest mpg, model A or model B?
   c) On average, which car model tested had better mpg, model A or B?
   d) Do you expect greater variability in the mpg of model A or model B? Explain.
   e) What percentage of model A cars have larger mpg than the best model B performer?

4.26 The 25th percentile for waiting time at a bank is 3 minutes. The 75th percentile is 17 minutes. Find the interquartile range.

4.27 You are given the following five-number summary for a sample: 2, 13, 22, 31,50. Find the *IQR*.

4.28 For the following data: 3, 4, 9, 11, 12, 17, 27, 29
   a) Find the five-number summary.
   b) Calculate the *IQR*.

4.29 Recall that in Example 4.8, the *IQR* was calculated for amount of time from passenger arrival at SFO to flight departure (in minutes).

Downloading[13] "2011_SFO_Customer_Survey" and using the "PASS-MIN" variable, compute $Q_1$ and $Q_3$. Note that the obtained quartile results may be slightly different than those in Example 4.8 depending on the computer software used.

4.30 Using the passenger data from Problem 4.29, create a boxplot of Total Landed Weight by GEOSUMMARY. Discuss.

4.31 Returning to the meteorite mass Problem 4.19, find the $IQR$ of the mass of the meteorites classified as Acapulcoite and compare to its standard deviation.

4.32 Many places establish rebate programs to promote energy savings. For example, Delaware has a rebate program for the purchase or lease of alternative fuel vehicles. Data is available to the public[14] including the amount of the rebate. Download "Delaware_State_Rebates_for_Alternative-Fuel_Vehicles." Construct a

a) Box plot of age by gender. Does it seem that the mean age of people getting rebates varies by gender?
b) Box plots of age by vehicle type. Interpret.

4.33 It is possible for a variable in a data set to be neither numerical nor categorical. One way this happens is that the instrument available to measure the variable can only reach a certain value, and below or above that level, the value cannot be known precisely. In the observations below, any value that is "< 10" is too low to be measured with precision.[15]

19.21  14.15  < 10  13.75  19.42  18.33  17.21  22.55  < 10  15.15  16.12

A summary of these type of data sets can still be useful.
a) Replace all <10 in the data set with 10.
b) Get $Q_1$ using the formula in this section.
c) Why is the $Q_1$ calculation reliable?
d) Calculate the median. Is the median reliable?
e) Would the mean be reliable?

4.34 Some studies suggest that bilingual children are better at problem solving than non-bilingual children. According to Figure 4.12, there is no evidence that students at bilingual service schools attain better average ACT scores. Comment on whether this contradicts that bilingual children are better at problem-solving skills.

---

13 Source: data.sfgov.org (accessed July 26, 2019).
14 Source: data.delaware.gov (accessed July 26, 2019).
15 This type of data is called censored data.

4.35 In the observations below, any value that is "<10" is too low to be measured with precision.

19.21  14.15  <10  13.75  19.42  <10  < 10  22.55  < 10  <10  <10

A summary of these type of data sets can still be useful.
a) Replace all <10 in the data set with 10 and get $Q_1$ using the formula in this section.
b) Why is the $Q_1$ calculation not reliable?
c) Is the median reliable?

4.36 In Toronto, Canada residents can pick up and drop off a water sample kit at one of six Toronto Public Health locations to sample tap water for lead. Download[16] "leadsampling_2016". The file has the sampling results per partial ZIP Code.
a) Note that the eighth observation of lead amount (parts per billion) is <0.05 and in fact, this occurs several times. Replace[17] all <0.05 in the data set with 0.05.
b) Use a computer to get the minimum and maximum.
c) With a computer get the mean, median, standard deviation, and *IQR* of lead amount.
d) What impact does replacing <0.05 by 0.05 have on obtaining the statistics in 4.36c?
e) Build a boxplot for lead amount (parts per billion) for the partial postal codes M6G. Interpret.
f) Build a boxplot for lead amount (parts per billion) for the partial postal codes M4L. Interpret.

## 4.5  Measuring the Association Between Two Variables

More often than not, databases contain multiple variables, not just one, and we have seen several examples. There are many instances when we want to evaluate the association between measured variables. How we study the association between variables depends on the mix of quantitative and qualitative variables that we have.[18]

---

16  Source: www.toronto.ca (accessed July 26, 2019).
17  If you open the file with Minitab, it will automatically replace <0.05 with missing values. You should replace these with 0.05.
18  Keep in mind that, on occasion, it will be preferable to turn quantitative variables into qualitative variables and then use methods of summarization for qualitative variables.

*Correlation and Covariance*   Let's look at how we numerically summarize the association between two numerical variables. Keep in mind that our aim is to measure "approximate" association, not perfect smoothed association as seen in math courses.[19] By looking into the association of the variables, we aim to answer the following two questions:

- When variable $X$ increases, what approximately happens to variable $Y$?
- If there is an approximate association, is it linear?

We can answer these questions with a statistic that measures how $X$ and $Y$ "covary;" how they move together. This is exactly what the statistic known as **covariance** does. When one variable increases, while the other (on average) increases, we say there is a positive association between the variables. When one variable increases, while the other (on average) decreases, we say there is a negative association between the variables.

Covariance makes sense as a metric of association between $X$ and $Y$ and is useful when studying things like the performance of financial assets. However, it has the drawback of depending on the scale of measurement. As a result, it is difficult to say how strong the linear association between $X$ and $Y$ is by looking at the covariance.

---

**Correlation**

The **correlation coefficient**, **r**, is a statistic that standardizes the covariance in such a way that it does not depend on the scale of measurement. $r$ is always between $-1$ and $1$.

---

Figure 4.13 demonstrates how the correlation coefficient works by partitioning a scatterplot into four mathematical quadrants. The quadrants are established by lines at $\bar{x}$ and $\bar{y}$:

- Points in quadrant I pertain to $x_i$ greater than $\bar{x}$ and $y_i$ greater than $\bar{y}$.
- Points in quadrant II pertain to $x_i$ less than $\bar{x}$ and $y_i$ greater than $\bar{y}$.
- Positive correlation indicates that most points are in quadrant I and quadrant III, as seen in the left panel of Figure 4.13.
- Negative correlation indicates that most points are in quadrant II and quadrant IV, as seen in the right panel of Figure 4.13.

The strength of the correlation will be related to the number of points in each quadrant. If almost all points are in quadrant I and III, then there is a strong positive correlation. If most points are in quadrant I and III, but there

---

19  When you see the word "association" in this book, it will always mean "approximate association."

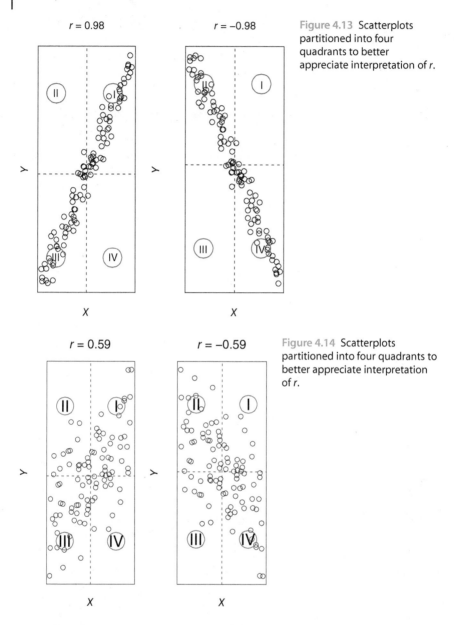

$r = 0.98$

$r = -0.98$

**Figure 4.13** Scatterplots partitioned into four quadrants to better appreciate interpretation of $r$.

$r = 0.59$

$r = -0.59$

**Figure 4.14** Scatterplots partitioned into four quadrants to better appreciate interpretation of $r$.

are some in quadrant II or IV, then there is a moderate positive correlation (Figure 4.14).

$r = 1$ occurs when there is a perfect positive linear association between $X$ and $Y$. When $r = 1$ or $r = -1$, it is possible to figure out the value of $Y$ exactly

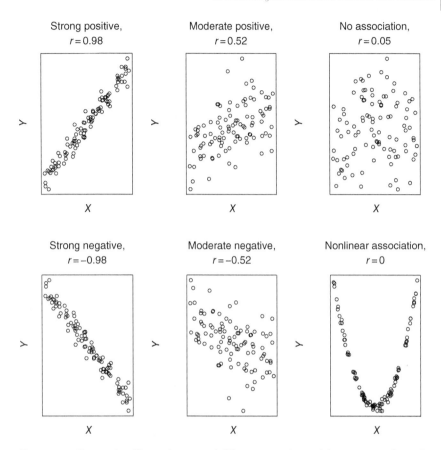

Figure 4.15 Scatterplots illustrating several different scenarios and the corresponding value of $r$.

if you know $X$. Generally, our interest is in scenarios where there is no perfect linear association between $Y$ and $X$.

The quadrants serve as scaffolding tools to define correlation, but they are not used explicitly in practice.

---

**Zero Correlation; But Variables Associatied**

A variety of associations among two variables is appreciated in Figure 4.15, with their corresponding $r$. The scenario in the lower right panel shows a nonlinear association, but $r = 0$. It is important to realize that $r$ measures linear association, not association overall. Therefore, associations that cannot be

Table 4.3 Generally used interpretations of ranges of *r*. Careful, weak linear association does not mean weak association overall.

| Range of *r* | Strength of linear association |
|---|---|
| $|r| \geq 0.80$ | Strong |
| $0.30 \leq |r| < 0.80$ | Moderate |
| $|r| < 0.30$ | Weak |

approximated well with a line (as in lower right panel of Figure 4.15) will result in *r* values close to zero. For this reason, association between two numerical variables should always be assessed with both a scatterplot and *r*.

Ranges of values of *r* that establish the strength of linear association are subjective, but Table 4.3 serves as a guideline. Moderate or better linear associations can be useful to predict one variable using another, a topic that we return to later on in the book.

We cannot solely rely on association assessments to argue that one variable, *X*, causes another, *Y*. Even strong associations may be caused by underlying variables.

**Summarizing Association: Qualitative Variables**  Correlation is useful when both variables are quantitative. When evaluating the association between two categorical variables, it is no longer appropriate to look at whether *Y* increases or decreases, when *X* is increasing, since categorical variables *X* and *Y* do not "increase" or "decrease." Instead, it is necessary to look differently at how *Y* depends on *X*. Informally, we can study if the counts in categories of *Y* depend on the counts in categories of *X*. This is often performed using "standardized" forms of counts and relative frequencies and presented in terms of cross-tabulated tables. We will return to this topic in later chapters.

When one variable is qualitative, the other quantitative, numerical summaries of the quantitative variable per category of the qualitative variable help assess dependence among the two variables. Alternatively, boxplots of the quantitative variable per category of the qualitative variable may suggest association among the variables (see Section 4.4).

## Practice Problems

4.37  In the two scatterplots below, state whether *r* has a small, medium, or large value, and interpret what is seen in the scatterplot.

(a)                              (b)

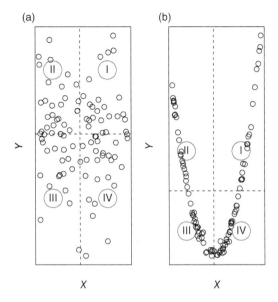

4.38 Recall that Figure 3.19 drew scatterplots of cheese fat percentage and moisture percentage using data from Canada, while also considering the type of cheese and adding a trend line. Below are the values of *r* by cheese type.

a) Which cheese type had the strongest linear association between fat percentage and moisture percentage?

b) Which cheese type had the weakest linear association between fat percentage and moisture percentage?

c) Does it seem like cheese type affects the linear association between fat percentage and moisture percentage?

| Cheese Type | r |
| --- | --- |
| Firm | −0.78 |
| Fresh | −0.64 |
| Hard | −0.47 |
| Semisoft | −0.62 |
| Soft | −0.70 |
| Veined | −0.38 |

**4.39** In their 2017 book *Teaching Statistics: A Bag of Tricks*, Andrew Gelman and Deborah Nolan provide a series of ideas for active student participation. In one of the activities, while taking an exam, students get to guess what score they will get. If their guess is close enough to their actual exam score, they get bonus points. This was implemented during an exam for the second semester of a business statistics course. Use the data "GuessexamscoresSecondPart":

a) To construct a scatterplot of the two variables. Interpret.
b) To find the sample correlation (using statistical software). Interpret.

**4.40** The file "VitalSignsBaltimore15wdictionarytoprows" includes an affordability index for mortgages, affordm15, and for rent, affordr15, for 55 communities in Baltimore.

*Hint:* The last row indicates overall measurement for the city and therefore should not be considered for computations.

a) Construct a scatterplot of the two variables. Interpret.
b) Find the sample correlation (using statistical software). Interpret.

## 4.6 Sample Proportion and Other Numerical Statistics

The **relative frequency** of a category is also known as a **sample proportion**. Table 4.4 summarizes the sample proportions of business programs studied by a sample of 272 students. These results estimate that students are three times more likely to study accounting than marketing.

Many other numerical summaries are available. **Skewness** measures the asymmetry of a distribution. High asymmetry can have important

Table 4.4 Sample proportions of business programs students are enrolled in.

| Business program | Sample proportion |
| --- | --- |
| Accounting | 0.46 |
| Finance | 0.13 |
| Management | 0.12 |
| Marketing | 0.13 |
| Other | 0.16 |
| Total | 1 |

ACCT, FINA, HUMR, INFS, MGMT, MKTG stand for accounting, finance, human resources, information systems, management, and marketing programs, respectively.

consequences in terms of statistical inference. **Kurtosis** measures the peakedness of the probability distribution or the heaviness of the tails of the distribution.

### 4.6.1  A Quick Look at Data Science: Murder Rates in Los Angeles

How violent is the city of Los Angeles? And how does it compare to other cities in the United States? One way to measure violence is through murders, a homicide with malicious intent. It is straightforward to realize that a comparison of murders among cities should not be based on the number of murders and that the demographics of the places to be compared play a role. It is best to compare murder rates per 100 000 people for a given year:

$$t = \frac{\text{Number of murders in the city}}{\text{City population size}} \times 100\ 000.$$

In this case study, murder rates for the city of Los Angeles are computed and then compared to several cities. The murder rates were obtained by processing and combining crime data from the Los Angeles open data portal and US Census population estimates for the city (see Figure 4.16). Crime data includes dates, times, and location of different types of crimes, among other variables. As of this writing, the data set had over 1.5 million rows and 26 columns. Only data from 2010 to 2016 were considered, since the 2017 data was incomplete. Vintage 2016 US Census annual population estimates for the period of interest were also retrieved. The quality of the crime data was assessed via checking for duplicated entries and comparison of annual number of murders found through other sources.

Table 4.5 presents the murder rates for the Los Angeles from 2010 until 2016. It shows that in 2010, there were 7.77 murders per 100 000 people in the city.

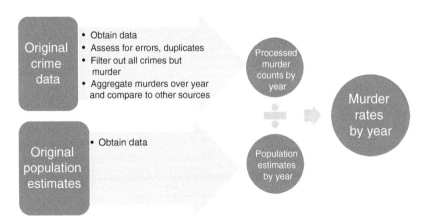

Figure 4.16  Steps required to calculate murder rate data for Los Angeles.

Table 4.5 Murder rates (per 100 000 people) for the city of Los Angeles from 2010 until 2016.

| 2010 | 2011 | 2012 | 2013 | 2014 | 2015 | 2016 |
| --- | --- | --- | --- | --- | --- | --- |
| 7.77 | 7.79 | 7.72 | 6.43 | 6.63 | 7.14 | 7.37 |

A reduction in the rate occurred in 2013 and 2014, but a bump in murders was recorded in the last two years available. On the positive side, the 2016 rate was not at the level of the beginning of decade, and the rates are much lower than what it was in the early 1990s, when the rate was over 20.

Without prior knowledge about murder rates, it is still hard to grasp the meaning of the Los Angeles murder rates. Thus, they are compared next to cities of at least 250 000 people in the United States. According to the FBI, the 2015 average murder rate in cities of 250 000 or more was[20] 10.1. Moreover, for that year, the cities with the top four murder rates[21] were St. Louis (59.2), Baltimore (55.37), Detroit (43.82), and New Orleans (41.68). In conclusion, Los Angeles does not have a high murder rate, although improvements can still be made. The city of New York has had a murder rate of around 3 for several years, while many European large cities have murder rates below 2.

---

**Want to Know More?**

If you want to know more about this case study, in the companion website, you will find the R code that reproduces the results presented here. This data set is rich, and you can modify the code to serve your own needs.

---

## 4.7 How to Use Descriptive Statistics

Statistics may not be weapons of mass destruction, but some people may disagree with this statement. It is easy to find examples where statistical methods are misused. We provide some guidelines to adequately make use of descriptive statistics.

- State objectives clearly. Not only this is important when presenting results to others, but it is also important for the person performing the summary. By often performing multiple tasks at work, this person can easily forget the objective of the analysis.
- Provide sample size. The number of observations used should always be mentioned. The more observations we have, the more reliable the results may be.

---

20  Source: ucr.fbi.gov (accessed July 26, 2019).
21  Murder and population numbers were obtained from ucr.fbi.gov (accessed July 26, 2019).

- Discuss the procedure to gather the data. Was a random sample used? A convenience sample? The procedure to gather the data also plays a role on how reliable the results are.
- Use visual aids in combination with descriptive statistics. Also, refer to these visual aids in the text of your report. The visual aids should be properly labeled and have good captions.

*Reminder: Parameter versus Statistic*   When possible to exactly quantify a parameter, it should always be preferred over a statistic. Defining the population and parameters needed is something that requires care. For example, suppose you have records for the students admitted to a program at a university, including gender, SAT score, and other information. Say you get the mean SAT score of new students admitted to the program at the university. Should you treat that as a statistic or a parameter? Before you continue reading, think about the question a bit and how would you answer it. It turns out that this is not an easy question to answer. First, we need to consider if the SAT scores we have correspond to all admissions. This is not always the case. The data may have information up to a certain date, say May, before school starts. But, some students get reconsidered, or sometimes schools admit students closer to the beginning of the school year. Let's assume for this argument that the data includes all admissions. Still, the response to whether the mean SAT score is a statistic or a parameter may depend of our question of interest. If our interest is in describing the mean SAT score of students entering that specific school program at the university, then one can argue that the mean SAT score is a parameter. However, one may wish to compare the admissions to a different program or university or may want to infer on future admissions to the program. In these cases, the mean SAT score should be interpreted as a statistic. The question that one wishes to answer is key in making the determination.

## Chapter Problems

4.41 Below are the yearly number of trees planted or funded by Seattle city programs from 2007 until 2016.[22]

1707   2345   2686   2187   1900   3258   2165   2385   2060   2410

a) Find the mean number of trees.
b) Find the median number of trees.

4.42 The file "VitalSignsBaltimore15wdictionarytoprows" includes an affordability index for mortgages, affordm15, for 55 communities in Baltimore.

---

22 Source: performance.seattle.gov (accessed July 26, 2019).

*Hint:* The last row indicates overall measurement for the city and therefore should not be considered for computations.
a) Find the mean of the index.
b) Find the median of the index.

4.43 A city manager has been given parking lot availability data. The data set provides parking spaces available every five minutes at parking lots (of all sizes, from small to very large) throughout the city. To assess the use of the parking lots in the city, the manager decides to take a sample from the large data set and find the sample mean number of lots taken. Explain why the sample mean is not a good summary to assess parking lot use throughout the city and recommend a more appropriate statistic.

4.44 Recall that a categorical scale with no logical order is known as nominal scale. Explain why for such data, the mode can be used, but not the mean or median.

4.45 University professors are evaluated by their students during each school term. The student survey answers can be turned into scores. In one question, students grade the performance of the professor. The table below shows the evaluations for one professor.

| Score | Category | Frequency |
| --- | --- | --- |
| 5 | Excellent | 5 |
| 4 | Good | 13 |
| 3 | OK | 4 |
| 2 | Deficient | 1 |
| 1 | Fail | 0 |

a) Find the mean score of the professor.
b) Find the standard deviation.
c) Find the median score of the professor.

4.46 Download the file "Air_Traffic_Landings_Statistics" with data on Air traffic landings for San Francisco.[23]
a) The mean Total Landed Weight, is it a statistic or a parameter?
b) What is the mean Total Landed Weight?
c) What is the median Total Landed Weight?

---

23 Source: data.sfgov.org (accessed July 26, 2019).

d) Which one do you think is a more appropriate measure of centrality of the landed weight, the mean or the median?

e) Create a boxplot for landed weight. Interpret.

4.47 2010 median house value by census group in Houston is available[24] in the file "Median_House_Value."

a) Get the minimum, median, mean, and maximum values. What is wrong with the minimum and how do you explain the issue?

b) What about the maximum, how do you explain it? (Hint: You can do a dotplot to observe the behavior of the maximum.)

c) Assuming the minimum is removed, what impact do you think the issue discovered in the maximum has on finding the median of the data?

d) Assuming the minimum is removed, what impact do you think the issue discovered in the maximum has on finding the mean of the data?

4.48 Return to Problem 4.47

a) Create a boxplot for landed weight by GEO Summary. Interpret.

b) Create a boxplot for landed weight by GEO Region, but only for landings that Landing.Aircraft.Type = "Freighter." Interpret.

c) Find the standard deviation for landed weight.

d) Find the interquartile for landed weight.

4.49 In a Washington Post article titled "A minimum-wage worker can't afford a two-bedroom apartment anywhere in the U.S." (June 13, 2018), it was highlighted how "… there is still nowhere in the country where someone working a full-time minimum wage job could afford to rent a modest two-bedroom apartment, according to an annual report[25] released Wednesday by the National Low Income Housing Coalition …." From the options below, choose the alternative that is closest to the correct interpretation of the statement. Explain your reasoning.

a) Apartments are too expensive in the United States.

b) Apartments are too cheap in the United States.

c) It is impossible for full-time minimum wage workers in the United States to find a modest two-bedroom apartment to rent.

d) It is not easy for full-time minimum wage workers in the United States to find a modest two-bedroom apartment to rent.

e) Full-time minimum wage workers in the United States can only rent less than modest two-bedroom apartment.

---

24 Source: data.houstontx.gov (accessed July 26, 2019).
25 nlihc.org/oor (accessed July 26, 2019).

**4.50** Recall from Example 4.9, that 2011 walking scores, "wlksc11" for 55 communities in Baltimore was available[26] at "Sustainability__2010-2012_"
*Hint:* Data set's last row is for all of Baltimore and must be removed.
a) Use a computer to get the mean, median, standard deviation, IQR, minimum, and maximum.
b) Build a histogram for the walking scores. Interpret.

**4.51** Download the file "Sanitation_2010Baltimore." One variable[27] is the rate of dirty streets and alleys in 2010 for Baltimore. The variable is called dirtyst10.
a) Use a computer to get the mean, median, standard deviation, IQR, minimum, and maximum.
b) Build a boxplot for these rates. Interpret.
c) Build a histogram for these rates. Interpret.

**4.52** Show that $\sum_{i=1}^{n}(x_i - \bar{x}) = 0$.
*Hint:* An example is not general enough. Rely on the mathematical definition of $\bar{x}$.

**4.53** According to Ireland's 2015 emergency records[28]:
a) There were 3315 instances that required ventilatory support. The mean length of stay of theses cases was 22, while the median was 9. What was the shape of the length of stay of cases that required ventilatory support?
b) From the companion website, download the table summarizing length of stays: "IrelandLengthofStay." (i) What can you say about the shape of length of stay for most of the cases summarized? (ii) How do the shapes found in (i) make sense?

**4.54** The third quartile for SAT math scores for region $X$ was 652. What is the MOST appropriate statement that can be made?
a) A third of test takers got a score of 652 or less.
b) 25% of test takers got a score of 652 or more.
c) The median score was over 700.
d) 75% of test takers got a score of 652.

**4.55** You are given the following five-number summary for a sample 2, 13, 22, 31, 50

---

26 Source: data.baltimorecity.gov (accessed July 26, 2019).
27 Source: data.baltimorecity.gov (accessed July 26, 2019).
28 Source: data.gov.ie (accessed July 26, 2019).

What is the LEAST appropriate statement that can be made?
a) The standard deviation of the sample is 18.26
b) The interquartile range is 18
c) 25% of values in the sample are smaller than 13.
d) The largest value in the sample was 50.

4.56 The following is the five-number summary of the traffic accidents last year in several intersections in a city.

36.00, 45.00, 54.00, 70.00, 108.00

What is the MOST appropriate statement that can be made?
a) The average of the sample is 62.6.
b) The average of the sample is 62.6 and the median is 54.
c) The data is highly skewed to the left.
d) 75% of the intersections have more than 45 accidents.

4.57 True or False
a) Approximately half of the data falls inside the box of a boxplot.
b) Large data sets tend to have smaller boxes in a boxplot.
c) A boxplot will always tell you what the mean of the data is.
d) A data set with a lot of large values can have a median outside the box of a boxplot.
e) The length of the box in a (horizontal) boxplot is a measure of data spread.

4.58 A production team must purchase a computer with some given specifications (processor speed, memory size, etc.), and a 25-inch monitor. Computers have a mean cost of $3000, with a standard deviation of $500, while monitors have a mean cost of $600 and a standard deviation of $100. They plan to purchase a computer costing $2500 and a monitor costing $800. Which cost is the highest considering other appliances of the same type? Justify your answer.

4.59 For a specific data set, $Q_1 = 250$ and $Q_3 = 250$. What is the value of the median?

4.60 The Canadian Dairy Information Centre compiles information from several companies on the cheeses made from cow, goat, sheep, and buffalo milk.[29] A random sample of 10 observations of moisture percent is

39   39   39   39   46   39   46   39   37   37

---

29 open.canada.ca (accessed July 26, 2019).

a) Find the *IQR*
b) Find *s*
c) Which of the two dispersion measurements calculated above makes more sense to summarize the data?

4.61 If there are $X_1, X_2, \ldots, X_n$ observations, the geometric mean, an alternative measure of centrality, is

$$\hat{x} = \sqrt[n]{X_1 \times X_2 \times \cdots \times X_n}$$

For example, beach water quality is frequently assessed through the geometric mean. Below are water quality sample results collected[30] by the Department of Health and Mental Hygiene at (Center) Manhattan Beach in New York on August 2017. Measurements are in most probable number per 100 ml (MPN / 100 ml):

60   320   100   110   48   4

a) Find the geometric mean.
b) Suppose the authorities have a standard stating that if the geometric mean of six consecutive measurements is 200 MPN/ 100 ml, that part of the beach should be closed. According to the result above, should they close the beach?

4.62 Sketch a scatterplot for which
a) $r = 1$;
b) $r = 0$, with no association between the two variables; and
c) $r = 0$, with strong association between the two variables.

4.63 2014 housing market data from 37 ZIP Codes in Austin, Texas, is available online.[31] Using the data file "2014_Housing_Market_Analysis_Data_by_Zip_Code," and with the assistance of statistical software, get the sample correlation between the following:
a) Median household income and median home value. Interpret.
b) Median household income and median rent. Interpret.

4.64 One of the Pew Research Center's survey topics is mobile phone use. The file "Riverside_Smartphone_Dependency" summarizes[32] the proportion of respondents who are smartphone dependent (use smartphones as their primary means of online access at home) in the County of Riverside, California.

---

30  data.cityofnewyork.us (accessed July 26, 2019).
31  Source: data.austintexas.gov (accessed July 26, 2019).
32  Source: data.countyofriverside.us (accessed July 26, 2019).

a) In 2016, what is the sample proportion of adults in the County who are smartphone dependent?
b) In 2016, which group showed the highest smartphone dependent?
c) How does income appear to affect smartphone dependency?

4.65 According to a recent report, in Northern Ireland, of 21 982 people previously found guilty of a crime, 4059 reoffended during a one-year observational period.[33] What is the proportion of reoffenders?

## Further Reading

Doane, D. P. and Seward L. E. (2011). *Applied Statistics in Business and Economics*, 3rd edition. McGraw-Hill.

Gelman, A. and Nolan, D. (2017). *Teaching Statistics: A Bag of Tricks*. Oxford.

Levine, D. M., Stephan, D. F., and Szabat, K. A. (2017). *Statistics for Managers: Using Microsoft Excel*, 8th edition. Pearson.

---

33 Source: Adult and Youth Reoffending in Northern Ireland (2015/2016 Cohort). Research and Statistical Bulletin.

# 5

# Introduction to Probability

## 5.1 Introduction

We need a measure that allows us to determine how feasible the occurrence of an event is. This will lead us closer to, among other things, being able to conduct statistical inference on population parameters. In this chapter, we will cover the topic of probability.

In practice, we encounter situations where it is convenient to use probabilities to make decisions. For example, if a company considers that a product price cut will result in an increase in sales of that product, the company can take a sample of days with sales and determine, through statistical inference, if the strategy works. Statistical inference requires probability. As another example, a shoe store knows by experience that 10% of the shoes from a certain brand are defective. The store wants to have a special sale of 100 pairs of shoes of that brand during a holiday. For the sale to succeed, management determines that they must sell 75 shoes with no defects. What is the probability that out of the 100 pair of shoes sold, 25 or more have defects? Furthermore, the store might be interested in finding if given that a customer buys a defective pair of shoes, what is the probability that the customer will return the shoes.

> **Why Probability?**
> Some general reasons why probability is necessary are as follows:
>
> - Probability allows us to determine decisions that will maximize revenue and/or minimize costs.
> - Probability allows risk control.
> - Probability establishes business policy.
> - Probability serves as a tool to conduct statistical inference.
> - Probability allows us to manage resources.

*Principles of Managerial Statistics and Data Science*, First Edition. Roberto Rivera.
© 2020 John Wiley & Sons, Inc. Published 2020 by John Wiley & Sons, Inc.
Companion website: www.wiley.com/go/principlesmanagerialstatisticsdatascience

## 5.2 Preliminaries

Before defining probability concepts, it is important to assure that the reader has a fair knowledge of basic set concepts. Let's begin with the necessary terms used to find probabilities.

**Data Collection Procedure:** a procedure that generates results that are not known before hand.

**Sample space:** a set that includes all possible outcomes of a data collection procedure. We denominate the sample space as $S$.

The sample space when a die is rolled is

$$S = \{1, 2, 3, 4, 5, 6\}$$

It should be noted that we are assuming that the data collection procedure generates random results, allowing any possibility in the sample space to occur.

**Example 5.1** *We are interested in verifying the strength of concrete blocks of a given dimension, that is made with a type of concrete a company manufactures. The company designs concrete blocks to hold more than 3000 pounds per square inch. If the blocks breaks before reaching that load, it is considered defective. If we want to do an experiment to test the strength of a concrete block of that specific type, we have a sample space with two possibilities: the concrete block is defective or it is not defective.*

Some sample spaces are too large to allow writing all possibilities. In these scenarios, a rule describing the sample space is established. For example, if $X$ represents the duration of a volleyball game, the sample space for the duration of a randomly chosen volleyball game would be

$$S = \{\text{all } X \text{ such that } X \geq 0\}$$

We are typically interested in the occurrence of specific outcomes from the sample space, not all outcomes.

**Event:** it is a subset of outcomes from the sample space. Typically, notation such as $A$ and $B$ are used to specify an event. We say that event $A$ occurs if at least one of its outcomes occurs. A specific outcome of the sample space is known as an **elementary event**. The sample space consists of all possible elementary events. Events may also consist of several elementary events.

**Example 5.2** *Suppose you roll a dice. An outcome of 2 is an elementary event. So is getting 6, or any other possible individual outcome of the experiment. If we define an event as "getting an even number," then we may say* A = {2,4,6}. *A occurs when the outcome of rolling the dice is 2, 4, or 6. It does not occur if the dice gives us any other value. Event A is not an elementary event, since it is based on more than one individual outcome of the experiment.*

When two dice are rolled, for an outcome to be considered well defined, the event must state something relating to the outcome of both dice. Specifically, $(1, 2)$ is an elementary event and so is $(6, 5)$. Thus, the sample space has 36 outcomes, namely

$$S = \begin{cases} (11), (12), \cdots (16), \\ (21), (22), \cdots (26), \\ \vdots \quad \ddots \quad \ddots \quad \vdots \\ (61), (62), \cdots (66), \end{cases}$$

**Example 5.3** *Suppose you roll two dice. If we define an event A as "getting a sum of 5," then*

$$A = \{(2,3), (3,2), (1,4), (4,1)\}$$

*Event A is not an elementary event. As another example, we could also define event B as "getting a 1 in at least one of the dice." Can you figure out which outcomes belong in B?*

*Special Events* Some events are particularly useful.

**Complement of an event:** The complement of an event, written as $A^c$, occurs when event $A$ does not occur. For example, if $A$ is the event of rolling an even number when rolling a die, $A^c = \{1, 3, 5\}$.

**Intersection of two events:** The intersection of events $A$ and $B$ occurs when these two events occur simultaneously, which is when an element that belongs to both $A$ and $B$ are a result of the data collection procedure.

---

**Toning Down the Math**

Many textbooks will use the notation $A \cap B$ to define the intersection of events $A$ and $B$. Although this notation is useful for theoretical work, for our purposes, we prefer to use $A$ *and* $B$ instead.

---

**Example 5.4** *When randomly choosing a card from a deck of 52 cards, if K = the chosen card is a king and R = chosen card is red; the event of choosing a red king is defined as*

$$K \text{ and } R = \{king \text{ } of \text{ } hearts, \text{ } king \text{ } of \text{ } diamonds\}$$

When events cannot occur simultaneously, they are said to be **mutually exclusive or disjoint**. As a result, when two events are mutually exclusive, $A$ *and* $B$ is an empty set,

$$A \text{ and } B = \{\}$$

Elementary events will always be mutuallyexclusive. That is, when you roll a dice once, you cannot get 1 and 6 at the same time. Also, still considering the roll of one dice, "getting an even number" and "getting an odd number" are mutually exclusive events. On the other hand, "getting an even number" and "getting a number greater than 3" are not mutually exclusive events.

**Union of two events:** The union of events $A$ and $B$ is defined as the event where $A$ occurs, $B$ occurs, or both occur simultaneously. Instead of using the traditional $A \cup B$ notation of union of two events, we will use $A$ *or* $B$. Another way of putting it is that for $A$ *or* $B$ to occur, at least one of the events $A$ and $B$ occurs.

**Example 5.5**  *When rolling two dice, if $O$ = roll one in first and $F$ = roll five in second, the event of roll one in first or five in second is defined as*

$$O \text{ or } F = \{(1,1),(1,2),(1,3),(1,4),(1,5),(1,6),(2,5),(3,5),(4,5),(5,5),(6,5)\}$$

We do not need to specifically call the union of two events $A$ *or* $B$; we could simply denominate a union as, say, $C$. For example, technically, event $A$ in Example 5.2 is the union of several events, but the definition of a union is more general than what we see in Example 5.2.

Later on in the textbook, we may not directly indicate that an event is a union, intersection, and so forth; but we will be applying these concepts indirectly. Also, these special events extend to more than just two events. The concept of the intersection of many events is straightforward to generalize, all of them must occur at the same. In the case of the union, we need to be a little bit more careful. For example, the union of events $A, B, C$ occurs when at least one of these events occurs. Notice that in this case, we cannot say that the union of $A, B, C$ occurs if $A$ occurs, $B$ occurs, $C$ occurs, or they all occur simultaneously, because other outcomes that make the union of $A, B, C$ occur are left out (e.g. that $A$ and $B$ occur but not $C$, that $B$ and $C$ occur but not $A$, and that $A$ and $C$ occur but not $B$).

## Practice Problems

5.1  Suppose we choose 2 balls from a bin with 100 red balls, 100 blue balls, and 100 green balls. If we define $R$ = a red ball is chosen, $B$ = a blue ball is chosen, and $G$ = a green ball is chosen,
   a) Write down the sample space S of the experiment.
   b) Let $C$ = a blue and red ball are part of the sample. Write the possible outcomes that form this event.

5.2  Define $A = \{1, 4, 6, 7\}$, $B = \{3, 6, 9\}$, $C = \{5, 8, 9\}$.
   a) What are the outcomes in the event $A$ *or* $B$?

b) What are the outcomes in the event *A and B*?
c) What are the outcomes in the event *A or B or C*?

5.3 Show that when rolling a dice, getting an even number and getting a number greater than 3 are not mutually exclusive events.

5.4 Determine what event would not be able to occur at the same time with the following events (the events are mutually exclusive):
a) You roll a die and get an even number.
b) Voter supports a candidate for president.
c) The maximum profit from three competing products are all equal.

5.5 Suppose a dice is rolled once.
a) Define $A$ = result is greater than 2 and even. What are the possible outcomes that form this event?
b) Define $B$ = result is greater than 5 and a prime number (only divisible by one and itself). What are the possible outcomes that form this event?
c) Define $P$ = result is an even number and $F$ = result is a number greater than four. Do both events contain the same results? If not, what would be the outcomes of an event $B$, that defines both events occurring at the same time?

5.6 A company wishes to conduct an analysis on the strength of some rods they produce. What is the sample space of the experiment?

5.7 You randomly choose a card from a deck of 52 cards. If $A$ = the chosen card is an ace and $B$ = chosen card is black, write down the possible outcomes that form $A$ *and B*.

## 5.3 The Probability of an Event

Once an event has been defined, we can perform different measurements of the event. For example, the probability of an event or the number of ways an event can occur are examples of measurements of the event. We designate the number of ways that $E$ can occur (i.e. the count), as $c(E)$. **Probability** is defined as the measurement of how likely a specific event is. To formally define the probability, some care is required, hence for now we are assuming that every possible result of a data collection procedure is equally probable. In mathematical terms, probability of an event $E$, $P(E)$, is defined as

$$P(E) = \frac{\text{Number of ways event } E \text{ occurs}}{\text{Total posibilities in the data collection procedure}} = \frac{c(E)}{c(S)} \quad (5.1)$$

This equation is an oversimplification, since it does not include situations where events may occur an infinite amount of ways and implies that every possible outcome of a data collection procedure is equally probable. But it helps in visualizing the probability as a measure of how likely any outcome in $E$ occurs relative to total possibilities in the sample space. An event with probability of the following:

- 1 means that the event is guaranteed to happen.
- 0 implies that the event is impossible.[1]
- 0.5 means that it is equally likely that the event will or will not occur.

The closer to 1 or 0, the more certain we feel that an event will occur or not occur: the more predictable the event is. In contrast, when $P(E) = 0.5$, then $E$ is unpredictable. Keep in mind that what is considered a high or low probability depends on the situation at hand. Expecting women over 35 are often told they have a high-risk pregnancy. This does not mean that the future mother has a high probability of complications, but that, relative to younger pregnant women, probability of complications is considerably higher.

**Example 5.6** *The past few presidential elections in the United States have seen the use of either polls or models (based on multiple polls or other information) to predict election winners. Just before the 2016 presidential election, most statistical models predicted Hillary Clinton to be the winner. However, it was Donald Trump who won the presidency. How did this happen?*

Multiple political pundits were quick to point out that polls and models were wrong. Now, there is plenty to say about this interpretation, but two arguments stick out as most important. First, it is not exactly true that polls and models missed it. Secondly, the misunderstanding is partly due to how poll and model summaries were communicated. For this example, the emphasis will be how poll and model output was communicated. Specifically, most models provided a winning probability for each candidate. That is, the New York Times upshot model gave Trump a 15% chance of winning, while FiveThirtyEight gave Trump a 29% chance of winning. Many took these probabilities to mean that a Clinton win was virtually guaranteed. However, 15% is slightly below the chances of getting a 4 when rolling a die. If you were to roll a die, do you think it is practically impossible to get a 4? Moreover, 29% is slightly below the chances of getting either a 3 or a 4 when rolling a die (more on this later). Thus, a Trump win was not as unlikely as many people thought.[2]

---

1 The interpretation of a probability of 1 and 0 here hold because the sample space is finite and countable. The meaning is slightly different in the case of uncountable or countable infinite sets. Strictly speaking, event $E$ is impossible when $E = \{ \}$, then $P(E) = 0$. But sometimes $P(E) = 0$ although $E \neq \{ \}$. In this second case, $E$ is not impossible, but very improbable.
2 Granted, some models did give substantially lower chances of winning to Trump, i.e. a 1% chance of winning. Such models may have been wrong.

Admittedly, the use of probability to summarize who may win an election can be hard to digest for the general public. Part of the problem is forgetting that a theoretical probability[3] always has a long-run interpretation. So when FiveThirtyEight gave about a 29% chance of winning to Trump, the interpretation is that over many elections between Trump and Hillary under the same political conditions, 29% of those elections would be won by Trump. Indeed, this is a strange summary for the model output since the probability gives us little information of the outcome of the 2016 election. Also problematic is our over reliance on one number summaries, such as a probability. In the case of the 2016 election outcome, an interval of the electoral college votes or an interval of proportion of popular votes would have been much more informative than probability of winning. FiveThirtyEight provided intervals for electoral college votes, popular votes, and expected margin of victory per state in their website, but only visually. Through any of those intervals, it was more apparent that the election was rather close. These intervals, known as prediction intervals, are discussed later in this book.

*Types of Probability* There are three ways to determine probabilities.

**Theoretical probability:** likeliness of an event is obtained based on logical/mathematical concepts. Theoretical probabilities are an indication of how likely an outcome is over a long-run of trials. Thus, the probability of getting a 4 when you roll a die is 1/6. Think about what this means for a second. If a die is rolled just once, you may get a 4 or you may get something else. But over a lot of attempts of rolling the dice, 1/6 of the time, 4 is the outcome.

**Empirical probability:** likeliness of an event is obtained based on observations or experimental results.

**Subjective probability:** this type occurs when the probability is based on what someone assumes or believes.

In some circumstances, like when we throw a coin and we want the probability of obtaining heads, we can directly calculate its theoretical probability. In other situations, the theoretical probability depends on parameters that we do not have; cases where we resort to empirical probabilities to either have a preliminary probability or to assess a conjectured theoretical probability. The subjective probability emerges from anecdotal experience. Sometimes, subjective probability is necessary when we do not have quantifiable information available. For example, when a brand-new product in a young industry field is developed. Still, it is best to rely on empirical or theoretical probabilities as soon as possible.

––––––

3 In this example, the models estimated the theoretical probability of candidates winning using polls and other information.

## Practice Problems

5.8 State whether the given probability is theoretical, empirical, or subjective.
   a) A poll of registered voters found that 38% of participants plan to vote for candidate $X$.
   b) In poker, the probability of a royal flush[4] is 0.000 001 54.
   c) From experience, the CEO of droneU determines there is a 15% that stricter drone regulations will be implemented in the next two years.

5.9 Yesterday, the local weatherman indicated that there was a 70% chance of rain. At the end of the day, it did not rain. Explain why the fact that it did not rain does not mean the local weatherman was wrong.

5.10 Based on the table below, evaluate each interpretation to determine if correct. Explain how the incorrect interpretations are wrong.
   a) The majority of plant managers consider the new machines to be noisy.
   b) It is more likely that a production employee works overtime most weeks than a plant manager considers the new machines to be noisy.
   c) It is equally likely that a worker has been working over 10 years for their current company than otherwise.
   d) We can virtually guarantee that an employee accesses their Facebook profile during work hours at least once a week.

| Event | Probability |
| --- | --- |
| Plant manager considers the new machines to be noisy | 0.44 |
| Production employee works overtime most weeks | 0.07 |
| Access their Facebook profile during work hours at least once a week | 0.78 |
| Has been working for their current company for over 10 years | 0.50 |

5.11 Before starting a new project, an analyst in a firm you work at determined that the probability the project would exceed its budget was 17%. Eventually, the project exceeded its budget. Was the probability given by the analyst certainly wrong?

5.12 A professional basketball team is one of the two remaining for the championship. The data analytics staff for the team has estimated that the probability of the team winning it all is 80%. The players interpret this as meaning there is nothing to worry about. Is the player's interpretation correct?

---

4 A hand of A, K, Q, J, 10 of the same suit.

5.13 A grocery store has calculated that the empirical probability of not running out of milk in any given day is 86%. Based on this probability, on average, how many times will they run out of milk in the next few days (assume the demand of milk is the same every day).

## 5.4 Rules and Properties of Probabilities

When working with theoretical or empirical probabilities, it is useful to apply certain rules of probability.

1) The probability of any event in a data collection procedure must be between 0 and 1:

$$0 \leq P(E) \leq 1$$

this result makes sense if we truly understand the concept of probability as defined in Eq. (5.1). We cannot get negative counts and an event can never occur in more ways than the total outcomes in the sample space.

2) The sum of probabilities[5] of all simple events that make up a data collection procedure will be 1. Implicitly, this way it is declared that the data collection procedure will occur with complete certainty; at least one of the possible simple events will occur. For example, if we roll a die, the probability that $E = $ a number between 1 and 6 is $P(E) = 1$. In this case, $E$ is defined as the event of the sample space. Mathematically, if we define the sample space $S$ based on elementary events so that $S = \{E_1, \ldots, E_n\}$, then:

$$\sum_{i=1}^{n} P(E_i) = 1$$

This rule can be generalized to the case of sample spaces with uncountable possibilities. Because of the second rule, $P(S) = 1$. This indicates that the data collection procedure is guaranteed to happen and therefore, if $A$ is a null event (nothing happens), $P(A) = 0$.

*Properties of Probabilities*   Several properties of probability can be established to facilitate calculation of probabilities. This is necessary because often it is not straightforward to intuitively determine the probability of an event. In what follows, what was learned about special events in Section 5.1 and about rules of probability is applied together.

---

5 This statement assumes a discrete sample space, but a similar rule applies for a continuous sample space.

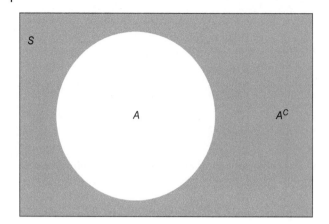

Figure 5.1 Venn diagram of the complement of *A*. Circle stands for *A*. Blue area stands for its complement: anything other than *A* occurs.

***Probability of the Complement of an Event*** A **Venn diagram** (Figure 5.1) illustrates how event *A* and its complement form *S*.

By definition of the complement, event *A* together with $A^c$ includes all the possibilities in the sample space. Therefore, considering rule 2 of probability we have

$$P(A^c) = 1 - P(A)$$

In Example 5.6, we stated that FiveThirtyEight gave Trump a 29% chance of winning. Thus, the probability of Trump not winning was

$$1 - 0.29 = 0.71$$

***Probability of the Intersection of Events*** The probability of the intersection of two events is denoted $P(A \text{ and } B)$ (see Figure 5.2).

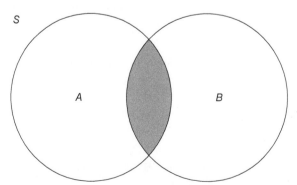

Figure 5.2 Venn diagram of the intersection of two events. Both occur at the same time.

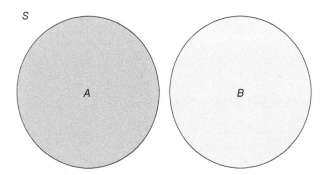

S

A

B

Figure 5.3 Venn diagram of two events that are mutually exclusive, they cannot occur simultaneously.

**Example 5.7** *What is the probability of getting a black ace when randomly choosing a card from a deck of 52 cards? If A = the chosen card is an ace and B = chosen card is black, the event of choosing a black ace is A and B. It can be shown (Problem 5.7) that this event has only two possibilities: ace of clubs and ace of spades. Therefore,*

$$P(A \text{ and } B) = \frac{2}{52}$$

or $1/26$. When $A$ and $B$ are mutually exclusive, $A$ *and* $B$ is an empty set. Since we know $P(S) = 1$, then $P(A \text{ and } B) = 0$ (see Figure 5.3).

***Probability of the Union of Events*** It can be shown that $P(A \text{ or } B)$:

$$P(A \text{ or } B) = P(A) + P(B) - P(A \text{ and } B)$$

Now, recall that $A$ *or* $B$ is defined as the event that at least one of these two events occur. So why is $P(A \text{ and } B)$ being subtracted on the right side of the equation? The best way to see what is going on is through a Venn diagram (see Figure 5.4). By adding $P(A)$ and $P(B)$, $P(A \text{ and } B)$ is being counted twice.

**Example 5.8** *A die is rolled once. What is the probability of getting a 3 or a 4?*

Let $T = $ a 3 is rolled and $F = $ a 4 is rolled. Then,

$$P(T \text{ or } F) = P(T) + P(F) - P(T \text{ and } F)$$
$$= P(T) + P(F)$$
$$= 1/6 + 1/6 = 1/3$$

The second equality holds because it is not possible to get 3 and 4 simultaneously when a die is rolled just once. Since $T$ and $F$ are mutually exclusive, $P(T \text{ and } F) = 0$.

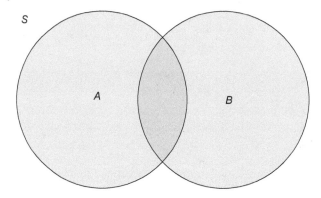

Figure 5.4 Venn diagram of the union of two events. Either event *A* occurs, *B* occurs, or both at the same time.

## Practice Problems

5.14 Define the following notation for the population of students at your local university:

$F$ = the event that a student is female.
$M$ = the event that a student is male.
$B$ = the event that a student is from the business program.
$E$ = the event that a student is from the engineering program.

Write the mathematical notation for the following:

a) The probability that a randomly chosen student is from the business program.
b) The probability that a randomly chosen student is from the business program and female.
c) The probability that a randomly chosen student is not male.
d) The probability that a randomly chosen student is from the engineering program or female.

5.15 Returning to Problem 5.14, write the mathematical notation for the following:

a) The probability that a randomly chosen student is neither in the business nor engineering program.
b) The probability that a randomly chosen student is from at least one of the two indicated programs.
c) The probability that a randomly chosen student is from the business or engineering program, but not both simultaneously.

5.16 If $P(A) = 0.15, P(B) = 0.44$, and $P(A \text{ or } B) = 0.29$. Find
   a) $P(A^c)$
   b) $P(B^c)$
   c) $P(A \text{ and } B)$

5.17 If $P(A) = 0.4, P(B) = 0.35$, and $P(A \text{ and } B) = 0.15$. Find
   a) $P(A^c)$
   b) $P(B^c)$
   c) $P(A \text{ or } B)$

5.18 One die is rolled. Find the probability of getting
   a) A one.
   b) An even number.
   c) An even and uneven number.
   d) A four and an even number.
   e) A one or an uneven number.

5.19 If $P(A) = 0.6, P(B) = 0.45$, and $P(A \text{ and } B) = 0.25$. Find
   a) at least one of the two events occurs.
   b) *A* occurs *B* occurs but not both simultaneously.

5.20 Given that $P(A) = 0.25$, $P(B) = 0.5$, and $P(A \text{ and } B) = 0.1$. Find $P((A \text{ or } B)^c)$.

5.21 Two dice are rolled. Find the
   a) number of possible outcomes in the experiment.
   b) probability of getting the same number in each die.
   c) probability of getting a 2 in one die or a 5 in one die.

5.22 $P(A) = 0.5, P(B) = 0.3$. It is known that if *B* occurs, then *A* will certainly occur. Find $P(A \text{ and } B)$ and $P(A \text{ or } B)$.

5.23 Concerned about the economy, the CEO of a marketing strategy company asks analysts to determine the probability of low demand of their services next year (*L*), of high demand (*H*), and of losses of at least $1 million due to unexpected events such as hurricanes (*U*). The analysts come up with $P(L) = 0.63, P(H) = 0.28, P(U) = 0.09, P(H \text{ and } U) = 0.02$, and $P(L \text{ and } U) = 0.07$.
   a) What is the probability that next year is not of high demand or low demand?
   b) What is the probability that next year has low demand for their services or losses of at least $1 million due to unexpected events?

    c) Are the events low demand for their services and losses of at least $1 million due to unexpected events mutually exclusive? Explain.

    d) Do you think that losses of at least $1 million due to unexpected events and the event that demand is not low are mutually exclusive? Explain.

## 5.5 Conditional Probability and Independent Events

Once again, we set a rhetorical situation. This time, a company that manufactures vehicles buys 100 000 radiators from a supplier. After completing the transaction, the company that manufactures cars notices that a considerable amount of radiators are defective, but they do not know specifically which ones. Even though the defect is not major, it can lead to overheating of a car in some instances. For this reason, management has two options: return all radiators and obtain new ones (at a discount), or continue manufacturing the cars. One way to decide what to do is answering the following question: given a consumer buys the vehicle with a defective radiator, what is the probability that their car overheats (and consequently the company is responsible for the cost)? Which is a conditional probability?

As another example, suppose a coin is tossed once, what is the probability that we get tails? In this case, most people can figure out that the probability of getting heads and of getting tails are the same: 0.5. Now, say we slightly modify the data collection procedure such that the coin is tossed 4 times. If we define $H$ as the event that we get heads on a toss and $T$ that we get tails on a toss, how do you think that the probabilities of the following events compare? $(H, H, H, H), (H, T, H, T), (T, H, T, H)$, and $(T, T, T, T)$. Take a few seconds to think about this problem before you continue reading. Most would answer that the events $(H, T, H, T)$, and $(T, H, T, H)$ have the highest probability. But it turns out that each one of these four events have the same probability of occurrence: $1/16$. This is related to the concept of independent events.

---

**Need a Refresher on Inequalities?**
Head to Section A.3 if you wish to review the notation for math inequalities.

---

    **Conditional probability:** the probability of event $A$ given that $B$ occurred is known as a conditional probability and is expressed as $P(A|B)$. Note that if we guarantee the occurrence of event $B$, then we can interpret conditional probability as the probability of $A$ and $B$ occurring simultaneously relative to the probability of $B$,

$$P(A|B) = \frac{P(A \text{ and } B)}{P(B)} \tag{5.2}$$

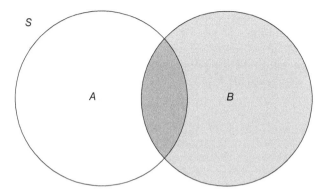

Figure 5.5 Venn diagram of the conditional probability of *A* occurring if *B* already occurred.

where $P(B) > 0$. Beware: $P(A|B) \neq P(B|A)$. The expression to the left of the inequality represents how likely *A* is given *B* occurred, while the expression to the right of the inequality represents how likely *B* is given *A* occurred (Figure 5.5).

Conditional probability is widely used in practice. An interpretation of $P(A|B)$ can be as follows: given some information (i.e. the occurrence of event *B*), what is the probability that an event (*A*) occurs? Conditional probabilities are also regularly used to conduct statistical inference. The way it works is that an analyst may not know if event *B* is true or not. But if it is true, then the probability of *A* must, say, be high. Therefore, she conjectures that *B* is true, and based on this, finds $P(A|B)$. If $P(A|B)$ is low, then this is evidence against *B* being true, and the analyst concludes that she has no evidence of *B* being true. This sort of argument is what is used in the statistical inference method known as hypothesis testing, which we turn to in Chapter 10.

**Example 5.9** *When randomly choosing a card from a deck of 52 cards, what is the probability of getting a five, given we know the card will be a 4, 5, or 6?*

If $F$ = the chosen card is a five and $G$ = chosen card is 4, 5, or 6;

$$P(F|G) = \frac{P(F \ and \ G)}{P(G)}$$

$$= \frac{4/52}{12/52} = \frac{1}{3}$$

Observe that $P(F) = 4/52 = 1/13$. Thus, $P(F|G) > P(F)$, knowing *G* occurred has increased the chances *F* occurs.

*Independent events*   Say you flip a coin twice. $H_1$ is the event of getting heads on the first try and $T_2$ is the event that you guess tails on the second try. Can you guess what $P(T_2|H_1)$ is without a calculation?

Two events are **independent** when the occurrence of one event does not affect the probability of occurrence of the other. When events $A$ and $B$ are independent, then

$$P(A|B) = P(A)$$

Although this equation represents the definition of independence, through Eq. (5.2) it is seen that independence also means

$$P(A \text{ and } B) = P(A)P(B)$$

When flipping a coin twice, the fact that we got heads on the first flip of a coin has no impact on the chances of getting tails on the second flip, $P(T_2|H_1) = P(T_2)$.

**Example 5.10**   *Returning to Example 5.7, what is the probability of getting an ace when randomly choosing a card from a deck of 52 cards, given we know the card will be black?*

If $A =$ the chosen card is an ace and $B =$ chosen card is black, the event of choosing a black ace is $A$ *and* $B$. It can be shown (Problem 5.7) that $A$ *and* $B$ has only two possibilities: ace of clubs and ace of spades. Therefore,

$$P(A|B) = \frac{P(A \text{ and } B)}{P(B)}$$
$$= \frac{2/52}{26/52} = \frac{1}{13}$$

But $P(A|B) = P(A)$, therefore, $A$ and $B$ are independent. The attribute of independence will be implied frequently in class. Often, this will be based on the argument that a sample was conducted randomly. This way, what we obtain as first observation would not provide information of what we can expect in the following observations. When **sampling without replacement**, the outcomes of a data collection procedure are not independent. For example, the card that you chose in the first attempt gives you (probabilistic) information about what you will get in the following attempts.

**Example 5.11**   *There are three left-handed workers and seven right-handed workers assigned to an assembly line. Two workers are randomly selected without replacement to test a new machine. What is the probability that both workers are right-handed?*

Note that the possibilities are that

- no right-handed workers are chosen, or
- one right-handed worker is chosen, or
- two right-handed workers are chosen.

Intuitively, both workers being right-handed should have a decent chance of occurring because the majority of workers assigned to the assembly line are right-handed. Let $R_i$ represent that the $i$th worker randomly chosen is right-handed. Then, the goal is to find $P(R_1 \text{ and } R_2)$. Implicitly, the scenario involves sampling without replacement. As a result, the chances of getting a right-handed worker on the second attempt depend on whether a right-handed worker was chosen on the first sample. Naturally, $P(R_1) = 7/10$. Moreover, if the first worker chosen is right-handed, then there are only six right-handed workers to be chosen from the remaining nine. This means

$$P(R_1 \text{ and } R_2) = P(R_1)P(R_2|R_1)$$
$$= \frac{7}{10}\left(\frac{6}{9}\right)$$
$$= \frac{42}{90}$$

*Bayesian Method* Conditional probabilities allow us to revise the probability of $A$ under a certain condition. We can also use conditional probability to modify our beliefs about an event through data. It works through what's known as **Bayes' Theorem**,

$$P(A|B) = \frac{P(B|A)P(A)}{P(B)}$$

where $P(B) > 0$. $P(B)$ is not always available, but it is possible to find using the law of total probability,

$$P(A|B) = \frac{P(B|A)P(A)}{P(B|A)P(A) + P(B|A^c)P(A^c)}$$

**Example 5.12** *Let's suppose that a company performs drug testing on its employees. Results from drug tests, and diagnostic tests of many diseases, cannot be guaranteed to be correct. A person may fail the test although they have not taken any illegal drugs recently or may pass the test although they have used illegal drugs recently. Say that if someone is an illegal drug user, then the test will come out positive 95% of the time. However, 10% of the time the test will come out positive although the person has not used any illicit drugs. According to the latest data, 9.4% of the US population have used illegal drugs in the past month.[6] The question is, given that a person in the company tests positive, what is the probability that the person used illegal drugs in the past month?*

---

6 www.drugabuse.gov (accessed July 26, 2019).

Let's define some events:

$I$ = person has taken illegal drugs,

$T$ = positive test.

The example provides $P(T|I) = 0.95$, $P(I) = 0.094$, and $P(T|I^c) = 0.1$, and the probability of interest is $P(I|T)$. According to Bayes' theorem,

$$P(I|T) = \frac{P(T|I)P(I)}{P(T|I)P(I) + P(T|I^c)P(I^c)}$$

since $P(I^c) = 1 - 0.094 = 0.906$, we have

$$P(I|T) = \frac{0.95 \times 0.094}{0.95 \times 0.094 + 0.10 \times 0.906}$$
$$= 0.496$$

Therefore, if someone tests positive for taking illegal drugs, the probability the person really used illegal drugs is about 50%. This probability at first looks surprisingly low. But keep in mind that only 9.4% of the population have used illegal drugs in the past month. So, by testing positive, it has become five times more likely that the person used illegal drugs. Nevertheless, it would be nice if the test gave us stronger evidence on drug use. However, the probability of the test giving a false positive when someone has not used drugs was fairly high (10%). If $P(T|I^c) = 0.01$ and all other arguments stay the same, then $P(I|T) = 0.91$, a marked improvement (you should check this is true). Also, before you go and tell your friends that drug testing is a farce, you should know that when someone tests positive for drugs, often another type of drug test is performed to confirm the results. The combination of these tests makes the results highly reliable.

In traditional statistical inference, a parameter of interest is fixed but unknown. It is estimated through a statistic, and this statistic is combined with probability to make conclusions about the parameter. Bayes' theorem can be used to apply inference in a nontraditional way, known as **Bayesian inference**. In Bayesian inference, any unknown is treated as random, including parameters. Using data and prior beliefs about the parameter, we can then adapt our conclusions about the parameter values. Let $H$ = some hypothesized event of interest and $D$ = data. A probability model is used for the data assuming the hypothesis is true, $P(D|H)$. Our beliefs on the hypothesis are modeled through a **prior distribution**, $P(H)$. Bayes' rule is used to obtain a **posterior distribution** for the hypothesized event given the data,

$$P(H|D) = \frac{P(D|H)P(H)}{P(D)}$$

Many people find the principle of Bayesian inference appealing, arguing that it is consistent with the principles of the scientific method, since it aims to answer the specific question of interest: What is the probability that a hypothesis or theory is correct given the data? Frequentist statistics on the other hand attempt to reach conclusions indirectly, by assuming that the hypothesis is true and getting the probability of the data. Although nowadays many people recognize the benefits of Bayesian inference, it still has not been broadly adopted. The are two main challenges. First, care is required when choosing a prior distribution, so that the results do not become subjective. It can be argued that the results could in fact be manipulated if a "convenient" prior distribution is selected. However, prior distribution sensitivity can always be conducted. Also, there is always a level of subjectivity in statistical models, even in frequentist methods: choosing a statistical model or probability distribution, or a cut off to reject a null is a subjective decision. Secondly, and most importantly, the computational requirements can be extensive. Often, the posterior distribution is not easy to work with and may not even have a well-known form. This problem is resolved through simulations which can be shown to eventually provide samples from the posterior distribution. However, the amount of simulations varies by problem. Sometimes 1 000 simulated samples will do. On occasion, a situation requires millions of samples which can take weeks to run. Software has been developed to make it easier to perform these simulations, yet expertise on the procedure is still required. The debate of whether Bayesian inference or frequentist inference is best can get intense. An important question in the debate is how often do conclusions drawn through frequentist methods vary from Bayesian methods? There is no general answer to this question, sometimes one method will be best, other times the other. Many data analysts nowadays will either perform both methods or a hybrid of both (known as **empirical Bayes method**).

## Practice Problems

5.24  Show that for two independent events $A$, $B$, $P(B|A) = P(B)$.
*Hint:* Use the meaning of independence for $P(A \text{ and } B)$.

5.25  If $P(A) = 0.4$, $P(B) = 0.35$, and $P(A \text{ and } B) = 0.05$. Find
  a)  $P(A|B)$
  b)  $P(B|A)$
  c)  Are events $A$ and $B$ independent? Explain

5.26  If $P(A) = 0.6$, $P(B) = 0.45$, and $P(A \text{ and } B) = 0.27$. Find
  a)  $P(A|B)$
  b)  $P(B|A)$
  c)  Are events $A$ and $B$ independent? Explain

5.27 Given that $P(A) = 0.10, P(B) = 0.35$, and $P(A \text{ and } B) = 0.07$.
  a) Are these two events independent? Explain.
  b) Calculate $P(A|B)$.
  c) Calculate $P(B|A)$.

5.28 What is the probability of getting a king when randomly choosing a card from a deck of 52 cards, given we know the card will be red?

5.29 In Example 5.7, verify that $P(A \text{ and } B) = P(A)P(B)$; the events are independent.

5.30 $P(B) = 0.35$ and $P(B|A) = 0.70$. If $B$ is something we hope occurs, do we want $A$ to occur or not? Explain.

5.31 $P(A) = 0.67$ and $P(B) = 0.52$, and $A$ and $B$ are mutually exclusive. Find $P(B|A)$.

5.32 The concept of conditional probabilities can be extended to more than two probabilities. If $P(A|B \text{ and } C \text{ and } D) = P(A)$
  a) Does it mean $A$ is independent from event $(B \text{ and } C \text{ and } D)$? Explain.
  b) Does it mean all events, $A, B, C, D$, are independent? Explain.

5.33 Referring to Example 5.11, what is the probability that both workers in your sample are left-handed?

5.34 Referring to Example 5.11, what is the probability that one worker in your sample is left-handed, and the other right-handed?(*Hint:* This probability is the union of two events.)

5.35 Referring to Example 5.12, show that $P(I|T) = 0.91$ if $P(T|I^c) = 0.01$ and all other arguments stay the same.

5.36 In multiple choice questions, there is always the possibility that a student luckily guesses the right answer. Suppose $K$ = student knows the right answer, $C$ = student answers correctly. If three possible choices are provided, from which only one is correct, assume $P(C|K^c) = 1/3$ (student guesses the right answer), $P(C|K) = 1$, and $P(K) = 3/4$. Find the probability the student knows the answer, given they answered correctly.

5.37 Continuing Problem 5.36, if there are four choices provided, then, $P(C|K^c) = 1/4$. Find the probability the student knows the answer, given they answered correctly.

## 5.6 Empirical Probabilities

Empirical probabilities rely on the frequency of events observed. The number of times an event occurs is counted and this value is divided by the total number of observations. Computer software can tally the occurrence of each category of a qualitative variable and summarize these counts in tabular form.

**Example 5.13** *Crime offenses in Orlando, Florida, from 2009 until 2015 are summarized in Table 5.1. When multiple crimes are committed, only the highest level of crime is given in the data set.[7] From the table, we can find that the probability that the highest offense was theft*

$$P(T) = \frac{78\,145}{160\,947} = 0.49$$

Table 5.1 Crime offenses in Orlando Florida from 2009 to 2015.

| Offense charge | Count |
| --- | --- |
| Arson $(R)$ | 244 |
| Assault $(U)$ | 17 026 |
| Bribery $(B)$ | 3 |
| Burglary $(Y)$ | 24 263 |
| Embezzlement $(E)$ | 48 |
| Fraud $(F)$ | 10 372 |
| Homicide $(H)$ | 109 |
| Kidnapping $(K)$ | 83 |
| Narcotics $(N)$ | 18 393 |
| Robbery $(O)$ | 4624 |
| Theft $(T)$ | 78 145 |
| Vehicle theft $(V)$ | 7637 |
| Total | 160 947 |

*Source:* data.cityoforlando.net (accessed July 26, 2019)

---

7 The table includes event notation for convenience. For example, event $R$ is when arson was the highest level of crime committed.

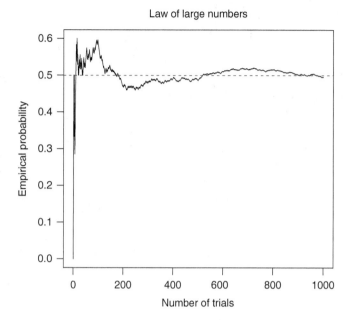

Figure 5.6 Law of large numbers: how the empirical probability tends to the theoretical probability (0.5 here) as the number of trials increases.

---

**Law of Large Numbers**

As observations increase, empirical probabilities will converge to theoretical probabilities, a theorem known as the **law of large numbers**. This theorem is at the core of statistical inference. Even when the theoretical probability is unknown, with enough empirical evidence we can statistically infer about it. An illustration of the law of large numbers is seen in Figure 5.6 for samples taken of an event with theoretical probability of 0.5. In the first few trials (small sample size), the empirical probability differs considerably from 0.5. But as the sample size continues to become larger, the empirical probability gets closer and closer to the theoretical probability. In the simulation results presented, by trial 17, the empirical probability never differs again from the theoretical probability by more than 0.06.

---

Recall from Section 3.4 that a contingency table allows us to visualize the relationship between two categorical variables, $A$, $B$.

**Example 5.14**  *Returning to Example 5.13, now crime offenses in Orlando, Florida from 2009 until 2015 are summarized in Table 5.2 by offense charge*

Table 5.2 Summary of highest offense charge and offense type in Orlando, Florida, from 2009 to 2015.

| Offense charge | Offense type | | Total |
| --- | --- | --- | --- |
| | Attempted (A) | Committed (C) | |
| Arson (R) | 7 | 237 | 244 |
| Assault (U) | 131 | 16 895 | 17 026 |
| Bribery (B) | 1 | 2 | 3 |
| Burglary (Y) | 3137 | 21 126 | 24 263 |
| Embezzlement (E) | 0 | 48 | 48 |
| Fraud (F) | 174 | 10 198 | 10 372 |
| Homicide (H) | 1 | 108 | 109 |
| Kidnapping (K) | 29 | 54 | 83 |
| Narcotics (N) | 29 | 18 364 | 18 393 |
| Robbery (O) | 481 | 4 143 | 4 624 |
| Theft (T) | 2 478 | 75 667 | 78 145 |
| Vehicle theft (V) | 960 | 6 677 | 7 637 |
| Total | 7 428 | 153 519 | 16 0947 |

*Source:* data.cityoforlando.net (accessed July 26, 2019).

*and offense type. A civil group would like to determine if the event of the highest offense charge being theft is independent of offense type.*

Recall from Example 5.13 that the probability that the highest offense was theft is $P(T) = \frac{78\,145}{160\,947} = 0.49$.

Now, what is the probability that the highest offense was theft, given that it was an attempted offense? And, what is the probability that the highest offense was theft, given that it was a committed offense?

$$P(T|A) = \frac{2478}{7428} = 0.33$$

$$P(T|C) = \frac{75\,667}{153\,519} = 0.49$$

From these results, we see that if we know that a crime was attempted, it is less likely that the highest offense was theft than if we knew that the crime was committed. Also, $P(T|C) = P(T)$ but $P(T|A) \neq P(T)$. If the highest offense charge being theft is independent of offense type, then $P(T|C) = P(T|A) = P(T)$. These are empirical probabilities though and the rule of independence, i.e. $P(T|A) = P(T)$, can only be ensured to hold exactly for theoretical probabilities. As an empirical probability, we can only expect $P(T|A)$ to be close to $P(T)$ if independence exists, a topic we return to later on in the book.

## 5.6.1   A Quick Look at Data Science: Missing People Reports in Boston by Day of Week

Are reports of a missing person in Boston more likely in some days than others? Police data for the city is available online.[8] Police data includes dates, times, location, and other variables for several types of incidents. At the time the data set was retrieved,[9] it had 237 074 rows and 17 columns. Little data wrangling is required for this data set: no errors were detected for our purposes, and incidents other than a missing person were identified as a non-missing person incident.

Let $M = $ a missing person report occurs and $D_i$ is the day of the week, $D_1 = $ Sunday, $D_2 = $ Monday, and so on. If the event of a missing person being reported is independent of day of week, then $P(M|D_i) \approx P(M)$ for all days of the week. Table 5.3 presents empirical probabilities of all 14 combinations of $D_i, M,$ and $M^c$. It can be seen that the probability of a missing person and the report be on a Friday was 0.0022. The empirical probability a missing person is reported in Boston regardless of day was 0.010 67.

Table 5.4 summarizes the empirical values of $P(M|D_i)$ and $P(M^c|D_i)$. At first, all values appear similar, but the probability of a missing person report given that it is a Friday, 0.0146, is 27% larger than the probability of a missing person report regardless of day; some evidence that a missing person report is more likely in some days than others.[10]

Table 5.3 Empirical probabilities of the 14 combinations of "day of week" and missing person report.

|             | Sunday | Monday | Tuesday | Wednesday | Thursday | Friday | Saturday |
|-------------|--------|--------|---------|-----------|----------|--------|----------|
| Missing     | 0.0014 | 0.0017 | 0.0017  | 0.0016    | 0.000 17 | 0.0022 | 0.0019   |
| Not missing | 0.1249 | 0.1416 | 0.1435  | 0.1450    | 0.1450   | 0.1496 | 0.1381   |

The cell probabilities add up to 1.

Table 5.4 Empirical conditional probability that there is a missing person report in Boston given the day of the week.

|             | Sunday | Monday | Tuesday | Wednesday | Thursday | Friday | Saturday |
|-------------|--------|--------|---------|-----------|----------|--------|----------|
| Missing     | 0.0113 | 0.0120 | 0.0121  | 0.0110    | 0.0118   | 0.0146 | 0.0136   |
| Not missing | 0.9886 | 0.9880 | 0.9879  | 0.9889    | 0.9882   | 0.9854 | 0.9864   |

---

8  data.boston.gov (accessed July 26, 2019).
9  November 5, 2017.
10  For a more formal approach to test independence, see Problem 15.2.

Table 5.5 Empirical conditional probability that there is a missing person report in downtown Boston given the day of the week.

|  | Sunday | Monday | Tuesday | Wednesday | Thursday | Friday | Saturday |
|---|---|---|---|---|---|---|---|
| Missing | 0.0020 | 0.0047 | 0.0041 | 0.0021 | 0.0018 | 0.0032 | 0.0031 |
| Not missing | 0.9980 | 0.9953 | 0.9959 | 0.9979 | 0.9982 | 0.9968 | 0.9969 |

To ensure adequate distribution of resources, it is also meaningful to police administrators to describe missing person probabilities by police district. It can be shown that the probability there is a missing person report in downtown Boston, regardless of day, was 0.0030. Table 5.5 shows $P(M|D_i)$ and $P(M^c|D_i)$ according to downtown incidents. Here, $P(M|D_i)$ is highest on Monday.

## Practice Problems

5.38 Returning to the missing person case study in Section 5.6.1, show that the empirical probability a missing person is reported in Boston regardless of day was 0.010 67, as stated.

5.39 Returning to the case study in Section 5.6.1, show that the empirical probability any police report in Boston was filed on a Monday is 0.1433.

5.40 The table below shows the type of incidents that lead to a call to the Fire Department in Detroit in 2015.[11]
a) What is the probability the incident is a false alarm?
b) What is the probability the incident is not a false alarm?
c) Find the probability the incident is a fire or good intent call.
d) Find the probability the incident is a good intent call and hazardous condition.
e) What is the probability that it was a weather incident given that it was a "fire, explosion" incident?

| False alarm | Fire, explosion | Good intent | Hazardous condition | Overpressure | Service call | Weather | Other |
|---|---|---|---|---|---|---|---|
| 6664 | 7023 | 2732 | 3151 | 96 | 265 | 4 | 24 |

---

11 Source: data.detroitmi.gov (accessed July 26, 2019).

**5.41** The following contingency table describes type of Cheese produced and whether it is organic based on 1296 companies from Canada.[12] Find each probability.

| Organic | Firm (F) | Fresh (R) | Hard (H) | Semisoft (E) | Soft (O) | Veined (V) | Total |
|---|---|---|---|---|---|---|---|
| | | | | **Type** | | | |
| Yes (Y) | 398 | 149 | 45 | 248 | 311 | 27 | 1178 |
| No (N) | 48 | 10 | 8 | 25 | 19 | 8 | 118 |
| Total | 446 | 159 | 53 | 273 | 330 | 35 | 1296 |

a) $P(O)$
b) $P(Y \text{ and } R)$
c) $P(O|Y)$
d) $P(V \text{ or } N)$
e) $P(R \text{ or } H)$
f) $P(E|F)$

**5.42** Delaware has a rebate program for the purchase or lease of alternative fuel vehicles. Data is available to the public.[13] Download "Delaware_State_Rebates_for_Alternative-Fuel_Vehicles." Construct a contingency table of gender and vehicle type.
a) What is the probability the rebate was for an electric vehicle?
b) What is the probability the rebate was for a plug-in hybrid?
c) What is the probability the rebate was for a plug-in hybrid given the customer was a female?
d) What is the probability the rebate was for a plug-in hybrid given the customer was male?

**5.43** In Section 5.5, we learned that if events $A$ and $B$ are independent, $P(A \text{ and } B) = P(A)P(B)$. This equality does not necessarily hold with empirical probabilities. However, when using empirical probabilities, do you think $P(A \text{ and } B)$ will be close to $P(A)P(B)$?

**5.44** Figure 3.16 suggested that race was related to whether a male driver was searched by police in San Diego in 2016. When exploring this question, it is important to consider the cause of the traffic stop; information that is available in the data.[14] Moving violations, equipment violation, and suspect information were some of the reasons for traffic stops. If we only

---

12 Source: the Canadian Dairy Information Centre, open.canada.ca. (accessed July 26, 2019)
13 Source: data.delaware.gov (accessed July 26, 2019).
14 data.sandiego.gov (accessed July 26, 2019).

consider data such that moving violations were the cause, the Table summarizes whether a male driver was searched according to his race.

| Searched | Race | | | | | |
|---|---|---|---|---|---|---|
| | Asian | Black | Hispanic | Other | White | Total |
| Yes | 43 | 292 | 526 | 63 | 373 | 1 297 |
| No | 2 609 | 3 687 | 10 575 | 3 693 | 16 770 | 37 334 |
| Total | 2 652 | 3 979 | 11 101 | 3 756 | 17 143 | 38 631 |

a) Find the probability that a White male driver was searched.
b) Find the probability that a male driver was not searched.
c) Find the probability that a male driver was searched given that he was Black.
d) Find the probability that a male driver was searched given that he was White.
e) Find the probability that a male driver was searched and was Hispanic.
f) Do you think the probability of a male driver being searched during a moving violation stop is independent of race? Explain.

5.45 Female drivers are searched less often than men during police stops. But is there evidence that race predicts whether a female driver was searched by police in San Diego in 2016? The Table summarizes whether a female driver was searched by police according to race of driver.

| Searched | Race | | | | | |
|---|---|---|---|---|---|---|
| | Asian | Black | Hispanic | Other | White | Total |
| Yes | 11 | 59 | 120 | 9 | 90 | 289 |
| No | 1 983 | 1 805 | 5 694 | 1 920 | 10 730 | 22 132 |
| Total | 1 994 | 1 864 | 5 814 | 1 929 | 10 820 | 22 421 |

a) Find the probability that a White female driver was searched.
b) Find the probability that a female driver was not searched.
c) Find the probability that a female driver was searched given that she was Black.
d) Find the probability that a female driver was searched given that she was Hispanic.
e) Find the probability that a female driver was searched given that she was White.
f) Do you think the probability of a female driver being searched during a moving violation stop is independent of race? Explain.

5.46 Many other cities around the US share police data. Statistics for 2015 and 2016 vehicle stops at Austin, Texas, are available to the public. Table 1 summarizes stops by race and Table 2 summarizes searches.[15] Use data from these tables to find the following:

**Table 1: Motor vehicle stops by race/ethnicity**

| Race/ethnicity | 2016 Stops Count | 2016 Stops % of total | 2015 Stops Count | 2015 Stops % of total |
|---|---|---|---|---|
| White | 48 743 | 46% | 59 699 | 50% |
| Hispanic | 36 006 | 34% | 37 702 | 31% |
| Black | 12 741 | 12% | 14 753 | 12% |
| Asian | 3 175 | 3% | 3 715 | 3% |
| Middle Eastern | 1 260 | 1% | 1 655 | 1% |
| Native American | 64 | 0% | 52 | 0% |
| Other | 2 939 | 3% | 2 480 | 2% |
| Total | 104 928 | 100% | 120 056 | 100% |

**Table 2: Searches by race/ethnicity**

| Race/ethnicity | 2016 Searches Count | 2016 Searches % of total | 2015 Searches Count | 2015 Searches % of total |
|---|---|---|---|---|
| White | 3 137 | 33% | 2 838 | 31% |
| Hispanic | 3 919 | 41% | 3 973 | 43% |
| Black | 2 243 | 23% | 2 228 | 24% |
| Asian | 119 | 1% | 109 | 1% |
| Middle Eastern | 50 | 1% | 41 | 0% |
| Native American | 7 | 0% | 5 | 0% |
| Other | 94 | 1% | 59 | 1% |
| Total | 9 569 | 100% | 9 253 | 100% |

a) The probability that in 2015 a driver was searched given that the driver was Black.

b) The probability that in 2015 a driver was searched given that the driver was Hispanic.

c) The probability that in 2015 a driver was searched given that the driver was White.

d) Any limitations you can think of when comparing these conditional probabilities to those found in Problems 5.45 and 5.46?

5.47 Returning to Problem 5.46 and its tables:

a) Find the probability that in 2016, a driver was searched given that the driver was Black,

b) Find the probability that in 2016, a driver was searched given that the driver was Hispanic.

c) Compare the calculated probabilities with those obtained in Problem 5.46.

## 5.7  Counting Outcomes

To determine how feasible an event of interest is, we must establish how many ways the event can occur. One way to do this is to write all the possibilities in a way we can count the number of ways an event happens. Let's say we throw a dice twice. As soon as we write all the possibilities (see Section 5.2), it is easier to count them and see that by throwing two dices we have 36 possible outcomes for the experiment. But, what if we throw the dice 10 times, or 100? Or if we conduct an experiment where we must select between 5 purse brands, 6 types

---

15  Source: 2016 Annual Racial Profiling Report, City of Austin, Austin Police Department: www.austintexas.gov (accessed July 26, 2019).

of stores, and 4 different counties in order to compare the price of each purse. How many possibilities can we encounter in this experiment?

It is not always practical to write every possibility of an experiment because it may be too complicated or we are only interested in a small number of possibilities.

*Multiplication Rule*   Let's suppose we can break down an event into $r$ steps and we can count the outcomes in each. The first step has $n_1$ possible outcomes and the second has $n_2$ possible outcomes until we reach step $r$ which has $n_r$ possible outcomes. The event will have

$$(n_1) \times (n_2) \times \cdots \times (n_r)$$

possibilities. Going back to the example where we needed to select 5 purse brands, 6 types of stores, and 4 different countries to compare the price of the purses, the experiment has three stages, select the

- purse brand,
- type of store,
- country.

Thus, $r = 3$, $n_1 = 5$, $n_2 = 6$, and $n_3 = 4$, and according to the multiplication rule the sample space consists of

$$(5)(6)(4) = 120$$

possibilities.

Sometimes, a possibility is allowed multiple times. For example, when you pick a card from a deck of 52 cards, then return the card to the deck, shuffle the cards, and pick another card. Recall that this type of situation is called **sampling with replacement**. Hence, if we choose 3 cards from a deck with replacement, how many sequences of 3 cards can we have? Well, for the first card, we have 52 possible cards. Since sampling is done with replacement, we also have 52 possible cards for the second and third outcomes. So according to the multiplication rule, there are $52^3 = 140\ 608$ possible sequences of 3 cards.

What about the amount of sequences of cards when we choose 3 cards from a deck without replacement? Now, for the first card, we have 52 possible cards. Since sampling is done without replacement, 51 possible cards are possible for the second and 50 for the third. So according to the multiplication rule, there are $52(51)50 = 132\ 600$ possible sequences of 3 cards.

When we perform sampling without replacement and the sampling is exhaustive (every possible outcome in the sample space will be eventually chosen in the sample), we get a special case of the multiplication rule. If a set has $n$ possible outcomes to choose from and we will choose $n$ times, then by the multiplication rule the sample space will have

$$(n) \times (n - 1) \times \cdots \times (1) = n!$$

possibilities.

**Example 5.15** *How many ways can we watch 3 movies, if we watch each movie only once. This is an example of sampling without replacement that is exhaustive. Therefore, the movies can be watched in* $3! = 3 \times 2 \times 1 = 6$ *ways.*

---

**Factorials**
Head to Section A.4 for more on factorials.

---

There are many instances when the sampling is done without replacement, but it is not exhaustive. Instead, a subset of size $k$ such that $k < n$ is chosen. When this is the case, it is necessary to distinguish between situations when order matters and when it does not. When order matters, then an outcome of $(A, B, C)$ is different than $(C, B, A)$. When order does not matter, then $(A, B, C)$ and $(C, B, A)$ are considered the same outcome. For our applications, we do not tend to care about order, so we focus on sampling without replacement when order does not matter. Since the sampling is not exhaustive, then the number of outcomes must be smaller than $n!$. It can be shown that the number of outcomes when sampling without replacement a subset $k$ from a set of size $n$ such that order does not matter is

$$\frac{n!}{k!(n-k)!}$$

This number of outcomes from this type of sampling, known as **combinations**, is also a special case of the multiplication rule. $k$ combinations tend to be expressed as "$n$ choose $k$."

*Using Counting Rules to Find Probability* Counting rules applied to ways an event can occur and ways a data collection procedure can occur lead to theoretical probabilities.

**Example 5.16** *A professor has 20 students and receives 20 compositions to grade. After grading 10 compositions, she checks the enrollment and sees 3 people have dropped the class. She now has 17 students. What is the probability she needlessly graded 3 compositions?*

Assuming the exams are randomly chosen to be graded, the sequence of gradings is 20 choose 10. That she graded the compositions of all three students that dropped the course is an event that can be broken into two steps: she chooses from the students that dropped the class, and then chooses from the students that did not. Using the multiplication rule, the first step can occur in 3 choose 3 ways, the second can occur in 17 choose 7.

$$P(T) = \frac{\frac{3!}{0!3!}\left(\frac{17!}{7!10!}\right)}{\frac{20!}{10!10!}}$$

$$= 0.1052$$

## Practice Problems

5.48  Find the numerical value of 3!.

5.49  Find the numerical value of 4!.

5.50  Solve the following combinatoric expressions:
   a) 10 choose 2.
   b) 5 choose 0.
   c) 7 choose 7.
   d) 5 choose 2.

5.51  Returning to Example 5.16, if all the numbers stay the same except that instead of 3 students dropping the class, 4 dropped that class:
   a) Intuitively, should the probability that all students who dropped the class be graded higher or lower than in Example 5.16?
   b) What is the probability she needlessly graded 4 compositions?

5.52  One version of the lottery chooses 6 numbers at random between 1 and 46, without replacement. Players choose their numbers and if their 6 numbers are selected in any order, they win the main price. What is the probability of winning the main price?

## Chapter Problems

5.53  Can events that are mutually exclusive and simultaneously independent exist? If so, provide an example.

5.54  Orlando building permits from 2010 to 2015 can be downloaded[16] (File "Orlando_Building_Permits_Issued").
   a) Find the probability that a building permit is of type "ROOF"?
   b) Construct a bar chart for type of building permit. What are the three top types of building permits?

5.55  Use Table 5.5 and $P(D_1) = 0.1263, P(D_2) = 0.1433, P(D_3) = 0.1452, P(D_4) = 0.1467, P(D_5) = 0.1467, P(D_6) = 0.1518$, and $P(D_7) = 0.1400$ to show that the probability there is a missing person report in downtown Boston regardless of day is 0.0030.

---

16  Source: data.cityoforlando.net (accessed June 6, 2018).

5.56 In an effort to increase the number of listeners, radio station WHACK-95.1 FM is doing sweepstakes for $50 000. The caller has to guess a four digit number to win the prize. The administration designed the sweepstakes in such a way that for every digit, any number from 0 to 9 is possible.

a) What is the probability that the first caller wins the prize if he chooses a number randomly?

b) Due to a miscommunication between administration and a technician, the sweepstakes were set up so that each digit can only occur once. What is the probability of the first caller winning the prize now?

c) The administrators did not inform the listeners well about how the digits were selected. What is the probability that the first caller wins the prize if they think that no number can be repeated in the sequence?

d) The administration intentionally misled the listeners to believe that no number can be repeated in the sequence and chose a number based on this restriction. What is the probability that the first caller wins the prize?

5.57 The following exercise is adapted from Chapter 5 of Steven Pinker's wonderful book, *The Better Angels of Our Nature*. Suppose there is a constant chance of your house being struck by lightning at any time throughout the year. Assume the strikes are random and independent and the rate is one strike a month. Your house is hit by lightning today, Monday. What is the most likely day for the next bolt to strike your house?

5.58 If a six-sided die is tossed two times and "3" shows up both times, what is the probability of "3" on the third trial? Explain.

5.59 When drawing two cards from a standard deck of 52 cards, find the probability of not getting an ace on the first card and getting an ace on the second when sampling is done

a) with replacement.

b) without replacement.

5.60 Given that an experiment has only outcomes associated to three events, $A, B, C$, while $P(A) = 0.5, P(B) = 0.4, P(C) = 0.1$, and $P(A \text{ and } B \text{ and } C) = 0$. Can any of these events happen simultaneously?

5.61 If $P(A) = 0.22, P(B) = 0.47$, and $P(A \text{ and } B) = 0.22$. Find

a) $P(B|A)$

b) $P(A|B)$

c) If you want $B$ to happen, is it preferable for $A$ to happen, not happen, or it does not matter?

d) If you want $A$ to happen, is it preferable for $B$ to happen, not happen, or it does not matter?

5.62 As you will see later on, one way to perform statistical inference is based on conditional probabilities. That is, a person wishing to infer on whether $\mu > 22$ may find the probability of some event $E$ given $\mu = 22$. If $P(E|\mu = 22) \leq 0.05$, then they decide $\mu > 22$. Suppose it was found that $P(E|\mu = 22) = 0.33$. Should they decide that $\mu > 22$?

5.63 A drug test has a false positive result – comes out positive given the person did not use illegal drugs – 5% of the time. Also, the drug test has a false negative result – comes out positive given the person did not use illegal drugs – 11% of the time. 9.4% of the US population have used illegal drugs in the past month.
   a) What is the probability that a person has used illegal drugs given that the test comes out positive?
   b) What is the probability that a person has not used illegal drugs given that the test comes out negative?

5.64 Let $A$ and $B$ be any two events. Assuming $P(A) > 0$ and $P(B) > 0$:
   a) prove that if $A$ is an event that is independent of $B$, then $A$ is independent of $B^c$. Must be proven GENERALLY (IN AN ABSTRACT WAY), NOT BY EXAMPLE!! (*Hint:* Use a Venn diagram to find a way to express $P(A \text{ and } B^c)$ in terms of $P(A)$ and $P(A \text{ and } B)$. Then proceed from this expression.)
   b) what do you think the general result in (a) means?

5.65 We have two independent events such that $P(A) = 0.35$ and $P(B) = 0.25$.
   a) What is the probability of the event: "$A$ occurs and $B$ does not occur" (*Hint:* Refer to Problem 5.64).
   b) What is the probability of the event: "$A, B$ do not occur together?"

5.66 If $P(A|B) > P(A)$, show that $P(B|A) > P(B)$. *Hint:* Start from the equation for the conditional distribution $P(A|B)$ and reexpress it.

5.67 At the beginning of 2017, managers and directors at Las Vegas were surveyed on their satisfaction with IT services. Their responses can be found in[17] the file "IT_Satisfaction_Survey." Use the data to answer the following:
   a) What is the probability that a manager or director will rate IT services as "Acceptable"?

---

17 opendata.lasvegasnevada.gov (accessed July 26, 2019).

b) What is the probability that a manager or director will rate IT services as "Very Unsatisfied"?

c) What is the probability that a manager or director will rate IT services as "Acceptable" or better?

5.68 A business is assessing the performance of their advertising campaign for healthy snacks. They conducted a survey of people that had not used the product before the ad campaign and the results are summarized below:

|  | Tried product | |
| --- | --- | --- |
| Saw any of the ads | Yes | No |
| Yes | 7.5% | 33% |
| No | 1% | 58.5% |

a) What is the estimated probability that a randomly chosen consumer would have seen the ad?

b) What is the estimated probability that a randomly chosen consumer would have seen the ad and tried the product?

c) Given that the randomly chosen consumer saw the ad, estimate the probability the consumer tried the product?

d) Do you think the event a consumer tries the product is independent of whether they have seen the ad? Justify your answer.

5.69 The table below summarizes alcohol-related deaths in England from 2001 to 2014, considering death classification and gender.[18]

|  | Gender | | |
| --- | --- | --- | --- |
| Category | Male | Female | Total |
| Mental/behavioral disorders | 339 | 150 | 489 |
| Alcoholic cardiomyopathy | 68 | 16 | 84 |
| Alcoholic liver disease | 2 845 | 1 488 | 4 333 |
| Fibrosis and cirrhosis of liver | 911 | 609 | 1 520 |
| Accidental poisoning | 242 | 127 | 369 |
| Other | 28 | 8 | 36 |
| Total | 4 433 | 2 398 | 6 831 |

---

18 Source of data: data.gov.uk (accessed July 26, 2019).

a) Find the probability that a death was classified as alcoholic liver disease.
b) Find the probability that a death was classified as alcoholic liver disease, given the person was male.
c) Find the probability that a death was classified as alcoholic liver disease, given the person was female.
d) Find the probability that a death was classified as Fibrosis and cirrhosis of liver and the person was female.
e) Find the probability that a death was classified as alcoholic liver disease or the person was male.

## Further Reading

Black, Ken. (2012). *Business Statistics*, 7th edition. Wiley.

Doane, D. P. and Seward L. E. (2011). *Applied Statistics in Business and Economics*, 3rd edition. McGraw-Hill.

Pinker, S. (2011). *The Better Angels of Our Nature: Why Violence Has Declined*. Penguin.

# 6

# Discrete Random Variables

## 6.1 Introduction

Up until now, to obtain probabilities for an event, we had to consider all the possibilities in the sample space and the ways that the event can occur. But, in many real-life situations, it is very difficult to do this in a practical way. In addition, usually our interest is in the characteristics of the data collection procedure that can be obtained based on some sort of counting or some measure. A **random variable** is a function of the outcomes of a data collection procedure. It turns outcomes of the data collection procedures into a number. If this number is representative of the count of something, then it is a **discrete random variable**. If this number is representative of the measure of something (e.g. the length, weight, duration, etc.), then it is a **continuous random variable**.

In this chapter, we will work with probabilities, expected value, and variance of discrete random variables.

---

**Random Variables and Their Nomenclature**

In algebra courses, variables are used to generalize situations. In statistics, random variables are slightly different to the variables used in algebra. Since random variables are a function of the data collection procedure, which has random outcomes, random variables involve uncertainty as well, and hence we must characterize this uncertainty. On the other hand, variables in algebra do not involve any randomness. Capitalized letters such as $X$, $Y$, or $Z$ are used to define a random variable, while specific values of the random variable are shown in lowercase $x$, $y$, or $z$. For example, $X = x$ means "When random variable $X$ is equal to the value $x$." The uncertainty in the value of a random variable is quantified through probability.

---

*Principles of Managerial Statistics and Data Science*, First Edition. Roberto Rivera.
© 2020 John Wiley & Sons, Inc. Published 2020 by John Wiley & Sons, Inc.
Companion website: www.wiley.com/go/principlesmanagerialstatisticsdatascience

## 6.2   General Properties

**Example 6.1**   *A coin is flipped. Arbitrarily, we define a variable X to have a value of 1 if heads come up and a value of 0 if tails come up.*

$X = 1$ and $X = 0$ are the only possible values of $X$. Because the values of $X$ are dependent on a random outcome of flipping a coin, $X$ is a random variable and before the experiment is done, neither its value of 1 nor 0 can be guaranteed. Can you guess what are $P(X = 0)$ and $P(X = 1)$ respectively?

**Example 6.2**   *Define a variable Y as the total number of times we get heads after flipping a coin twice.*

$Y$ is also a random variable. The data collection procedure implicitly gives the possible values of $Y$; since it's counting heads while flipping a coin twice, $Y$ can be 0, 1, or 2. These are the only possible values for $Y$. What about the probabilities of each possible value of $Y$? That requires more work than Example 6.1. But it is still doable. $Y = 0$ means that we got two tails when the coin was flipped twice. By independence of the events,

$$P(Y = 0) = \frac{1}{2}\left(\frac{1}{2}\right) = \frac{1}{4}$$

$P(Y = 2) = 1/4$ is also straightforward to figure out. $P(Y = 1)$ is a bit trickier, requiring us to consider getting heads in the first toss, tails on the second, or vice versa. A different way to calculate $P(Y = 1)$ will be applied later.

**Example 6.3**   *A marketing company claims that 1 in 6 people who click on an online ad will make a purchase transaction. If you randomly select from the history of 10 ad clicks views, what is the probability that 2 out of the 10 ad clicks resulted in a purchase transaction.*

Note that writing down all the possibilities of obtaining the 2 clicks that lead to a purchase (the first two lead to a purchase, the rest do not; the first eight do not lead to a purchase but the last two do; etc.) is very tedious.

**Example 6.4**   *If we roll a dice 10 times, what is the probability of getting 1 exactly 2 times?*

In this example, it would also be very tedious to figure out all the different ways that we can obtain this event to later find the probability.

Although discrete random variables may seem like an unnecessary abstract concept, in reality, it simplifies many problems in two ways:

● They make it unnecessary to write down all the different ways an event can happen so they can be counted.

- It generalizes problems in a way that different real-life situations can be solved with a specific formula to find probabilities, the expected value (mean) of the random variable, and its standard deviation.

Returning to the Examples 6.3 and 6.4, it turns out that both problems can be solved with a single equation to find the probability, and the same probability is obtained in both cases.

---

**Probability Distribution Function**

When a probability is assigned to each possible value of a discrete random variable, this is known as a **probability distribution** (or probability mass function). The probability properties for events established in Chapter 5 still apply. Considering each of the mutually exclusive possible values $x_i$ of the random variable,

- $0 \leq P(X = x_i) \leq 1$ for all $i$
- $\sum_{i=1}^{n} P(X = x_i) = 1$

The probability distribution for a discrete random variable tends to be represented in two ways: with a table that includes all possible values of the discrete random variable and their respective probabilities, or a mathematical equation to find these probabilities.

---

Returning to Example 6.1, since all probabilities add to 1,

$$P(Y = 1) = 1 - P(X = 0) - P(X = 2)$$
$$= 1 - 1/4 - 1/4$$
$$= 1/2$$

**Example 6.5** *A company decides to try to develop a rechargeable battery that lasts twice as long as the best battery in the market. It will try until it succeeds or until there are 3 failed prototypes. Assume that in each attempt the probability of success is 0.6 and that each attempt is independent. We need to construct the probability distribution of the number of attempts.*

Let

$X$ = number of attempts to attain the desired battery.
$A_i$ represents the event that the company succeeds in the $i$th attempt.
$F_i$ represents the event that the company fails in the $i$th attempt.

Table 6.1 shows how the possible outcomes relate to the values of $X$ and the probabilities of each value of $X$. According to the situation, the possible values of $X$ are 1, 2, 3:

- $X = 1$ when the company succeeds in their first try.

Table 6.1 Probability distribution function and possible outcomes.

| Possible outcomes | $x$ | $P(X = x)$ |
| --- | --- | --- |
| $A_1$ | 1 | 0.60 |
| $F_1, A_2$ | 2 | 0.24 |
| $F_1, F_2, A_3$ or $F_1, F_2, F_3$ | 3 | 0.16 |

- $X = 2$ when it fails in its first attempt and succeeds in its second attempt.
- $X = 3$ if it fails in all three attempts or fails in the first two tries and succeeds on the third.

For the first possible outcome, it is easily seen that $P(X = 1) = P(A_1) = 0.6$. For the next possible outcome, independence of the events is applied,

$$P(X = 2) = P(F_1 \text{ and } A_2) = P(F_1)P(A_2) = 0.4(0.6) = 0.24$$

The easiest way to find the remaining probability is by remembering that since the probabilities of all values of $X$ add to 1, $P(X = 3) = 1 - P(X = 1) - P(X = 2)$. One could also find this probability according to the possible outcomes that it entails (See Problem ??). Figure 6.1 presents the distribution function for this discrete random variable. As also seen from the table, the smallest two values of $X$ have the highest chances of occurring.

If $P(X \geq 2)$ is of special interest, then the probability distribution shows that

$$P(X \geq 2) = P(X = 2) + P(X = 3) = 0.24 + 0.16 = 0.48$$

For any random variable, $P(X \leq x)$ for all $x$ is called the **cumulative distribution function**. Table 6.2 shows the cumulative distribution function for Example 6.5.

*Expected Value* For a discrete random variable, the expected value is a mean, which assigns weight to the possible values of a random variable depending on the probability of occurrence. We represent the expected value of $X$ as $E(X)$. In general, for any discrete random variable with $N$ finite possible values,

$$E(X) = \sum_{i=1}^{N} x_i P(X = x_i) \tag{6.1}$$

$E(X)$ is a measure of centrality. If we had to guess what value we will get from a random variable, $E(X)$ would be a reasonable guess. Sometimes, however, it is not necessarily a value that can be obtained from the random variable. Also, as we can see from the equation, if all $N$ possible values of $X$ are equally likely, then each value is assigned the same weight ($P(X = x_i) = 1/N$) for the expected value.

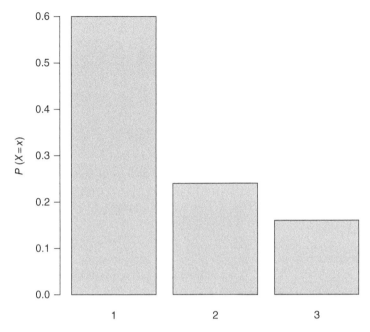

Figure 6.1 Probability distribution function for X.

Table 6.2 A cumulative distribution function.

| x | $P(X \leq x)$ |
| --- | --- |
| 1 | 0.60 |
| 2 | 0.84 |
| 3 | 1 |

**Example 6.6** *Returning to Example 6.5, what is the expected number of attempts for the company?*

Based on its probability distribution function (Table 6.1), some expected value computations are illustrated in Table 6.3.

That is

$$E(X) = 1 \times 0.60 + 2 \times 0.24 + 3 \times 0.16 = 0.60 + 0.48 + 0.48 = 1.56$$

Therefore, over the long run, companies attempting to develop a rechargeable battery under the same conditions set for Example 6.5 will perform on average, 1.56 attempts. Note that the expected value does not have to be one

Table 6.3 Expected value computations for rechargeable battery example.

| x | P(X = x) | xP(X = x) |
|---|----------|-----------|
| 1 | 0.60 | 0.60 |
| 2 | 0.24 | 0.48 |
| 3 | 0.16 | 0.48 |

of the possible values of $X$. Also, when $X$ has infinite values, $x_1, x_2, \ldots$, we can still define the expected value. Only that it will now be over infinite products of possible values and their probabilities. In these cases, mathematical trickery is needed to find an easier expression for $E(X)$. More on this later.

**Variance and Standard Deviation**   Variance, $Var(X)$, indicates the dispersion of the random variable or how the possible values are scattered around the expected value:

$$\sigma^2 = Var(X) = \sum_{i=1}^{N} (x_i - E(X))^2 P(X = x_i)$$

It is a kind of mean square distance from the expected value of each possible value of $X$. Sometimes the standard deviation, $\sigma = \sqrt{\sigma^2}$, is used as a measure of dispersion rather than the variance.

**Example 6.7**   *For Example 6.5, the variance worksheet is shown in Table 6.4. That is*

$$\sigma^2 = (1 - 1.56)^2 \times 0.60 + (2 - 1.56)^2 \times 0.24 + (3 - 1.56)^2 \times 0.16 = 0.57$$

So $\sigma = \sqrt{0.57} = 0.75$.

Table 6.4 Variance computations for rechargeable battery example.

| x | P(X = x) | xP(X = x) | x − μ | (x − μ)² | (x − μ)²P(X = x) |
|---|----------|-----------|-------|----------|------------------|
| 1 | 0.60 | 0.60 | −0.56 | 0.3136 | 0.188 160 |
| 2 | 0.24 | 0.48 | 0.44 | 0.1936 | 0.046 464 |
| 3 | 0.16 | 0.48 | 1.44 | 2.0736 | 0.331 776 |
|   |      | μ = 1.56 |       |          | σ² = 0.5664 |

### 6.2.1   A Quick Look at Data Science: Number of Stroke Emergency Calls in Manhattan

A New York City official wants to determine the distribution of stroke emergency calls in Manhattan on Sundays. This type of information can help ensure that enough emergency personnel is available at hospitals to attend this type of condition. Data from the computer-aided Emergency Management System for New York City is available online.[1] The raw data includes almost 5 million records and 32 columns for incidents from January 1, 2013 until June 30, 2016. After preprocessing the data (Figure 6.2), the probability distribution is determined (Table 6.5, Figure 6.3). There were 182 Sundays during the period of study, in which 1488 calls related to stroke occurred. In this case study, it is found that in 72% of the Sundays there were 4–11 emergency calls due to stroke, and in 16% of the Sundays there are 12–15 emergency calls due to stroke.

$E(X) = 2 \times 0.01 + 3 \times 0.06 + \cdots + 15 \times 0.01 = 8.18$. Thus, on average, a Sunday in Manhattan has 8.18 emergency calls due to stroke. Truth is, $E(X)$ is a theoretical number, a parameter. In this case study, sample data has been used, and it can be shown that 8.18 is indeed $\bar{x}$ (see Problem 6.36).

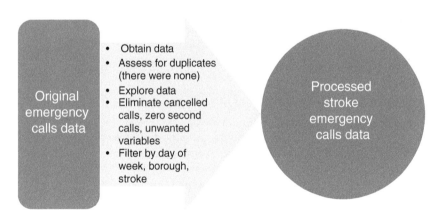

Figure 6.2   Steps required to process New York City emergency call data.

Table 6.5   Probability distribution of the number of emergency calls due to stroke on Sundays in Manhattan.

| $x$ | 2 | 3 | 4 | 5 | 6 | 7 | 8 | 9 | 10 | 11 | 12 | 13 | 14 | 15 |
|---|---|---|---|---|---|---|---|---|---|---|---|---|---|---|
| $P(X = x)$ | 0.01 | 0.06 | 0.05 | 0.08 | 0.09 | 0.16 | 0.12 | 0.08 | 0.09 | 0.10 | 0.08 | 0.04 | 0.03 | 0.01 |

---

1  data.cityofnewyork.us (accessed July 26, 2019).

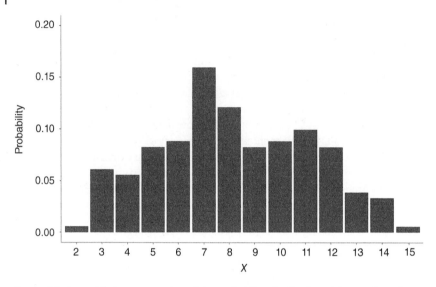

Figure 6.3 A graphical representation of the probability distribution of the number of emergency calls due to stroke on Sundays in Manhattan.

## Practice Problems

6.1 Returning to Example 6.2, write down the events that define $Y = 2$.

6.2 Returning to Example 6.2, show that $P(Y = 2) = 1/4$.

6.3 Which of the following cases does not define a proper probability distribution. Explain.
a) Case A

| $x$ | $P(X = x)$ |
| --- | --- |
| 0 | 0.60 |
| 1 | 0.14 |

b) Case B

| $x$ | $P(X = x)$ |
| --- | --- |
| 25 000 | 0.42 |
| 75 000 | 0.29 |
| 95 000 | 0.29 |

c) Case C

| $x$ | $P(X = x)$ |
|-----|------------|
| 1   | 0.10       |
| 5   | 0.25       |
| 7   | 0.50       |
| 7   | 0.10       |

6.4 Find $P(X = 3)$ in Example 6.5 according to the possible outcomes that it entails.

6.5 A woman decides to have children until she has her first girl or until she has four children, whichever comes first (do not try this at home people!). Let $X$ = the number of children she has. Find the probability distribution for $X$, given that the probability of having a girl is 0.488 for each birth and that the births are independent (Ignore the possibility of twins, triplets, etc. Just assume she will give birth to one child at a time).

6.6 There are three left-handed workers and seven right-handed workers assigned to an assembly line. Two workers are randomly selected without replacement to test a new machine. Let $X$ = the number of workers out of 2 selected who are left-handed. Write the probability distribution for $X$. *Hint:* See Example 5.11 in Chapter 5.

6.7 Returning to the case study in Section 6.2.1, what is the probability that there are no more than three emergency calls due to strokes in Manhattan?

6.8 A wide variety of data sets such as tax returns, population sizes, and power consumption tend to comply with what is known as Benford's law, which establishes a frequency distribution for the leading digits of data. For example, for the first leading digit, the law states the distribution should be

| First digit | 1 | 2 | 3 | 4 | 5 | 6 | 7 | 8 | 9 |
|-------------|------|------|------|------|------|------|------|------|------|
| Probability | 0.30 | 0.18 | 0.12 | 0.10 | 0.08 | 0.07 | 0.06 | 0.05 | 0.04 |

a) According to Benford's law, are all numbers from 1 to 9 equally likely to occur as first digits?
b) What is the probability that the leading first digit of a randomly chosen number is a 1 or a 2 according to Benford's law?

6.9 A die is rolled. If it rolls to a 1, 2, or 3, you win $3. If it rolls to a 4, 5, 6 you lose $2. Find the expected winnings.

6.10 A company is evaluating the potential of a future project. $Y$ is their profit, which depends on several uncertain factors. They have determined that their profit could be $1, $1.75, and $2.35 million with respective probabilities of 0.5, 0.28, and 0.22.
a) Find $E(Y)$.
b) Find $Var(Y)$.
c) Find the standard deviation of $Y$.
d) What is the probability that the profit is $1.75 million or more?

6.11 A company that sells air tickets online also provides a policy that pays $500 000 to a chosen beneficiary if the policy holder dies from a plane crash. The policy costs $5. There is 1 in 5 million chance of dying in a plane crash. If $Y$ = money received by beneficiary,
a) specify the probability distribution of $Y$.
b) find $E(Y)$. Interpret.

## 6.3 Properties of Expected Value and Variance

What if we need to work with a function of a random variable or of many random variables? The sample mean, sample standard deviation, and sample proportion are just some examples of functions of random variables. In this section, some properties of expected value and variance are defined that apply to linear functions of random variables. The properties are useful to find the expected value or variance of a linear function of random variables when the expected values or variances of these random variables are known.

If $a$ is a constant, and $X$ is a random variable, then
**Property 1:** $E(a) = a$
**Property 2:** $E(aX) = aE(X)$
Thus, the expected value of the constant 5 is 5. By being a constant, it has no variation. Similarly, if $E(X) = 2$, then

$$E(5X) = 5E(X) = 5(2) = 10$$

Furthermore, if $a, b$ are constants, then
**Property 3:** $E(aX \pm bY) = aE(X) \pm bE(Y)$

The sign $\pm$ on both sides of the equation indicates that if the sign on the left side is $+$, then it is $+$ on the right side of the equation and if the sign on the left side is $-$, then it is $-$ on the right side of the equation.

**Example 6.8** *One application of the properties of expected values presented above is to determine the expected value of investment portfolios. Suppose that there is an investment portfolio with two assets, fund 1 corresponds to 60% of the portfolio's value, and fund 2 accounts for the other 40% of the value of the portfolio. X is the yearly return of fund 1, and Y is the yearly return of fund 2. Therefore, the yearly return of the portfolio is $R = 0.6X + 0.4Y$. Also, suppose the yearly expected return of fund 1 is $E(X) = 0.10$ and for fund 2 $E(Y) = 0.03$. The goal is to determine the expected return, $E(R)$, of the investment portfolio.*

By Property 3,

$$E(R) = E(0.6X + 0.4Y)$$
$$= 0.6E(X) + 0.4E(Y)$$
$$= 0.6(0.10) + 0.4(0.03)$$
$$= 0.072$$

Property 3 extends to the sum of $n$ variables. It can be interpreted as meaning that the expected value is additive. That is, the expected value of $n$ random variables is the sum of the individual expected values of each individual random variable.

There are also properties that can be used when finding the variance of a function of random variables. If $a$ is a constant, and $X$ is a random variable, then

**Property 4:** $Var(a) = 0$

**Property 5:** $Var(aX) = a^2 Var(X)$

Thus, the variance of the constant 5 is 0. By being a constant, it has no variation. Similarly, if $Var(X) = 4$, then

$$Var(5X) = 5^2 Var(X) = 25(4) = 100$$

We have to be a little more careful to apply variance properties when combining random variables. The concept of independence carries over from independent events. We say that two random variables are **independent** when any value of one random variable, an event, does not affect the probability of the other random variable to have any specific value, another event. If $a$ and $b$ are constants and $X$ and $Y$ are independent random variables, then

**Property 6:** $Var(aX \pm bY) = a^2 Var(X) + b^2 Var(Y)$

Note that although we may be subtracting functions of $X$ and $Y$ on the left side of the equation, the sign on the right side is always $+$ (variances of functions of the variances are added).

**Example 6.9** *Returning to Example 6.8, if having only fund 1 has an expected return of 0.10, then why would someone combine those two assets to get a lower expected value? Suppose that for the same portfolio, $Var(X) = 0.04$ and $Var(Y) = 0$, and that the returns of fund 1 and fund 2 are independent.*

Then by Property 6,

$$Var(R) = Var(0.6X + 0.4Y)$$
$$= 0.6^2 Var(X) + 0.4^2 Var(Y)$$
$$= 0.36(0.04) + 0.4(0)$$
$$= 0.0144$$

The variability of an investment is considered a measure of its risk. By combining fund 1 (higher expected value, higher variance) with fund 2 (lower expected value, lower variance), the portfolio sacrifices some of the expected return for lower risk.

The last variance property can also be extended over a sum of more than two random variables and is frequently used in management and finance.

*Optional: Covariance* The property $Var(aX \pm bY) = a^2 Var(X) + b^2 Var(Y)$ requires that the random variables $X$ and $Y$ are independent. When the random variables $X$ and $Y$ are not independent, the property no longer applies. One way to measure the dependence between $X$ and $Y$ is through what is known as **covariance**, which measures how two variables covary or change together.

$$Cov(X, Y) = E(X - \mu_x)(Y - \mu_y)$$

where $\mu_x = E(X)$ and $\mu_y = E(Y)$. Covariance was seen in Section 4.5 as a measure of linear association between numerical random variables.

---

**Covariance Features**

- Unlike the variance of a random variable, covariance can be negative.
- If smaller values of $X$ generally occur with smaller values of $Y$ or if larger values of $X$ generally occur with larger values of $Y$, then the covariance is positive.
- When smaller values of one random variable generally occur with larger values of the other random variable, then the covariance is negative.
- Covariance is a measure of linear dependence between two random variables. Thus, if two random variables are independent, then $Cov(X, Y) = 0$, but not necessarily vice versa.

---

For two random variables $X$ and $Y$,

$$Var(aX \pm bY) = a^2 Var(X) + b^2 Var(Y) \pm 2abCov(X, Y)$$

When $X$ and $Y$ are independent, then[2] $Cov(X, Y) = 0$.

**Example 6.10** *Considering the context initiated in Example 6.8 and continued in Example 6.9, suppose that instead of independent funds, the funds are dependent with $Cov(X, Y) = -0.02$.*

Now,

$$Var(R) = Var(0.6X + 0.4Y)$$
$$= 0.6^2 Var(X) + 0.4^2 Var(Y) + 2(0.6)(0.4)(-0.02)$$
$$= 0.36(0.04) + 0.4(0) - 0.0096$$
$$= 0.0048$$

Hence, if the two funds would be inversely related, then the variance of the portfolio would be lower than a portfolio of two independent funds.

## Practice Problems

6.12 If $E(X) = 40, E(Y) = 25, Var(X) = 3, Var(Y) = 2$, and $X$ is independent of $Y$, find
  a) $E(5)$.
  b) $E(2X)$.
  c) $E(3X + 5Y)$.
  d) $E(7X - 3Y)$.
  e) $Var(12)$.
  f) $Var(7Y)$.
  g) $Var(3X + 5Y)$.
  h) $Var(3X - 5Y)$.

6.13 If $E(X) = 100, E(Y) = 225, Var(X) = 10, Var(Y) = 25$, and $Cov(X, Y) = 10$, find
  a) $E(70X - 35Y)$.
  b) $Var(70X + 35Y)$.
  c) $Var(70X - 35Y)$.
  d) If $Cov(X, Y) = -5$, find $Var(70X - 35Y)$.

6.14 If $X_1, X_2$, and $X_3$ each are independent and have an expected value of 10, find
  a) $E\left(\frac{X_1+X_2+X_3}{3}\right)$.
  b) $Var\left(\frac{X_1+X_2+X_3}{3}\right)$.

---

2 However, if $Cov(X, Y) \neq 0$, this does not mean $X$ and $Y$ are independent and they may have nonlinear dependence.

6.15 The expected revenue for a new product is $20 million. It is known that the total costs will be $15 million. What would the expected profit be?

6.16 A gas station sells two types of gasoline: regular and premium. Let $X$ be the amount (in gallons) of regular sold in a day and $Y =$ amount (also in gallons) of premium sold in a day. Regular is priced at $4/gallon and premium at $4.2/gallon. If $E(X) = 2000, E(Y) = 500, Var(X) = 2500, Var(Y) = 10000$, and $Cov(X, Y) = 4500$. Therefore, Revenue is $R = 4X + 4.2Y$. Determine
a) $E(R)$.
b) $Var(R)$.
c) what would happen if we wrongly assume $X$ and $Y$ are independent.

6.17 Two weeks ago, you were hired as a manager for one of the top companies headquartered in Houston, Texas. Unfortunately, a storm close by in the Gulf of Mexico has quickly gained strength and has become a powerful category 5 hurricane with sustained winds of 180 mph. According to the National Hurricane Center, it is estimated that there is a 20% chance that the plant you work in gets hit by the most powerful winds in 36 hours. The team has estimated that the cost of hurricane precautionary preparations (storm shutters, trimming tree branches, etc.) is $10 000. If the greatest winds do not hit, no major losses are estimated. On the other hand, if the greatest winds hit without any precautionary preparations, the losses are estimated to be $2.5 million. In summary, there are two alternatives, either precautionary preparations are done or not. If $Y =$ cost of hurricane with precautionary preparations and $N =$ cost of hurricane without precautionary preparations, choose one of these alternatives based on their expected values. (*Hint: Y* involves a fixed cost. *N* is a function of whether the strong winds hit or not.)

## 6.4  Bernoulli and Binomial Random Variables

We have already implied that often, the discrete random variable of interest can be seen in a generalized way so that different situations in practice can be solved with specific formulas to find probabilities, expected value (mean), and standard deviation. In this section, we start to explain how we can do this.

The simplest discrete random variable to explain is the **Bernoulli random variable**, characterized by only having two possible values: 0 and 1. It is used to determine if there was a success in a data collection procedure or not. Arbitrarily, let $X = 1$ for a "success" and $X = 0$ a "failure." If $P(X = 1) = \pi$, then,

$$P(X = 0) = 1 - \pi$$

since the probabilities should add to 1.

**Example 6.11**  *Suppose you flip a coin once and define $X = 1$, if you get heads, then $\pi = 0.5$ and $X$ follows a Bernoulli distribution.*

Since $0 < \pi < 1$, there are infinite examples of random variables that follow a Bernoulli distribution.

When $X$ has a Bernoulli distribution, it is easy to see that

$$E(X) = \sum_{i=1}^{2} x_i P(X = x_i) = 0 \times (1 - \pi) + 1 \times \pi = \pi$$

Therefore, a Bernoulli random variable will always have an expected value of $\pi$. $\pi$ is a parameter of the Bernoulli distribution.

Similarly, the variance of a Bernoulli random variable will be

$$\sigma^2 = Var(X) = \sum_{i=1}^{n} (x_i - E(X))^2 P(X = x_i)$$
$$= (0 - \pi)^2 (1 - \pi) + (1 - \pi)^2 \pi$$
$$= \pi(1 - \pi)$$

and $\sigma = \sqrt{\pi(1 - \pi)}$.

*The Binomial Random Variable*  Now suppose there are $n$ independent repetitions of Bernoulli random variables $X_1, X_2, \ldots, X_n$ and define $Y$ as the sum of these random variables.

$$Y = X_1 + \cdots + X_n$$

Then, $Y$ has a probability distribution known as the **binomial distribution**.

---

**Distribution Nomenclature**

We will use the notation $Y \sim$ binomial $(n, \pi)$ to mean "$Y$ follows a binomial distribution with parameters $n, \pi$." Similar notation will be used for other probability distributions.

---

Since $Y$ accumulates the results of all the Bernoulli iterations, its possible values are $0, 1, \ldots, n$. Moreover, for each trial you always have two possibilities, a success or a failure, which are independent. Thus, $\pi$ remains fixed for each trial.

---

**Properties of a Binomial Random Variable**
In summary, there are five essential properties for a discrete random variable to follow a binomial distribution:

- The random variable is discrete with possible values of $0, 1, \ldots, n$.
- The sample size ($n$) is fixed.
- The probability of "success" ($\pi$) is fixed for each observation.
- Independent observations.
- Per observation, we have two possible results: success or failure.

---

All these properties must be present to be able to state that a random variable follows a binomial distribution. $n$, $\pi$ are parameters of the binomial distribution. The assumption that $\pi$ is fixed and of independent observations are closely related. The independence of observations will often be implied by the fact that a random sample was taken. An implicit assumption is that the sample size is only a small fraction of the population size (no more than 10%). Otherwise, owing to the sampling being done without replacement, observations will not be independent: $\pi$ will no longer be constant.

*Expected Value of a Binomial Random Variable*    Like other discrete random variables, the expected value of a binomial random variable can be found with Eq. (6.1). But the equation requires finding the probabilities for each possible value of $Y$ which can be cumbersome, even for moderate sample size. Fortunately, when $Y \sim \text{binomial}(n, \pi)$, we have a shortcut to find $E(Y)$. There are several ways to reach the result of this shortcut, but the easiest is to use the fact that $Y$ is a sum of Bernoulli random variables and then apply the expected value property described in Section 6.3;

$$E(Y) = E(X_1 + \cdots + X_n)$$
$$= E(X_1) + \cdots + E(X_n)$$
$$= \pi + \cdots + \pi$$
$$= n\pi$$

a much quicker computation than using equation Eq. (6.1).

*Variance of a Binomial Random Variable*    To find the variance of a binomial variable, we can use the variance property described in Section 6.3

$$Var(Y) = Var(X_1 + \cdots + X_n)$$
$$= Var(X_1) + \cdots + Var(X_n)$$
$$= \pi(1 - \pi) + \cdots + \pi(1 - \pi)$$
$$= n\pi(1 - \pi)$$

and the standard deviation of $Y$ is $\sigma = \sqrt{n\pi(1 - \pi)}$.

***Probability for a Binomial Random Variable*** It is beneficial to conceptualize a simple experiment to understand the calculation of probabilities for a binomial distribution.

**Example 6.12** *Suppose we throw three dice once and we want to determine the probability of obtaining the number 1 twice.*

To find this probability through events, we can define
$O$ = obtain a 1.
$N$ = 1 is not obtained (any other number is obtained).

We work with $N$ because for the probability of interest, it is not necessary to distinguish between the possible outcomes that are not 1 when throwing the dice. Using these definitions of events, we find that to get 1 twice we must obtain $(O, O, N)$, or $(O, N, O)$, or $(N, O, O)$. Since the probability of obtaining 1 per throw is 1/6, the probability of obtaining any other number is 5/6, and the outcome of each die is independent of other dice, we get

$$P((O, O, N) \text{ or } (O, N, O) \text{ or } (N, O, O)) = P(O, O, N) + P(O, N, O) + P(N, O, O)$$
$$= P(O)P(O)P(N) + P(O)P(N)P(O)$$
$$+ P(N)P(O)P(O)$$
$$= 3\left(\frac{1}{6}\right)^2 \left(\frac{5}{6}\right)$$
$$= 0.069$$

Emphasizing the third equality of the example, we see that to find the probability of getting the number 1 twice, it is not necessary to distinguish between getting 1 on the first, the second, or third roll, and that they are treated with the same importance to calculate the probability of interest. Still, the procedure above is rather laborious. Furthermore, if there were 10 dice instead of 3, determining the events when 1 occurs exactly twice would take considerable more work. Fortunately, this is not the only way to solve this problem. If we say that $Y$ = number of times we get 1 out of 3 rolls of the dice, we can show that $Y$ follows a binomial distribution,

- The number of times we get the value 1 out of 3 rolls is discrete with possible values $0, 1, \ldots, n$.
- $n = 3$, $n$ is fixed.
- $\pi = 1/6$ for each roll of the dice.
- Each roll of the dice is independent of others.
- In every roll, we have two possible results: get a 1 (success) or some other number (failure).

When $Y \sim \text{binomial}(n, \pi)$, it can be shown that

$$P(Y = y) = \frac{n!}{y!(n-y)!} \pi^y (1 - \pi)^{n-y}$$

where $y = 0, 1, \ldots, n$. Compare this equation to the calculations used in Example 6.12. The term $\frac{n!}{y!(n-y)!}$ is responsible for counting all the different ways to get 1 twice without having to think about all the events of the experiment resulting in obtaining 1 twice. As we saw in the example, at each event we had $(1/6)^2$ and 5/6 which the remaining terms in the equation obtain. This equation of binomial probability may appear intimidating at first glance, but it hides an important advantage in its use. When $Y$ follows a binomial distribution, we do not need to determine the events that lead to the outcome of interest as we did in Example 6.12, which may become virtually impossible. Instead, we simply substitute the values of the parameters of the binomial and the value of $Y$ of interest to find the probability. Literally, the equation allows us to easily find probabilities for countless random variables that follow a binomial distribution. Table 6.6 summarizes the main characteristics of the random variable that follows the binomial distribution.

For any random variable $Y$ with a binomial distribution, it is useful to notice that there is only one way for $Y = 0$, no successes. Therefore by independence, $P(Y = 0) = (1 - \pi)^n$. The combinatorial term becomes 1 in this case. It is similar for $Y = n$ (no failures), only that $P(Y = n) = \pi^n$ (the combinatorial term also becomes 1 in this case). Hence, for any binomial random variable, the combinatorial term is not 1 when $0 < Y < n$.

**Example 6.13**   *Returning to Example 6.3, what is the probability that 2 out of 10 ad clicks result in a purchase transaction?*

Let $Y =$ number of ad clicks that result in a purchase transaction; $\pi =$ the probability an online ad click leads to a purchase.

- $Y$ is discrete with possible values $0, 1, \ldots, n$.
- $n = 10$, $n$ is fixed.
- $\pi = 1/6$, a constant.
- Each ad click is independent of others since they were randomly selected.
- In every trial, we have two possible results: purchase transaction (success) or no purchase transaction (failure).

Therefore, $Y$ follows a binomial distribution with sample size 10, $\pi = 1/6$ and from Table 6.6, the formula to find a probability for a binomial distribution can be applied:

Table 6.6 Main characteristics of a random variable $X$ that follows a binomial distribution.

| Property | |
| --- | --- |
| Probability mass function | $P(X = x) = \frac{n!}{x!(n-x)!} \pi^x (1 - \pi)^{n-x}$ for $x = 0, 1, \ldots, n$ |
| $E(X)$ | $n\pi$ |
| $Var(X)$ | $n\pi(1 - \pi)$ |

$$P(Y = 2) = \frac{10!}{2!(10-2)!}(1/6)^2(1-(1/6))^{10-2} = 45 \times (1/6)^2(5/6)^8 = 0.29$$

It we want $P(Y \leq 2)$, then the equation for binomial probabilities is applied with several values $y$,

$$P(Y \leq 2) = P(Y = 0) + P(Y = 1) + P(Y = 2)$$

Returning to the Example 6.4,

$$P(Y \leq 2) = \frac{10!}{0!(10-0)!}(1/6)^0(1-(1/6))^{10-0}$$
$$+ \frac{10!}{1!(10-1)!}(1/6)^1(1-(1/6))^{10-1}$$
$$+ \frac{10!}{2!(10-2)!}(1/6)^2(1-(1/6))^{10-2}$$
$$= (5/6)^{10} + 10(1/6)(5/6)^9 + 0.29$$
$$= 0.78$$

Figure 6.4 illustrates the binomial distribution and cumulative distribution function for $n = 10, \pi = 1/6$. In principle, $Y$ can have any value between 0 and

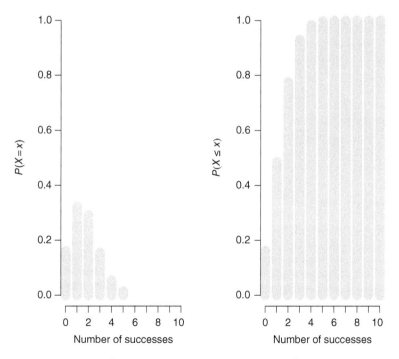

Figure 6.4 Binomial distribution and cumulative distribution function when $n = 10, \pi = 1/6$.

10, but the cumulative distribution function shows that most likely the random variable will have a value less than or equal to 5.

***The Shape of the Binomial Distribution*** It is informative to think about what happens to the shape of the binomial probability distribution as its parameters $\pi$ and $n$ change. In general, the shape of the binomial probability distribution will depend on these parameters:

- For small $\pi$, the binomial distribution is skewed to the right. This fact is expected, because when $\pi$ is small, then we have low probability of success on each iteration. Most likely, when we count the number of successes, we will get just a few.
- For large $\pi$, the binomial distribution is skewed to the left, since when $\pi$ is large, then we have a high probability of success in each iteration, and it is likely that when counting the number of successes we will get many.
- As $\pi$ tends to 0.5, the binomial distribution becomes more symmetric.

Regarding sample size, as $n$ increases the dispersion of the random variable tends to increase. In addition, if we keep $\pi$ fixed and increase $n$, the binomial distribution tends to become more symmetrical (Figure 6.5). This effect with increasing $n$ is tied to a very important concept that will be discussed later.

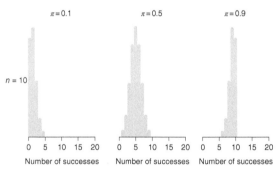

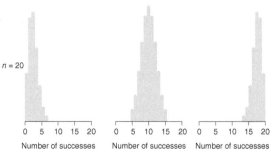

Figure 6.5 Binomial distributions. As we go down each column, the distribution becomes more symmetrical.

## Practice Problems

6.18 In each case, determine if a binomial random variable is in play. If the random variable does not follow a binomial distribution, state why.
   a) Waiting time of 10 people to be attended at an emergency room.
   b) Number of times you get an even number when you roll a dice 3 times.
   c) Possible sum obtained from a roll of a dice 5 times.
   d) Number of times a coin is tossed until we get a head.
   e) Number of times you get a green ball when randomly picking 4 times (without replacement) from a box with 30 green balls, 15 red balls, 50 black balls, and 5 blue balls.
   f) Number of times you get a green ball when randomly picking 4 times (with replacement by always adding a green ball) from a box with 30 green balls, 15 red balls, 50 black balls, and 5 blue balls.
   g) Number of times you get a green ball when randomly picking 4 times (with replacement: placing ball back and shaking box) from a box with 30 green balls, 15 red balls, 50 black balls, and 5 blue balls.

6.19 $X \sim$ binomial(10, 0.75). Without making a calculation, will $P(X = 2)$ be larger than, equal to, or smaller than $P(X = 8)$? Explain.

6.20 $X \sim$ binomial(10, 0.75). Find $P(X = 2)$ and $P(X = 8)$. Compare.

6.21 If $X$ follows a binomial(25, 0.40), determine
   a) $P(X = 1)$.
   b) $P(X = 10)$.
   c) $P(X \geq 1)$.
   d) $P(X > 1)$.
   e) $E(X)$.
   f) $Var(X)$.
   g) standard deviation of $X$.

6.22 If $X$ follows a binomial(5, 0.15), determine
   a) $P(X = 0)$.
   b) $P(X \leq 4)$.
   c) $P(X \geq 1)$.
   d) $P(X > 1)$.
   e) $E(X)$.
   f) $Var(X)$.
   g) standard deviation of $X$.

6.23 If $X$ follows a binomial$(10, 0.60)$ and $Y$ follows a binomial$(10, 0.10)$,
    a) without doing a calculation, would $P(X = 4)$ be smaller than, greater than, or equal to $P(Y = 4)$? Explain.
    b) find $P(X = 4)$ and $P(Y = 4)$.

6.24 Based on a true story! In Costa Rica, there are howling monkeys living in mango trees. People sometimes annoy them and they will throw mangoes at the people. Five tourists arrive in a car and start harassing the monkeys. The monkeys attack the tourists by throwing mangoes. Assume each attack on a tourist is independent, and that in every attempt there is a 0.15 chance that the monkey will hit a tourist.
    a) Show that the number of tourists hit follows a binomial distribution.
    b) Find probability that they hit 2 tourists.
    c) Find the expected number of tourists hit in the attack.
    d) Find the standard deviation of the number of tourist hit.
    e) Find the probability of hitting more than one, but no more than four tourists.

6.25 A student in a statistics course forgets to study for a multiple choice exam. Since the student does not understand the material, they must randomly guess the correct answer to each question (from four possible choices, only one correct answer). The student must answer at least half the questions correctly to pass the exam (all problems have same weight on exam grade). Find the probability that the student passes the exam if there are four questions.

6.26 Continuing with Problem 6.25,
    a) find the probability that the student passes the exam if there are ten questions.
    b) which type of multiple choice exam would be easier to pass, the one with four questions, or the one with ten?
    c) we have assumed the answers to the exam questions were independent. Why is this reasonable here?

## 6.5 Poisson Distribution

Imagine that on Thursday mornings, between 9 and 10 a.m., customers enter the queue at the local bank at a mean rate of 1.7 customers per minute. Let $X$ = number of customers that arrive at the given period. Note that $X$ does not follow a binomial distribution. For one, the number of customers that arrive at the local bank does not have an upper bound like a random variable with a binomial distribution does. There is another distribution called the **Poisson**

Table 6.7 Main characteristics of a random variable $X$ that follows a Poisson distribution.

| Property | |
|---|---|
| Probability mass function | $P(X = x) = \frac{\mu^x e^{-\mu}}{x!}$ for $x = 0, 1, \ldots$ |
| $E(X)$ | $\mu$ |
| $Var(X)$ | $\mu$ |

**distribution**, which takes its name from French mathematician Simeon Poisson. It is a probability distribution that describes the number of occurrences within a randomly chosen unit of time or space, and these occurrences happen independently. For example, number of occurrences within minutes, months, mile, etc. The Poisson distribution is also known as the model of arrivals.

When $X$ has a Poisson distribution, then

$$P(X = k) = \frac{\mu^k e^{-\mu}}{k!}$$

where $e$ is the exponential base $\approx 2.718$ and $\mu = E(X)$. Different than the binomial distribution, the Poisson distribution has only one parameter, $\mu$, the expected value and often referred to as the mean arrival rate. What is more, if $X$ has a Poisson distribution, then

$$E(X) = Var(X) = \mu$$

The Poisson distribution is appropriate for counting random variables with a range of values that have no upper bound. This is theoretically possible because as seen in Table 6.7, the probability mass function tends to zero as values become large.

Poisson distributions are generally right skewed. However, Figure 6.6 demonstrates how the Poisson distribution becomes less right skewed as $\mu$ increases.

There are certain conditions that indicate when the Poisson distribution is a good alternative for a random variable:

- $X$ counts the number of events within an interval of space or time. It has no upper bound.
- The counts are independent of each other.
- Expected arrival rate is constant.
- Number of occurrences are relatively low ($\mu$ is small) in small intervals of time or space.
- If data is available, the distribution of the values of the random variables is unimodal and right skewed.

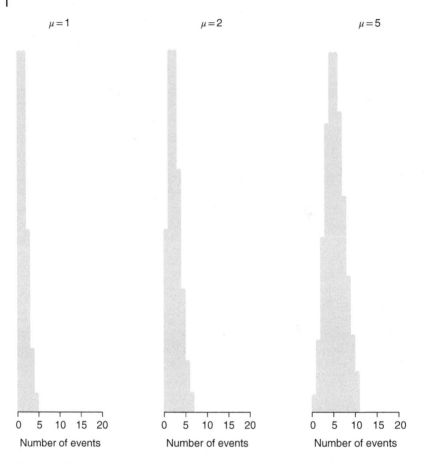

Figure 6.6 Poisson distributions. As we go right, $\mu$ increases and the distributions are less right skewed.

However, even when these conditions are all present, the Poisson distribution may not be an appropriate model. On occasion, data will suggest that it does not come from a distribution with equal expected value and variance. In practice, it is important to check if the Poisson distribution is an adequate model.

**Example 6.14** *The number of service requests to a company are canceled at an average rate of 1.5 requests a day. What is the probability that in a randomly chosen day, 3 service requests are canceled?*

Let $X$ = number of service requests to the company that are canceled. The expected arrival rate is constant, $\mu = 1.5$. It is reasonable to argue that the random variable has no upper bound. Also, we may assume that the counts are independent and the cancellations will be low. Thus, $X$ follows a Poisson distribution with $\mu = 1.5$ and

$$P(X = 3) = \frac{1.5^3 \times e^{-1.5}}{3!}$$

$$= 0.1255$$

***Optional: Poisson Approximation to a Binomial Random Variable*** A binomial random variable with a large $n$ may be hard to work with. When $n$ is large enough and $\pi$ small enough, the expected value of a random variable with a binomial distribution will be very similar to its variance. This suggests that the Poisson distribution can be used to approximate the binomial distribution by setting $\mu = n\pi$.

---

**A Rule of Thumb**

A rule of thumb to use the Poisson approximation to the binomial distribution is that the following three conditions are simultaneously satisfied:

- $n \geq 20$.
- $\pi \leq 0.05$.
- $n\pi < 7$.

---

**Example 6.15** *As of 2016, it is estimated that almost four million people live in the city of Los Angeles. Last year (365 days) records from an emergency room were used to evaluate the number of heart attack cases that were treated each day. Figure 6.7 illustrates the probability distribution of the collected data. Emergency staff would like to determine the probability of having 6 heart attack cases or more.*

Figure 6.7 Histogram for last year's number of heart attack cases treated each day.

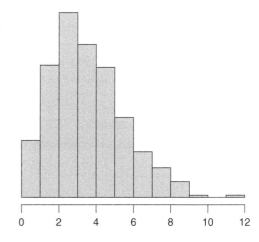

The number of heart attacks each day, $X$, can be treated as a binomial random variable. But it has a large sample size, 4 million. Also the data is right skewed. An alternative is to model $X$ with a Poisson distribution since $n > 20$, and based on historical data $\pi < 0.05$, and $n\pi < 7$.

The average of the data is 4.03. Therefore, we assume $X \sim Poisson(4.03)$ and find that $P(X \geq 6) = 0.22$.

## Practice Problems

6.27 Which of the following random variables are candidates to be modeled with a Poisson distribution:
   a) Number of car accidents at an intersection on a Tuesday night.
   b) Time it takes a machine to complete a task during a test run.
   c) Number of times you get a green ball when randomly picking 4 times (with replacement) from a box with 100 balls of four colors (including green).
   d) Number of callers to a company for service on a Monday from 10 to 11 a.m.

6.28 If $X$ follows a Poisson(5), determine
   a) $P(X = 1)$.
   b) $P(X = 10)$.
   c) $P(X \geq 1)$.
   d) $P(X > 1)$.
   e) $E(X)$.
   f) $Var(X)$.
   g) standard deviation of $X$.

6.29 You and your partner are the owners of a company that specializes in remodeling houses and interior decoration. You would like to track the amount of complaints you get a month, but do not know how. Your cousin currently goes to college and tells you "I can help out." Using the data of the past 20 years, she determines that the mean number of complaints a month is 1.5 and that it is reasonable to assume that it follows a Poisson distribution.
   a) What is the variance of the number of complaints in any given month?
   b) What is the probability that you will get exactly three complaints next month?

6.30 An employee is evaluating the number of arrivals at the drive-up teller window of a bank during a 15-minute period on weekday mornings. He assumes a Poisson distribution for the arrivals in this time period and that the mean number of arrivals in 15 minutes being 10.

a) What is the probability that 5 cars will arrive in a 15-minute period on weekday mornings?

b) What is the probability that at least 2 cars will arrive in a 15-minute period on weekday mornings?

c) What is the standard deviation of the number of cars that will arrive in this time period?

6.31    The number of lightning strikes in a year for a region in central United States has a Poisson distribution with a mean of 4.9. Find the probability that in a randomly selected year, the number of lightning strikes is 10.

6.32    Suppose we are interested in the number of major defects in a highway, one month after resurfacing. Assume a Poisson distribution for the major defects in every mile, with a mean number of two major defects per mile, one month after resurfacing. Also, assume the number of defects in every mile are independent.

a) What is the probability that 3 major defects are found in any given mile?

b) What is the variance of the number of defects in any given mile?

c) What is the probability that no major defects are found in a three mile-section?
    *Hint:* Express the number of defects in a three-mile section in terms of events.

d) What is the probability that at least one major defect is found in a three-mile section?

## 6.6    Optional: Other Useful Probability Distributions

There are a number of other probability distributions available for discrete random variables. For example, the binomial distribution can be extended to scenarios where for each outcome there are more than two possibilities, the **multinomial distribution**. Manual computations can be messy for these distributions. Void of any academic benefit in manual computations in this section Minitab is used instead.

*Hypergeometric Distribution*    The binomial distribution applied in instances of independent Bernoulli trials, which represented an instance of sampling with replacement. On occasion, when counting the number of successes, a sample of size $n$ is taken from a finite population without replacement. As a result, the

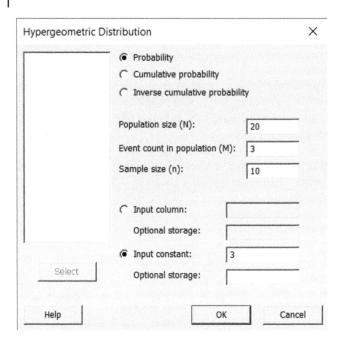

Figure 6.8 Minitab window to calculate (cumulative) probabilities using the hypergeometric distribution.

probability of success is no longer constant per trial. Furthermore, the number of total successes in the population to be sampled from is already known. If $X$ is the number of successes in a sample of size $n$, this random variable will follow a **hypergeometric distribution**.

**Example 6.16** *There are 20 employees in the actuarial department of a firm. Three of the 20 employees are women. 10 are randomly chosen to take a course for accreditation on an actuarial topic. What is the probability all 3 women are chosen?*

$X$ = number of employees that are women out of 10 employees randomly chosen from a total of 20. $X$ follows a hypergeometric distribution. In Minitab go to `Calc > Probability Distributions > Hypergeometric ....` Figure 6.8 shows the box that will pop up window and inputs to find the probability.

The answer is $P(X = 3) = 0.105$. A more interesting question is what is the probability that at least one woman is chosen? That probability is 0.895. Now, these results assume that the company randomly chose employees to take the course. Often preference will be given to some employees, like those considered most likely to get accreditation. When that is the case, the true probabilities may differ from those obtained here assuming randomly chosen employees. Conservatively, if 1 man is considered best in the company, then the probability that you chose at least 1 woman for the course (from the remaining 19 employees) is 0.913

*Negative Binomial and Geometric Distributions*   Suppose there is a scenario where it is necessary to characterize $X$, the occurrences of some event without an upper bound. But the requirement $Var(X) > E(X)$, while the Poisson distribution assumes $E(X) = Var(X)$. A good alternative is to use the **negative binomial distribution**. This distribution applies when independent Bernoulli trials occur until $r$ successes and these are therefore fixed. On the other hand $X$, the number of failures until $r$ successes is random. Thus, $X = r, r + 1, \ldots$. When $r = 1$, a special case arises that is known as the **geometric distribution**.

## Chapter Problems

6.33  Find the value of the missing probability (or probabilities) such that the table defines a proper probability distribution function:

a) Case 1

| $x$ | $P(X = x)$ |
| --- | --- |
| 250 | 0.35 |
| 750 | 0.25 |
| 950 | |

b) Case 2

| $x$ | $P(X = x)$ |
| --- | --- |
| 3 | 0.11 |
| 4 | |
| 5 | 0.29 |

c) Case 3

| $x$ | $P(X = x)$ |
| --- | --- |
| 0 | 0.05 |
| 1 | 0.65 |
| 3 | $a$ |
| 4 | $3a$ |

**6.34** Find the expected value for each Case in Problem 6.33.

**6.35** Find the standard deviation for each Case in Problem 6.33.

**6.36** In Section 6.2.1, $E(X)$ was determined from open data. By using sample data, an estimate of $E(X)$ was found, not its true value. Suppose that in a random sample of 10 observations, we find the relative frequencies below. Show that the $E(X)$ estimated from the table below is equal to $\bar{x}$.

| $x$ | $P(X = x)$ |
| --- | --- |
| 0 | 2/10 |
| 1 | 7/10 |
| 2 | 1/10 |

**6.37** State what are the five main properties of a random variable that follows a binomial distribution.

**6.38** Why is it reasonable in Example 6.15 to assume that the number of daily heart attacks follows a binomial distribution?

**6.39** Every fall, managers at a bank run a marketing campaign to promote their credit card to customers. The marketing pitch will be presented to a random sample of 20 customers visiting the bank. Historically, the probability a customer accepting the credit card has been 0.15.
a) What is the probability that two customers accept the credit card offer?
b) What is the expected number of customers that will accept the credit card offer?
c) Find the standard deviation of the number of customers that will accept the credit card offer.

**6.40** You are the coach of a team playing against the Golden State Warriors in the NBA finals. With 1.5 seconds left in the game, the score is 115–112 in your team's favor; Golden State called a time out, and you are pretty sure Stephen Curry will take the last shot. The two coaching choices are to foul him while taking the shot or let him take the shot. If you foul him while taking a three-point shot, he would get three free throws. During the season, Curry makes 90% of his free throws and 44% of his three point shots.
a) Find the probability he would make all three free throws. Assume free throw outcomes are independent.[3]

---

3 Studies support this assumption.

b) Which choice makes more sense: fouling him for three free throws or guarding against a three-point shot?

6.41 Management at a bank have developed a new marketing strategy for their credit card. A random sample of 20 customers visiting the bank are chosen to present the marketing pitch, and seven accepted the offer. Historically, the probability a customer accepts the credit card has been 0.15.

    a) What is the probability that seven customers or more accept the credit card offer if the new marketing strategy did not improve the acceptance rate?

    b) Managers have decided that if the probability 6.41a is below 0.05, then they should conclude that the new marketing strategy was able to increase the proportion of people that accept the credit card offer. Based on the result in 6.41a, what should they conclude?

6.42 In Problem 6.41, assume 5 out of the 20 customers accepted the credit card offer:

    a) What is the probability that 5 customers or more accept the credit card offer?

    b) Managers have decided that if the probability in 6.42a is below 0.05, then they should conclude that the new marketing strategy was able to increase the proportion of people that accept the credit card offer. Based on the result in 6.42a, what should they conclude?

6.43 For Example 6.15, the keen reader may have wondered why use a Poisson distribution to find probabilities, instead of the empirical probabilities. The table below presents the empirical probabilities.

| $x$ | 0 | 1 | 2 | 3 | 4 | 5 | 6 | 7 | 8 | 9 |
|---|---|---|---|---|---|---|---|---|---|---|
| $P(X = x)$ | 0.0137 | 0.0548 | 0.1589 | 0.2219 | 0.1836 | 0.1562 | 0.0959 | 0.0548 | 0.0356 | 0.0192 |

    a) What is the empirical probability that $X = 3$?
    b) What is the empirical probability that $X \geq 6$?
    c) If $X \sim Poisson(4.03)$, find $P(X = 3)$ and compare with (a).
    d) If $X \sim Poisson(4.03)$, find $P(X \geq 6)$ and compare with (b).

6.44 A firm will randomly draw credit data of 500 households at once in a town of 1 000 households. If $X =$ number of households in the random sample with credit debt of at least $5 000, does this random variable follow a binomial distribution?

6.45 Of the students applying to a university, 63% are accepted. Assuming independent applications, what is the probability that among the next 18 applicants?
- a) At least one will be accepted.
- b) Fifteen or more will be accepted.
- c) Three will be rejected.

6.46 Returning to Problem 6.45, determine:
- a) the expected number of acceptances.
- b) the standard deviation.

6.47 A professor receives 20 exams from his 20 students. After grading 10 exams, she checks the enrollment and sees 3 people dropped the class. She now has 17 students. If she grades the exams at random,
- a) what is the probability she needlessly graded 3 exams?
- b) what is the probability she needlessly graded at least one exam?

## Further Reading

Black, Ken. (2012). *Business Statistics*, 7th edition. Wiley.

Levine, D. M., Stephan, D. F., and Szabat, K. A. (2017). *Statistics for Managers: Using Microsoft Excel*, 8th edition. Pearson.

Wackerly, Dennis D., Mendenhall III, William, and Scheaffer, Richard L. (2002). *Mathematical Statistics with Applications*, 6th edition. Duxbury.

# 7

# Continuous Random Variables

## 7.1 Introduction

With discrete random variables, we were often able to establish an equation to find the probability of specific values of the random variable. These equations had to be such that the sum of the probabilities of all possible values of the discrete random variable was 1. Special cases of distributions for discrete random variables such as the Binomial and the Poisson were discussed.

In this chapter, some properties of[1] **continuous random variables** are introduced. There are many situations in real life where continuous random variables are needed. For example, when measuring something, like the weight of a product being manufactured or the operation time of a machine to complete a procedure, continuous random variables are defined. With a continuous random variable, we still need to work with its expected value, variance (or standard deviation), and probabilities. But these traits are calculated differently than with discrete random variables. Specifically, continuous random variables have an infinite amount of possible values that cannot be written down in an orderly fashion. The alternative approach is using a **probability density function,** $f(x)$, or pdf for short. A pdf is a function, which defines a curve such that the following two conditions hold:

- $f(x_i) \geq 0$ for all $x_i$ (it is a nonnegative function).
- The area under the entire curve must be 1.

Thus, $f(x) = -3x$, where $0 \leq x \leq 1$ is not a pdf, because the function is negative. Moreover, $f(x) = 0.1$ where $0 \leq x \leq 1$ is nonnegative, but the area under the entire curve is not 1, while $f(x) = 2x$ where $0 \leq x \leq 1$ is a pdf because it is nonnegative and the area under the entire curve is 1 (the area of a triangle).

---

[1] Continuous random variables are those that, given an interval, its possible values are infinite and uncountable.

*Principles of Managerial Statistics and Data Science*, First Edition. Roberto Rivera.
© 2020 John Wiley & Sons, Inc. Published 2020 by John Wiley & Sons, Inc.
Companion website: www.wiley.com/go/principlesmanagerialstatisticsdatascience

---

**Probability as an Area**
The area under the pdf curve for a given interval $(c, d)$ is the probability of obtaining a value of the random variable in that interval (Figure 7.1).

---

Strictly speaking, for a continuous random variable $X$, integrals are required to find probabilities;

$$P(c < X < d) = \int_c^d f(x)dx$$

By definition, the expected value of a continuous random variable is the mean of all possible values of the variable. The mean is influenced by the weight

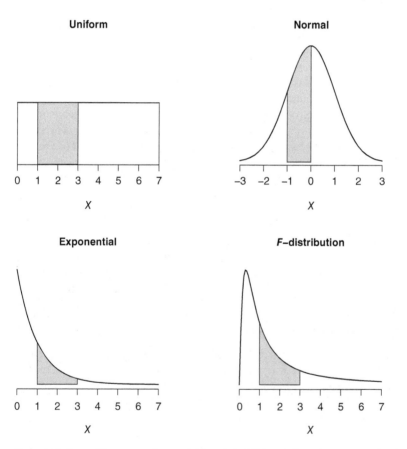

**Figure 7.1** For continuous random variables, probabilities are found as areas under the probability density function. Here are examples of four probabilities using different density functions.

given to each possible value, which is dictated by the probability distribution of the variable. Although the definition of the expected value is the same as for a discrete random variable, it is calculated differently. Again, this calculation is obtained through an integral, as is the calculation of the variance. Fortunately, there are special cases of continuous random variables which do not require calculus. In Section 7.2, we discuss one of these continuous random variables.

## Practice Problems

7.1 Which of the following functions represent a probability density function for a continuous random variable? *Hint:* Check if both rules of a proper probability density function hold.

a) $f(x) = 0.5$ where $0 \leq x \leq 2$.
b) $f(x) = x/2 - 1$ where $0 \leq x \leq 2$.

7.2 Which of the following functions represent a probability density function for a continuous random variable? *Hint:* Check if both rules of a proper probability density function hold.

a) $f(x) = 0.25$ where $0 \leq x \leq 8$.
b) $f(x) = x/2$ where $0 \leq x \leq 2$.

## 7.2 The Uniform Probability Distribution

If a random variable follows a **uniform probability distribution** in some interval $(a, b)$, then any subinterval of possible values of equal length will be equally likely. For example, if a random variable is continuous and has a uniform distribution with possible values between 5 and 10, the curve used for obtaining probabilities is seen in Figure 7.2.

As one can see, the uniform probability density function has a familiar shape, a rectangle. The area of a rectangle is the height multiplied by its length. The height of this rectangle is such that when multiplied by the length of the rectangle, the result is 1. Recall that this is a must for any pdf. The uniform distribution can be defined in general terms for any interval $(a, b)$,

$$f(x) = \begin{cases} \frac{1}{b-a}, & a \leq x \leq b \\ 0 & \text{otherwise} \end{cases}$$

This equation defines the height of the rectangle to any uniform distribution with parameters $(a, b)$, the lowest and highest possible values of the random variable $X$. This uniform probability density function has a constant value for

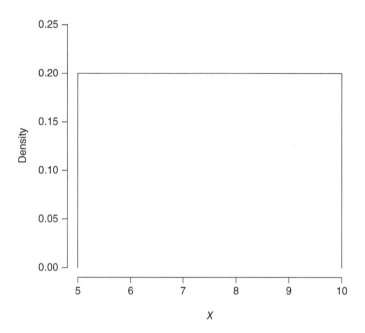

Figure 7.2 Uniform distribution for a continuous random variable with possible values between 5 and 10.

any $x$, except values outside the interval $(a, b)$, where $f(x)$, and hence any probability, is zero.

**Example 7.1** *Suppose you plan to take a bus to a meeting. Say the waiting time of this bus follows a uniform distribution between 5 and 10 minutes in the morning. What is the probability that you have to wait between 7 and 9 minutes? What about between 20 and 30 minutes?*

According to what we have covered, to find the required probability, one must obtain the area of a sub-rectangle within the uniform distribution (see Figure 7.3).

As we can see from Figure 7.3, the height of the sub-rectangle is $\frac{1}{10-5} = 0.2$ since the variable follows a uniform distribution, while its length is $9 - 7 = 2$. Therefore, the area of interest is $2 \times 0.2 = 0.4$. In addition, the definition of $X$ in this problem implicitly tells us that $P(20 < X < 30) = 0$, since it is a region that falls completely outside the rectangle which defines the variable, $(5, 10)$. You think you could find $P(6 < X < 20)$?

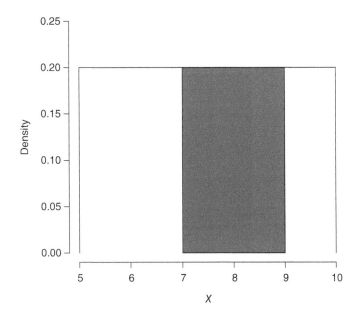

Figure 7.3 The uniform distribution for Example 7.1. Shaded in red is the area that we must calculate to obtain $P(7 < X < 9) = 0.4$.

---

**$P(X = k)$ When $X$ is Continuous**

Since probabilities for continuous random variables are based on the area under a curve, $P(X = k) = 0$ for any $k$, since $X = k$ defines an interval of length zero. Consequently, if $X$ is a continuous random variable,

$$P(X \leq k) = P(X < k)$$

Another way of explaining this is that irrelevant of whether we wish to find $P(X \leq k)$ or $P(X < k)$, so the limits of integration do not change. Similarly, we can conclude the same for other inequalities, $P(X \geq k) = P(X > k)$, $P(c < x < d) = P(c \leq X \leq d)$, and so forth.

These equalities, DO NOT APPLY for discrete random variables. Overall for discrete random variables $P(X \leq k) \neq P(X < k)$, etc.

---

*Expected Value and Variance of a Uniformly Distributed Random Variable* The expected value of a continuous random variable, $X$, with possible values ranging from $(a, b)$ and probability density $f(x)$ is

$$\mu = E(X) = \int_a^b xf(x)dx$$

Think of it as a form of weighted average. The weight is based on $f(x)$ for that value $x$. Assume a uniform random variable $X$ with possible values ranging from $(a, b)$. Then, the expected value can be shown to be,

$$\mu = E(X) = \frac{b+a}{2}$$

Returning to the example of the bus, since $X$ is a uniformly distributed random variable with values between 5 and 10 minutes, the expected value is

$$\mu = E(X) = \frac{10+5}{2} = 7.5$$

If we think carefully about this result, we see that it is no surprise, 7.5 minutes falls right in the middle of the uniform distribution with values between 5 and 10. In this case, the average is equal to the median, since the distribution is symmetrical.

For the variance of a uniform random variable $X$, we have

$$\sigma^2 = Var(X) = \frac{(b-a)^2}{12}$$

while the standard deviation is $\sigma = \sqrt{Var(X)}$. Returning to the example of the bus, the standard deviation is $\sigma = \sqrt{(10-5)^2/12} = 1.44$.

Although the uniform distribution is easy to use, it has limited applications in real life. In business, applications will often be related to conducting simulations. There is a continuous distribution that applies widely in practice. It also has properties that allow estimation and inference very efficiently. It is the topic of Section 7.3.

---

**$P(X = k)$ When $X$ is Continuous: part 2**

Suppose that $X$ has a uniform distribution (0,10). We take a sample and obtain $X = 5$. However, based on the fact that $P(X = k) = 0$ for continuous random variables, $P(X = 5) = 0$. So how did we get that result during sampling? This statement applies for any possible value of $X$, not just 5. Our "area" explanation only sort of helps in understanding how we can get $X = 5$, while $P(X = 5) = 0$. Another explanation is that all continuous variables have an uncountable amount of possible values. Hence, if we want to be able to get probabilities of possible values of continuous variables, we need $P(X = k) = 0$ for any constant $k$. This discussion also implies an important distinction: an impossible result has zero probability, but a zero probability event is not impossible. This theoretical result is useful to state so we understand how probability works for continuous random variables. From a practical point of view, we may sometimes need to obtain something close to $P(X = k)$ for any continuous random variable. We can approximate this probability by finding $P(k - 0.5 \leq X \leq k + 0.5)$, the probability of a rounded value of $X$. By adding and subtracting 0.5, we now have a length that is greater than zero and hence an event of positive probability. In the bus example, $P(6.5 \leq X \leq 7.5) = 0.2$.

## Practice Problems

7.3 When $X \sim U(75,125)$. Show that
   a) $P(75 \leq X \leq 100) = 0.5$.
   b) $P(80 \leq X \leq 111) = 0.62$.
   c) $P(X \geq 112.5) = 0.25$.
   d) $P(X < 60) = 0$.
   e) $E(X) = 100$.
   f) $Var(X) = 208.33$.

7.4 If $X \sim U(15, 50)$. Find
   a) $P(15 \leq X \leq 25)$
   b) $P(20 \leq X \leq 30)$
   c) $P(X < 37)$
   d) $P(X \leq 37)$
   e) $P(X > 42)$
   f) $E(X)$
   g) $Var(X)$

7.5 Find the expected value and standard deviation of $X$ for each case.
   a) $U(10, 50)$
   b) $U(100,500)$
   c) $U(200,400)$
   d) $U(0,100)$

7.6 If $X \sim U(15, 50)$, determine the value $x$ such that it is the 75th percentile.

7.7 If $X \sim U(103,225)$, determine the value $x$ such that only 5% of values of the random variable exceed it.

7.8 $X \sim U(0,800)$ has the same expected value as $Y \sim U(200,600)$. Check if $P(X \leq 250) = P(Y \leq 250)$. Also check if $Var(X) = Var(Y)$.

7.9 Assume the weight of cargo for a company truck follows a uniform distribution ranging between 300 and 450 pounds.
   a) What is the expected value of the weight of the cargo?
   b) What is the standard deviation of the weight of the cargo?

c) Draw a plot of the pdf.
d) Find the probability that the weight is between 325 and 375 pounds.
e) What is the probability that the cargo's weight is at least 325 pounds?
f) Suppose that an efficiency study was conducted and found that the probability that the cargo weight is greater than the expected value and less than $x$ is 0.3. Furthermore, weight $x$ should be the maximum weight allowed for cost efficiency. Find what this weight is.

7.10 Past records of the temperature in Miami show an average temperature, for the months from September to October, which follows a uniform distribution with temperatures between 82 and 87. Suppose that you own a yogurt producing company and you know the best temperature for organic yogurt production is between 84 and 86°. What is the probability tomorrow's temperature will NOT be best?

## 7.3 The Normal Distribution

The uniform random variable is ideal for introducing probability distributions for continuous variables, since it is easy to understand and allows us to easily visualize how to find the probability of $X$ being within a given interval for a continuous random variable, by finding the subarea under the curve established by the interval of interest. But there is a probability distribution that applies very often in practice and has a number of properties that are useful for inference. It is known as the **normal distribution** and it is the most important probability distribution of all. We will use the notation $X \sim N(\mu, \sigma)$ to read as follows: "X follows a normal distribution (or is normally distributed) with mean $\mu$ and standard deviation $\sigma$."

The normal pdf forms a bell-shaped curve. As seen in Figure 7.4, the "top of the bell" is located at $x = \mu$. As the values of $X$ go further below or above $\mu$, they become less probable because the area under the normal pdf becomes smaller. The normal distribution is symmetric around its mean and therefore, the mean of a normally distributed random variable is equal to its median.

In its mathematical form, the normal probability density function is

$$f(x) = \frac{1}{\sqrt{2\pi\sigma^2}} e^{-\frac{(x-\mu)^2}{2\sigma^2}} \quad -\infty < x < \infty$$

where $\pi \approx 3.14$ and $e \approx 2.71$. This pdf defines the curve seen in Figure 7.4. From the $f(x)$ equation we see that in theory, the normal random variable $X$ has a range of $(-\infty, \infty)$. Still, the pdf has a total area under the curve of one, thanks to the exponential term which allows a rapid decrease in probability as values of $X$ head to the extreme right or left of the mean. More specifically, the probability

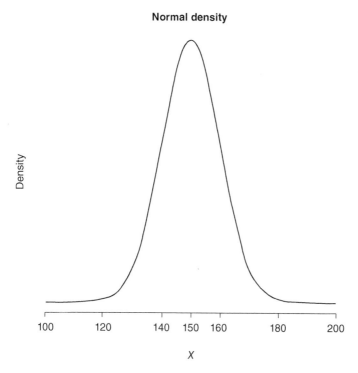

**Normal density**

Density

100 120 140 150 160 180 200

X

Figure 7.4 A continuous random variable that follows a normal distribution with mean $\mu = 150$ and standard deviation $\sigma = 10$.

that a value of a normally distributed $X$ with mean $\mu$ and standard deviation $\sigma$ is within $3\sigma$ of $\mu$ (i.e. $P(\mu - 3\sigma < X < \mu + 3\sigma)$) is 0.9997.

---

**Empirical Rule**
If data follows a normal distribution, then approximately (Figure 7.5)

- 68% of the observations fall within 1 standard deviation from the mean, $\mu \pm 1\sigma$.
- 95% of the observations fall within 2 standard deviations from the mean, $\mu \pm 2\sigma$.
- Virtually all observations fall within 3 standard deviations from the mean, $\mu \pm 3\sigma$.

---

Also, observe that the normal distribution has two parameters, the mean $\mu$ and the variance $\sigma^2$. As $\mu$ changes the peak of the bell shape shifts. That is, if we keep $\sigma$ fixed, reducing the mean shifts the peak of the bell-shaped curve to the

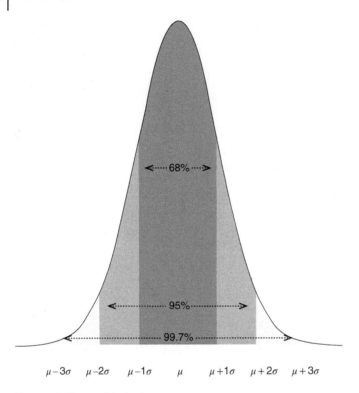

$$\mu-3\sigma \quad \mu-2\sigma \quad \mu-1\sigma \quad \mu \quad \mu+1\sigma \quad \mu+2\sigma \quad \mu+3\sigma$$

Figure 7.5 The empirical rule.

left, while if we increase $\mu$ then the peak of the bell-shaped curve shifts to the right. Figure 7.6 shows the effect of reducing the mean in red and increasing the mean in blue.

Although the curve may shift by changing $\mu$ this way, it retains its bell shape. On the other hand, if we keep $\mu$ fixed and change $\sigma$, the scale of the normal curve changes (Figure 7.7). It is no surprise that the scale changes with $\sigma$ if we recall its definition. The standard deviation measures the dispersion of $X$ around $\mu$. Therefore, a high value of $\sigma$ means more spread out values of $X$. Changing $\sigma$ does not affect the bell shape form of the curve either. As we can imagine, by changing both, $\mu$ and $\sigma$, we shift the bell-shaped curve and change its scale. Yet, the curve retains its bell shape. When shifting and rescaling the curve, we are subtracting one constant from $X$, and dividing that difference by another constant. Therefore, we are defining a new random variable. Now that this new random variable also has a bell-shaped curve does not automatically guarantee that it also follows a normal distribution (in fact, there are pdf curves other than the normal that form bell-shaped curves). However, we can formally make the following statement.

**Normal density**

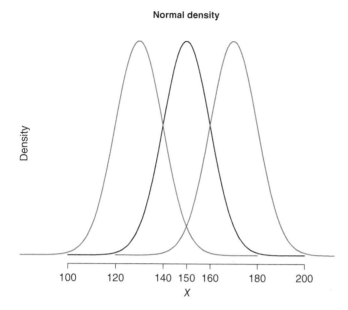

Figure 7.6 Three normally distributed continuous random variables with $\mu = 130, 150, 170$, and $\sigma = 10$ in red, black, and blue, respectively.

---

**Normal Transformations**

For constants $a$ and $b$ such that $b > 0$, if $X \sim N(\mu, \sigma)$, and we define a new random variable, $Y = \frac{X-a}{b}$, then $Y$ is also normally distributed.

---

We need to obtain probabilities of values of normally distributed $X$. As mentioned before, for continuous random variables, probabilities consist of an area under the pdf curve. For a normally distributed random variable, this sounds intimidating, since the pdf has a complicated mathematical form. We would need to perform calculus each time we have a different $\mu$ or $\sigma$. But it turns out that we can use the results above when shifting the mean of a normal and rescaling the bell curve to obtain probabilities.

***Standardized Normal Random Variable*** A normal distribution with mean zero, and standard deviation 1 is known as a **standardized normal distribution**. Mathematically, the standardized normal probability density function is

$$f(z) = \frac{1}{\sqrt{2\pi}} e^{-\frac{z^2}{2}} \quad -\infty < z < \infty$$

The important thing to notice about this equation is that it no longer depends on $\mu$ or $\sigma$. Tables of cumulative probabilities for this distribution are widely available (see Appendix B), providing $P(Z \leq z)$, for many different values $z$.

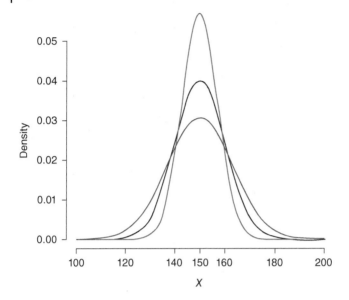

Figure 7.7 Three normally distributed continuous random variables with mean $\mu = 5$ and standard deviations $\sigma = 7, 10, 13$ in red, black, and blue, respectively.

**Example 7.2** *$Z$ has a standardized normal distribution (with $\mu = 0$, $\sigma = 1$). We want to find $P(Z < 1)$.*

Figure 7.8 illustrates what the probability of interest is as an area under the standard normal pdf. Considering that the standard normal has a mean of zero and it is a symmetric distribution,[2] Figure 7.8 suggests that $P(Z < 1)$ is greater than 0.5.

Figure 7.9 illustrates how we find $P(Z < 1) = 0.84$ using the table in Appendix B.

What if we need to find $P(Z > 1)$? This probability represents the white area under the pdf curve in Figure 7.8. The standardized normal table provides $P(Z \leq z)$. Therefore, we can use the complement rule for probabilities, $P(Z > 1) = 1 - P(Z \leq 1) \approx 1 - 0.84 = 0.16$.

What if we need to find $P(-0.5 < Z < 0.5)$? In this case, it is convenient to appeal to geometry. Say you divide a rectangle into two overlapping sub-rectangles, 1 and 2. It is easy to deduce that the area of the rectangle minus the area of sub-rectangle 2 is equal to the area of sub-rectangle 1. This fact applies in general regardless of the geometrical shape of interest. Based on this reasoning, Figure 7.10 shows what we propose to do to find $P(-0.5 < Z < 0.5)$. Using the standardized normal table, we first find the total area seen in red

---

2  Half of the values of $Z$ are to the left of 0, and the other half to the right.

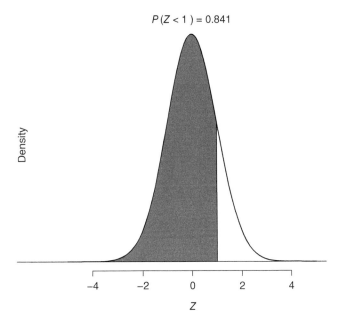

Figure 7.8 Area under the curve representing the probability of interest, $P(Z < 1)$.

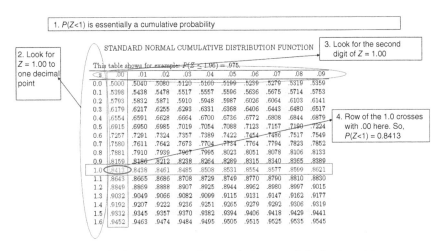

Figure 7.9 Steps to find $P(Z < 1)$ using the cumulative normal distribution table.

and blue, which corresponds to $P(Z < 0.5)$. Then we find the blue area, which is $P(Z < -0.5)$. Finally, the area that interests us, in red, is

$$P(-0.5 < Z < 0.5) = P(Z < 0.5) - P(Z < -0.5) \approx 0.383$$

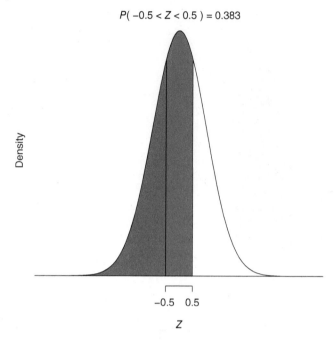

$P(-0.5 < Z < 0.5) = 0.383$

Figure 7.10 Areas under the standard normal pdf curve to find, $P(-0.5 < Z < 0.5)$.

This last example can be generalized the following way. When we need to find the probability that a standardized normal random variables $Z$ is between $z_1$ and $z_2$, we use the formula,

$$P(z_1 < Z < z_2) = P(Z < z_2) - P(Z < z_1)$$

The following steps help find probabilities of values of a standard normal random variable:

1) Sketch the pdf of $Z$ with the area (probability) of interest.
2) Determine if the probability of interest is a cumulative probability. If not, express the probability as a function of a cumulative probability.
3) Find the value $z$ to one decimal point in the first column of the cumulative normal distribution table.
4) Determine which column has the remaining decimal point of $z$.
5) Cross the row in step 3 with the column in step 4 to find $P(Z \leq z)$. Apply steps 3–5 more than once if necessary.
6) Use the probability or probabilities in step 5 to find the probability of interest.

**Example 7.3**  *Z has a standardized normal distribution (with $\mu = 0$, $\sigma = 1$), we want to find, $P(Z \geq 4.55)$.*

Figure 7.11 Area under the standard normal pdf curve to find $P(Z \geq 4.55)$. The area under the standard normal pdf is so small that a zoomed-in window of that portion of the bell-shaped curve was required.

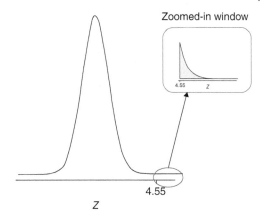

Let us follow the steps proposed above. A sketch of the standard normal pdf and the area of interest is seen in Figure 7.11. Note that the area below the curve was so small in this case that a zoomed-in panel of the right portion of the standard normal pdf was required (right panel). Thus, this probability must be very small. Next, note that the probability of interest is not a cumulative probability. Therefore, we state that[3] $P(Z \geq 4.55) = 1 - P(Z \leq 4.55)$. Next is to look up 4.5 in the first column of the $Z$ table. However, the largest possible value found is 3.49, smaller than 4.5. This does not prevent us from determining the probability of interest. According to the $Z$ table, $P(Z \leq 3.49) = 0.9998$. Therefore, $P(Z \leq 4.55)$ is even closer to 1 that $P(Z \leq 3.49)$, and we can say that $P(Z \geq 4.55) = 1 - P(Z \leq 4.55) = 1 - 1 = 0$. As expected, $P(Z \geq 4.55)$ was very small.

Recall that when $X$ is normally distributed, $P(\mu - 3\sigma < X < \mu + 3\sigma) = 0.9997$. Therefore, in the case of the standard normal random variable, $P(-3 < Z < 3) = 0.9997$. So it is easier to determine if values of $Z$ are rare compared to other normal random variables with different parameter values. That is why values of the standard normal random variable are often called $z$-**scores**. They define a number of standard deviations that an observation falls from the mean when the observations are normally distributed.

---

**Tips to Use the Z Table**
Students commonly encounter a question to find a probability and jump immediately to the $Z$ table to draw a value without carefully thinking of the problem. The temptation to jump to the final answer should be avoided. As always, we must not forget to use common sense to solve problems. We have already

---

3 Remember that $Z$ is a continuous random variable, and therefore $P(Z \leq 4.55) = P(Z < 4.55)$.

abided to common sense when finding probabilities, verifying that the result is realistic (between 0 and 1) and sometimes corroborating that we get a result that makes sense according to the defined problem. We can do the same in the case of probabilities from normal distributions. Assume that $P(z_1 < Z < z_2)$ is needed and $Z$ has a standard normal distribution.

- If the interval strictly falls to the left of 0, then $P(z_1 < Z < z_2) < 0.5$. This is because the normal distribution is symmetric, and therefore its mean is equal to the median. If 50% of the values of $Z$ fall to the left of $\mu$, then any interval $(z_1, z_2)$ strictly to the left of $\mu$ cannot have an area greater than 0.5. An analogous statement can be made when an interval is strictly to the right of $\mu$.
- Open intervals $(-\infty, a)$ or $(b, \infty)$ that include the peak of the bell curve will always be greater than 0.5. We can use this fact to corroborate if the probabilities we obtain make sense.
- The exception is if you have an interval $(z_1, z_2)$ such that $z_1$ is to the left of $\mu$ and $z_2$ is to the right of $\mu$, the interval includes the top of the bell curve. Then, we cannot automatically determine if the probability is less than or greater than[4] 0.5.

***Finding Values of Z for Given Probabilities*** There is no reason to only use the $Z$ table to find probabilities. For a given probability, we can find a $Z$ value. For example, what $z$ value corresponds to the 10th percentile of the standard normal distribution? Since the standard normal has a mean of 0 and is a symmetric distribution, the 10th percentile must be a negative value (Figure 7.12). Our

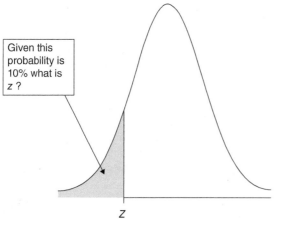

Figure 7.12 Area under the normal curve defining $P(Z \leq z) = 0.10$.

Given this probability is 10% what is z ?

Z

---

4 Unless the values $(z_1, z_2)$ permit applying the empirical rule.

1. Given probability is a cumulative probability, $P(Z \leq z) = 0.10$

| z | .00 | .01 | .02 | .03 | .04 | .05 | .06 | .07 | .08 | .09 |
|---|---|---|---|---|---|---|---|---|---|---|
| -3.4 | .0003 | .0003 | .0003 | .0003 | .0003 | .0003 | .0003 | .0003 | .0003 | .0002 |
| -3.3 | .0005 | .0005 | .0005 | .0004 | .0004 | .0004 | .0004 | .0004 | .0004 | .0003 |
| -3.2 | .0007 | .0007 | .0006 | .0006 | .0006 | .0006 | .0006 | .0005 | .0005 | .0005 |
| -3.1 | .0010 | .0009 | .0009 | .0009 | .0008 | .0008 | .0008 | .0008 | .0007 | .0007 |
| -3.0 | .0013 | .0013 | .0013 | .0012 | .0012 | .0011 | .0011 | .0011 | .0010 | .0010 |
| -2.9 | .0019 | .0018 | .0018 | .0017 | .0016 | .0016 | .0015 | .0015 | .0014 | .0014 |
| -2.8 | .0026 | .0025 | .0024 | .0023 | .0023 | .0022 | .0021 | .0021 | .0020 | .0019 |
| -2.7 | .0035 | .0034 | .0033 | .0032 | .0031 | .0030 | .0029 | .0028 | .0027 | .0026 |
| -2.6 | .0047 | .0045 | .0044 | .0043 | .0041 | .0040 | .0039 | .0038 | .0037 | .0036 |
| -2.5 | .0062 | .0060 | .0059 | .0057 | .0055 | .0054 | .0052 | .0051 | .0049 | .0048 |
| -2.4 | .0082 | .0080 | .0078 | .0075 | .0073 | .0071 | .0069 | .0068 | .0066 | .0064 |
| -2.3 | .0107 | .0104 | .0102 | .0099 | .0096 | .0094 | .0091 | .0089 | .0087 | .0084 |
| -2.2 | .0139 | .0136 | .0132 | .0129 | .0125 | .0122 | .0119 | .0116 | .0113 | .0110 |
| -2.1 | .0179 | .0174 | .0170 | .0166 | .0162 | .0158 | .0154 | .0150 | .0146 | .0143 |
| -2.0 | .0228 | .0222 | .0217 | .0212 | .0207 | .0202 | .0197 | .0192 | .0188 | .0183 |
| -1.9 | .0287 | .0281 | .0274 | .0268 | .0262 | .0256 | .0250 | .0244 | .0239 | .0233 |
| -1.8 | .0359 | .0351 | .0344 | .0336 | .0329 | .0322 | .0314 | .0307 | .0301 | .0294 |
| -1.7 | .0446 | .0436 | .0427 | .0418 | .0409 | .0401 | .0392 | .0384 | .0375 | .0367 |
| -1.6 | .0548 | .0537 | .0526 | .0516 | .0505 | .0495 | .0485 | .0475 | .0465 | .0455 |
| -1.5 | .0668 | .0655 | .0643 | .0630 | .0618 | .0606 | .0594 | .0582 | .0571 | .0559 |
| -1.4 | .0808 | .0793 | .0778 | .0764 | .0749 | .0735 | .0721 | .0708 | .0694 | .0681 |
| -1.3 | .0968 | .0951 | .0934 | .0918 | .0901 | .0885 | .0869 | .0853 | .0838 | .0823 |
| -1.2 | .1151 | .1131 | .1112 | .1093 | .1075 | .1056 | .1038 | .1020 | .1003 | .0985 |
| -1.1 | .1357 | .1335 | .1314 | .1292 | .1271 | .1251 | .1230 | .1210 | .1190 | .1170 |
| -1.0 | .1587 | .1562 | .1539 | .1515 | .1492 | .1469 | .1446 | .1423 | .1401 | .1379 |

2. Find closest value to given probability in table

3. Find value of z from row and column that cross probability, $P(Z \leq -1.28) = 0.10$

Figure 7.13 Steps to find z using the cumulative normal distribution table such that $P(Z \leq z) = 0.10$.

task is to find the probability in the cumulative normal distribution table that is closest to 0.10 (see Figure 7.13). Thus, $P(Z \leq -1.28) \approx 0.10$.

## Practice Problems

7.11 Refer back to Figure 7.4. What do the values in the $x$ axis of a normal probability density function represent? What do the values in the $y$ axis represent?

7.12 Suppose that $Z$ follows an $N(0, 1)$. In the following, state whether the probability is greater than 0.5, lower than 0.5, equal to 0.5, or if it is not possible to tell just by common sense, without computing the actual probabilities. Explain your reasoning.
a) $P(Z < -2.44)$.
b) $P(Z \geq 1.44)$.
c) $P(Z < 0)$.
d) $P(-0.28 \leq Z << 0.14)$.

7.13 $Z$ follows an $N(0, 1)$. In the following, state whether the probability is greater than 0.5, lower than 0.5, equal to 0.5, or if it is not possible to

tell just by common sense, without computing the actual probabilities.
Explain your reasoning.
a) $P(Z > 1.33)$.
b) $P(Z \leq 1.75)$.
c) $P(Z \geq 0)$.
d) $P(-0.75 \leq Z \leq< 0.21)$.

7.14 $Z$ follows an $N(0, 1)$. Show that
a) $P(Z < -1.57) = 0.058$.
b) $P(Z \geq 0.18) = 0.43$.
c) $P(Z < -5.34) = 0$.
d) $P(-0.12 < Z < 0.33) = 0.18$.

7.15 $Z$ follows an $N(0, 1)$. In the following, state whether the given probability
is correct or incorrect. If incorrect, explain why.
a) $P(Z \geq 1.28) = 0.8997$.
b) $P(Z < -0.11) = 0.4602$.
c) $P(Z < 7.37) = 1$.
d) $P(Z \leq 0.92) = 0.8212$.
e) $P(Z \geq 2.18) = 0.0146$.

7.16 If $Z$ follows an $N(0, 1)$, compute the following probabilities:
a) $P(Z < 2.44)$.
b) $P(Z \leq 2.44)$.
c) $P(Z \geq 15.32)$.
d) $P(Z \leq -0.15)$.
e) $P(Z \geq -1.96)$.
f) $P(Z > -7.43)$.
g) $P(Z \leq -9.32)$.
h) $P(-0.28 \leq Z \leq< 0.14)$.

7.17 If $Z$ follows an $N(0, 1)$.
a) show that $P(-3 \leq Z \leq 3) = 0.9997$.
b) find $P(-2 < Z < 2)$.
c) find $P(-1 < Z < 1)$.
d) find $P(-0.75 < Z < 0.75)$.

7.18 Find the $z$-value or $z$ values:
a) Such that it is the 25% percentile.[5]
b) Such that it is the 75% percentile.[6]

---

5 The theoretical first quartile.
6 The theoretical third quartile.

c) Such that at most 15% of $z$-scores are higher.
d) Such that the area under the middle of the curve is 0.75.
e) So that there is no more than 40% probability of exceeding that value.

7.19 Find the $z$-value or $z$ values:
   a) Such that there is a probability of 0.025 of a lower $z$-score or 0.025 probability of a higher $z$-score.
   b) $P(Z \geq z) = 0.80$.
   c) $P(Z < z) = 0.33$.
   d) $P(-z \leq Z \leq z) = 0.75$.

## 7.4   Probabilities for Any Normally Distributed Random Variable

Section 7.3 demonstrated how easy it is to find probabilities for a standardized normal random variable. Additionally, Figures 7.6 and 7.7 displayed how changes in normal distribution parameters still result in a bell-shaped probability density function of the random variable. And at the end of Section 7.3, we stated that this is because when you shift a normal random variable (change its mean) or scale it (changing the standard deviation), the "new" random variable also has a normal distribution. All this together allows us to state the following:

---

**Standardizing a Normal Random Variable**
If $X \sim N(\mu, \sigma)$ and we define a new variable $Z$,

$$Z = \frac{X - \mu}{\sigma} \qquad (7.1)$$

Then this new variable $Z$ follows a standard normal distribution with $\mu = 0$ and $\sigma = 1$.

---

So for any normal random variable, no matter what the value of $\mu$ or $\sigma$ are, we can calculate probabilities through transforming it into a standardized normal random variable, $Z$. The method literally allows us to solve an infinite amount of different situations, a unique distinction of the normal family of distributions.

**Example 7.4**    *A local company that produces canned juice knows that the weight of the cans follows a normal distribution with a mean of 12.02 oz and standard deviation of 1.2 oz. What is the probability that the weight of a randomly chosen can is between 11.5 and 12.5 oz?*

We can summarize the example as follows,

- $X$ = the weight of a randomly chosen can of juice from the company.
- $X \sim N(12.02, 1.2)$.

This problem is solved by finding the bounds $(z_1, z_2)$ for a standardized normal random variable such that $P(11.5 < X < 12.5) = P(z_1 < Z < z_2) = P(Z < z_2) - P(Z < z_1)$. First, we have to calculate $z_2$ to find $P(Z < z_2)$:

$$z_2 = \frac{12.5 - 12.02}{1.2}$$
$$= 0.4$$

Second, we must calculate $z_1$ to find $P(Z < z_1)$:

$$z_1 = \frac{11.5 - 12.02}{1.2}$$
$$= -0.43$$

Therefore,

$$P(11.5 < X < 12.5) = P(-0.43 < Z < 0.4)$$
$$= P(Z < 0.4) - P(Z < -0.43)$$
$$= 0.32$$

Now, suppose that in this example the company is concerned about the production of cans with too much juice. Specifically, management determines that cans with 14 oz or more result in considerable cost. What is the probability that a can of juice from this company has 14 oz or more juice? We want to find z such that $P(X > 14) = P(Z > z)$:

$$z = \frac{14 - 12.02}{1.2}$$
$$= 1.65$$

Since the normal probability table is cumulative, we use the property:

$$P(Z > 1.65) = 1 - P(Z < 1.65) = 0.049$$

Hence, approximately 5% of the cans contain too much juice. If management believes that this number is too high, they can take steps to reduce the likelihood that a can contains a lot of juice. Possibilities include optimizing the procedure to minimize variability of the process or redesigning the machinery or equipment.

**Normal Percentiles** What if we want to find a limit for a normal random variable with a given probability? This value of the normal random variable is a percentile.

**Example 7.5**   *Returning to Example 7.4, what is the weight of a can such that only 10% of the cans have a weight equal to or less than?*

For this example, we must find the value $x$ of $X$ such that $P(X \le x) = 0.1$. One way to find this value is using the standardized distribution table. According to this table, we find that $P(Z \le -1.28) = 0.1$. That is, there is a 0.1 probability of obtaining a value less than or equal to $-1.28$. Now, we get the limit $x$ of interest as follows:

$$z = -1.28 = \frac{x - 12.02}{1.2}$$
$$\Rightarrow x = 12.02 - 1.28(1.2) = 10.48$$

The meaning of this is that $P(X \le 10.48) = 0.1$ and that 10% of cans have weight of 10.48 or less.

### 7.4.1   A Quick Look at Data Science: Normal Distribution, A Good Match for University of Puerto Rico SATs?

By design, SAT scores of all test takers will follow a normal distribution. This is not to say that the SAT scores of recently admitted students to a school follow a normal distribution. In this case study, our aim is to informally determine if the SAT scores of new admissions at the University of Puerto Rico, Carolina campus, are normally distributed. Establishing what the unknown SAT scores distribution can help calculate probabilities or model the scores. Data wrangling was applied to the original admissions data from 10 University of Puerto Rico campuses. Data set included academic year of admission, student's gender, high school of origin, campus the student was admitted to, GPA, admission index (*ai*), and place of residence. Data covers almost 70 000 undergraduate admissions for the academic years from 2009–2010 through 2013–2014. SAT scores[7] were calculated from the admission index.[8] Figure 7.14 recaps the data wrangling.

Our task requires probability density function estimation. A histogram is a very basic method to perform this estimation. Although simple, histograms have their disadvantages (see Chapter 3). We will also use a more sophisticated, smooth method, to estimate the pdf. In a nutshell, the method uses our sample to estimate an unknown pdf. Then, we will visually determine if the normal distribution is a reasonable candidate for this unknown pdf. Figure 7.15 shows both pdf estimates: the histogram and the smooth method. They indicate that the normal distribution indeed is a good candidate to the unknown pdf for SAT scores of students admitted to the University of Puerto Rico, Carolina campus.

---

7  Note: these are pre-2016 total SAT scores, with a minimum of 400 and a maximum of 1600.
8  The admission index of each student is $ai = 50gpa + \frac{0.5}{3}(sat - 400)$.

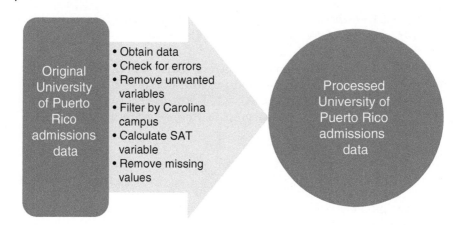

Figure 7.14 Steps taken to create a data set of University of Puerto Rico, Carolina campus, SAT scores for density estimation.

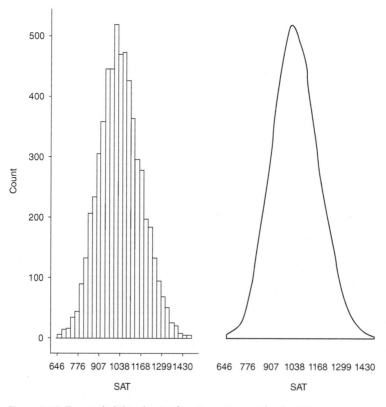

Figure 7.15 Two probability density function estimates for the SAT scores.

## Practice Problems

7.20 In the following, without computing the actual probabilities, state whether the probability is greater than 0.5, lower than 0.5, equal to 0.5, or if it is not possible to tell just by common sense. Explain your reasoning.[9]
a) $P(X \leq 612)$ when $X \sim N(725, 100)$.
b) $P(X > 11.44)$ when $X \sim N(15.05, 3.45)$.
c) $P(252 < X \leq 277)$ when $X \sim N(245, 25)$.

7.21 In the following, without computing the actual probabilities, state whether the probability is greater than 0.5, lower than 0.5, equal to 0.5, or if it is not possible to tell just by common sense. Explain your reasoning.
a) $P(X > 113)$ when $X \sim N(50, 3)$.
b) $P(59 \leq X \leq 81)$ when $X \sim N(65, 2)$.
c) $P(X \leq 7)$ when $X \sim N(30, 2)$.

7.22 The 2015 average commute time (in minutes) in a West Hills, Los Angeles, census tract was[10] 29.9. Let $X =$ commute time to work and $Z =$ the usual standardized random variable. Assuming a normal distribution, and a standard deviation of 5 minutes, to find the probability of taking at least 25 minutes, which is the right notation: $P(X \geq 25)$, or $P(Z \geq 25)$?

7.23 Determine if the probability notation is adequate in each case when $Z \sim N(0, 1)$. For
a) $X \sim N(32.5, 6.3)$, the probability that the random variable is less than 37.48 is $P(Z < 37.48)$.
b) $X \sim N(177, 9)$, the probability that the random variable is at least 164 is $P(Z \geq 164)$.
c) $X \sim N(12.25, 0.75)$, the probability that the random variable is greater than 11.90 is $P(X > 11.90)$.
d) $X \sim N(1500, 115)$, the probability that the random variable is no more than 1380 is $P(Z \leq -1.04)$.

7.24 Calculate each probability.
a) $P(X \leq 612)$ when $X \sim N(725, 100)$.
b) $P(X > 11.44)$ when $X \sim N(15.05, 3.45)$.
c) $P(252 < X \leq 277)$ when $X \sim N(245, 25)$.

---

9 $X \sim N(\mu, \sigma)$ stands for $X$ follows a normal distribution with mean $\mu$, standard deviation $\sigma$.
10 usc.data.socrata.com (accessed July 26, 2019).

7.25 Calculate each probability.
   a) $P(X > 113)$ when $X \sim N(50, 3)$.
   b) $P(59 \leq X \leq 81)$ when $X \sim N(65, 2)$.
   c) $P(X \leq 7)$ when $X \sim N(30, 2)$.

7.26 For each scenario, determine the value of $X$.
   a) The 10th percentile when $X \sim N(275, 10)$.
   b) The 90th percentile when $X \sim N(25, 2)$.

7.27 For each scenario, determine the value of $X$.
   a) Such that 70% of values of $X$ are greater and $X \sim N(30, 5)$.
   b) Such that there is a 66% chance that $P(-x \leq X < x)$ when $X \sim N(55, 3)$.

7.28 A company owns two stores. The monthly profit of the first store follows a normal distribution with mean \$10 000 and standard deviation \$1 000. The monthly profit of the second store follows a normal distribution with mean \$10 000 and standard deviation \$2 000. For both stores, the monthly profits are independent. Which store has a higher probability of having a profit of \$12 000 next month? (Computation not necessary here) Do you think your answer to the first question implies one store is better than the other? Why or why not?

7.29 A company owns two stores. The monthly profit of the first store follows a normal distribution with mean \$10 000 and standard deviation \$1 000. The monthly profit of the second store follows a normal distribution with mean \$8 000 and standard deviation \$1 000. For both stores, the monthly profits are independent. Which store has a higher probability of having a profit of \$12 000 next month? (Computation not necessary here) Do you think your answer to the first question implies that one store is better than the other? Why or why not?

7.30 The pressure of a type of tire follows a normal distribution with a mean of 30 pounds per square inch (psi) and a standard deviation of 2 psi. When the tire pressure is below 28 psi, it is underinflated. When the tire pressure exceeds 32 psi, it is overinflated. If the tire is under- or overinflated, it requires adjustment.
   a) Should the probability that the tire is underinflated be greater than 0.5 or lower than 0.5? (Computation not necessary here)
   b) What is the probability that the tire is underinflated?
   c) What is the probability that the tire is overinflated?

d) What is the probability that the tire has a correct inflation, $P(28 \leq X \leq 32)$?

e) What is the probability that the tire requires adjustment of its pressure?

7.31 The 2016 combined (math and critical reading) mean SAT scores for college-bound females was 987 and the standard deviation was[11] 162.64. The scores were normally distributed. If a college program includes a minimum combined score of 900 on the SAT among its requirements, what percentage of females do not satisfy that requirement?

7.32 If $X \sim N(80, 10)$, show that
a) the 25% percentile is 73.25.
b) the 90% percentile is 92.82.
c) 65% of $X$ values are smaller than 83.85.
d) 85% of $X$ values are greater than 69.64.

7.33 The exam scores of a course follow a normal distribution with a mean of 78 and a standard deviation of 11.
a) We are interested in the probability of a student getting a C or better (score of 70 or more). Based on the description of the problem, should this probability be greater than 0.5 or lower than 0.5? (Computation not necessary here)
b) Find the probability that a student gets a C or better in the exam.
c) What would be the grade of a student such that she is the 20th percentile?
d) What would be the grade of a student such that only 5% of the students obtain a better grade.

7.34 Refer to Example 7.4. The management determines that any can with a weight less than or equal to the 10% percentile is considered defective. What is the weight of a can such that 10% of the cans have the same or less weight?

7.35 A company extracts oil at a rig where the monthly weight of oil follows a normal distribution with mean 10 tons and standard deviation of 1.2 tons. Any monthly weight falling below the 25th percentile is considered not to be cost-effective. Find the weight that establishes the limit for cost-effectiveness.

---

11 Source: secure-media.collegeboard.org (accessed July 26, 2019).

## 7.5 Approximating the Binomial Distribution

When $X$ follows a binomial distribution and the sample size is large, it can be cumbersome to obtain probabilities. This is particularly true when the probability is over many values of $X$. Section 6.4 discussed how the shape of the binomial distribution is a function of $n$ and $\pi$. Figure 7.16 presents the shape of a two binomial distributions. For the first case, when $X \sim binomial(10, 0.25)$, the normal approximation to $P(X \leq 3)$ (in blue) misses the mark a bit, in part because this binomial distribution has too much right skewness. On the other hand, for the second case, when $X \sim binomial(60, 0.25)$, the normal approximation to $P(X \leq 15)$ (in blue) is good, in part because the binomial distribution is less skewed.

Generally, the normal approximation will be good as long as the binomial distribution is not too skewed to the left or the right. A preferred way to check this is by evaluating the expected successes and expected failures.

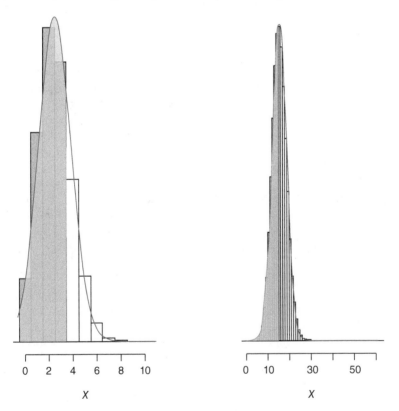

Figure 7.16 Binomial and normally approximated probabilities. On the left $P(X \leq 3)$ when $X \sim binomial(10, 0.25)$, on the right $P(X \leq 15)$ when $X \sim binomial(60, 0.25)$.

---

**Rule of Thumb**

Let $X \sim binomial(n, \pi)$. We will use the rule of thumb[12] $n\pi \geq 15$, and $n(1 - \pi) \geq 15$. When both these conditions hold, then we will say that $X$ is approximately normally distributed, with $\mu = n\pi$ and $\sigma = \sqrt{n\pi(1 - \pi)}$.

---

**Example 7.6**  *Suppose that for a large drug company, 30% of drugs undergoing clinical tests will be approved for the market. Sixty drugs (that are unrelated to each other) are being tested. What is the probability that less than 20 will be approved for the market?*

Let $X$ be the number of drugs that approved for the market out of the 60 being tested. $X \sim binomial(60, 0.3)$ because

- $X$ is discrete with possible values $0, 1, \ldots, n$.
- $n$ is a constant.
- $\pi$ is also fixed.
- drug approvals are independent.
- two possible outcomes per test: approval or non-approval.

Also, $n\pi = 18 > 15$ and $n(1 - \pi) = 42 > 15$, so the normal approximation should be good. Since $E(X) = 18, \sigma = \sqrt{18 * (1 - 0.30)}$

$$Z = \frac{20 - 18}{\sqrt{18 * (1 - 0.30)}} = 0.56$$

Therefore, $P(X < 20) = P(Z < 0.56) = 0.71$. Keep in mind that the insinuation here is that a distribution meant for continuous random variables, the normal, can be applied to find probabilities for a discrete random variable. The actual implementation requires what is called a **continuity correction**, which involves subtracting 0.5 or adding 0.5 to the limit in the probability, depending on the type of inequality. Returning to Example 7.6, the true probability (using the binomial distribution) is 0.669. Using the continuity correction, the normal approximation probability would be 0.66, closer than the 0.71 found above. Computer software can easily implement the continuity correction for us, and computing it manually has no educational value and is thus not discussed any further.

---

12  Other textbooks have the rule of thumb $n\pi \geq 5$, and $n(1 - \pi) \geq 5$ or $n\pi \geq 10$, and $n(1 - \pi) \geq 10$. However, some researchers have found these rule of thumbs are not conservative enough.

## Practice Problems

7.36 Among residents of a region, 35% live in and around a big city. If 50 residents are chosen at random, the exact probability that no more than 19 live in and around the big city can be found to be 0.7264. Find the normal approximation of this result. (Do not use continuity correction. With continuity correction, it would be 0.723 409 2.)

7.37 Suppose the owner of a store knows that 15% paying costumers buy clothing from a well-known brand. If he chooses 150 paying customers at random, what is the probability that no less than 45 customers purchased clothing from the well-known brand?

7.38 Among employed women, 25% have never been married. Select 500 employed women at random.
a) What is the approximate probability that at least 150 have never been married?
b) What is the probability that 120 or less have never been married?

7.39 An investment company claims that only 10% of their portfolios give a negative return on investment in a 1-year span. You take a random sample of 150 of their portfolios and find that 20 had a negative 1-year return:
a) Show that the number of portfolios that have a negative 1-year return follow a binomial distribution.
b) Find the probability that at least 20 portfolios have a negative 1-year return using a normal approximation.
c) Your team has decided that if $P(X \geq 20)$ is lower than 0.01, then the investment company's claim that only 10% of their portfolios give a negative return to investment in a 1-year span should be rejected. What should your team decide based on the probability found in 7.39b? Explain.

7.40 A company that develops motorbike batteries knows that 10% of the batteries are defective. 300 batteries are randomly sampled:
a) Show that $X$ = number of defective batteries out of 300 follows a binomial distribution.
b) What is the probability that of the next 300 batteries, 20 or less will be defective?

## 7.6 Exponential Distribution

The uniform and normal distribution are symmetric distributions. Some continuous random variables have **asymmetric distributions**. The **exponential**

**distribution** has high right skewness and is used for strictly positive random variables. Specifically, the exponential distribution is often used to model time between events. Say the time between the arrival of two customers at a store or times between failure of a part of a product. In fact, recall that to model arrivals at independent time intervals we used the Poisson distribution in Chapter 6. It can be shown that the time between arrivals in a Poisson model follows an exponential distribution. Other applications of the exponential distribution is to model distance between events.

Fortunately, when $X$ is exponentially distributed, there is an easy expression to find probabilities,[13]

$$P(X > x) = e^{-\lambda x}$$

(see Figure 7.17). From properties of probability distribution functions, note that $P(X > x) + P(X \leq x) = 1$.

Figure 7.17 Probability when $X$ is exponentially distributed. The left panel shows the equation and the area under the exponential curve for $P(X > x)$, the right panel shows the equation and the area under the exponential curve for $P(X \leq x)$.

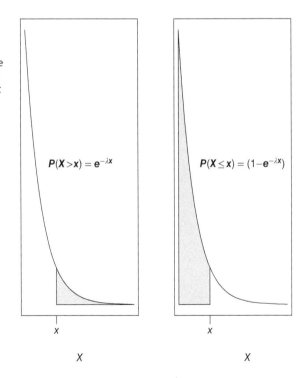

$$P(X > x) = e^{-\lambda x}$$

$$P(X \leq x) = (1 - e^{-\lambda x})$$

---

13 The exponential distribution is sometimes parametrized using $\mu = \frac{1}{\lambda}$. Then, $P(X > x) = e^{-x/\mu}$ and $E(X) = \mu$. $\mu$ is the mean time per arrival. In practice, both parameterizations are used and we should check if we are given mean arrival per unit of time ($\lambda$ which is what we use in this book) or mean time per arrival ($\mu$) to determine how to find probabilities.

Table 7.1 Main properties of a random variable X that follows an exponential distribution.

| Property | |
| --- | --- |
| Cumulative distribution function | $P(X \le x) = 1 - e^{-\lambda x}$ for $x > 0$ |
| $E(X)$ | $\dfrac{1}{\lambda}$ |
| $Var(X)$ | $\dfrac{1}{\lambda^2}$ |

**Example 7.7**  *The waiting time when calling a company for service follows an exponential distribution with mean of 5 minutes, $\lambda = 1/5$. What is the probability a customer will have to wait over 15 minutes before their call is attended?*

According to the Example, if $X$ = waiting time for service call to be attended (in minutes), then $X \sim exp(1/5)$. So $P(X > 15) = e^{-15/5} = 0.05$.

Table 7.1 summarizes some properties for a random variable that follows an exponential distribution. Percentiles are found using the same premise as the uniform and normal distributions: the probability is known and we must find the value of $x$ that gives the probability.

There are far too many probability distributions for continuous random variables to cover them all in this book. Drawing inference on the variance of the population, comparing two population variances, and inferring on the population mean without knowing $\sigma$ are some situations where new continuous distributions will be introduced later on.

## Practice Problems

7.41  If $X \sim exp(1/2)$, show that
a)  $P(X > 1) = 0.61$.
b)  $P(X \le 0.5) = 0.22$.
c)  $P(0.5 \le X \le 1) = 0.17$.
d)  The ‰ is 1.39.

7.42  The time waiting in line at a bank follows and exponential distribution with mean of 2 minutes, $\lambda = 1/2$.
a)  What is the probability a customer will have to wait over 10 minutes to be attended?

b) What is the probability a customer will have to wait no more than 2 minutes to be attended?

c) Calculate the variance of the waiting time while at the bank.

7.43 You and your partner are the owners of a company that specializes in remodeling houses and interior design. You would like to track the time between complaints (in months) but do not know how. Your daughter (thanks to her Business Stats class) tells you it turns out the time intervals between complaints can be modeled using an exponential distribution with $\lambda = 0.57$.

a) Find the probability that it will be at least one month from today's complaint until the next complaint appears.

b) Find the probability that it will take between 0.5 and 0.75 months for the next complaint.

7.44 A laptop manufacturer produces laptops with a mean life of seven years. What warranty should they offer such that no more than 25% of the laptops will fail before the warranty expires? Assume laptop failure times follow an exponential distribution ($\lambda = 1/7$).

## Chapter Problems

7.45 Determine if the following defines a discrete or continuous random variable:

a) Time it takes to complete a product in an assembly line.

b) Tire pressure after a stress and safety test.

c) Number of students taking an exam.

d) Total monthly desk sales at a store.

7.46 Which of the following functions represent a probability density function for a continuous random variable?

a) $f(x) = 0.25$ where $0 \le x \le 4$.

b) $f(x) = 3 - x^3$ where $-1 \le x \le 5$.

7.47 In Example 7.1, find $P(6 < X < 20)$.

7.48 The travel time for a university student between her home and her favorite nightclub is uniformly distributed between 15 and 25 minutes. Find the probability that she will finish her trip in 22 minutes or more.

7.49 From the following illustrations, determine if the probability is greater than 0.5, lower than 0.5, or cannot be sure.

(a)                              (b)                              (c)

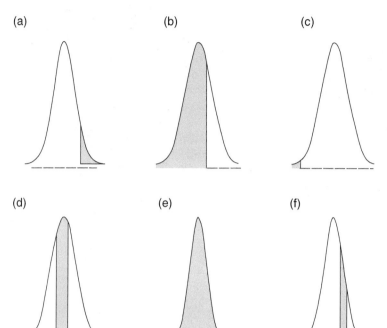

(d)                              (e)                              (f)

7.50 Show that for a normal distribution, the probability within
   a) 1.28 standard deviations from the mean is 0.80.
   b) 1.96 standard deviations from the mean is 0.95.
   c) 2.57 standard deviations from the mean is 0.99.

7.51 For a normally distributed random variable, find the probability of falling
   a) more than 2.33 standard deviations above the mean.
   b) within 0.33 standard deviations from the mean.

7.52 For a normally distributed random variable, find the $z$-score that is 0.75 standard deviations from the mean.

7.53 The weight of a model of computer laptops, $X$, is normally distributed with a mean of 2.8 pounds and a standard deviation of 0.1 pounds. What argument would you use to state that $P(X \leq 2.6) = P(X \geq 3.0)$ without doing the calculation?

7.54 Suppose you are given the option of choosing from two production procedures. Both have possibilities that follow a normal distribution. One has mean profit of $280 and standard deviation $10. The other has mean $280 and standard deviation $40.

a) Which one would you choose if you want the lowest chance of a profit of less than $200?

b) Say that the first production plan has mean profit $280 and standard deviation $10, but that the second has mean $300 and standard deviation of $80. Which one would you choose, if you prefer the production plan with the lowest probability of earning less than $240.

7.55 Scores on the Wechsler Adult Intelligence Scale (a standard "IQ test") for the 20–34 age group are approximately normally distributed with a mean of 110 and a standard deviation of 25. What score does a person aged 20–34 need to be the 98th percentile?

7.56 There is a situation where you wish to find $P(X < 3)$ for some continuous random variable $X$. You assume that the random variable $X$ follows a normal distribution with mean 10 and standard deviation of 1. However, it turns out that $X$ has a distribution with high right skewness with the peak of the density curve at $X = 3$ (a truth that you don't know). Explain how mistakenly using the normal distribution will affect the calculation of $P(X < 3)$.

7.57 Lightning strikes are a forest fire hazard. Therefore, it is important to model lightning strike patterns within a region. If time between lightning strikes for a region follows an exponential distribution with a rate of 5 per hour:

a) What is the probability that the amount of time that passes between two consecutive lightning strikes is more than two minutes?

b) What is the probability that the amount of time that passes between two consecutive lightning strikes is less than one minute?

c) Find the probability that the amount of time that passes between two consecutive lightning strikes is between one and two minutes.

7.58 The time to produce a ring for an engine is exponentially distributed with a $\lambda = 0.1$ minutes.

a) Find the probability that it will take more than 12 minutes to produce the ring.

b) Find the probability that it will take less than 2 minutes to produce the ring.

## Further Reading

Agresti, Alan and Franklin, Christine. (2016). *Statistics: The Art and Science of Learning from Data*, 4th edition. Pearson.

Ruppert, D. (2004). *Statistics and Finance*. Springer.

Wackerly, Dennis D., Mendenhall III, William and Scheaffer, Richard L. (2002). *Mathematical Statistics with Applications*, 6th edition. Duxbury.

# 8

# Properties of Sample Statistics

## 8.1   Introduction

Several statistics have been presented in the previous chapters. The main ones are $\overline{X}$, $s$, and the sample proportion, $\hat{p}$. Recall that $\overline{X}$ is used to estimate the population mean $\mu$, $s$ is used to estimate the population standard deviation $\sigma$, and $\hat{p}$ is used to estimate the probability of success at the population level $\pi$. Also, contrary to a parameter, such as $\mu$, which is fixed, a statistic, such as $\overline{X}$, is random: we obtain different results whenever a different sample is used.

For statistical inference, a probabilistic characterization of the values of the statistic is needed. To accomplish this, we must determine the expected value, standard deviation, and distribution of these statistics. Since the probability distribution represents many potential samples, it is referred to as the **sampling distribution**.

Also, how do we decide to use a statistic among "many alternative statistics"? That is, if we are in a situation where we can use $\overline{X}$ or the first observation of a random sample $X_1$ to estimate $\mu$. Which one of the two should we choose? To answer these questions, we will need to determine the properties that we would like our statistic to have. We establish arguments here that help explain why some statistics are frequently used in practice. We will return to the situation of choosing between $\overline{X}$ and $X_1$ several times. Before diving into these topics, we define some necessary terminology.

**Estimator**: Sometimes referred to more precisely as a point estimator; it is the abstract notation of a sample statistic. For example, $\overline{X} = \frac{1}{n} \sum_{i=1}^{n} x_i$, $\hat{p} = \frac{Y}{n}$ (where $Y$ is the number of successes in an experiment with $n$ observations), and $s = \sqrt{\frac{\sum_{i=1}^{n} (X_i - \overline{X})^2}{n-1}}$ are three estimators. Note that, being a statistic, estimators are random.

**Estimate**: It is the result of applying an estimator to observations from a sample. For example, $\overline{x} = 5$, $s = 2$, and $\hat{p} = 0.27$ are three estimates obtained by

*Principles of Managerial Statistics and Data Science*, First Edition. Roberto Rivera.
© 2020 John Wiley & Sons, Inc. Published 2020 by John Wiley & Sons, Inc.
Companion website: www.wiley.com/go/principlesmanagerialstatisticsdatascience

applying the respective estimators to data from some sample. Different than estimators, an estimate is a specific number.

**Population Distribution:** The probability distribution from where the sample comes from. Usually, the goal is to reach a conclusion about the parameters of the population distribution. Frequently, the population distribution is unknown.

**Data Distribution:** The probability distribution of the sampled data. The larger the sample size $n$, the closer the data distribution resembles the population distribution.

**Sampling Distribution:** The probability distribution of a sample statistic. It represents how the sample statistic varies over many equivalent studies of same sample size. The sampling distribution plays a crucial role in inferring about parameters. The standard deviation of a sampling distribution is known as the **standard error**.

## 8.2  Expected Value and Standard Deviation of $\overline{X}$

$\overline{X}$, a random variable, is an estimator of $\mu$, a constant. The meaning of this is that the value of $\overline{X}$ will change for each sample of size $n$.

**Example 8.1**  *The new SAT exams (with a total back down to 1600) are designed to have an expected score of 1000. Assume the population standard deviation is $\sigma = 100$. Suppose a random sample of 10 SAT scores is taken. The sample results in an average of 989.42. Hence, this sample estimates that the true mean SAT score of the population is 989.42. Now, say that 4 additional random samples (each of sample size $n = 10$) are taken which give the following results: $\overline{x}_2 = 1042.10$, $\overline{x}_3 = 950.22$, $\overline{x}_4 = 995.69$, and $\overline{x}_5 = 989.50$. The different estimates of $\mu$ are due to different students being part of every sample. It is appropriate to ask, what is the expected value of $\overline{X}$ for each sample of size n?*

We can answer this question easily using the properties of expected value (Section 6.3).

$$E(\overline{X}) = E\left(\frac{1}{n}\sum_{i=1}^{n} X_i\right)$$

$$= \frac{1}{n}E\left(\sum_{i=1}^{n} X_i\right)$$

The last equality is due to $E(aY) = aE(Y)$ for any constant $a$ (in this case, $a = 1/n$ and $Y = \sum_{i=1}^{n} X_i$). Now, recall that one property of expected values is that the expected value of a sum of random variables is equal to the sum of the expected values. Therefore,

$$\frac{1}{n}\left(E(X_1 + X_2 + \cdots + X_n)\right) = \frac{1}{n}\left(E(X_1) + E(X_2) + \cdots + E(X_n)\right)$$

Finally, aware that each observation comes from the same population, the expected value of each observation is $\mu$,

$$\frac{1}{n}\left(E(X_1) + E(X_2) + \cdots + E(X_n)\right) = \frac{1}{n}(\mu + \mu + \cdots + \mu)$$

$$= \frac{1}{n}(n\mu) = \mu$$

Similarly, through the property $Var(aY) = a^2 Var(Y)$ (in this case, $a = 1/n$ and $Y = \sum_{i=1}^{n} X_i$) and the fact that observations are independent,[1] if the population has a standard deviation $\sigma$, we can show that

$$\sigma_{\overline{x}} = \frac{\sigma}{\sqrt{n}}$$

$\sigma_{\overline{x}}$ is the standard error of the sample mean. In summary, we know that

- $\overline{X}$ will have the same expected value ($\mu$) of the population where the sample comes from.
- The standard deviation of the sample mean will be a fraction of the standard deviation of the population.
- $\sigma_{\overline{x}} = \frac{\sigma}{\sqrt{n}}$ indicates that as the sample size increases, the variability of $\overline{X}$ decreases.

Figure 8.1 helps appreciate this effect. Moreover, it is also seen that at first, unit increments in $n$ drastically decrease $\sigma_{\overline{x}}$, but increase the sample size after 30 observations or so have a smaller effect in decreasing $\sigma_{\overline{x}}$.

Figure 8.1 $\sigma_{\overline{x}}$, the standard deviation of $\overline{X}$, as $n$ increases and $\sigma = 1$.

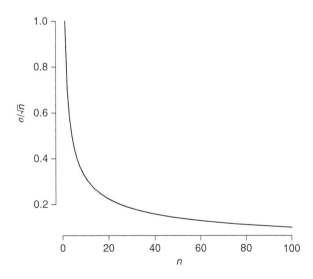

---

1 When the population is small or the sample size is large relative to the population size, it can no longer be argued that all elements from the population are equally likely to be selected. Then, a finite population correction is necessary.

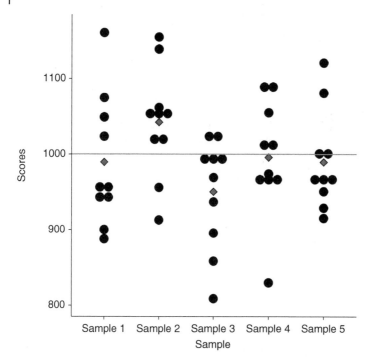

Figure 8.2 Dotplots for 5 samples of 10 observations each. Mean of each sample appears as a blue diamond, and the population mean, $\mu = 1000$, appears in red.

Figure 8.2 illustrates the results of the five samples in Example 8.1. Note that observations (black dots) are approximately as likely to be above $\mu$ (red line) and as below $\mu$. The same can be said for the mean of each sample (blue diamond). More importantly, although sometimes individual observations may be very close to $\mu$, on other occasions they can be far above or far below $\mu$. In contrast, sample means tend to be relatively closer to $\mu$. That is, $\overline{X}$ has a smaller variance. Figure 8.2 presents empirical evidence of the difference between the standard deviation of $\overline{X}$, which is $\sigma/\sqrt{n}$, and the standard deviation of an individual observation, $\sigma$. The larger the sample size, the smaller the standard deviation of $\overline{X}$.

## Practice Problems

8.1 If a population follows a normal distribution with mean 10 and standard deviation 2, show that

 a) if $X$ is the value of a randomly sampled observation from this population, it has an expected value of 10.

b) if a sample of 25 observations is taken, its sample mean also has an expected value of 10.
c) if $X$ is the value of a randomly sampled observation from this population, it has a standard deviation of 2.
d) if a sample of 25 observations is taken, its sample mean has a standard deviation of 0.4.
e) if a sample of 36 observations is taken, its sample mean has a standard error of 0.33.

8.2 For a population with mean 25 and standard deviation 4, which sample size would lead to the smallest variance for $\overline{X}$, $n = 100$ or $n = 400$?

8.3 Find the standard error of $\overline{X}$ for each random sampling situation.
a) $\sigma = 6, n = 10$
b) $\sigma = 6, n = 40$
c) $\sigma = 6, n = 60$
d) Do you need to assume the population is normally distributed for the computations in this problem?

8.4 The company BEEFJERK produces 100 g of beef jerky with a mean saturated fat of 10 g and a standard deviation of 0.5 g. For
a) one serving of 100 g, what is the expected value of $\overline{X}$?
b) one serving of 100 g, what is the standard error of $\overline{X}$?
c) a random sample of 25 servings of 100 g, what is the expected value of $\overline{X}$?
d) a random sample of 25 servings of 100 g, what is the standard error of $\overline{X}$?

## 8.3 Sampling Distribution of $\overline{X}$ When Sample Comes From a Normal Distribution

Besides $E(\overline{X})$ and $\sigma_{\overline{x}}$, we must also characterize the distribution of $\overline{X}$. The sampling distribution allows probability calculations on values of $\overline{X}$. First, let us state a well-known statistics theorem,

**Theorem 8.1** *If a random sample comes from a population of interest with a normal distribution,[2] with mean $\mu$, and standard deviation $\sigma$, then* $\overline{X} \sim N\left(\mu, \frac{\sigma}{\sqrt{n}}\right)$. □

---

2 It can be shown that any linear combination of random variables that follow a normal distribution will also follow a normal distribution.

As remarked in Section 8.2, although $\overline{X}$ has an expected value of $\mu$, its standard deviation is smaller than the one from the population. $\overline{X}$ having a smaller standard deviation was implied by Example 8.1; it means that $\overline{X}$ values are generally closer to $\mu$ than single observations.

**Example 8.2** *Let's return to Example 8.1 where SAT scores follow a normal distribution with $\mu = 1000$ and $\sigma = 100$. Five thousand samples, each of size 10 are taken.*

Figure 8.3 shows the theoretical and empirical distributions of the sample mean and the population distribution. The empirical distribution (red frequency bars) follows closely the theoretical distribution (black curve) according to Theorem 8.1. Moreover, the arguments stated in Section 8.2 are still apparent, as the theoretical and empirical distributions of $\overline{X}$ both have smaller variance than the population distribution (blue line). Thus, any single observation from the population is more likely to be further above or below $\mu$.

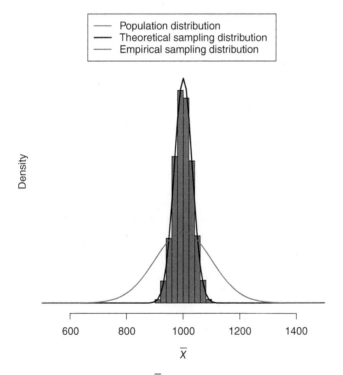

**Figure 8.3** The distribution of $\overline{X}$ according to Theorem 8.1 is presented in black. Also, 5000 samples with $n = 10$ are simulated and the empirical distribution of all the sample means is seen in red. The distribution of the population from where all the samples were taken is seen in blue.

> **Different Normal?**
> Many are often confused on why $\overline{X}$ does not have the exact same normal distribution as the population where the sample came from. One way of looking at it is that the normal distribution of the population is representative of all the possible values of the population, while the normal distribution of $\overline{X}$ is representative of all the possible sample means estimating $\mu$.

Theorem 8.1 is extremely convenient because given that $\overline{X} \sim N(\mu, \frac{\sigma}{\sqrt{n}})$, when $\sigma$ is known, if we define

$$Z = \frac{\overline{X} - \mu}{\sigma/\sqrt{n}} = \frac{\sqrt{n}(\overline{X} - \mu)}{\sigma} \tag{8.1}$$

then $Z \sim N(0, 1)$. The first version of the formula in Eq. (8.1) is the theoretical version for standardization. However, when performing the computations by hand, it may be easier to use the second formula shown. This transformation allows us to easily find probabilities of values of $\overline{X}$ through the standardized normal distribution. We can compare the $Z$ Formula (8.1) with the one used in Section 7.4. Basically, Formula (7.1) is a special case of Formula (8.1) above, when $n = 1$, so Eq. (8.1) generalizes the use of $Z$ for standardizing a sample mean of any size $n$.

**Example 8.3** *Continuing with Example 8.1, where SAT scores follow a normal distribution with $\mu = 1000$, $\sigma = 100$, and a random sample of 10 scores is obtained. Find $P(\overline{X} \leq 984)$.*

Based on Theorem 8.1, the average SAT from a random sample follows an $N(1000, \frac{100}{\sqrt{10}})$. This information allows us to determine if a probability is relatively large (greater than 0.5) or relatively small (smaller than 0.5). Considering that this normal distribution has a mean of 1000 and it is a symmetric distribution, Figure 8.4 illustrates as an area under the normal pdf what the probability of interest is, and suggests that $P(\overline{X} \leq 984)$ is smaller than 0.5.
Now,

$$Z = \frac{\sqrt{10}(984 - 1000)}{100} = -0.506$$

and from Appendix B, we find $P(\overline{X} < 984) = P(Z < -0.506) = 0.306$ (Figure 8.5), confirming that the requested probability is smaller than 0.5.

Let's return to the comparison of $X_1$ and $\overline{X}$ as estimators of $\mu$. Both will have the same expected value $\mu$, but $X_1$ has a standard deviation of $\sigma$, while $\overline{X}$ has a smaller standard deviation, $\sigma/\sqrt{n}$. Naturally, if two estimators have the same

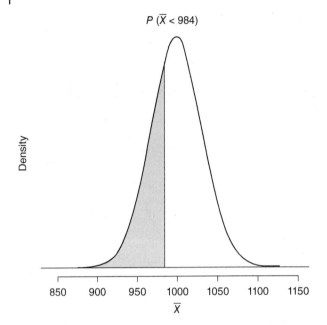

Figure 8.4 Area under the curve representing the probability of interest, $P(\overline{X} < 984)$ when $\overline{X} \sim N(1000, \frac{100}{\sqrt{10}})$

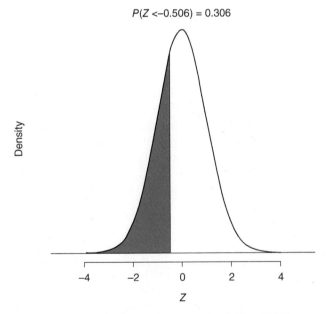

Figure 8.5 Area under the curve representing $P(Z < -0.506)$.

expected value, we prefer the one that has the smallest variability. Note that if the random sample comes from a population with a normal distribution, we need $\sigma$ to be known to apply the $Z$ standardization procedure. If we do not know $\sigma$, it must be estimated. Although a standardization statistic can still be obtained, the estimation of $\sigma$ will add uncertainty to it, increasing the chances of values further away from $\mu$.

## Practice Problems

8.5   Given a normal population with mean 250 and standard deviation 10,
   a) should $P(\overline{X} \leq 259)$ be greater or smaller than 0.5?
   b) for which sample size, $n = 30$ or $n = 20$ would $P(\overline{X} \leq 259)$ be greater? (*Hint:* it can be explained without computing both probabilities.)
   c) for which sample size, $n = 30$ or $n = 20$, would $P(\overline{X} > 265)$ be greater? (*Hint:* it can be explained without computing both probabilities.)

8.6   Determine if the probability notation is adequate in each case when $Z \sim N(0, 1)$. For a random sample from
   a) $N(32.5, 6.3)$, where $n = 32$, the probability that the sample mean is less than 37.48 is $P(Z < 37.48)$.
   b) $N(12.25, 0.75)$, where $n = 50$, the probability that the sample mean is greater than 11.90 is $P(\overline{X} > 11.90)$

8.7   Determine if the probability notation is adequate in each case when $Z \sim N(0, 1)$. For a random sample from
   a) $N(177, 9)$, where $n = 32$, the probability that the sample mean is at least 164 is $P(Z \geq 164)$.
   b) $N(1500, 115)$, where $n = 60$, the probability that sample mean is no more than 1480 is $P(Z \leq -1.35)$.

8.8   According to the National Center for Education Statistics, the mean costs (fees, room, and board rates) for a four-year public university in the United States was \$15 640 for the 2013–2014 academic year. Assume the costs follow a normal distribution with $\sigma = 2000$. If a random sample of 60 costs for four-year public university was taken that year,
   a) what is the expected value of the sample mean?
   b) what is the standard deviation of $\overline{X}$?
   c) what is the probability that the sample mean is over \$17 000?
   d) what is the probability that the sample mean is within \$1000 of the population mean?

8.9  Returning to Problem 8.8, suppose a researcher wanted to determine if the four-year public university mean cost for 2015–2016 was higher than $15 640. She takes a sample of 100 public schools and gets an average cost of $15 850.

 a) Find $P(\overline{X} > 15\ 850)$.

 b) The researcher has determined that if $P(\overline{X} > 15\ 850)$ is smaller than 0.1, then she will determine the true mean cost in 2015–2016 was higher than in 2013–2014. What should they decide based on the result of 8.9a?

8.10  The kilowatt/hour (kWh) generated by a solar system of 12 panels from (12–1 p.m.), during the summer months, has a normal distribution with mean 2.25 and a standard deviation of 0.5. If 10 solar systems of this kind are randomly sampled, find

 a) the probability that the sample average is 3 kWh or more.

 b) the probability that the sample average is between 2.09 and 2.41 kWh.

8.11  Continuing with Example 8.1, where SAT scores follow a normal distribution with $\mu = 1000$ and $\sigma = 100$ and a random sample of 10 scores is obtained, find

 a) $P(\overline{X} > 984)$.

 b) $P(\overline{X} > 1007)$.

 c) $P(984 \leq \overline{X} < 1007)$.

 d) $P(\overline{X} < 1030)$.

 e) $P(\overline{X} < 800)$.

8.12  Continuing with Example 8.3, where SAT scores follow a normal distribution with $\mu = 1000$ and $\sigma = 100$ and a random sample of 10 scores is obtained,

 a) find the 25‰.

 b) $\overline{x}$ such that 5% of the sample means could be greater.

## 8.4   Central Limit Theorem

In Section 8.3, we established that when a random sample comes from a normally distributed population with mean $\mu$ and standard deviation $\sigma$, $\overline{X}$ also has a normal distribution with mean $\mu$ and standard deviation $\sigma/\sqrt{n}$. However, typically we either do not know the distribution of the population or the population distribution is not normally distributed. The central limit theorem provides us with another reason to use $\overline{X}$ to estimate $\mu$.

**Theorem 8.2 (*Central Limit Theorem*)**   *Suppose we have a random sample from a population with an arbitrary distribution, with mean $\mu$ and standard* deviation $\sigma$. Then, the distribution of $\overline{X}$ will tend to an $N(\mu, \sigma/\sqrt{n})$ as $n$ increases. □

Essentially, the theorem tells us that as long as the sample size is large enough, $\overline{X} \sim N(\mu, \sigma/\sqrt{n})$, no matter what the distribution of the population is. In fact, the central limit theorem applies even for discrete random variables. The central limit theorem is the reason why probabilities of random variables with a binomial distribution can be approximated using the normal distribution.[3]

**Example 8.4** *A simulation of many random samples from a uniform distribution with an expected value of 7.5 and from an exponential distribution with an expected value of 1 are obtained. In each random sample of sizes* n = 2, 10, 50, *the sample mean is obtained. Figure 8.6 elucidates what happens to the distribution of $\overline{X}$ as the sample size increases.*

On the left panel, Figure 8.6 presents the sampling distribution of $\overline{X}$ for the samples taken from the uniform distribution. The histogram is the empirical distribution obtained from all the $\overline{X}$ from samples of size $n$, the red curve is the normal distribution. When $n = 2$, the histogram displays a symmetric form but not exactly a normal distribution, there is too much variability. When $n = 10$, the histogram closely follows the shape of the normal distribution, and the normal approximation is even better when $n = 50$. On the right panel of Figure 8.6, the corresponding sampling distribution for samples from the exponential distribution are presented. These results also give support to the central limit theorem. However, the rate at which the normal approximation is good is shown to be slower than in the case of samples from the uniform distribution. At $n = 2$, the histogram is very skewed. At $n = 10$, there is some improvement but still some visible right skewness. By $n = 50$, the skewness has been reduced but it is still apparent.

As we can see from the empirical distributions, in both cases, $\overline{X}$ tends to have a more normal distribution as $n$ increases. But Example 8.4 indicates that care is needed. If you take a careful look at what the central limit theorem says, you will notice that it does not say when the sample size is big enough for $\overline{X}$ to be normally distributed in general. The central limit theorem cannot make such a statement, because the adequate sample size will depend on the distribution of the population and recall that we may not even know what type of probability distribution the population has.

---

3 The central limit theorem also applies to the sum of random variables, not just their mean.

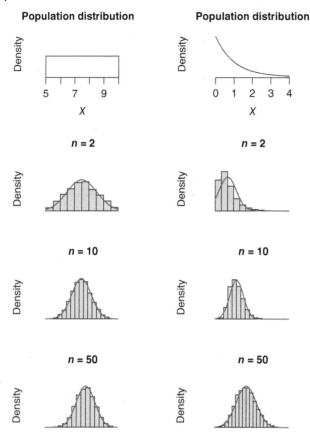

Figure 8.6 The left column shows the empirical distribution of the sample mean obtained from a random sample from a uniform distribution for different sample sizes. The right column presents the same, but for random samples obtained from an exponential distribution.

---

**Central Limit Theorem's Rule of Thumb**

As a rule of thumb, $n \geq 30$ is often sufficient to make the central limit theorem applicable. However, the more skewed the population distribution, the larger the sample size needed for the central limit theorem to apply.

---

Considering the central limit theorem, $\overline{X}$ will frequently follow approximately a normal distribution and therefore the trick of transforming to $Z$, which has a standard normal distribution applies. Figure 8.7 illustrates an approach that can be used when determining if the $Z$ table can be used to find probabilities or percentiles associated to $\overline{X}$.

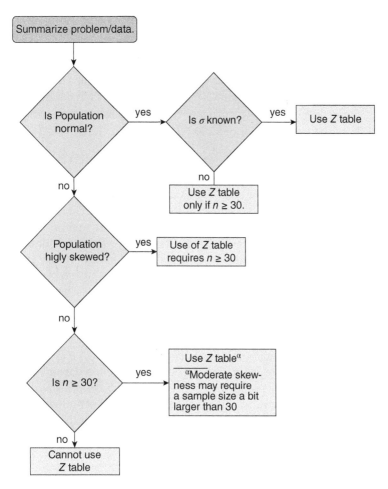

Figure 8.7 A flowchart for the steps to determine if use of the standard normal is possible to find probabilities or percentiles associated to $\overline{X}$.

$Z$ is a function of the standard deviation of the population. If the population is normal and $n \geq 30$, for the moment, we will rely on the central limit theorem to use the $Z$ table. However, we will discuss how to deal with this scenario more precisely later on.

**Example 8.5** *Joe's restaurant offers all you can eat Thursdays for $15 per customer. Management is aware that the expenses per customer, which accounts for the amount of food customers eat and other costs, is right skewed with a mean of*

*$14.95 and a standard deviation of $2.50. If 300 customers come in next Thursday and can be assumed to be a random sample what is the probability that the restaurant will make a profit?*

The restaurant will make a profit if the sample mean expense is less than $15. Let $\overline{X}$ be the sample mean expense of the 300 customers next Thursday. By the central limit theorem, $\overline{X}$ follows approximately a normal distribution with a mean of $14.95 and a standard error of $2.50/\sqrt{300} = 0.144$. Thus,

$$z = \frac{15 - 14.95}{0.144} = 0.35$$

$P(\overline{X} < 15) = P(Z < 0.35) = 0.64$.

**Central Limit Theorem in Practice**   The central limit theorem is a fascinating theoretical result that is used to argue over the general use of the $\overline{X}$ to draw inference on $\mu$. It has been applied successfully countless times. The principles of the central limit theorem were first introduced in the eighteenth century and generalized at the beginning of the twentieth century. With the advent of computers, these days the need of the central limit theorem is somewhat diminished. Yet, the theorem is still regularly used in practice. Moreover, recall that the central limit theorem does not say when the sample size is big enough for $\overline{X}$ to be normally distributed in general, and we have shown that the rate at which the distribution of $\overline{X}$ tends to follow a normal distribution as $n$ increases depends on the how skewed the population distribution is. Hence, normality of $\overline{X}$ should always be double-checked in practice, even when $n \geq 30$. A histogram can be used to evaluate the skewness of the data or the presence of outliers. The higher the skewness, the larger the $n$ required for the central limit theorem to apply.

The central limit theorem is generally applied to find probabilities and percentiles. If we rely on the central limit when it does not apply, calculated statistics may be incorrect (see Section 8.4.1).

Do not forget that we are assuming that the observations are independent. For this assumption to hold, a random sample must be drawn and the sample must account for no more than 10% of the population size. A more subtle assumption is that the population mean is constant.

---

**Central Limit Theorem and Big Data**

Lastly, a final warning is appropriate in the age of big data. With some data sets having millions of observations, it is tempting to assume that the central limit theorem will immediately apply. This may be true, but one must keep in mind that although it is often helpful to work with the expected value of a random variable, as a measure of centrality, $\mu$ is not always best. Again, we do not always know the distribution of the population. If

- the distribution of the sample observations is highly skewed,
- centrality of the population is the main interest, and
- there are no arguments for extreme values to be discarded (i.e. arguments such as they are erroneous or not representative of the population of interest),

then we should also question the use of the sample mean over the sample median. As we have seen in Chapter 4, $\overline{X}$ and the median will be quite different when the data distribution is asymmetric and the median will be a better descriptor of centrality or typical values of the random variable.

***Approximating the sampling distribution of $\hat{p}$*** If $Y$ counts the number of successes out of $n$ attempts and $Y$ follows a *binomial*$(n, \pi)$, the statistic $\hat{p} = \frac{Y}{n}$ is used to estimate the probability of success. Then,

$$E(\hat{p}) = \pi$$

and

$$Var(\hat{p}) = \frac{\pi(1 - \pi)}{n}$$

The standard deviation of $\hat{p}$ is also called the **standard error** of the sample proportion. Although $\hat{p} = Y/n$ does not follow a binomial distribution,[4] fortunately we can still use the normal approximation as long as $n\pi \geq 15$ and $n(1 - \pi) \geq 15$. Essentially, this is a special application of the central limit theorem. In summary, if $n\pi \geq 15$ and $n(1 - \pi) \geq 15$, then $\hat{p}$ follows an $N\left(\pi, \sqrt{\frac{\pi(1-\pi)}{n}}\right)$.

### 8.4.1  A Quick Look at Data Science: Bacteria at New York City Beaches

Suppose an administrator from New York is interested in assessing the percentage of times water around Manhattan Beach has levels of Enterococci, a bacteria, that exceed 200 most probable number per 100 ml (MPN/100 ml). Specifically, an interval estimate will be obtained. In contrast to just one numerical statistic, an interval estimate incorporates the uncertainty of the statistic. Results from over 20 000 beach water quality samples taken across New York are available[5] online, with measurements taken as far back as 2005. We filter the original data by beach location. Also, only samples from 2017 will be considered and missing values from measurements below the detection limit are removed (Figure 8.8). Through this procedure, it is found that out of 60 Manhattan Beach samples in 2017, two exceeded 200 MPN/100 ml. Thus, $\hat{p} = 2/60$.

---

4  See Problem 8.29.
5  data.cityofnewyork.us (accessed July 26, 2019).

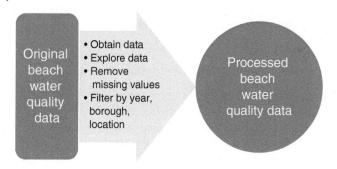

Figure 8.8 Steps required to extract 2017 Manhattan Beach (New York) Water Quality data.

Note that $\pi$ is not known. However, the standard error $\hat{p}$ can be estimated by

$$\sqrt{\frac{\hat{p}(1-\hat{p})}{n}} = \sqrt{\frac{2/60 \times 58/60}{60}} = 0.0232$$

Since $\pi$ is unknown, we can estimate the conditions to use the normal approximation using $n\hat{p}$ and $n(1-\hat{p})$. Now that $n\hat{p} = 2 < 15$, the normal approximation would not be good. One way to deal with this is to use resampling, also known as bootstrapping. In this method, the results from one sample are used to construct the sampling distribution of a statistic, in this case $\hat{p}$. The method proceeds with the following steps[6]:

- Randomly select an observation from your sample of size $n$.
- Place the "resampled" observation back into the original sample, then randomly choose another observation (sampling is performed with replacement).
- Perform the steps above until you get a new sample of size $n$.
- Calculate the statistic of interest.
- Generate new statistics $B$ times. $B$ in practice tends to be large: $\geq 50\ 000$.
- Use the percentiles of your "resampled" statistics, say the 0.5 and 99.5% to determine the bounds of your interval.

The bootstrapping method is applied using estimated proportions from $B = 50\ 000$ simulated samples. Figure 8.9 shows the bootstrap sampling distribution. It shows right skewness, an expected consequence of a small $\hat{p}$, and indicative of how the normal distribution approximation should not be trusted. The 0.5 and 99.5% percentiles of the resamples result in the interval (0, 0.10).

So far, one reason why $\overline{X}$ is commonly used as a measure of centrality was presented: $\overline{X}$ often follows a normal distribution. Still, convenience is just not enough to argue for one statistic being better than others. In Section 8.5, a series of other properties that do indicate when one statistic is better to estimate a

---

6 There are many flavors of bootstrapping. The provided steps are just for one version.

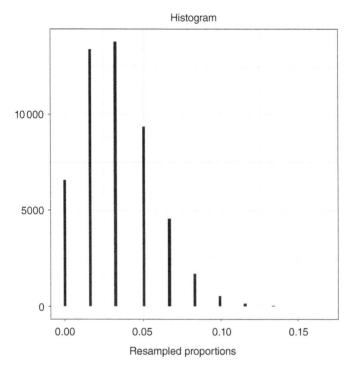

Histogram

**Figure 8.9** The sampling distribution of 50 000 resampled proportions generated from the original data. It indicates right skewness.

parameter than other statistics are discussed. These properties are the main arguments when determining which estimator is best to estimate a parameter.

## Practice Problems

8.13 State in which of these cases it is more reasonable to assume that $\overline{X}$ follows a normal distribution and explain why or why not
  a) sample of size 2 comes from a population with normal distribution (known variance).
  b) sample of size 10 comes from a population with normal distribution (unknown variance).

8.14 State in which of these cases it is more reasonable to assume that $\overline{X}$ follows a normal distribution and explain why or why not
  a) sample of size 300 comes from a population with some unknown distribution.

b) sample of size 15 comes from a population with some unknown distribution.

8.15 Two of your employees are conducting statistical analysis. Each one is working with a random sample of 30. The first employee took the sample from a population with a symmetric distribution. The second employee took the sample from a highly skewed distribution. Both based their conclusions on the sample mean following approximately a normal distribution. For which of the two employees do you think the normal assumption is more appropriate?

8.16 Average commute times (in minutes) to work at locations across Oregon are found online.[7] With the variable "COMMUTING TO WORK - Mean travel time to work (minutes)," in the file "OregonAverage_Commute_Times,"
a) draw a histogram. Interpret its shape.
b) could the mean of the variable follow a normal distribution? Explain.
c) filter the variable "COMMUTING TO WORK - Mean travel time to work (minutes)" by Region "2." Draw the histogram of these commute times and interpret its shape.
d) could the mean of the variable "COMMUTING TO WORK - Mean travel time to work (minutes)" for Region "2" follow a normal distribution? Explain.

8.17 Returning to Example 8.5,
a) If next Thursday there are 500 customers instead of 300, will the probability of a profit increase or decrease?(*Hint:* find the probability that the sample mean expense is less than $15 when there are 500 customers.)
b) What if only the mean expense per customer changed to $14.75 instead of $14.95 will the probability of a profit increase or decrease?

8.18 Based on the findings in Problem 8.17, what would you recommend to management to ensure the all you can eat event brings a profit?

8.19 A machine produces bolts for construction. The length of the bolts has an expected value of 5 in and a variance of 0.05 in$^2$. A random sample of 100 bolts is taken.
a) What is the mean (expected value) and standard deviation of the random sample average?
b) Find the probability that the sample mean is greater than 5.1 in.

---

7 Source: 2014 American Community Survey, data.orcities.org (accessed July 26, 2019).

    c) What is the probability that the sample mean is smaller than 4.9 in or greater than 5.1 in?

    d) What argument did you use to support applying the normal distribution to solve parts (b) and (c)?

8.20 If $\pi = 0.20$, find the mean and the standard error of $\hat{p}$ when

    a) $n = 100$

    b) $n = 200$

    c) $n = 50$

8.21 Free-throw proportion in basketball is[8] the number of free throws made divided by the number of attempts. A player has a 0.86 career free-throw proportion. Explain why any free-throw proportion between 0.79 and 0.93 for his next 25 free throws would not be unusual for the player.

8.22 Batting averages in baseball are sample proportions;[9] the number of hits divided by the number of at-bats. A player has a 0.31 career batting average. Explain why any batting average between 0.29 and 0.33 for the player's next 500 at-bats would not be unusual for the him.

8.23 Determine if normality can be assumed for $\hat{p}$ when

    a) $n = 25, \pi = 0.50$

    b) $n = 100, \pi = 0.97$

    c) $n = 1000, \pi = 0.001$

    d) $n = 1500, \pi = 0.21$

## 8.5 Other Properties of Estimators

The preceding sections of this chapter help illustrate the sampling distribution of some estimators, an important property to draw inference. Admittedly, the estimators $\overline{X}$ and $\hat{p}$ were preselected. Sure, intuitively it makes sense to use a sample mean to estimate a population mean. Similarly, it is reasonable to use a sample proportion to estimate a population proportion. But intuition is just not enough. Recall from Chapter 4 that there are multiple measures of centrality, with the sample mean and sample median being the most common. The sample median was less affected by outliers that are the sample mean. Thus, the sample median may also be a candidate to estimate $\mu$, especially if there is knowledge that the population distribution is symmetric. In the case of the sample variance, $s^2$ divides by $n - 1$ to make it unbiased. However, as a consequence,

---

8 The percentage used in professional basketball is just the proportion times 100.

9 At-bats is the total times the player has attempted to hit the ball with a bat during the season.

the variance of $s^2$ is higher than $\sum_{i=1}^{n}(X_i - \overline{X})^2/n$. Does the unbiasedness of $s^2$ compensate for its higher variance? These arguments raise the question: what makes us in general decide for one estimator over others? We can think about this in terms of what properties does the estimator satisfy. To do this, we must consider the following:

- Is the expected value of the estimator equal to the parameter it is trying to estimate?
- Is the variability of the estimator smaller than any competitor?
- Does it become more likely that the estimator is arbitrarily close to the parameter as $n$ increases?
- Does the estimator include all the information available from the sample to estimate the parameter?

Our choice of estimator will be based on which estimator adequately answers most of these questions simultaneously. For example, from Section 8.4, it can be deduced that, if we compare two estimators with the same expected value, it is preferable to use the one that has the smallest variance (standard deviation). In what follows, we formalize important estimator properties:

**Unbiasedness:** an estimator is unbiased when estimating a parameter if it is expected value is exactly equal to the parameter it is estimating. For example, since

$$E(\overline{X}) = \mu$$

$\overline{X}$ is an unbiased estimator of $\mu$. Also, the reason why the sample variance computation divides by $n - 1$ is to make it an unbiased estimator of $\sigma^2$. Unbiasedness is not necessarily essential, and there are situations when biased estimators are best, but it is a very useful property because of its intuitive appeal and because it becomes easier to establish other properties of estimators (see below) when restricting comparisons to unbiased estimators.

**Efficiency:** An estimator, say $\hat{\theta}_1$, is more efficient in terms of another estimator, say $\hat{\theta}_2$, in estimating $\theta$ if $Var(\hat{\theta}_1) < Var(\hat{\theta}_2)$. Each time we use an estimator to estimate a parameter, we want as little variability as possible in the estimates. Sometimes we can determine that an estimator has the smallest variance within a family of estimators. This is the case with $\overline{X}$ to estimate $\mu$; no other unbiased estimator has a smaller variance than $\overline{X}$. The same can be said about $s^2$ and $\hat{p}$ that they are the unbiased estimators with the smallest variance when estimating $\sigma^2$ and $\pi$, respectively.

**Consistency:** An estimator is consistent when estimating a parameter if the probability that the estimate is arbitrarily close to the parameter tends to 1 as $n$ increases. It can be shown that an unbiased estimator, with a variance that tends to 0 as $n$ increases, is consistent. Estimators $\overline{X}$, $s^2$, and $\hat{p}$ are unbiased and have variances that tend to zero as $n$ increases, therefore they are also consistent.

**Sufficiency:** An estimator is sufficient in estimating a parameter if it includes all the information available from the sample in relation to that parameter. An estimator is **minimal sufficient** if it resumes the information available from the sample without losing the necessary information of the parameter of interest. Taking this definition literally, the entire sample $X_1, \ldots, X_n$ is sufficient to estimate $\mu$ since it contains all the information available from the sample. But the entire sample is not minimal sufficient, since it does not summarize the information about $\mu$ from the sample. In contrast, $\overline{X}$ is minimal sufficient to estimate $\mu$.

Now, we return to the comparison of $\overline{X}$ and $X_1$ to estimate $\mu$, but based on the properties just defined. Table 8.1 resumes the properties of $\overline{X}$ and $X_1$. Both estimators are unbiased, $E(\overline{X}) = E(X_1) = \mu$. But $X_1$ is not sufficient or consistent estimating $\mu$, while $\overline{X}$ is. In addition, $\overline{X}$ is efficient in relation to $X_1$, since it has a smaller variance. Moreover, no other unbiased estimator has a smaller variance than $\overline{X}$ when estimating $\mu$.

And what about the median of a sample, $X_{(\frac{n}{2})}$, as an estimator of $\mu$ for a normal population? Recall that the sample median has the advantage of not being affected by extreme values in a sample. So when we take samples from distributions that are highly skewed, we expect the median to be a better measure of centrality than the average. However, the standard deviation of $X_{(\frac{n}{2})}$ is $\sqrt{\frac{\pi}{2}}(\frac{\sigma}{\sqrt{n}})$, greater than $\frac{\sigma}{\sqrt{n}}$, the standard deviation of $\overline{X}$. That is why in the case of a normally distributed population, we prefer to use $\overline{X}$ to estimate $\mu$ instead of $X_{(\frac{n}{2})}$, $X_1$, or other estimators.

In addition, when populations with skewed distributions are sampled, the central limit theorem tells us that $\overline{X}$ will have an approximately normal distribution with an expected value of $\mu$, a convenient result allowing us to easily find probabilities. Hence, the reasoning to use the sample mean to estimate the population mean is that $\overline{X}$ is

- unbiased
- consistent

Table 8.1 Comparison of $\overline{X}$ and $X_1$ as estimators of $\mu$.

| Property | $\overline{X}$ | $X_1$ |
|---|---|---|
| Unbiased | √ | √ |
| The most efficient | √ | — |
| Consistency | √ | — |
| Sufficiency | √ | — |

- sufficient
- the most efficient (No other unbiased estimator has a smaller variance.)

Similarly, $\hat{p}$ is unbiased, consistent, and sufficient in estimating the parameter $\pi$ of a population with a Binomial distribution, while $s^2$ is unbiased, consistent, and sufficient in estimating[10] $\sigma^2$.

**The Importance of the Properties of an Estimator**   The more of these properties an estimator has, the better. Sometimes, estimators may be missing some of these properties and they can still be useful. For example, on occasion, minimizing the bias of an estimator will lead to higher variance. This is especially the case when we need to estimate several parameters. On other instances, such as the case of data with spatial dependence, estimators of parameters may not be consistent, but prediction procedures of a process of interest using these estimators will lead to useful results.

Perhaps the cautious reader has realized there is a bit of a contradiction in how probabilities are calculated in this chapter. The argument that may arise goes as follows: if we use statistics to estimate parameters of interest, then why do the probability problems we have solved here give us a known $\mu$? If we know $\mu$, isn't it unnecessary to work with $\overline{X}$? In practice, $\mu$ is often unknown or we only know about its past value and we would like to see if it has changed in value under different population conditions. The properties of the estimator discussed here allow us to reason about the value of $\mu$. The idea works like this: given the properties of $\overline{X}$, we should feel quite confident that $\overline{x}$ should be "relatively close" to $\mu$, regardless of whether $\mu$ is unknown or not. Therefore, one alternative is that we can conjecture about the value of $\mu$ before gathering data. Then, using results of a random sample, we can probabilistically assess the value of $\overline{x}$ assuming the conjectured value of $\mu$, and from this determine how reasonable the value of $\mu$ we guessed is. This type of procedure will be implemented formally through hypothesis testing in Chapter 10.

## Chapter Problems

8.24  An employee from a company's new products department is conducting research on the number of product parts a machine can assemble in an hour. Unbeknown to him, the population distribution is left skewed with a mean of 100 and a standard deviation of 10. The researcher takes a random sample of 200 hours and counts the number of product parts completed in each hour.

a)  What shape will the data distribution probably have?

---

10  It turns out that technically $s$ is not an unbiased estimator of $\sigma$. However, if the sample size is large enough, the bias is negligible.

b) What shape will the sampling distribution of $\overline{X}$ have? Why?

c) What is the expected value of the sample average number of product parts completed in each hour?

d) What is the standard error of the sample average number of product parts completed in each hour?

8.25 For a population with population mean 50 and standard deviation 7,

a) if the population is normally distributed, which sample size would lead to the smallest variance for $\overline{X}$, $n = 100$ or $n = 300$?

b) if the population is NOT normally distributed, which sample size would lead to the smallest variance for $\overline{X}$, $n = 100$ or $n = 300$?

8.26 Suppose you are given the option of choosing from two production procedures. Both have possibilities that follow a normal distribution. One has mean profit of $280 and standard deviation $10. The other has mean $280 and standard deviation $40.

a) Which one would you choose if you want the lowest chance of a profit of less than $200?

b) Say that the first production plan has mean profit $280 and standard deviation $10, but that the second has mean $300 and standard deviation of $80. Which one would you choose, if you prefer the production plan with the lowest probability of earning less than $240?

8.27 Scores on the Wechsler Adult Intelligence Scale (a standard "IQ test") for the 20–34 age group are approximately normally distributed with a mean of 110 and a standard deviation of 25. For a random sample of 50 people, what would be $\overline{x}$ such that 98% of means from equally sized samples are less than or equal to $\overline{x}$?

8.28 A December 2017 *CNN article*[11] established that life expectancy in the United States, on average, was 78.6. Assume that the standard deviation of life expectancy was 12 years. If a random sample that year of the age at the time of death of 52 individuals is selected, what is the probability that the sample mean will be between 76.5 and 79 years?

8.29 In Section 8.4, it was stated that when $Y \sim binomial(n, \pi)$, $\hat{p} = Y/n$, does not follow a binomial distribution. Explain why this is the case. (*Hint:* evaluate the properties of a random variable with the binomial distribution to see if they all hold.)

---

11 Source: edition.cnn.com (accessed July 26, 2019).

8.30 A company is about to randomly choose customers to fill surveys evaluating the company's service. One of the questions asks whether customers will use or recommend the company's services in the future. The potential answers are yes or no. When estimating the proportion that answers no, which sample would have a smaller standard error, $n = 1000$ or $n = 1250$?

8.31 The hours a week college students spend on social media is believed to follow a normal distribution with a mean of 8.2 hours and a standard deviation of 2 hours. A study found that anyone who spends more than 12 hours on social media tends to have mediocre grades (D average or lower). Suppose that in reality, hours spent on social media by college students really do follow a normal distribution BUT with mean of 9.3 hours and a standard deviation of 2 hours. By using a mean of 8.2 hours that we believe to be true (and it is not), would the WRONG probability of a student getting mediocre grades (spends 12 hours or more on social media) be smaller or bigger than the TRUE probability obtained when a population mean of 9.3 hours is used to calculate the probability? Explain. (*Hint:* plotting both situations together may help.)

8.32 The monthly returns on a stock have a standard deviation of 0.02 and are suspected of having a population mean of 0.01. If all monthly returns are considered independent,
   a) what is the probability that for the next 60 months we will get a sample mean return of 0.02 or higher?
   b) what argument did you rely on to use the normal distribution to find the requested probability above?
   c) suppose that a manager decides that if the probability found in (a) is less than or equal to 0.05, then the company decides that the population mean return is actually higher than 0.05. What should they decide from your result in (a)?

8.33 Suppose $X$ has a highly asymmetric distribution (skewed to the right to be precise). Should we use the sample mean to infer on the centrality of the distribution of $X$ (a typical $X$ value in the population)? Explain.

8.34 It has been stated that the standard error of $\overline{X}$ is $\sigma/\sqrt{n}$. This assumes that the population is infinite. If $n$ is a significant proportion of the population size, $N$, a finite population correction to the standard error is needed:

$$\frac{\sigma}{\sqrt{n}}\sqrt{\frac{N-n}{n-1}}$$

Find the correction to the standard error of $\overline{X}$ when

a) $n = 20, N = 40$.
b) $n = 50, N = 100\ 000$.

8.35 Continuing Problem 8.34, find the standard error of $\overline{X}$ when $n = N$, and explain why the result makes sense.

8.36 For two sample sizes $n_1 > n_2$ from a population with standard deviation $\sigma$, show that the sample mean of size $n_1$ will always have a smaller standard error than the sample mean of size $n_2$.

8.37 Show that (*Hint:* See Section 6.3.)
a) $E(\hat{p}) = \pi$.
b) $Var(\hat{p}) = \frac{\pi(1-\pi)}{n}$

8.38 Mention at least three properties that we want an estimator to have.

## Further Reading

Agresti, Alan and Franklin, Christine. (2016). *Statistics: The Art and Science of Learning from Data*, 4th edition. Pearson.
Casella, G. and Berger, R. L. (2001). *Statistical Inference*, 2nd edition. Duxbury.
Wackerly, Dennis D., Mendenhall III, William, and Scheaffer, Richard L. (2002). *Mathematical Statistics with Applications*, 6th edition. Duxbury.

# 9

# Interval Estimation for One Population Parameter

## 9.1 Introduction

For the most part, we have been assuming the values of the parameters of a distribution are known. Often, we wish to reach managerial decisions, and we can express a problem at hand in terms of an unknown parameter value. By drawing inference on the value of the parameter, we can solve the managerial problem. To perform inference effectively, the best estimator of a parameter among many choices must be used. Chapter 8 covered the properties that help determine what the best estimator for a parameter is. However, once the best estimator is found, it is not enough to get an estimate from it to perform inference.[1]

---

**$P(\bar{X} = k) = 0$ for Any Constant $k$**

Suppose you take a random sample of size 60 from a normally distributed population with mean 10 and standard deviation 2.5. $\mu$ is unknown and inference on its value is needed. We already know that $\bar{X}$ is the best estimator of $\mu$. Suppose, the sample average turned out to be 8.5. Now, the probability that the sample mean would be 8.5 is zero. So, how did that value happen? The implication is that we should not expect the sample mean to be exactly equal to the population mean, only close based on its properties as an estimator. So just using a sample mean value is not enough to infer about the population mean.

---

Traditionally, statistical inference is conducted in two ways: using confidence intervals or using hypothesis testing. With confidence intervals, the sample estimate of the parameter of interest is combined with the uncertainty of its value to infer on the true value of the parameter. With hypothesis testing, we speculate on the value of the parameter, then see if the empirical data supports that

---

1 Remember for a continuous random variable $P(Y = k) = 0$ for any $k$.

*Principles of Managerial Statistics and Data Science*, First Edition. Roberto Rivera.
© 2020 John Wiley & Sons, Inc. Published 2020 by John Wiley & Sons, Inc.
Companion website: www.wiley.com/go/principlesmanagerialstatisticsdatascience

parameter value. This chapter introduces the concept of confidence intervals to conduct inference on the parameter of a population. In Chapter 10, we introduce hypothesis testing to conduct inference.

## 9.2 Intuition of a Two-Sided Confidence Interval

Suppose a professor wants to infer on the mean score of exams in some course. She wishes to obtain an interval to draw inference from. If exams are worth 100 points, naturally, the mean score will be between 0 and 100. But of course, this is not a very good interval, but it is too wide, providing virtually no information (other than the obvious) on what the mean exam score is. More precision is needed, in such a way that the lower and upper limits of the interval are "as close as possible" (i.e. the interval is as short as possible). The idea of a **two-sided confidence interval** is that we find two values, a lower bound and an upper bound, such that we feel pretty confident that the true value of the parameter is between those two bounds and the interval is as short as possible.

---

**Point Estimator versus Interval Estimate**

- A point estimator (e.g. $\overline{X}$) gives us one estimate of the parameter.
- A **confidence interval** gives you an interval estimate: two bounds between which, according to the data, the parameter of interest falls.

---

In general, for a confidence interval, we want to use not only the estimate of the parameter of interest, but also the uncertainty around that estimate, and have a certain degree of confidence of inferring the true parameter value from the interval estimate. To accomplish this, we take the point estimator of the parameter of interests and add/subtract a **margin of error,**

$$\text{Point estimator} \pm \text{Margin of error} \qquad (9.1)$$

$L$ = point estimator − margin of error will be the lower bound of the confidence interval, and $U$ = point estimator + margin of error will be the upper bound. Thus, the confidence interval will be $(L, U)$. The margin of error quantifies the uncertainty and our confidence on the resulting interval. It is the error in estimation that is deemed acceptable. This is the intuition for all two-sided confidence intervals. The chosen point estimator and margin of error will vary by the parameter one wishes to draw inference on. It is important to point out that the lower and upper bounds of the interval estimator obtained from Eq. (9.1) are a function of the point estimator, a random variable. Therefore, the interval estimator itself is random. This has implications on the interpretation of the confidence interval.

## 9.3 Confidence Interval for the Population Mean: σ Known

So how do we apply the intuition of Section 9.2 to build a confidence interval of the population mean? Let's assume there is a sample from a population with mean $\mu$ and standard deviation $\sigma$. $\sigma$ is known, while $\mu$ is not and we must infer on its value. In Chapter 8, we demonstrated how $\overline{X}$ has the properties of unbiasedness, consistency, efficiency, and sufficiency, making it the point estimator of $\mu$ of choice. Note that we are assuming that $\overline{X}$ follows a normal distribution. Referring back to Eq. (9.1), we already have part of the confidence interval,

Point estimator $= \overline{X}$

What is left is to figure out the margin of error component. For that, we rely on the trusty standard normal random variable $Z$ once more. Recall that

$$Z = \frac{(\overline{X} - \mu)}{\sigma/\sqrt{n}}$$

has a normal probability distribution that does not depend on $\mu$ or $\sigma$, and that any normal distribution is symmetric. Thus, we may write

$$P(-z_{\alpha/2} < Z < z_{\alpha/2}) = 1 - \alpha \tag{9.2}$$

$1 - \alpha$ is a probability value that has been determined to be adequate and it is known as the **confidence level** of the confidence interval. For the moment, think of $\alpha$ as a number that controls the confidence level. Informally, $\alpha$ is the probability that the confidence interval does not include $\mu$ when it should. Because the normal distribution is symmetric, we split $\alpha$ equally into the areas of the tails of the sampling distribution of $\overline{X}$ as seen in Figure 9.1. A more formal definition of $\alpha$ will be discussed in Chapter 10. Substituting $Z = \frac{(\overline{X}-\mu)}{\sigma/\sqrt{n}}$ in Eq. (9.2) and isolating $\mu$ in the middle of the inequality, we find that the $1 - \alpha$ confidence level interval estimate is

$$\overline{X} \pm z_{\alpha/2}\frac{\sigma}{\sqrt{n}} \tag{9.3}$$

where $z_{\alpha/2}$ is the value of $Z$ such that $P(Z < -z_{\alpha/2}) = \alpha/2$. Thus, in this scenario, $z_{\alpha/2}(\frac{\sigma}{\sqrt{n}})$ is the margin of error.

To infer on $\mu$, an interval estimate is better than a point estimate because it incorporates uncertainty around the point estimate and it measures how confident we are about the result. It is highly unlikely that a point estimate will be exactly equal to the parameter being estimated. In contrast, it is very likely that the confidence interval includes the true parameter value. Instead of inferring the value of $\mu$ by only using $\overline{x} = 10$ from a sample of size 100 with $\sigma = 10$, we can rely on, say the 95% interval estimate, $(8.04, 11.96)$.

**Confidence level for a two–sided confidence interval**

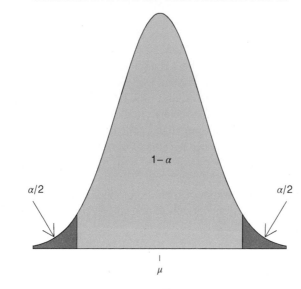

Figure 9.1 The sampling distribution of $\overline{X}$. The area under the curve in green is the confidence level. The areas under the curve in red are $\alpha/2$.

**Example 9.1** *According to the 2013 Canadian General Social Survey, the average hours a year volunteers contribute is*[2] *154. Suppose there were 100 volunteers and the standard deviation of the population is known to be 75. Find a 99% confidence interval for the true mean volunteer hours a year.*

Since $n = 100$, we assume that $\overline{X}$ is approximately normally distributed by the central limit theorem. To apply Eq. (9.3), the only thing that is not directly given to us is $z_{\alpha/2}$. For a 99% confidence interval, it can be shown that $z_{.005} = 2.576$. Therefore, the confidence interval is obtained as

$$154 \pm 2.576 \left( \frac{75}{\sqrt{100}} \right)$$

which is (134.68,173.32). That is, we are 99% confident that the true mean hours worked by volunteers is between (134.68,173.32).

Table 9.1 summarizes the $z$-scores for the confidence levels most commonly used in practice.

***Effect of Sample Size and Confidence Level on Interval Estimate*** It is constructive to look at what happens to the length of the confidence interval when we change

---

2 Source: www.statcan.gc.ca (accessed July 26, 2019).

Table 9.1 z-scores for the most commonly used confidence levels.

| Confidence level | z |
| --- | --- |
| 90 | 1.645 |
| 95 | 1.960 |
| 99 | 2.576 |

$1 - \alpha$ or $n$. The higher the confidence level, the larger the length of the interval. This makes sense, for us to be more confident that an interval includes a parameter, the interval should be longer. We can also see why from Eq. (9.3).

- As the confidence level increases, so does the value of $Z$, making the margin of error larger which leads to an increase in length of the confidence interval. More confidence that the interval includes the parameter comes at a price: larger margin of error means loss of precision (Figure 9.2). From a managerial perspective, the choice of confidence level can be determined from the problem at hand.
- On the other hand, by increasing $n$ and leaving everything else fixed, then the length of the confidence interval should decrease. This also makes sense,

Figure 9.2 Three confidence intervals from one sample from an $N(50, 10)$ distribution. Confidence levels are 90% (green), 95% (blue), and 99% (red). The dots in each line are $\bar{x}$ for each sample and each line extends to the confidence interval's lower and upper bound.

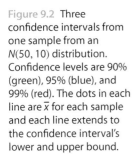

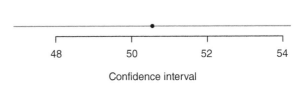

Confidence interval

larger $n$ results in smaller variability of $\overline{X}$ making the margin of error smaller, which leads to a decrease in length of the confidence interval (see Eq. 9.3).

The length of the interval estimate also depends on $\sigma$, but this is something we cannot control.

---

**Effect of Sample Size and Confidence Level Sidenote**
Importantly, in this section, the impact of sample size and confidence level on length of the confidence interval has been presented. However, this should not be misconstrued as meaning that if the confidence level leads to unwanted conclusions, the confidence level should be changed. Once a confidence level value is chosen, before constructing the interval, and based on what is considered appropriate for the problem at hand, it is kept fixed. The sample size is either what is available, or it is determined from a desired outcome, a topic discussed in Section 9.4.

---

**Interpretation of the Confidence Interval**
At this stage of the book, it is hard to count how many times we have distinguished between a parameter and a statistic, but we must again remind the reader about the distinction.

- A parameter comes from a population and is a constant.
- A statistic comes from a sample and varies according to the observations in the sample.

Since $\mu$ is constant:

- It is either inside the confidence interval, or it is not.
- For a 95% confidence interval, we cannot say that there is a probability of 0.95 that $\mu$ is between (8.04, 11.96). The right interpretation is that out of a large amount of interval estimates over many different samples, 95% include the true $\mu$ while 5% of them do not.
- The middle point of the confidence interval, $\bar{x}$, is not more likely to be close to $\mu$ than any other value in the confidence interval.

---

Specifically, recall that since the point estimator $\overline{X}$ is a random variable, and the lower and upper bounds of the interval estimator obtained from Eq. (9.3) are a function of it, the interval estimator itself is random. For any $n$ and $\alpha$, every different sample will give us a different interval estimate. This makes the interpretation of a confidence interval a bit tricky.

A simulation exercise helps illustrate the interpretation of a confidence interval. The following steps are necessary:

1) Take a random sample of size 50 from a normally distributed population with $\mu = 50$, $\sigma = 10$.
2) Construct a 95% confidence interval using Eq. (9.3).
3) Repeat the steps above using 100 independent samples from the population.

Figure 9.3 summarizes the outcome of this simulation. Each of the horizontal lines represents an interval estimate using the respective sample. The dots in each line are $\bar{x}$ for each sample. So from the first sample, the 95% confidence

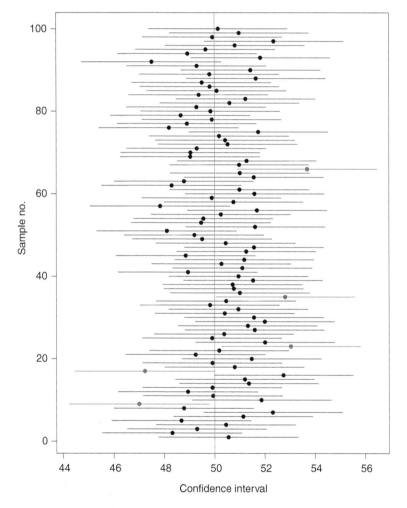

**Figure 9.3** Samples from an $N(50, 10)$ distribution were simulated 100 times. For each sample, a confidence interval using Eq. (9.3) was constructed. Red confidence intervals indicate they do not include $\mu = 50$. Dots are the mean of each sample.

interval at the bottom of Figure 9.3 was (47.77, 53.32), which encloses $\mu = 50$. All in all, 95 of the 100 confidence intervals included $\mu = 50$, while 5 of the 100 confidence intervals (in red) did not include $\mu$. In contrast, sometimes the sample means were close to 50 but not a single time was the estimate exactly equal to the parameter value. A confidence level of 99% would have led to more confidence intervals enclosing $\mu$, but at a cost of wider intervals. In the real world, you obtain the confidence interval just once and will not know if it includes the population mean or not. But from the properties of $\overline{X}$ and the simulation results, we see it is likely the interval will include $\mu$.

It is preferable to make the interpretation of the confidence interval in the following way: "we are $1 - \alpha$% confident, that the true mean of the population is between the lower limit and the upper limit." Using the word confident instead of probability is intentional, to avoid misinterpretation of the confidence interval.

## Practice Problems

9.1 Construct a confidence interval for $\mu$ using each random sample. Assume $\overline{X}$ is normally distributed.
   a) $\overline{x} = 58.5, n = 55, \sigma = 9.2$, 99% confidence.
   b) $\overline{x} = 11.11, n = 150, \sigma = 1.25$, 90% confidence.
   c) $\overline{x} = 1032.85, n = 15, \sigma = 29.4$, 95% confidence.
   d) $\overline{x} = 93.61, n = 70, \sigma = 4.2$, 75% confidence.

9.2 In each case, state whether the interpretation of a confidence interval is correct or incorrect. If incorrect, explain why.
   a) There is a 99% probability that the true mean repair cost for a machine is between ($15 232.11, $17 667.89).
   b) We are 99% confident that the true mean repair cost for a machine is between ($15 232.11, $17 667.89).
   c) The sample mean repair cost is inside of the 99% confidence interval and thus the range is correct.
   d) There is a 99% probability that the interval ($15 232.11, $17 667.89) includes the true mean repair cost.
   e) We are 99% sure that the true mean repair cost for a machine is between ($15 232.11, $17 667.89).

9.3 A 90% confidence interval is built for the population mean using 32 observations.
   a) How likely is it that the confidence interval includes the sample mean?
   b) How likely is it that the confidence interval includes the population mean?

c) If the chosen confidence level was instead[3] 95% at the same sample size, which confidence interval would be wider, the 90% confidence interval or the one with 95%?

d) Had the sample size been 25 yet confidence level remained at 90%, which confidence interval would be wider, the one with $n = 32$ or the one with $n = 25$?

9.4 A study of the effects of running on a random sample of adults found a 95% confidence interval for the true mean heart rate to be (68.74, 111.14), using a normal approximation procedure (the $Z$ equation).

a) What is the value of the point estimate (sample mean) for the true mean heart rate of senior citizens?

b) What is the value of the margin of error for the confidence interval that was constructed?

c) If the sample size was 25, what was the population standard deviation?

9.5 According to www.fueleconomy.gov, a random sample of seven 2016 Subaru Forester AWD 2.5 L, 4 cyl, automatic cars gave a mean of 26.4 miles per gallon (MPG). Assume MPG for this vehicle follow a normal distribution with a standard deviation of 3.23 MPG.

a) A rental company wishes to infer on the mean MPG for this vehicle. Construct a 90% confidence interval.

b) Interpret the confidence interval found in part (a).

c) Suppose that before constructing the confidence interval, the rental company's boss decided that if the true mean MPG of the 2016 Subaru Forester was 32 MPG, he would add some of those vehicles to their fleet. Should he instruct the staff to simply increase the confidence level to simply ensure it includes that mean value?

9.6 The weight of Holstein cows in a farm has a standard deviation of 66.70 pounds. A random sample of 50 cattle has a mean weight of 1420.90 pounds.

a) Why can we assume that the sample mean follows a normal distribution?

b) Construct the 99% confidence interval for the population mean.

c) Interpret the confidence interval found in 9.6b.

9.7 A random sample of 40 grass-fed beef prices in the United States gave an average of $3.70 (per pound). If the prices are not highly skewed and historically the standard deviation is $0.70,

---

3 This goal of this exercise is to understand the impact that the confidence level has on the width of the confidence interval. It does not imply that confidence levels should be altered, in practice, they shouldn't.

a) construct the 95% confidence interval for the population mean and
b) interpret the confidence interval found in 9.7a.

9.8 Show that for a 99% confidence interval, $z_{\alpha/2} = 2.576$.

## 9.4 Determining Sample Size for a Confidence Interval for $\mu$

Commonly, our aim is to infer on $\mu$ within an acceptable margin of error, $\pm m.e$, and confidence level. If we can assume that $\overline{X}$ follows a normal distribution and $\sigma$ is known, then from Eq. (9.3)

$$m.e = z_{\alpha/2}\frac{\sigma}{\sqrt{n}}$$

All terms are known except $n$. Solving for $n$ gives,

$$n = \left(\frac{z_{\alpha/2}\sigma}{m.e}\right)^2 \tag{9.4}$$

since sample size must be a whole number, we conservatively round up the result. Equation (9.4) makes intuitive sense. The more confident you want to be with your confidence interval, the larger $n$ should be. Moreover, the larger the variability of your population and the smaller the margin of error, the larger the sample size needed.

**Example 9.2**  *A solar power company must determine the sample size to infer on the mean utility bill of residents. From previous studies, it was established that the population standard deviation is 25 kilowatts/hour (kWh). They would like a margin of error of 2kWh and a confidence level of 95%. What is the sample size needed?*

For a confidence level of 95%, from Table 9.1, we get that $z_{0.025} = 1.96$. According to the equation above,

$$n = \left(\frac{1.96(25)}{2}\right)^2$$
$$= 600.25$$

Therefore, 601 observations should be made to obtain the required margin of error at the desired confidence level. When $\sigma$ is unknown, alternatives to use Eq. (9.4) are relying on preliminary or historical data to estimate $\sigma$, or with established lower and upper limits for the data, $a$ and $b$, set $\sigma = \sqrt{\frac{(b-a)^2}{12}}$

## Practice Problems

9.9 In the past, a car company had to face a burst of bad publicity when faulty brake parts led to car accidents. They would like to ensure their new car models do not have similar problems. To do so, company staff wants to test the strength of a certain component of the brakes. Previous work has determined the standard deviation is 2.5 psi. They would like a 99% confidence for their experiment and a margin of error of 0.5.
   a) Find the sample size for the experiment.
   b) Find the sample size for the experiment if they wanted a margin of error of 0.25.

9.10 A solar power company must determine the sample size to infer on the mean utility bill of residents. From previous studies, it was established that the population standard deviation is 40 kilowatts/hour (kWh). They would like a margin of error of 5 kWh and a confidence level of 90%. What is the sample size needed?

9.11 A university is concerned with its student's first year mean GPA. They implement new guidelines in the hopes of improving the overall students performance. To verify the guidelines' effectiveness, they would like to sample a number of students and infer on their first year mean GPA. Previous work has determined that the standard deviation is 0.3. They would like a 98% confidence for their experiment and a margin of error of 0.15. What is the sample size required for this experiment?

9.12 To construct a 99% confidence interval on the MPG of a type of vehicle, researchers are debating whether they want a margin of error of 1, or of 2.
   a) Which margin of error will require a bigger sample size and why?
   b) If a 95% confidence interval is to be constructed, will a margin of error of 1 or of 2 require a bigger sample size?

## 9.5 Confidence Interval for the Population Mean: σ Unknown

When $\sigma$ is unknown, we can no longer rely on the confidence interval based on the standard normal distribution. Although $\sigma$ is not directly of interest, it must be estimated as well, to account for uncertainty in the point estimator and to obtain the margin of error of the confidence interval. This leads to the use of another distribution, known as the **Student's *t*-distribution or *t*-distribution** for short. It applies to instances when samples come from a normally distributed population. The $t$-distribution arises from a similar random variable to the $Z$ statistic used to find probabilities of values of $\overline{X}$.

- First, take the sample mean and subtract its expected value, $\mu$.
- Next, since now $\sigma$ is unknown, divide the difference by $s/\sqrt{n}$.

Like in the case of the $Z$ statistic, this new random variable has an expected value of zero, but on account of having to use $s$, there is more variability in its values, and it follows a $t$-distribution. Fortunately, the $t$-distribution is also bell-shaped, yet there is a higher probability of falling in the tails. Since they standardize a random variable to mean zero just like $z$-scores, these $t$ statistics are known as $t$-**scores**. The $t$-distribution has a parameter known as[4] the **degrees of freedom**. Often, the expression $df$ is used for degrees of freedom. When building a confidence interval for the population mean using a random sample from a normal distribution and $\sigma$ is unknown, the degrees of freedom is

$$df = n - 1$$

Figure 9.4 compares several $t$-distributions to the standard normal distribution. It can be seen that for all values of $df$, the $t$-distribution has a bell shape, similar to the standard normal, but with heavier tails (and hence, higher probability of

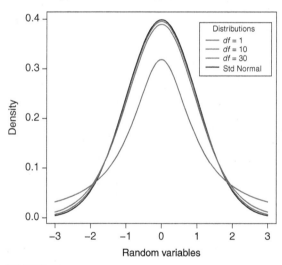

**Comparison of t and standard normal distributions**

Distributions
— $df = 1$
— $df = 10$
— $df = 30$
— Std Normal

Figure 9.4 Comparison of three $t$-distributions with 1, 5, and 30 degrees of freedom, respectively, to the standard normal distribution.

4 Degrees of freedom can be thought as measuring the "loss of freedom" in values of a sample when one must estimate a parameter using estimates of other parameters. Although degrees of freedom can be defined well mathematically, it cannot be defined precisely in words. The definition provided in the first sentence of this footnote does not apply in other situations when degrees of freedom are used. More generally, it is best to think of it as a parameter required for certain probability distributions such as the $t$-distribution.

falling away from the mean). Also, as $n$ increases, the distribution tends toward the standard normal distribution, a feature that can be explained through the central limit theorem.

Therefore, to construct a confidence interval for $\mu$ based on a random sample from a normal distribution when $\sigma$ is unknown, one can use

$$\overline{X} \pm t_{df,\alpha/2} \frac{s}{\sqrt{n}} \tag{9.5}$$

where $t_{df,\alpha/2}$ is the value of $t$ such that $P(t < -t_{df,\alpha/2}) = \alpha/2$. With the $t$-distribution, in contrast to the standard normal distribution, one must still work with a parameter, $df$, that identifies the $t$-distribution. Appendix C provides the value of $t$ based on the confidence level of the interval and the degrees of freedom. Figure 9.5 illustrates how to use Appendix C for a 95% confidence interval and $df = 15$. Then, $t = 2.131$. That is, $P(-2.13 \leq t \leq 2.131) = 0.95$ when the statistic follows a $t$-distribution with 15 degrees of freedom (Figure 9.6).

Some further remarks:

- The $t$ table differs from the $Z$ table; in that, it presents information for many $t$ distributions and very few probabilities.
- The value of $t$ is larger than an analogous $z$-based confidence interval.
- When $df = \infty$, Appendix C gives the value of $z$ for the pertinent confidence level. Observe how for a fixed confidence level, as $n$ increases ($df$ increases), the values of $t$ tend to $z$ values.
- Figure 9.7 illustrates an approach that can be used when determining if the $Z$ or $t$ table can be used to find probabilities or percentiles associated to $\overline{X}$.

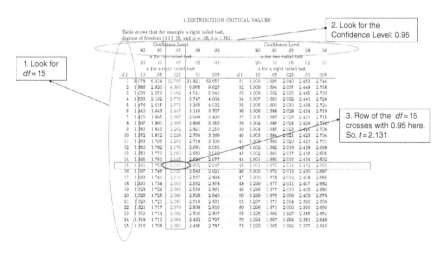

Figure 9.5 Steps to find the critical value $t$ for $df = 15$ and a 95% confidence level using the $t$-distribution table.

**Confidence level for a two–sided confidence interval**

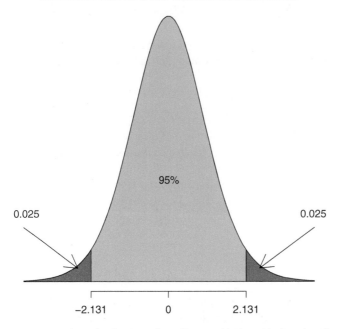

Figure 9.6 The *t*-distribution when *df* = 15 with the critical *t* values for a 95% confidence interval. The area under the curve in green is the confidence level. The areas under the curve in red are $\alpha/2$.

- When the population distribution is skewed, the confidence interval can still be obtained, but the true coverage of the confidence interval may be lower than the established confidence level because the confidence interval will be shorter than what it should be. When in doubt, simulations could help determine the empirical coverage of the confidence interval.

---

**Confidence Intervals in Practice**

In almost all practical situations when building a confidence interval for $\mu$, $\sigma$ is not known.

- When the sample size is small, use of the *t*-distribution will result in wider confidence intervals than using a *z* confidence interval.
- Although theoretically the *t*-distribution applies to samples from a normal population, in practice the confidence interval is fairly robust to a violation of this assumption as long as the sample does not come from a population with a highly skewed distribution.

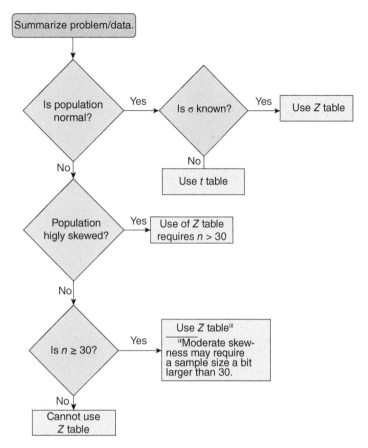

Figure 9.7  A flowchart for the steps to determine if use of the standard normal or $t$-distribution is possible to find probabilities or percentiles associated to $\overline{X}$.

**Example 9.3**  *In Example 4.9, the Baltimore 2011 walking score, calculated as a function of distance to amenities, was introduced. The score is normalized from 0 to 100 (with 100 being best). Figure 9.8 illustrates the histogram of the scores for 55 communities. The random sample resulted in an average of 61.94 and standard deviation of 20.54. We will construct a 95% confidence interval.*

According to Figure 9.8, the distribution is not too skewed. Thus, we will use the $t$-statistic confidence interval Eq. (9.5) with[5] $t_{54,0.025} = 2.004$ (Appendix C),

$$61.94 \pm 2.004 \left( \frac{20.54}{\sqrt{55}} \right)$$

---

5  Actually, Appendix C does not include $df = 54$. Fortunately, at 95% confidence, $t_{50,0.025}$ and $t_{55,0.025}$ are very similar. For simplicity, we treat $t_{54,0.025} = t_{55,0.025}$.

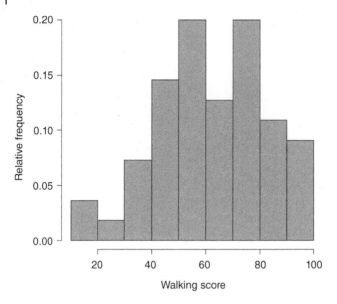

Figure 9.8 Histogram of the 2011 walking scores for 55 communities within Baltimore.

So we are 95% confident that the mean walking score in Baltimore is between (56.39, 67.50).

## Practice Problems

9.13 If $t$ follows a $t$-distribution with 10 degrees of freedom, and $Z$ a standard normal distribution, determine which probability will be greater (no computation is necessary).
a) $P(Z > 1.12)$ versus $P(t > 1.12)$
b) $P(Z < -0.71)$ versus $P(t < -0.71)$
c) $P(Z < 2.4)$ versus $P(t > -2.4)$
d) $P(-0.75 \leq Z \leq 0.75)$ versus $P(-0.75 \leq t \leq 0.75)$

9.14 If $t_1$ follows a $t$-distribution with 10 degrees of freedom and $t_2$ a $t$-distribution with 5 degrees of freedom, determine which probability will be greater (no computation necessary).
a) $P(t_2 > 1.12)$ versus $P(t_1 > 1.12)$
b) $P(t_2 < -0.71)$ versus $P(t_1 < -0.71)$
c) $P(t_2 < 2.4)$ versus $P(t_1 > -2.4)$
d) $P(-0.75 \leq t_2 \leq 0.75)$ versus $P(-0.75 \leq t_1 \leq 0.75)$

9.15 Use Appendix C to find $t$ in the following situations

a) $df = 5$, 99% confidence.
b) $df = 75$, 90% confidence.
c) $df = 23$, 98% confidence.
d) $df = 205$, 95% confidence.

9.16 A company is in the early stages of developing an eco-friendly laundry machine that will use less water than the competition. A team has conducted a study to determine the mean number of gallons of water used by washing machines when completing a washing cycle. The number of cycles is approximately normally distributed. A sample of 8 washing machines leads to an average of 15.37 gallons and a standard deviation of 1.30. With this information, a team member incorrectly built a confidence interval using a $z$-score. Will the $Z$ confidence interval have a longer or shorter length than a $t$ confidence interval? Explain.

9.17 Construct a 90% confidence interval for Problem 9.16.

9.18 Construct a confidence interval for $\mu$ using each random sample. Assume a normally distributed population.
a) $\bar{x} = 79.3, n = 12, s = 9.2$, 95% confidence.
b) $\bar{x} = 5.81, n = 20, s = 0.75$, 90% confidence.
c) $\bar{x} = 312.11, n = 5, s = 19.62$, 99% confidence.
d) $\bar{x} = 93.61, n = 50, s = 10.74$, 95% confidence.

9.19 As an owner of a local accounting firm, you would like to know the mean tax refund people get in your town. This way you can advertise how customers can benefit from your services to get a better refund. You take a random sample of 16 tax refunds last year from people you surveyed by mail (who did not file with you), and find a sample average of $576.41 and a standard deviation of $150.39
a) Construct a 95% confidence interval for the true mean tax refund.
b) Interpret the confidence interval in 9.19a.
c) Are you assuming anything that is not directly given to you in the problem to construct the confidence interval in 9.19a? Explain.

9.20 Suppose authorities in Chicago, Illinois, would like to estimate the mean number of days to complete graffiti removal request from private property owners. Weekly mean number of days to complete requests from 2011 to 2012 are available[6] in the file "ChicagoPerformance:

---

6 Source: data.cityofchicago.org (accessed July 26, 2019).

Metrics_-_Streets___Sanitation_-_Graffiti_Removal." Authorities determined the data is representative of the current environment in the city.

a) Construct a 99% confidence interval for the mean number of days to complete graffiti removal these days.
b) Construct a histogram of the weekly mean number of days to complete requests. Is it approximately normal?

9.21 In an analysis of a random sample of 75 claims to an insurance company, collision repair costs are found to have a mean of $1386. From previous studies, the standard deviation is known to be $735.

a) Construct a 95% confidence interval for the true mean repair cost.
b) Interpret the confidence interval in 9.21a.
c) Notice that to complete 9.21a it was implied that the sample mean follows a normal distribution. Is this a reasonable assumption based on how the problem is stated? Explain why or why not.

9.22 According to official Kansas City January 2018 records,[7] the monthly average pay of 16 Accountant II positions was $4053. The minimum pay was $3353 and maximum pay was $5757. Someone would like to use this information to infer on the mean salary of Accountant II positions in similar locations to Kansas City. Assuming the salaries are normally distributed and the provided data represents a random sample, estimate the standard deviation of the salaries as one-fourth the range of the data and find a 90% confidence interval.

9.23 The Canadian Dairy Information Centre compiles information on the cheeses made from cow, goat, sheep, and buffalo milk.[8] Use the data in "canadianCheeseDirectory" to construct a 95% confidence interval for the mean fat percentage of cheese. State any necessary assumptions.

## 9.6   Confidence Interval for $\pi$

The principle to construct a confidence interval for $\mu$ also applies in the case of $\pi$: a margin of error is added and subtracted to a point estimator. It was argued in Chapter 8 that $\hat{p}$ has a series of properties that makes it the best point estimator of $\pi$. Also from Section 8.4, it is known that $E(\hat{p}) = \pi$ and $Var(\hat{p}) = \frac{\pi(1-\pi)}{n}$. As a rule of thumb, if $n\pi \geq 15$, and $n(1 - \pi) \geq 15$, then we will say that $\hat{p}$ is

---

7  Source: data from data.kcmo.org (accessed February 12, 2018).
8  Source: open.canada.ca (accessed July 26, 2019).

approximately normally distributed[9] by the central limit theorem. When $\pi$ is unknown, we estimate the rule of thumb, $n\hat{p} \geq 15$ and $n(1 - \hat{p}) \geq 15$. Therefore, it can be shown that the $1 - \alpha$ confidence level interval estimate for $\hat{p}$ is

$$\hat{p} \pm z_{\alpha/2} \sqrt{\frac{\hat{p}(1 - \hat{p})}{n}} \tag{9.6}$$

Equation (9.6) helps us understand the influence of $n$ and confidence level on the length of the confidence interval.

- Larger confidence level means higher value of $Z$, making the margin of error larger, which leads to an increase in length of the confidence interval.
- On the other hand, by increasing $n$ and leaving everything else fixed, then the length of the confidence interval should decrease. A larger $n$ results in smaller variability of $\hat{p}$, making the margin of error smaller, which leads to a decrease in length of the confidence interval.

The length of the interval estimate also depends on the number of successes in the sample, but this is something we cannot control.

**Example 9.4**   *A 2016 survey conducted in New York asked 9831 adults whether they had been binge drinking.[10] 1315 people responded affirmatively.[11] Find a 90% confidence interval for the true proportion of people that have been binge drinking.*

Now $n\hat{p} = n(x/n) = x$, the number of people that answered yes, 1315 and which is greater than 15. Meanwhile $n(1 - \hat{p}) = 8516$, also greater than 15. So a normal approximation is appropriate. We find that $\hat{p} = 1315/9831 = 0.134$, and the confidence interval is obtained through:

$$0.134 \pm 1.645 \sqrt{\frac{0.134(1 - 0.134)}{9831}}$$

We are 90% confident that the proportion of adults that have been binge drinking is between $(0.128, 0.140)$.

## Practice Problems

9.24   Construct a confidence interval for $\pi$ using each random sample and a normal approximation.

a)   $x = 34, n = 55$, 95% confidence.

---

9 The use of this approximation is for the sake of simplicity. Alternative confidence intervals exist for the population proportion that are more robust to the one presented here but, mathematically more complicated.

10 Binge drinking was defined as five or more drinks on one occasion for men and four or more drinks on one occasion for women in the past 30 days.

11 Source: a816-healthpsi.nyc.gov (accessed July 26, 2019).

b) $x = 17, n = 149$, 90% confidence.
c) $x = 214, n = 555$, 99% confidence.
d) $x = 4, n = 104$, 95% confidence.

9.25 In Problem 9.24, check for each case if the conditions for a normal approximation hold.

9.26 A 95% confidence interval is built for the population proportion using 200 observations.
a) If the chosen confidence level was instead 90% at the same sample size, which confidence interval would be wider, the 90% one or the 95% one?
b) If the sample size would have been 250 and confidence level remained at 95%, which confidence interval would be wider: the one with $n = 200$ or the one with $n = 250$?

9.27 Suppose the administration at a local university claims that 65% of their college graduates obtain jobs no longer than a month after graduation. For 150 randomly chosen college graduates of the school, it is found that 76 of them had job offers right after graduation.
a) Construct a 90% confidence interval for the true proportion of the school graduates who have job offers right after graduation.
b) Interpret the Confidence interval.
c) Is the school administration's claim justified? Explain.

9.28 From the 2013 New York Youth Behavioral Risk Survey, we can be 95% confident that the proportion of youth who drank five or more alcoholic drinks in the past 30 days is between[12] (0.098, 0.118). If a researcher believed 15% of youth who drank five or more alcoholic drinks in past 30 days, does the confidence interval support this belief?

9.29 According to the CDC, in 2015, of 208 595 kids younger than six years old tested for lead in the blood in Massachusetts, 5889 had levels[13] between 5 and 9 µg/dl. Calculate a 95% confidence interval for the true proportion of children with levels of lead between 5 and 9 µg/dl. Interpret.

## 9.7 Determining Sample Size for $\pi$ Confidence Interval

Sometimes it is necessary to determine $n$ when constructing a confidence interval of $\pi$. Based on the margin of error for the proportion confidence interval in Eq. (9.6), it can be shown that

---

12 Source: data.cityofnewyork.us (accessed July 26, 2019).
13 Source: www.cdc.gov (accessed July 26, 2019).

$$n = \left(\frac{z_{\alpha/2}}{m.e}\right)^2 \hat{p}(1 - \hat{p}) \tag{9.7}$$

This equation poses a problem, however it depends on $\hat{p}$, which is not available. There are a few ways to deal with this quandary:

- Rely on a past estimate of $\pi$. A favorite in practice, since it involves no cost. Of course, for this alternative to work, it is assumed that the proportion has not changed much.
- Conduct a pilot study to estimate $\pi$. Unfortunately, this involves additional cost and resources, which may make the alternative unfeasible.
- Judgment estimate. The most subjective option, an estimate based on opinion, is used. There is a risk of the judgment estimate being way off, resulting in an inadequate sample size.
- Use $\hat{p} = 0.5$. Equation (9.7) is maximized at this value when all other terms are held fixed, resulting in a conservative $n$. Problem is, that if $\hat{p}$ differs considerably from 0.5, the extra cost of a bigger $n$ is not justified.

If $p_n$ is the proportion determined from any of the alternatives above, then we can find $n$ through

$$n = \left(\frac{z_{\alpha/2}}{m.e}\right)^2 p_n(1 - p_n) \tag{9.8}$$

Occasionally, a combination of the options above is used. For instance, it is easy to set up a table or worksheet with several variables of $p_n$ and the resulting $n$. Then, based on available resources and opinion of likely value of $\hat{p}$ (perhaps relying on past data), a final selection of sample size can be made.

**Example 9.5** *A survey is in the works to determine the proportion of female executives that have been sexually harassed by subordinates. How large is the required sample size to construct a 90% confidence level interval with a margin of error of 2%?*

With $m.e = 0.02$ and conservatively setting $p_n = 0.5$, the required sample size is

$$n = \left(\frac{1.645}{0.02}\right)^2 0.5(1 - 0.5) = 1691.27$$
$$= 1692$$

To estimate $\pi$ within $\pm 2\%$ with 90% confidence, 1692 female executives would need to participate in the survey. It is possible that the research budget does not allow for such a large $n$. The team prefers not to alter the confidence level nor the margin of error. After some digging, it is found that a similar survey was conducting two years ago and the proportion of female executives that who had been sexually harassed by subordinates was 17%. Using $p_n = 0.17$, the required sample size is

$$n = \left(\frac{1.645}{0.02}\right)^2 0.17(1 - 0.17) = 954.55$$
$$= 955$$

This sample size is much smaller than the conservative estimate.

## Practice Problems

9.30 A reporter wants to gauge the opinion of college undergraduates about legalizing marijuana,
   a) using $p_n = 0.65$ (from a similar study), and a 95% confidence level, determine what the sample size is for a margin of error of 5%.
   b) using $p_n = 0.5$ (from a similar study), and a 95% confidence level, determine what the sample size is for a margin of error of 5%.
   c) using $p_n = 0.65$ (from a similar study), and a 95% confidence level, determine what the sample size is for a margin of error of 1%.
   d) using $p_n = 0.5$ (from a similar study), and a 99% confidence level, determine what the sample size is for a margin of error of 1%.

9.31 A nonprofit wishes to build a 95% confidence interval on the proportion of uninsured Americans in the United States. In 2010, 16% of Americans had no insurance. In 2016, it was 8.6%.
   a) If the nonprofit prefers to estimate the sample size conservatively, which of the two historical proportions of uninsured Americans should they use?
   b) Using the most conservative historical proportion and a margin of error of 1%, determine the required sample size.

9.32 To infer on the proportion of new business that fail within five years, it is found that a random sample of 1877 businesses is needed. The boss considers this sample size to be too large and suggests to place a question on their website, which tends to have very large daily traffic. Is this a good idea?

## 9.8    Optional: Confidence Interval for $\sigma$

Inference on the standard deviation or variance of the population may also be required. For example, when an assembly procedure is designed, it is useful to verify if $\sigma$ is in line with specifications. Trouble with machinery, change in personnel, and wear can all distort the variability in the production line. When a random sample is taken from a normal population, the **chi-squared distribution** (pronounced "kai" squared distribution) can be used to build a confidence interval for $\sigma^2$. We will not delve into the details of this distribution, other than to say that it is useful for random variables that can only have positive values, and that the shape of the chi-squared distribution varies according to its degrees of freedom, in this case $n - 1$, but it is generally a right-skewed distribution. Specifically, if the population follows a normal distribution, the chi-squared distribution can be used to find a confidence interval for the variance of the population. Then the square root is taken to get the bounds of the interval for the standard deviation.

**Example 9.6**   *Returning to Example 9.3, suppose we wish to find a 95% confidence interval for the population standard deviation of a walking score of communities in Baltimore. Figure 9.9 shows the necessary[14] Minitab commands (left panel) and the result (right panel).*

The sample standard deviation is found to be 20.5. The interpretation is that we are 95% confident that the true Baltimore walking score standard deviation is between (17.3, 25.3). In the case of wanting a confidence interval for the variance, one can square the bounds for the standard deviation confidence interval: (299.29, 640.09).

---

**Limitation of the Chi-Squared Distribution**
Be aware that contrary to the $t$-distribution, the chi-squared distribution is not robust to violations of the normal distribution assumption. In case this assumption does not hold, the chi-squared distribution method cannot be used. An alternative is to use some bootstrapping method, a type of resampling procedure.

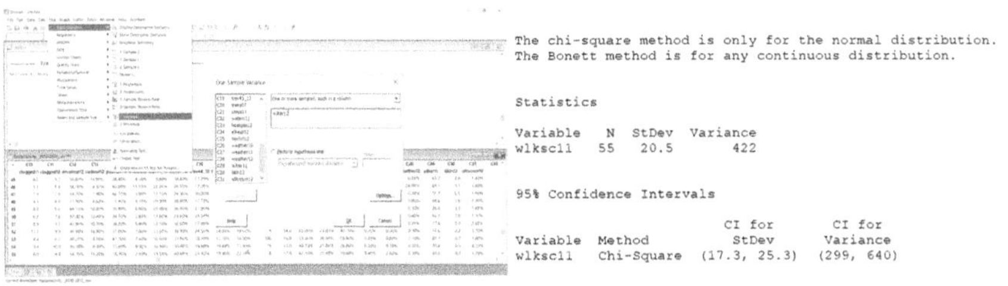

---

Figure 9.9   Window grabs to construct a 95% confidence interval for σ using Minitab.

---

14  Last row had data measured over all Baltimore and was removed.

### 9.8.1 A Quick Look at Data Science: A Confidence Interval for the Standard Deviation of Walking Scores in Baltimore

Suppose we wish to find a 95% confidence interval for the population standard deviation of a walking score of communities in Baltimore, and that the assumption of a normal distribution is not trusted. One flexible option is to use **bootstrapping** (see Section 8.4.1). Recall that in this method, the results from one sample are used to construct the sampling distribution of a statistic, in this case $s$. We apply the bootstrapping method using estimated standard deviations from $B = 50\,000$ simulated samples. The histogram of the resampling distribution shows an approximately normal distribution (Figure 9.10).

The 2.5 and 97.5% percentiles of the "resampled" standard deviations are the bounds of our 95% confidence intervals: (17.14, 23.40). For very small data sets (say $n < 30$), bootstrap methods should be avoided. From a practical point of view, for the confidence interval of the walking score standard deviation, it does not matter whether the chi-squared method or bootstrap method is used (see Example 9.6).

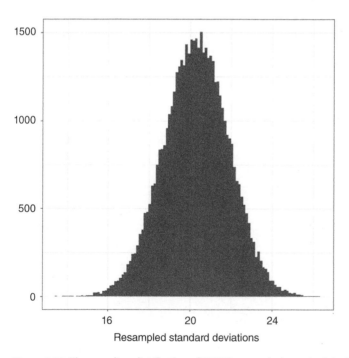

Figure 9.10 The sampling distribution of 50 000 resampled standard deviations generated from the walking score data.

# Chapter Problems

9.33 Researchers studying criminal domestic violence rates per 100 000 people in country $X$ took a random sample of 39 cities with at least 250 000 residents. Below, a graphical summary is presented.

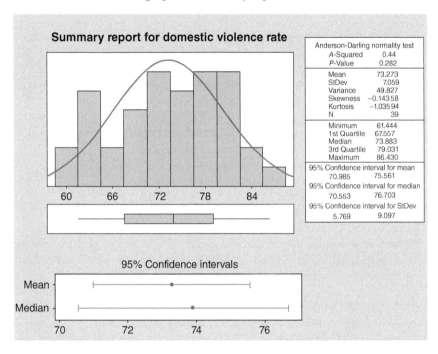

a) What is the 95% confidence interval?
b) Interpret the 95% confidence interval.

9.34 After much effort, you are convincing your "old school" boss to construct a confidence interval as part of the decision-making process for a new customer service program. Regarding the confidence level you tell him "The confidence level is a number between 0 and 100, chosen based on how confident we want to be that the interval includes the population mean." To which he responds "Fine then. Set the confidence level to 100%." How would you explain to your boss that it is not reasonable to set the confidence level to 100%?

9.35 A company will conduct an experiment to determine whether the mean time it takes to make a type of earring is within a certain limit. Assuming that the times to make the earrings have a standard deviation of 10 seconds, how many earrings should be tested so that we can be 95% confident that the estimate of the mean will not differ from the true mean time of production by more than two seconds (margin of error)?

9.36 In Problem 9.35, if the margin of error were 4 instead of 2 seconds, would the sample size be smaller?

9.37 A university is concerned with the second-year student's mean GPA. They implement new guidelines in the hopes of improving the overall students' performance. To verify the guidelines' effectiveness, they would like to sample a number of students and check their second-year GPA's. Previous work has determined the standard deviation is 0.1. They would like a 98% confidence for their experiment and a margin of error of 0.5. Find the sample size required for this experiment and explain why it is so small.

9.38 Use Appendix C to find $t$ for a
   a) 90% confidence interval using 10 observations.
   b) 95% confidence interval using 10 observations.
   c) 95% confidence interval using 50 observations.

9.39 The Southern California authorities are 95% confident that the rate of people killed in traffic accidents in 2010 was[15] between (4.27, 4.89). Find the point estimate of the rate of people killed.

9.40 As part of the decision about opening a new store, the management of a company wanted to infer about the average amount of daily sales the store would have. They knew for a fact that the standard deviation was $3500 and that the daily sales followed a normal distribution. To conduct the analysis, they took a random sample of 15 days from the records of the store at a location with similar demographics to the one considered and found a sample average of $25 500.
   a) Construct a 95% confidence interval for the true mean daily sales.
   b) Interpret the confidence interval from 9.40
   c) If in the problem you were only to change the confidence level to 90%, which confidence interval would be smaller, the 95% one or the 90% one. Explain (no need to obtain the actual new confidence interval, but simply look at confidence interval equation and explain).
   d) If in the problem you were only to change the sample size to 30, which confidence interval would be smaller, the one with $n = 15$ or the one with $n = 30$. Explain (no need to obtain the actual new confidence interval, but simply look at confidence interval equation and explain).

───────

15 Source: data.chhs.ca.gov (accessed July 26, 2019) (before last row).

e) If the management wants mean daily sales of $30 000 to bring the store to the location, what will they decide according to the confidence interval obtained in 9.40? Explain.

9.41 The owner of an oil change shop would like to know the mean time it takes to do an oil change at the shop. He takes a random sample of 16 cars that were serviced and finds a sample average of 20.5 minutes and a standard deviation of 7.4 minutes.

a) Construct a 90% confidence interval for the true mean time to do an oil change to a car.
b) Interpret the confidence interval.
c) Are you assuming anything that is not directly given to you in the problem to construct the confidence interval in 9.41a? Explain.

9.42 Current weekly sales of regular ground coffee at a supermarket average 354 pounds and vary according to a normal distribution. The store wants to reduce the price to sell more ground coffee. After the price reduction, sales in the next three weeks averaged 398 pounds, with a standard deviation of 17. Is this good evidence that average sales are now 360 pounds? Construct a 99% confidence interval to answer the question.

9.43 In 2013, it was estimated that out of 5395 births in the Australian Capital Territory, the mean rate of perinatal deaths[16] per 1000 babies was 7.6, with a standard error of 1.18. Find a 95% confidence interval using a $t$ statistic.

9.44 An indicator of the reliability of an estimate is known as the Relative Standard Errors (RSE). $RSE = s.e./estimate \times 100$, where $s.e.$ is the standard error of the estimate. In 2013, with 95% confidence, it was found that the rate of babies in Australia, born at less than 37 weeks gestation, was between (8.47, 8.67). Find the $s.e.$ if the $RSE = 0.6$.

9.45 A manufacturer of automobile tires claims that the average number of trouble-free miles given by one line of tires made by his company is 45 000 miles. When 16 randomly picked tires were tested, the mean number of miles was 50 000 with a standard deviation of 8200 miles. Assume normal distribution for the population of interest.

a) Construct a 95% confidence interval for the true mean number of trouble-free miles by this line of tires.
b) Determine from the interval whether the manufacturer's claim is justified.

---

16 Source: www.data.act.gov.au (accessed July 26, 2019).

9.46 Construct a 99% confidence interval for the true proportion of adolescents in California who consumed five or more servings of fruits and vegetables a day, when in a 2012 random sample of 1143 teenagers, 40.2% answered they did.[17]

9.47 The spokesperson of a company claims that over 88% of Americans either own a drone or know somebody who does. In a random survey of 1036 Americans, 856 said they or someone they know owned a drone. Test the spokesperson's claim by answering the questions below.
   a) Construct a 80% confidence interval.
   b) Interpret the confidence interval.
   c) Summarize the assumptions taken to construct the confidence interval.

## Further Reading

Casella, G. and Berger, R. L. (2001). *Statistical Inference*, 2nd edition. Duxbury.
Efron, Bradley and Hastie, Trevor (2016). *Computer Age Statistical Inference: Algorithms, Evidence, and Data Science*. Cambridge.

---

17 Source: archive.cdph.ca.gov (accessed July 26, 2019), page 5.

# 10

# Hypothesis Testing for One Population

## 10.1   Introduction

Three adventurers are in a hot-air balloon. Soon, they find themselves lost in a canyon in the middle of nowhere. One of the three says, "I've got an idea. We can call for help in this canyon and the echo will carry our voices far." So he leans over the basket and yells out,

"Hello! Where are we?"

They hear his voice echoing in the distance. Fifteen minutes pass. Then they hear this echoing voice:

"Hello! You're lost!!"

One of the three says, "That must have been a statistician." Puzzled, his friend asks, "Why do you say that?" He replied, "For three reasons. One – he took a long time to answer, two – he was absolutely correct, and three – his answer was absolutely useless."[1]

This joke has some truth to it, in the sense that statisticians do not always communicate well their results. But, we have all heard it, communication is a two-way street. The issue is also due to the lack of understanding of statistics in part of the receiver of the "statistical message." We have previously seen multiple concepts that are not straightforward to understand, such as the interpretation of the results from confidence intervals and the meaning of the correlation coefficient. Hypothesis testing is one of the most used statistical methods, yet the most misunderstood topic in statistics.

In practice, we often have to make decisions in situations involving uncertainty. As a result of this uncertainty, we cannot guarantee a good final result. The inferential process requires a series of steps:

- Determine what is the objective of the study.

---

1  Source: www.amstat.org (accessed July 26, 2019).

*Principles of Managerial Statistics and Data Science,* First Edition. Roberto Rivera.
© 2020 John Wiley & Sons, Inc. Published 2020 by John Wiley & Sons, Inc.
Companion website: www.wiley.com/go/principlesmanagerialstatisticsdatascience

- Develop a question that helps attain the objective. This is a crucial part of the process, and often it is given little to no credit in the statistics courses. We must determine how feasible it will be to answer the question through data, while considering the cost of the study, the complexity of the procedure to answer the question, the type of data that we can obtain, and the type of measurements necessary to answer the question, among other things.
- Determine how to answer the question with data. This will lead to establishing a hypothesis.
- Collect the data.
- Perform data analysis.
- Reach a statistical conclusion.
- Make a managerial decision.

**Example 10.1** *Imagine that you work for a company that has just manufactured a new gasoline additive to increase miles per gallon (mpg) of vehicles. Before its release to the market, it is appropriate to determine if the additive works as expected. The question of interest is: Does this additive increases miles per gallon of a car?*

How can we answer this question? That is, how do we measure miles per gallon, and what type of hypothesis should we establish? In theory, there are various ways to proceed. You could take into consideration every factor you know affects the miles per gallon of a car: the driver, type of car, type of tires, type of surface on which it is being driven, weather, etc. Then, according to those factors, determine what the miles per gallon are for a driver. However, this level of detail can become costly. Moreover, it could be impossible to determine adequately due to the fact that in reality, many factors will have an impact in the miles per gallon for a vehicle, some which are yet unknown. Another alternative is to focus on MEAN miles per gallon of every driver under certain conditions. That is, we answer the question in terms of mean miles per gallon of every driver under certain standard conditions and state the hypothesis in terms of the mean miles per gallon.

As seen in the example above, it can be handy to solve problems according to the mean value of a measure. In Chapter 9, confidence intervals were used as a tool to conduct statistical inference. In this chapter, the procedure for statistical inference most used in practice is introduced: hypothesis testing. In this method, an assumption is made about the value of the parameter, then the data is used to determine if the assumption is reasonable. At first, our focus is hypothesis tests about the mean of a population. We later discuss hypothesis tests for proportions. Important principles of statistical inference will also be covered in more detail.

## 10.2    Basics of Hypothesis Testing

Our aim is to reach a conclusion about a parameter from the population or about several parameters. Hypothesis testing starts by making two statements about the world. The two statements ares

- mutually exclusive, and therefore only one of the statements must be true.
- collectively exhaustive, together they cover all the possible values of the parameter(s).

Since it is not known what the truth is, each statement is a hypothesis. One of them is called the **null hypothesis**, $H_o$. The second statement with an opposite argument than the one established in $H_o$ is called the **alternative hypothesis**, $H_1$. Both hypotheses are usually presented in the following format,

$H_o$: Null statement about parameter(s).

$H_1$: Alternative statement about parameter(s).

For example, if in a given situation the null hypothesis is that the population is greater than or equal to 5.4, then the alternative hypothesis is that the population is less than 5.4 and they are presented together as follows:

$H_o$: $\mu \geq 5.4$

$H_1$: $\mu < 5.4$

As can be seen, $H_o$ and $H_1$ together cover all the possible values of $\mu$ and are mutually exclusive, so only one of them is true (and both of them cannot be simultaneously false). Similarly, if in a given situation $H_o$: $\mu \leq 10$, then $H_1$: $\mu > 10$, and if $H_o$: $\mu = 100$, then $H_1$: $\mu \neq 100$. These principles also apply when performing hypothesis testing on other parameters. For example, if $H_o$: $\pi \leq 0.65$, what is $H_1$? In later chapters, we will conduct hypothesis testing comparing multiple parameters. If $H_o$: $\mu_1 = \mu_2$, what is $H_1$? And what about $H_o$: $\mu_1 = \mu_2 = \mu_3$, what is $H_1$? When $H_o$: $\beta_1 = \beta_2 = \beta_3 = 0$, what is $H_1$? In practice, we are not told explicitly what $H_o$ or $H_1$ is. How to determine both hypotheses will be discussed later.

Hypothesis testing works by speculating that $H_o$ is true, before data is observed. This is followed by collecting data which either provides evidence against $H_o$ or it does not. The evidence is through the estimator of the parameter of interest. Recall from Chapter 8 that estimators have a series of properties such that although we do not know the true parameter value, we feel confident that the estimator is relatively "close" to the parameter value.

*Errors in Hypothesis Testing*    Hypothesis testing has been designed in such a way that we reach a conclusion which has the highest probability of being correct. Despite this, since the true value of the parameter is unknown, it is possible that the wrong hypothesis is chosen as true. Viewing the decision in terms of the null and alternative hypotheses, we can classify our decision to choose a hypothesis

Table 10.1 The decision that we make about $H_o$ versus the true statement about the population.

|  | Truth about population | |
| --- | --- | --- |
| Decision about $H_o$ | $H_o$ is true | $H_o$ is false |
| Reject $H_o$ | Type I error | Right decision |
| Do not reject $H_o$ | Right decision | Type II error |

into two correct decisions and two erroneous decisions. Most likely, the null will be rejected when false, and the null will not be rejected when true. For the errors, we can reject the null hypothesis (a decision relying on data) when the null hypothesis is in fact true. This is known as **type I error** (see Table 10.1). Or, the second type of error that one can make is NOT to reject the null hypothesis when in fact it is false. This is known as **type II error**. The repercussions of incurring in these errors tend to have different tolls on the company and the customer, with which error being worse (and to whom) depending on the situation at hand.

**Example 10.2** *If our null and alternative are,*

$$H_o: \mu \geq 5.4$$
$$H_1: \mu < 5.4$$

*Then, incurring in type I error means that we conclude $\mu < 5.4$, while in reality the true population mean is greater than or equal to 5.4. On the other hand, incurring in type II error means that we conclude $\mu \geq 5.4$, while in reality the true population mean is lower than 5.4.*

Naturally, it is not desirable to incur in these errors. Our choice of a hypothesis as true is dependent on random statistics gathered from a sample. Therefore, our aim is to keep the probabilities of incurring both types of errors as small as possible. The probability of incurring in type I error is often written as

$$\alpha = P(\text{type I error}) = P(\text{Reject } H_o | H_o \text{ is true})$$

In practice, this probability is known as the **significance level or level of significance**. The probability of incurring type II error is often written as

$$\beta = P(\text{type II error}) = P(\text{Do not reject } H_o | H_o \text{ is false})$$

**A Compromise with Hypothesis Testing Errors**
Unfortunately, it is not straightforward to reduce simultaneously the probabilities of both types of errors because they are associated with one another. For example, all things equal, if $\alpha$ is chosen to be small, it is harder to reject the null. However, a side effect is that now it becomes more likely not to reject the null

by mistake; the probability of incurring type II error has increased. Only increasing the sample size will simultaneously reduce $\alpha$ and $\beta$. Statisticians decided to fix $\alpha$ at an "adequate value" and develop a procedure of hypothesis testing where $\beta$ is minimized. This implies that the value of $\alpha$ is known BEFORE gathering data and analyzing the results. The adequate value of $\alpha$ is dependent on how acceptable it would be to incur in type I error. Typical values for $\alpha$ are 0.01, 0.05, and 0.10.

Why not use $\alpha = 0$? See Problem 10.67. From a managerial perspective, the negative impact in profit when incurring in type I error is one thing that is used to determine $\alpha$. A considerable reduction in profit when incurring in type I error can be decreased by setting a low $\alpha$.

**Example 10.3** *An old product is to be compared to a newer version that the company is evaluating to put on the market. Expenses to the company would be lower with the new version of the product; but higher supplier costs would affect the selling price of the new version, making it considerably more expensive to the consumer.*

$H_o$ : *Both versions of the product are just as effective on what they do for the consumer.*

$H_1$ : *New version of the product is more effective on what they do for the consumer.*

In this example, type I error means concluding that the latest version of the product is more effective when in fact both versions are just as effective. The consumer would then have to pay a higher price. From the consumer's standpoint, $\alpha$ should be low. The burden of type I error may not be as bad for the company however, unless type I error affects considerably its profit (say, if customers find out both versions of the product are just as effective). On the other hand, a type II error means concluding that both versions of the product are as effective on what they do for the consumer when in reality the new version is more effective. The company would have missed out on an opportunity to make more money. Thus, the company would like to keep $\beta$ low.

**Example 10.4** *An old product wants to be compared to a newer version that the company is evaluating to put on the market. Expenses to the company would be lower with the new version of the product. Both versions will have the same performance and cost the same to the consumer. The old version never had any real serious defects after many years on the market.*

$H_o$: *A recall is less or equally likely with the new version than with the old one.*

$H_1$: *A recall is more likely with the new version than with the old one.*

This time, type I error is of no consequence to the consumer, but the company may miss an opportunity to reduce expenses by not placing the new version of the product in the market. On the other hand, type II error would lead to the market release of a product with a higher chance of being defective. Both the company and the consumer could be affected. A low $\beta$ is preferable.

In summary, it is important to recognize the possibility of incurring in any of these errors. Traditionally, focus is on $\alpha$ though incurring in type II error can lead to serious consequences. To minimize the probability of incurring type II error does not mean that the probability is low. It may mean that in certain situations, the lowest $\beta$ could be 0.2, that is, one in five hypotheses are incurring in type II error.

*Power of a Test*   Another important concept is

**Power of a test** $= 1 - \beta = P(\text{Reject } H_o | H_o \text{ is False})$

As we can see from this equation, the event "Reject $H_o | H_o$ is False" is the complement of type II error. The higher the power of a test, the more likely we made the right decision when rejecting $H_o$. Also, since $H_o$ is False has the same meaning as $H_1$ is True, the power of the test supposes the alternative hypothesis is true. Typically, we desire the power of a test to be 0.8 or greater. The calculation of power of a test requires knowing the true mean parameter value or an effect size (a departure from the null parameter value deemed practically important). Since the true parameter value is unknown, often a power curve is displayed instead of the actual power of a test. The power curve calculates the probability of rejecting the null over many potential values of the parameter that are part of the alternative hypothesis set of values.

**Example 10.5**   *To illustrate how* $\alpha$, $\beta$ *and power of a test work, suppose a sample mean is used to draw inference on* $\mu$ *in a situation where it is known that* $\overline{X}$ *follows a normal distribution. Specifically,*

$H_o$: $\mu \le 20$

$H_1$: $\mu > 20$

*The population mean is assumed to be 20 when, unknown to us, in reality* $\mu = 23.5$. *Under* $H_o$, *the peak of the normal curve should be at 20 (Figure 10.1). The darker blue shade represents the probability of rejecting the null when it is true,*

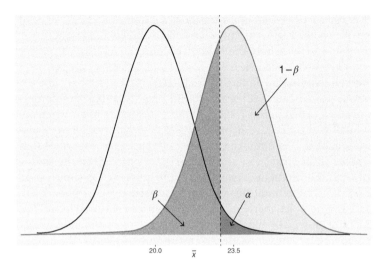

Figure 10.1 How $\alpha$, $\beta$, and the power of a test interact in a scenario where the null states that $\overline{X}$ is normally distributed with population mean 20, when in reality $\overline{X}$ is normally distributed with a population mean of 23.5.

$\alpha$. *Moreover, the brown-shaded region represents the probability of not rejecting the null when it is false, $\beta$. Note that if $\alpha$ were bigger, then $\beta$ would be smaller and vice versa. The probability of rejecting the null when it is false is shown in light blue, the power of the test.*

The further the true value (or effect size) from the parameter value under the null, the higher the power. Also, the greater the sample size, the greater the power of the test and the smaller $\alpha$. It is possible to determine the required sample size so that the test has some preset power.

---

**Choosing the $H_o$ Statement**
The null hypothesis argument is determined from a verbal problem, and this tends to create confusion among students. In some cases, the problem establishes directly what the null hypothesis is and from this we deduce the alternative hypothesis. In other circumstances, the problem establishes the alternative hypothesis directly and from that we deduce the null hypothesis. We need to recognize that the null hypothesis argument always has an equality symbol. That is, it will have $\geq$, $\leq$, or = sign in it.

---

The alternative hypothesis on the other hand is based on arguments of strict inequalities $<$, $>$, and $\neq$. Why are the hypotheses established this way? Because the way the hypothesis test is completed is by finding probabilities, assuming the argument in $H_o$ is true. When inferring about a parameter, say $\mu$, we need to assign a specific value, $\mu = \mu_0$, to find what will be known as a test statistic. It is exactly in this equality where the hypothesis test has the highest power to reject the null hypothesis when it is not true, which is why we always need equality in the null hypothesis. Taking this into account and that $H_o$ and $H_1$ should be mutually exclusive, that leaves us with only three types of combinations of $H_o$ and $H_1$ when performing hypothesis testing. Earlier, when we stated that the null and alternative hypothesis were mutually exclusive, some examples were provided. If you return to those examples, you will notice that the null statements always had $\geq$, $\leq$, or $=$ sign in it. The sign of the alternative may sometimes require careful understanding of the problem. On some cases, the use of the word "increase" indicates a $>$ should be used in $H_1$. In other cases, the use of the very same word "increase" indicates a $<$ should be used in $H_1$. The statements can be expressed in many different ways, and using different words than increase, and thus, practice is needed to ensure one can identify the proper $H_1$ inequality.

## 10.3   Steps to Perform a Hypothesis Test

In what follows, we define some intuitive steps necessary to perform hypothesis testing. As students will see through the rest of this book, there is a very wide range of hypothesis tests that can be performed. All of these different types of hypothesis tests will involve the following five steps.[2] Steps for hypothesis testing are part of the inferential process discussed in Section 10.1.

***Step 1: Define $H_o$ and $H_1$, and choose the significance level***   We only have three possible combinations of $H_o$ and $H_1$, but this step tends to be problematic to students. Remember, $H_o$ must always have an equality sign, regardless of whether we obtain the argument directly or indirectly from the problem at hand. Also, at this stage, a significance level is chosen. In practice, $H_o$, $H_1$, and $\alpha$ are all selected before observing the data.

***Step 2: Choose and calculate the test statistic to be used***   $H_o$ is assumed to be true until empirical evidence suggests otherwise. According to the parameter being estimated, a test statistic must be selected. For convenience, this test statistic should be as general as possible. It must depend on a statistic that estimates the parameter of interest yet should apply regardless of the value of the parameters. For all the hypothesis tests performed in this book, we can rely on some quantity that is a function of the parameter of interest, is a function of the statistic estimating the parameter of interest, maybe is also a function of an additional (nuisance) parameter or its estimate, and whose probability distribution does

---

2  Keep in mind that there are some initial steps required before one reaches the point of hypothesis testing.

not depend on these parameters. This quantity is known as a **pivotal quantity**. In fact, confidence intervals are also constructed based on pivotal quantities.

**Step 3: Find the value of the decision rule**   In the context of inference on $\mu$, given the properties of $\overline{X}$ as an estimator of $\mu$, we anticipate that the sample average is close to the true population mean, regardless of whether $\mu$ is the value established in $H_o$ or not. The **decision rule** is the rule used to decide whether to reject or not reject $H_o$.

**Step 4: Make a decision about $H_o$**   Is the evidence against or in favor of $H_o$? The significance level helps in determining a threshold for the decision rule.

**Step 5: Statistical conclusion**   In this step, we interpret, in the context of the problem at hand, the decision to reject or not reject $H_o$. Keep in mind that the statistical interpretation of the results may not automatically imply practical importance.

## Practice Problems

10.1   Determine the appropriate mathematical symbol that corresponds to each statement: $=, \geq, \leq, \neq, >$, or $<$.
   a)  Mean SAT score of high school students is 625.
   b)  The proportion of batteries produced that are defective is at least 0.05.
   c)  The proportion of houses with solar panels is no more than 0.20.
   d)  The variance in weight of cement bags is less than 0.5 pounds.
   e)  Mean weight of new laptops is greater than 3 pounds.
   f)  Variance in pipe circumference is not equal to 5 inches.
   g)  The difference between mean systolic blood pressure in men and mean systolic pressure in women is at least zero.

10.2   Determine if each statement in Problem 10.1 pertains to $H_o$ or $H_1$.

10.3   In what follows, write the expression for $H_o$ according to the given $H_1$.
   a)  $H_1 : \mu > 103.2$,
   b)  $H_1 : \mu < 547$,
   c)  $H_1 : \mu \neq 706.44$,
   d)  $H_1 : \pi \neq 0.75$,
   e)  $H_1 : \mu_1 - \mu_2 > 0$,
   f)  $H_1 :$ at least one coefficient is nonzero.

10.4   State what is wrong in each combination of null and alternative hypotheses:
   a)  $H_o : \mu = 17; H_1 : \mu \geq 17$.

b) $H_o: \mu \leq 29; H_1: \mu > 35$.
c) $H_o: \pi > 0.60; H_1: \pi \leq 0.60$.
d) $H_o: \mu \neq 105; H_1: \mu = 105$.
e) $H_o: \sigma^2 < 28; H_1: \sigma^2 \geq 28$.
f) $H_o: \mu_1 - \mu_2 - \mu_3 = 0; H_1:$ the means are not different.

10.5  Suppose a researcher sets $\alpha = 0.01$. Interpret the meaning of this value of $\alpha$.

10.6  Suppose $\beta = 0.20$. Interpret the meaning of this value of $\beta$.

10.7  In the following, find the power of the test and the probability of not rejecting the null when the null is true.
a) $\alpha = 0.01, \beta = 0.10$.
b) $\alpha = 0.05, \beta = 0.20$.
c) $\alpha = 0.10, \beta = 0.05$.
d) $\alpha = 0.01, \beta = 0.50$.

10.8  Determine in the following whether type I, type II, or no error occurs
a) $H_o: \mu \leq 12$ is not rejected. True population mean is 15.
b) $H_o: \mu \geq 27$ is rejected. True population mean is 35.
c) $H_o: \mu \leq 90$ is not rejected. True population mean is 90.
d) $H_o: \mu \geq 75$ is rejected. True population mean is 80.

10.9  A firm was testing whether a new gadget design allowed an increase in its battery life. Specifically, if $\mu$ is the expected battery life in hours when the gadget is new, $H_o: \mu \leq 10.7$ and $H_1: \mu > 10.7$. The power of the test was found to be 0.84. Interpret the meaning of the power in the context of the situation presented.

10.10  List the five steps of hypothesis testing.

10.11  Define
a) What is a pivotal quantity?
b) What is a decision rule?

## 10.4   Inference on the Population Mean: Known Standard Deviation

It is time to consider hypothesis testing on $\mu$ when $\sigma$ is known.

For now, let's assume that $\overline{X}$ follows a normal distribution and the population standard deviation $\sigma$ is known. In addition, we speculate that the mean of the population has a specific value, say $H_o: \mu = \mu_o$. Later, we mention ways to verify the normality and known standard deviation assumptions. Our goal is to determine if the value $H_o$ is appropriate based on data, in a probabilistic way. $H_1$ would present an argument such that $H_o$ and $H_1$ are mutually exclusive.

**Sample Mean Properties as an Estimator of $\mu$**   Recall from Section 8.5 that $\overline{X}$ has many properties that allow us to feel confident that it is estimating $\mu$ adequately. It is

- unbiased
- consistent
- efficient
- sufficient

Due to these properties, if we assume that the null hypothesis is correct, $\mu = \mu_o$, $\overline{X}$ must be "close" to $\mu_o$. Since $\overline{X}$ follows a normal distribution and the population standard deviation $\sigma$ is known, we can transform to the standard normal distribution. Thus, after standardization, assuming the null is true, $Z$ should be close to its expected value, 0. Moreover, the probability of $Z$ being equal or as extreme as the value obtained in the sample should not be "too" small. If the probability is too small, this indicates that the premise on the null hypothesis is not justified, it is unlikely that $\overline{X}$ will be far away from the true population mean, and thus we should reject $H_o$.

---

**Other Estimators Have Good Properties Too**
The same intuition applies in the case of hypothesis testing for other parameters, the distribution of the estimators are known. Additionally, common estimators have good properties, leading us to conclude that the estimator should be, statistically speaking, "close" to what it is attempting to estimate, even when that value is unknown.

---

**Type of Hypothesis Tests (and Determining $H_o$ Statement)**   From "Choosing the $H_o$ Statement" in Section 10.2, it is known that the null must always have $\geq$, $\leq$, or $=$ signs. When inferring about $\mu$, and when $\overline{X}$ follows a normal distribution, we need to assign a specific value, $\mu = \mu_o$ to find the test statistic, $Z$. Recall that it is exactly in this equality where the hypothesis test has the highest power to reject the null hypothesis when it is not true, which is why we always need equality in the null hypothesis. Taking this into account and that $H_o$ and $H_1$ should be mutually exclusive, that leaves us with only three type of hypothesis tests on $\mu$.

**Lower Tail Test**

$$H_o: \mu \geq \mu_o$$
$$H_1: \mu < \mu_o$$

In this case, if $\bar{x}$ is far enough to the lower tail of the sampling distribution, then $H_o$ would be deemed unlikely, hence $H_o$ is rejected. On the other hand, an $\bar{x}$ close enough to $\mu_o$ or greater than $\mu_o$ would be considered weak evidence against $H_o$ and hence it would not be rejected.

### Upper Tail Test

$$H_o: \mu \leq \mu_o$$
$$H_1: \mu > \mu_o$$

This case works the other way around, if $\bar{x}$ is far enough to the upper tail of the sampling distribution, then $H_o$ would be deemed unlikely, hence $H_o$ is rejected. On the other hand, an $\bar{x}$ close enough to $\mu_o$ or lower than $\mu_o$ would be considered weak evidence against $H_o$ and hence it would not be rejected.

### Two-Tailed Test

$$H_o: \mu = \mu_o$$
$$H_1: \mu \neq \mu_o$$

In the two-tailed test, to reject $H_o$, $\bar{x}$ must be far enough to the left of or far enough to the right of $\mu_o$. On the other hand, an $\bar{x}$ close enough to $\mu_o$ would be considered weak evidence against $H_o$ and hence the hypothesis would not be rejected.

Now, although $H_o$ and $H_1$ should be mutually exclusive, recall that it is at $\mu = \mu_o$ where the hypothesis test has the highest power to reject the null hypothesis when it is not true. That is, if $H_o: \mu \leq \mu_o$, at $\mu = \mu_o$, the hypothesis test procedure has the highest power. Therefore, from a practical point of view, the $H_o$ and $H_1$ pairings

$$H_o: \mu \geq \mu_o \qquad H_o: \mu = \mu_o$$
$$\text{and}$$
$$H_1: \mu < \mu_o \qquad H_1: \mu < \mu_o$$

are equivalent. Similarly,

$$H_o: \mu \leq \mu_o \qquad H_o: \mu = \mu_o$$
$$\text{and}$$
$$H_1: \mu > \mu_o \qquad H_1: \mu > \mu_o$$

are equivalent.

---

### Interpreting the Decision About $H_o$

The equivalences above bring to the surface a conundrum: how should we word the decision about $H_o$? Specifically, is there a difference between "accept the $H_o$," and "do not reject $H_o$"? The answer is yes, there is a difference, and the correct wording of the decision on $H_o$ is the latter.

Suppose we have

$H_o: \mu \le \mu_o$

$H_1: \mu > \mu_o$

---

**Can I Accept $H_o$?**

In hypothesis testing, the data gives weak evidence against $H_o$ when $\overline{X}$ is close to $\mu_o$. However, for the null and alternative above, $\mu$ may be much smaller than $\mu_o$, yet the data has simply given us evidence to not reject $H_o$ based on using $\mu_o$. Does this mean that we can only be capable of accepting the null when conducting a two-tailed test? The answer is no. From the definition of $\alpha$, the confidence level of a confidence interval is

$$1 - \alpha = P \text{ (Do not Reject } H_o | H_o \text{ is True)}$$

Therefore, we have "confidence" in not rejecting the null, not on accepting it. Other arguments can be used to show that hypothesis testing does not allow accepting $H_o$. Some believe that this is simply a wording issue, but making the distinction helps us recognize the limitations of hypothesis testing.

---

In this book, we prefer to provide the expressions of the null and alternative more in line with how we defined them at the beginning of this chapter (that they are together exhaustive and mutually exclusive). That way, it will be more natural to determine $H_1$ from the null and to reach the appropriate conclusion from hypothesis testing.

---

**The Hypothesis Testing and Confidence Interval Connection**

At a fixed $\alpha$, the same conclusion about $\mu$ is reached whether a two-sided hypothesis test is performed, or a two-sided confidence interval is calculated. It is tempting to think that hypothesis testing is more general on account of there being three type of tests. However, it is also possible to construct **one-sided confidence intervals** (one bound is finite and the other is infinite) which at a fixed $\alpha$ have equivalent conclusions to lower tail and upper tail hypothesis tests. In one-sided confidence intervals, you either subtract or add a margin of error to the point estimate

- if you subtract the margin of error, the upper limit is open (infinite);
- if you add the margin of error, the lower limit is open (infinite).

Table 10.2 Equivalence between hypothesis tests and confidence intervals (assuming the same significance level).

| Type of hypothesis test | Type of confidence interval | Reasoning |
|---|---|---|
| Upper tail | Lower bound: $\overline{X} - z_{1-\alpha}\frac{\sigma}{\sqrt{n}}$ | Estimating the lowest possible value for $\mu$ according to data |
| Lower tail | Upper bound: $\overline{X} + z_{1-\alpha}\frac{\sigma}{\sqrt{n}}$ | Estimating the highest possible value for $\mu$ according to data |
| Two-sided | Two-sided: $\overline{X} \pm z_{\alpha/2}\frac{\sigma}{\sqrt{n}}$ | Estimating the highest and lowest possible value for $\mu$ according to data |

Table 10.2 summarizes the equivalence between hypothesis tests and confidence intervals. For example, instead of an upper tail test, one could obtain a lower bound confidence interval – estimating the lowest possible value for $\mu$ according to data.[3]

From Chapter 8, we know that $\overline{X}$ has a normal distribution when the population is normal. Also, if the sample size $n$ is large, $\overline{X}$ is approximately normally distributed according to the central limit theorem. Defining $H_o$ and $H_1$ will depend on the problem at hand under these conditions, but let's focus on obtaining the test statistic. We want the test statistic to be as general as possible, something that depends on a statistic that estimates the parameter of interest yet applies regardless of the value of the parameters, $\mu$ and $\sigma$, a pivotal quantity. Thus, we must obtain a test statistic to draw inference on $\mu$ when $\sigma$ is known, and $\overline{X}$ is $N(\mu, \sigma/\sqrt{n})$. The test statistic in this case is

$$Z = \frac{\overline{X} - \mu_o}{\sigma/\sqrt{n}} = \frac{\sqrt{n}(\overline{X} - \mu_o)}{\sigma}$$

We already encountered $Z$ in Chapters 7 and 8, and from there we know $Z$ has an $N(0, 1)$. That is, the probability distribution of $Z$ does not depend on $\mu$ or $\sigma$. To obtain the test statistic, note that $\mu = \mu_o$. This is the value of the population mean such that the hypothesis test has the most power. In the current scenario – $\overline{X}$ is normally distributed and $\sigma$ is known – hypothesis testing is also called a **Z-test**.

Figure 10.2 presents decision rules for the three types of possible hypothesis tests. Specifically, to reject the null in a lower tail test, the value of $Z$ must be "extremely" below 0. In contrast to reject the null in an upper tail test, the value of $Z$ must be "extremely" above 0, and to reject the null in a two-tailed test, the value of $Z$ must be either "extremely" below or "extremely" above 0.

---

3 For an upper tail test, the analogy with a one-sided confidence interval is to estimate the lowest possible value that the population mean could have, it is a lower bound confidence interval.

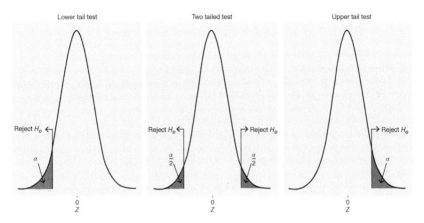

Figure 10.2 Sampling distribution of $Z$ under the null. If $H_o$ is true, in each of the three types of tests seen here, the observed test statistic should not be too extreme. What is considered too extreme from the expected value of $Z$ depends on the alternative and is the region presented in red.

Next in line, step 3 of hypothesis testing: find the value of the decision rule. $H_0$ must be either rejected or not rejected. This decision will depend on what the data tells us about $H_0$, information provided through the test statistic. In hypothesis testing, there are two equivalent decision rules. Our emphasis will be on the rule most commonly used: the **$p$-value**.

**The p-value** It is the probability of obtaining a test statistic value at least as extreme as the observed result given the null hypothesis is true. Hence,

- if the $p$-value is low, this implies that our statement that the population mean has the value $\mu_0$ appearing in $H_0$ is likely to be incorrect. That is, considering the $\bar{x}$ should be close to the true value of the population mean and our sample result is unlikely assuming $\mu = \mu_0$, the most reasonable explanation is that the null hypothesis is incorrect.
- if the $p$-value has a large value, then $\bar{x}$ does not contradict $\mu = \mu_0$, and thus the null is not rejected.

The limit assigned to the $p$-value is $\alpha$, because this has been chosen as an acceptable type I error. In summary, if $p$-value $\leq \alpha$, we reject $H_0$, otherwise we do not. Now care is needed when quantifying if the test statistic is at least as extreme than the observed result when the null hypothesis is true. To quantify this, we must pay attention to what is stated in $H_1$. Let $z_0$ be the test statistic observed in the sample,

If $H_1$: $\mu < \mu_0$, then $p$-value $= P(Z \leq z_0)$.
If $H_1$: $\mu > \mu_0$, then $p$-value $= P(Z \geq z_0)$.
If $H_1$: $\mu \neq \mu_0$, then $p$-value $= 2P(Z \geq |z_0|)$.

These equations to calculate the $p$-value come straight from its definition, the last equation exploiting the fact that normal distributions are symmetric and hence we obtain one probability and multiply it by 2. A subtle consequence of the $p$-value formulas is that if the null is rejected in a two-sided test, a one-sided test performed with the same data, at the same significance level, would certainly reject the null too. On the other hand, rejecting the null with a one-sided test does not guarantee rejection of the null when performing a two-sided test since the $p$-value is always bigger. The two-sided test is a more conservative test than the one-sided test.

**Example 10.6** *Suppose*

$$H_o: \mu \geq 13.5$$

$$H_1: \mu < 13.5$$

*and* $\bar{x} = 13.34$, $\sigma = 2.75$, *and* n = 80. *What is the* p-*value?*

By the central limit theorem, the sample mean follows approximately a normal distribution. Therefore,

$$Z = \frac{13.34 - 13.5}{2.75/\sqrt{80}} = \frac{\sqrt{80}(13.34 - 13.5)}{2.75} = -0.52$$

The $p$-value is the probability that $Z$ is $-0.52$ or more extreme. According to the alternative, more extreme means lower

$$P(Z \leq -0.52) = 0.30$$

Figure 10.3 helps us visualize how the $p$-value works when $H_1: \mu < \mu_o$. Under $H_o$, $Z$ should have a standard normal distribution with an expected value of 0. The $p$-value was 0.3. If $\alpha = 0.05$, then $0.3 \nless 0.05$ and therefore the null hypothesis is not rejected.

Currently, to conduct hypothesis testing on $\mu$, we need

- $\bar{X}$ to follow a normal distribution,
- independent observations, and
- known population standard deviation.

The validity of our conclusions depends on these three statements being true. In Section 8.4, we stated that for academic purposes we would assume that $\bar{X}$ is approximately normally distributed if $n \geq 30$, although in practice the skewness of the population distribution must always be assessed.

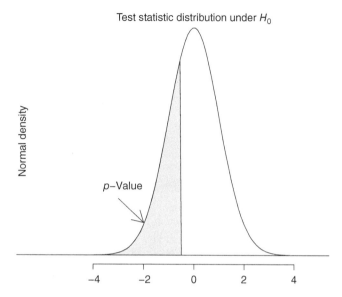

Test statistic distribution under $H_0$

Normal density

$p$–Value

-4  -2  0  2  4

Figure 10.3  Under $H_o$, $Z$ should have a standard normal distribution. The $p$-value is shaded in blue and leads us to not reject $H_o$ in this example, since $\alpha = 0.05$.

> In terms of inference, incorrectly using a normal approximation for $\overline{X}$ is of less concern in the case of two-sided hypothesis tests or confidence intervals because the probability error of one tail is often compensated by the opposite error in the other tail. For one-sided inference, the normality of $\overline{X}$ should be more carefully evaluated.

**Example 10.7**  *Returning to Example 10.1, suppose that without the gasoline additive, the mean is 25 mpg with a standard deviation of 1.5. A random sample of 30 cars using the gasoline additive results in an average of 26.4 mpg. The miles per gallon are approximately normal and its variability is unaffected by the additive. We would like to see if there is evidence that the gasoline additive was able to achieve higher miles per gallon at 5% significance.*

**Step 1: State null and alternative**  Let $\mu$ = mean miles per gallon with the additive. The company wants to determine if the gasoline additive is able to increase miles per gallon. This is a strict inequality statement and thus goes in the alternative hypothesis,

$$H_o: \mu \le 25$$
$$H_1: \mu > 25$$

$\alpha = 0.05$.

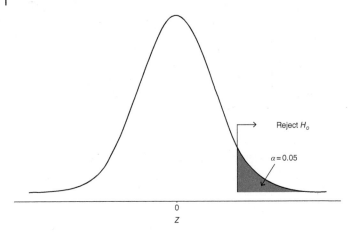

Figure 10.4 If $H_o$ is true, it is unlikely the observed test statistic will be too far above 0. The region presented in red is defined by $\alpha = 0.05$.

**Step 2: Choose and calculate the test statistic to be used**  Since the standard deviation is known and miles per gallon are normally distributed, a $Z$-test will be performed. Figure 10.4 illustrates the decision rule,

$Z$ is found as

$$Z = \frac{26.4 - 25}{1.5/\sqrt{30}} = \frac{\sqrt{30}(26.4 - 25)}{1.5} = 5.11$$

**Step 3: Find the value of the decision rule, the p-value**  To reject the null, the $Z$ value must be "extreme" enough (Figure 10.5), which in this example, according to $H_1$, means that it must be far enough above 0.

$$p\text{-Value} = P(Z \geq 5.11) = 1 - P(Z \leq 5.11) = 0$$

**Step 4: Make a decision about $H_o$**  Since $0 < 0.05$, the null is rejected.

**Step 5: Statistical conclusion**  At 5% significance, we conclude that there is evidence that the gasoline additive was able to achieve an increase in miles per gallon.

**Using Minitab**  A $Z$-test can be performed using Minitab. Like many statistical software, hypothesis testing tends to come together with confidence intervals. To solve Example 10.7 using Minitab, go to `Stat > Basic Statistics > 1-Sample Z`. Figure 10.6 illustrates the steps (since the data is summarized, the `Summarized data` option must be chosen). Pressing the `Options` button, a new window will open where one can select the type of test

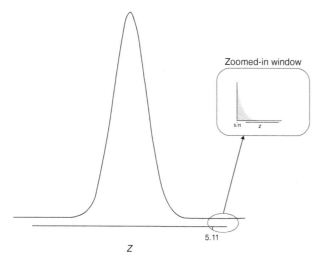

Figure 10.5  Under $H_o$, Z should have a standard normal distribution. The *p*-value is shaded in blue.

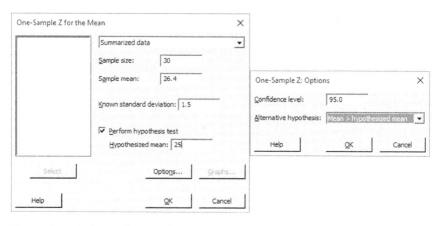

Figure 10.6  Window grabs to perform a Z-test using Minitab.

and change the confidence level[4] (default 95%). The output (Figure 10.7) leads to the same conclusion we reached by manually performing the computations.

Even though practical importance is paramount, this determination will be dependent on factors that are separate from the statistical aspects discussed here. In Example 10.7, statistically we have determined that the gasoline additive was successful in increasing miles per gallon. However, the practical relevance of the results will depend on the cost of gasoline and the cost of the

_____
4  Recall that the confidence level is $1 - \alpha$.

**One-Sample Z**

```
Test of μ = 25 vs > 25
The assumed standard deviation = 1.5

 N    Mean   SE Mean   95% Lower Bound     Z      P
30   26.400   0.274             25.950   5.11   0.000
```

Figure 10.7 *Z*-test Minitab output.

additive among other things. Earlier, we stated that instead of an upper tail test, one could obtain a lower bound confidence interval – estimating the lowest possible value for $\mu$ according to data. The one-sided 95% confidence interval for this example would have a lower bound of 25.95 (Figure 10.7). The interpretation of this confidence interval is that we are 95% confident, that mean miles per gallon of cars with the gasoline additive is 25.95 or higher. This one-sided confidence interval indicates the true mean miles per gallon for the additive may only be slightly above 25.

**Example 10.8** *A company produces steel rods for construction. Every once in a while, the staff inspects if steel rods produced by their machine fulfill specifications, among them, that their diameter is 14 mm. The staff has already determined that the standard deviation of rod diameters is 1 mm, as it should. A random sample of 60 rods results in a mean diameter of 13.98 mm. Test at 0.01 significance if the machine is producing adequate steel rods.*

**Step 1: State null and alternative** Let $\mu$ = mean diameter of the steel rods. The company wants to determine if the produced steel rods have the diameter that they should. This is an equality statement and thus goes in the null hypothesis,

$$H_o: \mu = 14$$
$$H_1: \mu \neq 14$$

$\alpha = 0.01$.

**Step 2: Choose and calculate the test statistic to be used** Since the standard deviation is known and by the central limit theorem we assume the sample mean is normally approximated, our procedure is a $Z$-test. Figure 10.8 illustrates the decision rule,

$Z$ is found as

$$Z = \frac{\sqrt{60}(13.98 - 14)}{1} = -0.15$$

**Step 3: Find the value of the decision rule, the p-value** To reject the null, the $Z$ value must be "extreme" enough, which in this example, according to $H_1$, means that

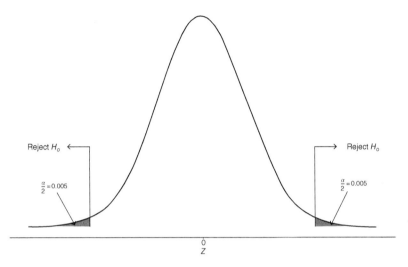

Figure 10.8 If $H_o$ is true, the observed test statistic should likely be sufficiently close to zero. Each region presented in red is defined by $\alpha = 0.01/2$.

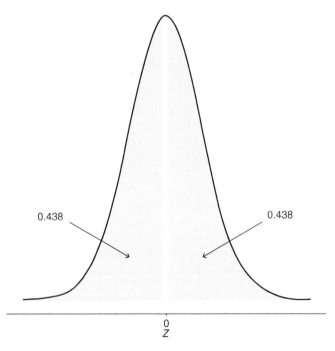

Figure 10.9 Sampling distribution of the test statistic. Each shaded area corresponds to $p$-value/2.

it must be far enough above or below 0. This could be simplified to finding (Figure 10.9)

$$p\text{-Value} = 2P(Z \geq |-0.15|) = 2(1 - P(Z \leq 0.15)) = 0.877$$

**Step 4: Make a decision about $H_o$** Since $0.877 \nless 0.01$, the null is not rejected.

**Step 5: Statistical conclusion** At 1% significance, we find no evidence that indicates that the machine is producing faulty steel rods.

## Practice Problems

10.12 For a study, $H_o: \mu = 50$ and $H_1: \mu \neq 50$. A two-sided confidence interval at confidence level $1 - \alpha$ was found to be (35.7, 52.3). Do you think that conducting the hypothesis testing at the same significance level will reject the null or not? Explain.

10.13 For a study, $H_o: \mu = 175$ and $H_1: \mu \neq 175$. A two-sided confidence interval at confidence level $1 - \alpha$ was found to be (113.57, 126.83). Do you think the $p$-value will be smaller than $\alpha$? Explain.

10.14 A computer company wishes to install a fan that cools the equipment so that the temperature of the mother-board does not exceed a 115°F. A company inspector wishes to test the design of the computer fan. It chooses at random 36 computers. If the computers are found to perform well, they are placed in the market, if they are found not to perform well, then the fans are redesigned.
   a) Write the null and alternative hypothesis.
   b) According to the described problem, explain in your own words what it would mean to incur in a type I error. What is the consequence of incurring in a type I error?
   c) According to the described problem, explain in your own words what it would mean to incur in a type II error. What is the consequence of incurring in a type II error?

10.15 A firm is considering to use a new machine to reduce the current production time of 5.25 minutes for a product. Since they can try out the machine before purchase, they will take a random sample of production times of items completed with the new machine. What is the null and alternative for their analysis?

**10.16** In an attempt to increase daily sales of their coffee, Company C just implemented a large marketing campaign. A manager believes the marketing campaign was a failure. To test this, she decides to treat the daily sales in the past 90 days as a random sample. Before the campaign, the amount of coffee sold daily was 300 pounds. What would $H_o$ and $H_1$ be if the manager wants to test that the mean amount of coffee sold daily (in pounds) after the marketing campaign does not exceed the amount before the campaign?

**10.17** For over 25 years, Fishy company has been selling canned sardines, mainly with a net (mean) weight of 425 g. To your surprise, management has not tested the actual mean weight of cans in the last three years, and you wonder if the net mean weight of 425 g is still appropriate. What would be the null and alternative hypothesis to test this?

**10.18** Assuming a $Z$ test statistic will be used, sketch the rejection region under the test statistic:
a) $H_o: \mu = 250; H_1: \mu \neq 250; \alpha = 0.05.$
b) $H_o: \mu \leq 181; H_1: \mu > 181; \alpha = 0.01.$
c) $H_o: \mu \geq 343; H_1: \mu < 343; \alpha = 0.10.$

**10.19** Determine whether $H_o$ should be rejected
a) $p$-Value $= 0.32, \alpha = 0.01.$
b) $p$-Value $= 0.11, \alpha = 0.05.$
c) $p$-Value $= 0.99, \alpha = 0.10.$
d) $p$-Value $= 0.001, \alpha = 0.05.$
e) $p$-Value $= 0, \alpha = 0.01.$
f) $p$-Value $= 0.0101, \alpha = 0.01.$

**10.20** Say that
$H_o: \mu \geq 25.5$
$H_1: \mu < 25.5$
If $n = 50$, $\sigma = 5$, and $\alpha = 0.01$, and $\overline{X}$ is normally distributed, compute the $p$-value and state if you reject the null or not when
a) $\overline{x} = 26.$
b) $\overline{x} = 24.7.$
c) $\overline{x} = 23.6.$
d) $\overline{x} = 22.$

**10.21** Suppose that
$H_o: \mu \leq 173.8$
$H_1: \mu > 173.8$

If $n = 42$, $\sigma = 21$, and $\alpha = 0.05$, and $\overline{X}$ is normally distributed, compute the $p$-value and state if you reject the null or not when
a) $\overline{x} = 181.4$.
b) $\overline{x} = 152.3$.
c) $\overline{x} = 176.8$.
d) $\overline{x} = 190.3$.

**10.22** A researcher is working with

$$H_o: \mu \leq 25.5$$
$$H_1: \mu > 25.5$$

when $n = 50$, $\sigma = 5$, $\alpha = 0.01$, and $\overline{X}$ is normally distributed.
a) Determine whether the researcher rejects the null when $\overline{x} = 22$.
b) Determine whether the researcher rejects the null when $\overline{x} = 26$.
c) Now suppose that the researcher did a mistake stating $H_o$ and $H_1$, and in reality they should have been stated as in Problem 10.20. Compare the decision about the null in 10.22a with the one reached in Problem 10.20(d).
d) Again suppose the researcher did a mistake stating $H_o$ and $H_1$ and in reality they should be stated as in Problem 10.20. Compare the decision about the null in 10.22a with the one reached in Problem 10.20(a).

**10.23** If a researcher by mistake wrote
$$H_o: \mu > 25.5$$
$$H_1: \mu \leq 25.5$$
Under this erroneous statement, what are the consequences for the rest of the hypothesis testing procedure?

**10.24** A company is attempting to determine if the content of chemical $X$ in their product is below a certain threshold. Specifically,
$$H_o: \mu \geq 30 \text{ mg}$$
$$H_1: \mu < 30 \text{ mg}$$
A random sample was taken from which they did not reject the null using $\alpha = 0.05$. Based on this decision, they conclude the following: "At 5% significance, we do not find evidence against the mean content of chemical $X$ in our product exceeding 30 mg." What is wrong with the conclusion statement?

**10.25** If $H_o: \mu \leq 40.5$ and $H_1: \mu > 40.5$, which of the following is an equivalent statement?
a) $H_o: \mu < 40.5$; $H_1: \mu \geq 40.5$.

b) $H_o: \mu = 40.5; H_1: \mu \neq 40.5.$
c) $H_o: \mu \geq 40.5; H_1: \mu < 40.5.$
d) $H_o: \mu = 40.5; H_1: \mu > 40.5.$

10.26 If $H_o: \mu \geq 73; H_1: \mu < 73$, which of the following is an equivalent statement?
a) $H_o: \mu = 73; H_1: \mu < 73.$
b) $H_o: \mu \leq 73; H_1: \mu > 73.$
c) $H_o: \mu = 73; H_1: \mu \neq 73.$
d) $H_o: \mu > 73; H_1: \mu \leq 73.$

10.27 An engineer was proud to have installed a new solar system at his house. The system was economically feasible because the generated surplus power was sold to the power company. To the engineer's dismay though, right after installation, every single day from around 11 a.m. to about 3 p.m. the system stopped selling back to the power company. After some observations, he noticed that the voltage in the power grid seemed to be higher than the expected value, 120 V. If the voltage in the power grid is too high, due to federal regulations, he will not be able to sell power back to the grid. He took 30 random voltage measurements during the time of the day in question and got an average of 130 V. The engineer was able to find out that the power grid voltage for the area is normally distributed with a standard deviation of 1 V. Conduct the five steps of hypothesis testing to determine if mean voltage in his neighborhood is higher than 120 V at 5% significance.

10.28 February 2018 daily page views for the Boston Archives[5] website are available[6] ("digitalrecords-pageviews-2018-02").
a) Find the mean and standard deviation of daily pageviews for February from the data set.
b) Assuming the pageviews are accurate, the values found in 10.28a are $\mu$ and $\sigma$ for February 2018 daily pageviews. Use a computer to take a random sample of 15 daily pageviews from the data set. Then, conduct a hypothesis test to see if the mean daily pageviews for February 2018 was 160 at $\alpha = 0.1$ (use $\sigma$ found in 10.28a).
c) How did the results from the hypothesis tests compare with what was found in 10.28a? Explain.

10.29 A one-sided 90% confidence interval for the mean monthly cost of supplies for a printing service shop gave an upper bound of $714.

---

5 cityofboston.access.preservica.com (accessed July 26, 2019).
6 Source: data.boston.gov (accessed July 26, 2019).

a) Interpret the result of this confidence interval.
b) Would the equivalent hypothesis test be an upper tail test or a lower tail test?

10.30 A one-sided 99% confidence interval for the mean number of visits to a theme park in July gave a lower bound of 11 321.44.
a) Interpret the result of this confidence interval.
b) Would the equivalent hypothesis test be an upper tail test or a lower tail test?

10.31 Recall from Problem 10.15 that a firm is considering to use a new machine to reduce the current production time of 5.25 minutes for a product. They can try out the machine before purchase. From a random sample of production times of items completed with the new machine, they got a one-sided 90% confidence interval upper bound of 5.75 minutes.
a) Do you think the null hypothesis in Problem 10.15 was rejected?
b) Do you think that the results are of practical importance? Explain.
c) Suppose the one-sided 90% confidence interval upper bound was instead 5.10 minutes. Do you think that the results are of practical importance? Explain.
d) Suppose the one-sided 90% confidence interval upper bound was instead 2.38 minutes. Do you think that the results are of practical importance? Explain.

10.32 If $H_o$: $\mu \leq 11.4$ but $\mu$ is in fact 13, $\sigma = 4$, $n = 60$, $\alpha = 0.05$, and the value of $\overline{X}$ to reject the null is 12.25, determine what is the power of the test.

10.33 (Determining sample size accounting for power of the test) In Section 9.4, sample size was determined to construct a confidence interval for a fixed confidence level and a set margin of error. In that technique of determining $n$, $\alpha$ is accounted for. But the power of the test, rejecting the null when it is false, is not taken into account. Of course, there are many situations where it is important to also determine $n$ according to values of $\alpha, \beta$, and effect size, the difference between $\mu_o$ and some $\mu_a$ value that belongs in $H_1$.
a) Just controlling $\alpha$. Find $n$ for a two-sided hypothesis test on $\mu$ when $\alpha = 0.01$, $\sigma = 7$, and a margin of error of 5 is desired.
b) Controlling $\alpha$ and $\beta$. Use Minitab to find $n$ for a two-sided hypothesis test on $\mu$ when $\alpha = 0.01$, $\beta = 0.05$, $\sigma = 7$, $H_o$: $\mu = 45$, and the true population mean is 50 (effect size is 5).

c) Use Minitab to find $n$ for a one-sided hypothesis test on $\mu$ when $\alpha = 0.01, \beta = 0.05, \sigma = 7, H_o: \mu \le 45$, and the true population mean is 50 (effect size is 5).

## 10.5  Hypothesis Testing for the Mean ($\sigma$ Unknown)

Section 10.4 focused on hypothesis testing for the mean when $\sigma$ is known. However, most of the time when inference on $\mu$ is needed, $\sigma$ is unknown. Thus, we now consider scenarios where there is a random sample from a normal population with mean $\mu$ and standard deviation $\sigma$, yet both are unknown. It is informative to look at what was previously done to draw inference on $\mu$ when the population standard deviation was known. It was then convenient to determine $Z$, a test statistic with a standardized normal distribution of mean 0 and standard deviation 1. The fact that it has a standardized distribution makes inference on $\mu$ easy regardless of the values of $\mu$ and $\sigma$. However, to calculate $Z$, we must know $\sigma$, hence a different test statistic is now needed. In Chapter 4, the use of $s$ to estimate $\sigma$ was introduced. Perhaps an analogous statistic to $Z$ can be used, using $s$ instead of $\sigma$ for standardization. It is already known that when a random sample comes from a normally distributed population, then $\overline{X} \sim N(\mu, \sigma/\sqrt{n})$. The analogous statistic to $Z$ in this scenario is

$$t = \frac{\overline{X} - \mu_o}{s/\sqrt{n}} = \frac{\sqrt{n}(\overline{X} - \mu_o)}{s}$$

When $H_o$ is true, instead of following a standard normal distribution, $t$ will follow a $t$-distribution with $n - 1$ degrees of freedom (Figure 10.10). The $t$-statistic

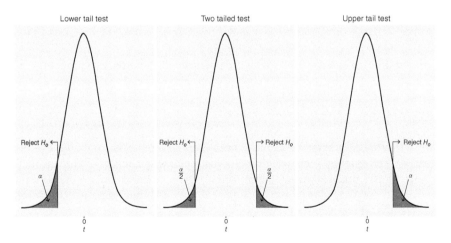

Figure 10.10  If $H_o$ is true, $t$ will follow a $t$-distribution with $n - 1$ degrees of freedom.

is a pivotal quantity, and a hypothesis test procedure on the population mean that uses the $t$-statistic is called a **$t$-test**.

There are five steps to perform a $t$-test. The main difference from the $Z$-test is in obtaining the $p$-value, step 3. $t$ tables are not able to give exact $p$-values, instead providing an interval of $p$-values. However, this is enough since the $p$-value interval can still be compared to $\alpha$. Besides, in practice instead of $t$-tables, statistical software capable of finding exact $p$-values is used.

**Example 10.9** *Unsatisfied with the mean Scholastic Assessment Test (SAT) scores in his high school, the principal implemented a series of measures to improve SAT mean scores. The administration now claims that the SAT is higher than the past mean of 500.71. A student wonders if the measures worked, but the school administration is not willing to provide the latest SAT results. With the help of some teachers, she is able to randomly choose 23 senior students to anonymously answer what SAT score they got (in reality she got data from 25 students, but one dude claimed to have gotten 5000 and another 75 000, so she had to remove their scores). The sample mean was 514.5 with a standard deviation of 76.11. Her teacher told her that SAT scores tend to follow a normal distribution. Does the data support the administration's mean SAT claim at 5% significance?*

**Step 1: State null and alternative**   The administration's claim is that mean SAT scores ($\mu$) are now greater than 500.71. This is an inequality statement and thus goes in the alternative hypothesis,

$$H_o: \mu \leq 500.71$$
$$H_1: \mu > 500.71$$

$\alpha = 0.05$.

**Step 2: Choose and calculate the test statistic to be used**   Since the standard deviation is unknown and the SAT scores are normally distributed, our procedure is a $t$-test,

$$t = \frac{\sqrt{23}(514.5 - 500.71)}{76.11} = 0.87$$

**Step 3: Find the value of the decision rule, the p-value**   To reject the null, the $t$ value must be "extreme" enough, which in this example, according to $H_1$, means that it must be far enough above 0 (Figure 10.11). Of course, with the $t$ table, we can not find the exact value of the $p$-value $= P(t > 0.87)$, but statistical software reveals[7] that the $p$-value $= 0.197$.

---

7 The $t$-table does show that, for 22 degrees of freedom, the $p$-value $> 0.10$, enough for us to decide about the null.

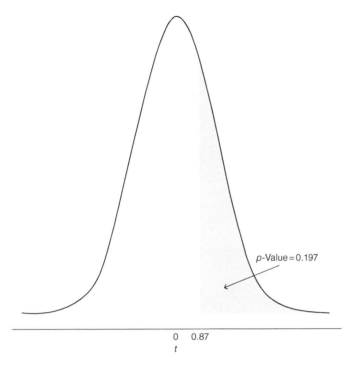

*p*-Value = 0.197

0    0.87
*t*

Figure 10.11  If $H_o$ is true, *t* will follow a *t*-distribution with 22 degrees of freedom. The *p*-value is shaded in blue.

***Step 4: Make a decision about H₀***  Since 0.197 ≮ 0.05, the null is not rejected.

***Step 5: Statistical conclusion***  At 5% significance, the student cannot conclude that the mean SAT score is now higher than 500.71.

***Using Minitab***  To solve Example 10.9 using Minitab, go to Stat > Basic Statistics > 1-Sample t. Figure 10.12 illustrates the steps (since the data is summarized, the Summarized data option must be chosen). Pressing the Options button, a new window will open where one can select the type of test and change the confidence level. The output (Figure 10.13) leads to the same conclusion we reached by manually performing the computations.

The *t*-test is developed under two assumptions:

- Independent observations
- The population is normally distributed

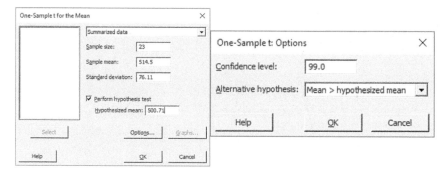

Figure 10.12 Window grabs to perform a *t*-test using Minitab.

## One-Sample T

```
Test of μ = 500.71 vs > 500.71

 N    Mean   StDev   SE Mean   99% Lower Bound     T      P
23   514.5   76.1     15.9                      474.7   0.87   0.197
```

Figure 10.13 *t*-test Minitab output.

The normal population assumption is theoretically needed to ensure that $\overline{X}$ follows a normal distribution and thus the test statistic follows a *t*-distribution.[8]

- The normal assumption is of most concern in the case of a small random sample available for a one-sided test.
- If the sample size is large enough, say $n \geq 30$, we rely on the central limit theorem coming into play to argue that $\overline{X}$ follows approximately a normal distribution.
- As suggested in Section 8.4, in practice the skewness of the population must always be assessed, the higher the skewness, the larger the sample size needed for the central limit theorem to apply. However, when $n \geq 30$, the high skewness is mainly a problem when a one-tail test is conducted. The reason is that when conducting a two-tailed test, calculation of the *p*-value requires finding areas from both tails. Thus, although the test statistic no longer follows a *t*-distribution, the smaller area under one tail will be symmetrically compensated by the larger area under the other tail. For a one-tail test, there is no such compensation for the skewness issue.

---

8 More specifically, when the population is normally distributed, $\overline{X}$ and $S$ are independent and $t$ follows a $t$-distribution. The more skewed the population, the more dependent $\overline{X}$ and $S$ are, and $t$ no longer will follow a $t$-distribution.

The skewness of the population is often assessed graphically through a histogram or boxplot of the data.

- Regardless of sample size, one should always assess the data. Large asymmetry or presence of outliers would contradict the normality assumption. When the assumption of a normal population is worrisome, it might be better to use a nonparametric test (see Chapter 16). If an observation appears suspect or potentially influential in your results, the analysis can be redone without that observation to see if the conclusions change and if they do, then a closer assessment of the observation is warranted.

## Practice Problems

10.34 Use the table in Appendix C to find a $p$-value interval in the following situations.
a) $t = 2.31$, two-sided test, $df = 5$.
b) $t = 2.57$, one-sided test, $df = 23$.
c) $t = 4.74$, one-sided test, $df = 19$.
d) $t = -12.33$, two-sided test, $df = 250$.

10.35 Use the table in Appendix C to find a $p$-value interval and decide if the null is rejected in the following situations.
a) $t = 2.31$, two-sided test, $df = 27, \alpha = 0.1$.
b) $t = 3.77$, right-sided test, $df = 39, \alpha = 0.05$.
c) $t = -4.14$, left-sided test, $df = 16, \alpha = 0.01$.
d) $t = 1.35$, two-sided test, $df = 35, \alpha = 0.05$.

10.36 Say that
$H_o: \mu \geq 25.5$
$H_1: \mu < 25.5$
If $n = 50$, $\alpha = 0.01$, and the population is normally distributed, use statistical software to compute the $p$-value and state if you reject the null or not when
a) $\bar{x} = 22, s = 3.22$.
b) $\bar{x} = 26, s = 6.47$.

10.37 Suppose that
$H_o: \mu \leq 173.8$
$H_1: \mu > 173.8$
If $n = 42$, $\alpha = 0.05$, and the population is normally distributed, use statistical software to compute the $p$-value and state if you reject the null or not when

a) $\bar{x} = 181.4, s = 25.93$.
b) $\bar{x} = 152.3, s = 17.31$.

10.38 Say that

$H_o: \mu = 33.7$
$H_1: \mu \neq 33.7$

If $n = 30$, $\alpha = 0.05$, and the population is normally distributed, compute the $p$-value and state if you reject the null or not when
a) $\bar{x} = 34, s = 12.55$.
b) $\bar{x} = 32.2, s = 8.44$.

10.39 A company claims that its food snacks contain an average of 25 g of sugar per serving. A researcher takes a random sample of 25 snack servings and finds an average of 32 g of sugar and a standard deviation of 1.4 g. Conduct a hypothesis test to determine if the company's claim is valid at 10% significance.

10.40 Recently, a friend of yours who works at the local pizzeria stated that she is skeptical of the manager's claim that they will deliver pizza in 30 minutes or less, anywhere in the local neighborhood. Anxious to put your new hypothesis testing skills to the test, you convince her to request that starting next week, employees record the time of each order and delivery boys record the time of each delivery. Your friend gives you the times. To eliminate the influence of work ethic of some employees (and other factors), you take a random sample of 15 times and find a mean of 31 and a standard deviation of 5. You determine that a 1% significance is adequate in this case.
a) State the null and alternative.
b) Determine the value of $t$.
c) Calculate the $p$-value (or a range of values from the $t$ table).
d) Make a decision about the null.
e) Make a statistical conclusion.

10.41 After evaluating customer satisfaction survey data, a computer company finds that one of the main complaints of customers was how long it took to repair computers. They make some changes with the objective of reducing the average time to 5 days. A random sample of 30 computer repairs found a mean of 4.7 days with a standard deviation of 2.3 days. Conduct all five steps of the hypothesis testing process to determine if the company was successful in attaining the proposed mean repair time at 5% significance.

**10.42** According to public health officials, the 2007 US mortality rate of breast cancer was[9] 22.9 per 100 000 women. Use Chicago data found[10] in "Selected_public_health_indicators_by_Chicago_community_area" to determine if the breast cancer rate there is different at 5% significance.

**10.43** A news show reports that every year a monthly mean of 570 miles of sewer lines are cleaned. Monthly miles of cleaned sewage is reported by the Los Angeles authorities for the Fiscal years[11] of 2014–2017 ("LASAN__Miles_of_Sewer_Cleaned").
a) Test the news media claim at $\alpha = 0.01$.
b) Construct a 99% confidence interval for the monthly mean miles of sewer cleaned in LA.
c) What assumptions are being made to draw inference?

**10.44** Interpret the confidence interval in Figure 10.13. Does it support the conclusion reached with the hypothesis test?

## 10.6   Hypothesis Testing for the Population Proportion

In the case of hypothesis testing for the population proportion, $\pi$, we still have three possible scenarios:

*Lower Tail Test*

$$H_o: \pi \geq \pi_o$$
$$H_1: \pi < \pi_o$$

In this first case, if $\hat{p}$ is far enough to the lower tail of the sampling distribution, then $H_o$ would be deemed unlikely, hence $H_o$ is rejected. On the other hand, a $\hat{p}$ close enough to $\pi_o$ or greater than $\pi_o$ would be considered weak evidence against $H_o$ and hence it would not be rejected.

*Upper Tail Test*

$$H_o: \pi \leq \pi_o$$
$$H_1: \pi > \pi_o$$

This case works the opposite way of a lower tail test, if $\hat{p}$ is far enough to the upper tail of the sampling distribution, then $H_o$ would be deemed unlikely, hence $H_o$ is rejected. On the other hand, a $\hat{p}$ close enough to $\pi_o$ or lower than $\pi_o$ would be considered weak evidence against $H_o$ and hence it would not be rejected.

---

9  Source: data table at seer.cancer.gov (accessed July 26, 2019).
10  Source: data.cityofchicago.org (accessed July 26, 2019).
11  data.lacity.org (accessed June 8, 2018).

**Two-Tailed Test**

$$H_0: \pi = \pi_0$$
$$H_1: \pi \neq \pi_0$$

In this scenario, if $\hat{p}$ is either too far above, or too far below $\pi_0$, $H_0$ is rejected. On the other hand, a $\hat{p}$ close enough to $\pi_0$ would be considered weak evidence against $H_0$ and hence it would not be rejected.

---

**A Normal Approximation**

Recall that if the distribution of $\hat{p}$ is not too skewed, then it will follow approximately a normal distribution according to the central limit theorem. For hypothesis testing on the proportion of a population $\pi$, the condition for approximate normality becomes $n\pi_0 \geq 15$ and $n(1 - \pi_0) \geq 15$.

---

If the conditions above holds, to get a test statistic, the idea is the same as conducting hypothesis testing for the mean: work with a pivotal quantity. We take an estimator of the population proportion, subtract its expected value under the null, and divide by the estimators standard error. "Approximating the Sampling Distribution of $\hat{p}$" in Section 8.4 provided guidance to do this. Under $H_0$, $\hat{p}$ has an expected value of $\pi_0$ and a standard error of $\sqrt{\pi_0(1 - \pi_0)/n}$. Thus, the test statistic

$$Z = \frac{\hat{p} - \pi_0}{\sqrt{\frac{\pi_0(1-\pi_0)}{n}}}$$

will follow an $N(0, 1)$. Lower tail, upper tail and two-tailed hypothesis tests are still possible, and the procedure will also be broken down into five steps.

**Example 10.10** *The Pew Research Center reported that 31.2% of young adults (aged 18–34) in the United States, lived with their parents in 2014.[12] A manager at a real estate firm in Santa Barbara California thinks that there the rate of young adults living with their parents is lower. From a random sample of 125 young adults in the area, 26 respond that they live with their parents. Check at 5% significance, if the proportion of young adults in Santa Barbara living with the parents is lower than the proportion nationwide.*

**Step 1: State null and alternative**  Let $\pi$ = proportion of young adults (aged 18–34) in Santa Barbara, living with their parents. The belief is that $\pi$ is lower than for the country overall. This is an inequality statement and thus goes in the alternative hypothesis,

---

12 Source: Pew Research Center.

$H_o: \pi \geq 0.312$

$H_1: \pi < 0.312$

$\alpha = 0.05.$

**Step 2: Choose and calculate the test statistic to be used** $\hat{p} = 26/125 = 0.208$. Since $n\pi = 39$ and $n(1 - \pi) = 86$ are both greater than 15, it is assumed that $\hat{p}$ is approximately normally distributed, therefore,

$$Z = \frac{\sqrt{125}(0.208 - 0.312)}{\sqrt{0.32(1 - 0.312)}} = -2.51$$

**Step 3: Find the value of the decision rule, the p-value** To reject the null, the $Z$ value must be "extreme" enough, which in this example, according to $H_1$, means that it must be far enough below 0 (Figure 10.14).

p-Value $= P(Z \leq -2.51) = 0.006$

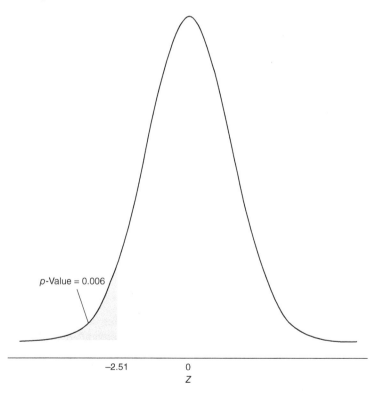

Figure 10.14 If $H_o$ is true, $Z$ will follow approximately a standard normal distribution. The p-value is shaded in blue.

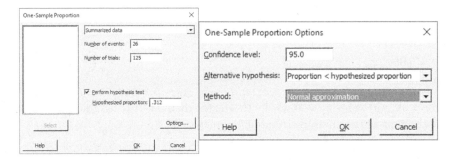

Figure 10.15 Window grabs to perform a Z-test for a proportion using Minitab.

## Test and CI for One Proportion

```
Test of p = 0.312 vs p < 0.312

Sample   X    N Sample p 95% Upper Bound Z-Value P-Value
1       26 125 0.208000        0.267713   -2.51   0.006

Using the normal approximation.
```

Figure 10.16 Minitab output for a proportion test using a normal approximation.

**Step 4: Make a decision about $H_o$** Since $0.006 < 0.05$, the null is rejected.

**Step 5: Statistical conclusion** At 5% significance, it is concluded that the proportion of young adults in Santa Barbara living with their parents is less than 0.312.

**Using Minitab** To solve Example 10.10 using Minitab, go to Stat > Basic Statistics > 1 Proportion. Figure 10.15 illustrates the steps (the Summarized data option must be chosen for this example). Pressing the Options button, a new window will open where one can select the type of test, the confidence level, and the method: normal approximation or "exact" method.[13] The output (Figure 10.16) leads to the same conclusion we reached by manually performing the computations. Choosing the method would have not affected the results either.

---

### When to Avoid the Normal Distribution Approximation
When $n\pi_o < 15$ or $n(1 - \pi_o) < 15$, the sample proportion does not approximately follow a normal distribution. It turns out, that this may be only a problem when a one-tail test is conducted, analogous to previous hypothesis tests presented in this chapter. For a one-tail test, usually statistical software has an option to find the p-value exactly (using the binomial distribution). This "exact test" is also called a **binomial test**.

---

13 The exact method uses the binomial distribution for the successes.

One last thing to add. When performing inference on $\mu$, we learned that hypothesis tests procedures lead to the same conclusions as confidence intervals (at same significance). With $\pi$ this is not necessarily true. The reason is that the test statistic for the hypothesis test uses $p_0$ to find the standard error, while the confidence interval uses $\hat{p}$. Since $\hat{p}$ and $p_0$ are not necessarily close, confidence intervals and hypothesis tests on $\pi$ could lead to different conclusions.

### 10.6.1   A Quick Look at Data Science: Proportion of New York City High Schools with a Mean SAT Score of 1498 or More

Suppose that the National Center for Education Statistics is studying the SAT scores of students by location, like New York City. We will help them perform their analysis! It is still rather early to have enough data from the new SAT,[14] so we will work with pre-2016 SAT scores. The aim is to determine if the proportion of schools in NYC that have a mean total score of 1498 or more is greater than 10%. Test is to be performed at 5% significance.

Let $\pi$ = proportion of New York City High Schools with a mean SAT Score of at least 1498. Our hypotheses are

$H_o: \pi \leq 0.10$

$H_1: \pi > 0.10$

For New York City, we will use the 2012 mean total SAT score per school.[15] The original data set provides location, number of test takers, scores on Critical Reading, Math, and Writing. Out of 478 schools, some have very few test takers. Arbitrarily, we will only consider schools if they had at least 30 test takers. After

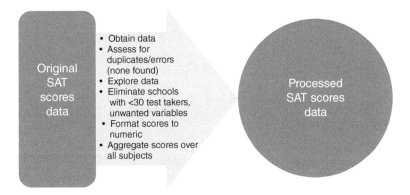

Figure 10.17  Procedure to preprocess NYC SAT scores.

---

14  Since March 2016, a new SAT is based on 1600 points instead of 2400.
15  Source: data.cityofnewyork.us (accessed July 26, 2019).

**Test and CI for One Proportion**

```
Test of p = 0.1 vs p > 0.1

                                                Exact
Sample    X    N   Sample p   95% Lower Bound   P-Value
1         26  359  0.072423       0.051240      0.971
```

Figure 10.18 Minitab output for a proportion test using exact proportion test.

some data wrangling (Figure 10.17), we find that 26 out of 359 schools had mean total score of 1498 or more.

Using Minitab (Figure 10.18), the $p$-value $= 0.97$ is not less than 0.05. Therefore, the null is not rejected. At 5% significance, we find no evidence against stating that the proportion of NYC schools mean total score of 1498 or more is less than or equal to 10%.

## Practice Problems

10.45 Write the expressions for $H_o$ and $H_1$.
 a) The proportion of batteries produced that are defective is no less than 0.05.
 b) The proportion of houses with solar panels is greater than 0.15.
 c) The majority of cars driven in a county comply with environmental standards.

10.46 Assuming a $Z$-test statistic will be used, sketch the rejection region under the test statistic:
 a) $H_o: \pi = 0.4; H_1: \pi \neq 0.4; \alpha = 0.05$
 b) $H_o: \pi \leq 0.75; H_1: \pi > 0.75; \alpha = 0.01$
 c) $H_o: \pi \geq 0.60; H_1: \pi < 0.60; \alpha = 0.10$

10.47 Say that
 $H_o: \pi \geq 0.25$
 $H_1: \pi < 0.25$
 If $n = 150$ and $\alpha = 0.01$, compute the $p$-value and state if you reject the null or not when
 a) $X = 37$.
 b) $X = 44$.
 c) $X = 12$.
 d) $X = 23$.

10.48 Suppose that
 $H_o: \pi \leq 0.80$
 $H_1: \pi > 0.80$

If $n = 400$ and $\alpha = 0.05$, compute the *p*-value and state if you reject the null or not when
a) $X = 315$.
b) $X = 350$.
c) $X = 394$.
d) $X = 320$.

10.49 Say that
$H_o$: $\pi = 0.33$
$H_1$: $\pi \neq 0.33$
If $n = 250$ and $\alpha = 0.10$, compute the *p*-value and state if you reject the null or not when,
a) $X = 70$.
b) $X = 85$.
c) $X = 83$.
d) $X = 200$.

10.50 You are the owner of a local women's clothing store. Blouses, hats, and purses are among some of the items you sell. There are two sort of blouses of a certain brand, regular cut style and skinny cut style. To be able to supply demand you are trying to determine, if the proportion of skinny styles sold in your store this year has gone down from 25% of all blouses last year. A random sample of 200 blouse sales this year are taken from the records and 39 sales are found to be for skinny style blouses:
a) State the null and alternative hypothesis.
b) Find the value of the test statistic.
c) Calculate the *p*-value.
d) Should $H_o$ be rejected at $\alpha = 0.025$?
e) Write the statistical conclusion.
f) Check the necessary conditions to see if the normal approximation used is adequate.

10.51 In a survey of 90 participants from the LGBT community from across Riverside and San Bernandino counties in California, 74.4% of respondents indicated that they had seriously considered committing suicide.[16] 6.1% of the general population have seriously considered committing suicide. Test at 10% significance if the proportion of members of the LGBT community, who seriously considered committing suicide, is higher than for the general population.

---

16 Source: Gardner, Aaron T. and Curlee, Erin (2015). *Inland Empire Transgender Health & Welness Profile*, page 29.

a) State the null and alternative hypothesis.
b) Find the value of the test statistic.
c) Calculate the $p$-value.
d) Should $H_0$ be rejected at $\alpha = 0.10$?
e) Write the statistical conclusion.
f) Check the necessary conditions to see if the normal approximation used is adequate.

10.52 With the assistance of statistical software, build a 90% confidence interval for Problem 10.51, and check the same conclusion is reached.

10.53 According to experts, by mid-2016, US shoppers made 51% of their purchases online (*Wall Street Journal*, June 8, 2016), marking the first time that shoppers made more than half their purchases online. A researcher believes that shoppers in a small US community make less than half of their purchases online. Their random sample of 150 residents found that 80 stated they make more than half their purchases online. Test the researchers belief at 10% significance.

10.54 A political candidate claims that from 2008 to 2012 the percentage of people aged 25+ in Chicago without a high school diploma exceeded 25%. Socioeconomic indicators for the city of Chicago during this period are available.[17] Use the data "Census_ Data_Selected_socioeconomic_indicators_in_Chicago_2008_2012" to test the researchers claim at 1% significance.
*Hint:* Pay attention to the last row in the data set.

10.55 A political candidate claims that from 2008 to 2012 the percentage of people aged 16+ in Chicago who were unemployed was 15%. Socioeconomic indicators for city of Chicago during this period are available.[18] Use the data "Census_Data_Selected_socioeconomic_indicators_ in_Chicago_2008_2012" to test the researchers claim at 10% significance.

10.56 Interpret the confidence interval in Figure 10.16. Does it support the conclusion reached with the hypothesis test?

---

17 Source: data.cityofchicago.org (accessed July 26, 2019).
18 Source: data.cityofchicago.org (accessed July 26, 2019).

## 10.7 Hypothesis Testing for the Population Variance

It is also possible to perform formal hypothesis testing on the population variance. A manufacturing plant may need to assess the variability in some numerical property of a product. For example, the intention could be to test

$$H_o: \sigma^2 \geq \sigma_o^2$$
$$H_1: \sigma^2 < \sigma_o^2$$

Recall that the pivotal quantity is about convenience and its probability distribution does not depend on the population parameters. The form of the pivotal quantity to perform hypothesis testing for the variance is different than for $\mu$ or $\pi$. If $H_o$ is true, then it can be shown that

$$\chi^2 = \frac{(n-1)s^2}{\sigma_o^2}$$

will follow a chi-squared distribution with $n - 1$ degrees of freedom when the random sample comes from a normal distribution.

**Example 10.11**   *We want to determine if the variance of the Baltimore walking scores is 425 minute$^2$. 2011 walking scores from 55 communities resulted in a sample variance of 421.89. Test is to be performed at 5% significance.*

Let $\sigma^2$ = the variance of community walking scores. Assuming the community walking scores follow a normal distribution, then the $\chi^2$-test proposed above can be used.

$$H_o: \sigma^2 = 425$$
$$H_1: \sigma^2 \neq 425$$

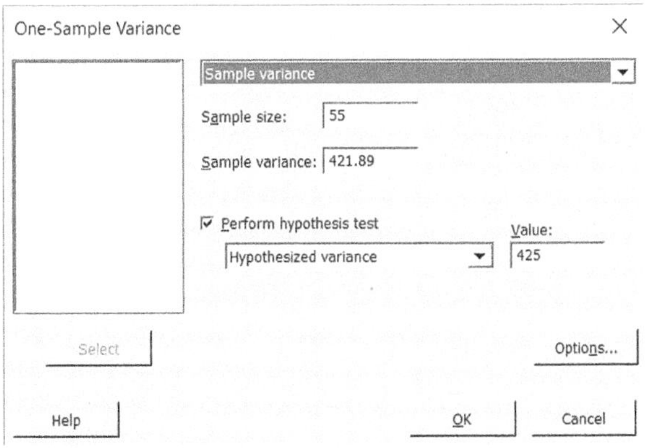

**Figure 10.19** Minitab Window grab to perform a test for a population variance.

One-Sample Variance ✕

Sample variance

Sample size: 55

Sample variance: 421.89

☑ Perform hypothesis test

Hypothesized variance

Value:

425

Select

Options...

Help

OK

Cancel

```
Statistics

 N   StDev   Variance
55    20.5     422
```

Figure 10.20 Minitab output for a variance test.

```
95% Confidence Intervals

                CI for      CI for
Method          StDev       Variance
Chi-Square    (17.3, 25.3) (299, 640)


Tests

                   Test
Method          Statistic  DF  P-Value
Chi-Square         53.60   54   0.979
```

In Minitab, go to Stat > Basic Statistics > 1 Variance. Figure 10.19 illustrates the steps (the Summarized data and Hypothesized variance options must be chosen for this example). The output (Figure 10.20) shows a $p$-value = 0.979, which is greater than 0.05. So the null is not rejected. This way, at 5% significance we find no evidence against the variance 425 minute$^2$.

## 10.8  More on the $p$-Value and Final Remarks

Recall that a $p$-value is the probability of obtaining a test statistic value at least as extreme than the observed result when the null hypothesis is true.[19] The $p$-value is a useful tool in conducting statistical inference. It has a certain intuitive logic to it, allows us to reach conclusions by just looking at one number, is taught in school as part of introductory statistics courses, and is easily accessible using statistical software. It is widely used in practice. But the $p$-value does have drawbacks. It is often misinterpreted and even abused. Here, we try to give some background on the use of the $p$-value and discuss some issues with the measure and provide alternatives when conducting statistical inference.

*Do Political Moderates See Shades of Gray More Accurately Than Either Left-Wing or Right-Wing Extremists?* In 2014, there was an article in the journal *Nature* about a study of 2000 people to determine if political moderates see shades of gray more accurately than either left-wing or right-wing extremists. The null was that there was no difference in the shades of gray of the groups, while the alternative was that they were not equal. The researchers found a $p$-value of 0.01, less than their 5% significance, which is considered pretty strong evidence to reject the null hypothesis and suggests that extremists are more likely to see the world in black and white. Although they were excited about this result, they decided

---

19 Technically, we should say when the null hypothesis is true, at sample size $n$, and all model assumptions hold.

to collect more data and run the analysis again. Their result in the second try was a *p*-value of 0.59, which leads to a completely different conclusion when comparing these populations. So, what happened?

*A Bit of History* The concept of a *p*-value was made popular in the 1920s by Ronald Fisher, a scientist who made enormous contributions in the field of statistics. His intention was to use the *p*-value as one component of the process of scientific inquiry and he acknowledged the importance of the consideration of other facts. That is, he never proposed that the *p*-value be used as definitive evidence to build decisions upon. But back in the 1920s and 1930s, there were two competing statistical procedures: the one proposed by Fisher using *p*-values and the work of Newman and Pearson based on hypothesis testing and the power of these tests. These scientists did not like each other and often were very critical of each other's work. Although the philosophies of the two procedures were different, in some ways they were reconcilable. Eventually, statistical inference became a hybrid of both Fisher and Neyman and Pearson, and this hybrid procedure is wildly applied today.

## 10.8.1  Misunderstanding the *p*-Value

In the case of the political leanings example, some may interpret the *p*-value = 0.01 as the probability of rejecting the null erroneously. But this would be the wrong interpretation. *p*-Values are sometimes also misconstrued as a probability that the null hypothesis is true. This is not only incorrect, but in fact a *p*-value may often be way off from the probability of the null being true. Recall that we may define a *p*-value as the probability of getting the observed statistic or a more extreme value given the null hypothesis is true. Therefore, a small enough *p*-value indicates that the alternative hypothesis makes more sense according to the data. Strictly speaking, the *p*-value does not tell us anything about the chances of the null being true or false. Hence, a *p*-value does not allow us to accept a hypothesis (see Section 10.4). At best, it tells us we should not reject it which is different. Finally, we must remember that a *p*-value is a function of observed data. As so, it is a statistic with its own sampling distribution. Therefore, measurements obtained from an additional sample may result in a (very) different *p*-value.

*A Coin Flip Experiment* Suppose we have a two-headed coin. We flip it 4 times to test if the coin is biased (e.g. the null is that the coin is fair) at 5% significance and count the number of times we get heads. Of course, since it is a two-headed coin, we get all successes. When flipping a coin 4 times in this scenario, the *p*-value would never be smaller than 0.05 and in fact in this case *p*-value = 0.0625. Therefore, we do not reject the null hypothesis that the coin is fair, although clearly, the coin is biased.

**Another coin flip experiment**    Let's say we are doing a coin experiment again, but now with a normal coin. We flip it 12 times while testing if the coin is biased (e.g. the null is that the coin is fair) at 5% significance, and we get heads 9 times. Now here comes the tricky part, were the 12 flips fixed by design? If so, then the number of heads follows a binomial distribution and therefore the $p$-value is 0.075. However, if the flips were done until we got 3 tails, the distribution used would be different and would result in a $p$-value of 0.0325. That is, in the first case, we reject the null at 5% significance, while in the second we do not reject the null. The implication is that our conclusion about the null depends on how the experiment was designed and not just the results.

Now, the discussion from these examples is not to say that the hypothesis test procedure has failed. A power calculation would tell us that for the initial coin example, the test was not too powerful in the first place. But the example serves to drive the point about why it is important to say "do not reject the null" as opposed to "accept the null." The second example presents another challenge, more philosophical in nature, of using the $p$-value.

### Properly Using p-Values and Alternatives

Considering the drawbacks discussed above, what is one to do with $p$-value methods?

**How to Interpret the $p$-Value?** Remember, the $p$-value is the probability of obtaining a test statistic value at least as extreme than the observed result when the null hypothesis is true. For example, suppose that a company had to recall a certain product five years ago and wanted to check if the recall decreased monthly mean sales of the product. The null hypothesis is that the true monthly mean sales are greater than or equal to what they were five years ago. If we use monthly sales from the recall until now to obtain a sample mean, the interpretation of the $p$-value is the probability of getting a sample mean at least as extreme as the one observed, if the recall did not decrease monthly mean sales. Note that the $p$-value does not directly answer the question on whether the recall decreased monthly mean sales of the product, and that is somewhat disappointing. But this brings us to our next point.

**Recall the Properties of Your Statistical Estimator.** In reality, the $p$-value is not used on its own to decide to reject or not reject the null. We must remember that many estimators have a series of properties that make us "confident" that the estimator should be "close" to the parameter it is estimating (see Section 8.5). Hence, $\overline{X}$ should be close to the true mean of the population, and if the null is stating the true population mean, then we expect the $p$-value to not be too low. If the $p$-value is too low, then it is reasonable to argue that based on the properties of $\overline{X}$, the null hypothesis is not appropriate.

**Report Sample Size.** This will put into context how reliable the results are. When the sample size is small, statistics used for inference, such as the sample mean, are more variable. Furthermore, when the sample size is small, there is a good chance of incurring in type I and type II errors when conducting a hypothesis test.

**Consider Effect Size**, and remember that statistical importance does not imply practical importance. As suggested from the first coin flip example, be careful of using the $p$-value as definitive evidence to build decisions. Some researchers believe that the smaller a $p$-value, the more important the result, and the larger a $p$-value the less important a result. Yet, the $p$-value is a function of sample size. If the sample size is large enough or standard deviation is small enough, even very small effects will produce small $p$-values. In contrast, truly important results may be found statistically insignificant if the sample size is small enough or variability of the statistics is high. To sum things up, data may support the null or data may not be enough to reject it, but the $p$-value is not able to distinguish between these two possibilities.

**Choose the Significance Level Based on the Problem at Hand (and Consider Power of the Test Too).** By definition, the $p$-value is obtained assuming the null is true. If the null is false, the $p$-value is no longer reliable.[20] In contrast, the power of the test is a measure dependent on the alternative being true. Thus both, the $p$-value and the power, must be assessed in any given situation.

---

**Rely on Confidence Intervals**

As we have seen, a $p$-value is a statistic, one that does not account for uncertainty in its value. On the other hand, confidence intervals do account for uncertainty while providing us with information about the value of a parameter. By providing a range of plausible parameter values, confidence intervals can be more informative of the practical importance of the results. Although we emphasized two-sided confidence intervals, one-sided confidence intervals can be constructed as well. Computer software can easily find any of these kind of confidence intervals. Just remember that confidence intervals must be interpreted carefully too. Consider a scenario where $H_o: \mu = 20$, $H_o: \mu \neq 20$; $\alpha = 0.05$, and the $p$-value was 0.000 000 12. Thus, $p$-value is much smaller than $\alpha$ and the null is rejected. Yet with the same data, the 95% confidence interval is found to be (19.15, 19.64), indicating the true mean is close to 20. So although there was statistical significance, there may be no practical significance. In contrast, had the 95% confidence interval been (7.34, 9.45), strong practical importance would have been indicated.

---

**Alternative: Use Bayesian Methods.** Essentially, the $p$-value method reaches a conclusion about $H_o$ by using $P(\text{Data}|H_o \text{ is true})$. This is indirect evidence of the validity of $H_o$. In math, things are often proven, unequivocally,

---

20 Some consider how far below the $p$-value is from $\alpha$ as strength of evidence against $H_o$. Since the $p$-value assumes the null to be true, this is not exactly true.

by negation: assuming something is true, which then leads to a contradiction. With the *p*-value this is not what is being done, our conclusion cannot be guaranteed to be correct. Without being able to guarantee that our decision about the null is the correct one, a reasonable expectation is to quantify how likely it is that the null is true based on data[21]: $P(H_0$ is true|Data). Unfortunately, the hypotheses testing procedures discussed in this chapter assume parameters are fixed yet unknown, therefore it is not possible to say anything about $P(H_0$ is true|Data). Bayesian methods for inference treat any unknown value as a random variable and hence apply a probability distribution to it, even parameters. The Bayesian method is not as widely taught as the *p*-value method, but it has been gaining popularity because it can quantify $P(H_0$ is true|Data). Say you are interested in drawing inference on the mean of a population. Before observing the data, Bayesian statistics assigns a probability distribution, called a prior distribution, to the possible values of $\mu$. Then this probability distribution belief is "adjusted" based on data. As a consequence of this, under a Bayesian perspective, one can determine the probability of the null being true given the data, $P(H_0$ is true|Data), something the *p*-value method is unable to do. The Bayesian method does come with some drawbacks. Among them, the prior distribution may be argued to be subjective, and computations may become complicated in many real-life scenarios. There are ways to deal with these drawbacks, but they may lead to unacceptable "complexity" to some. Since *p*-value based methods often do lead to useful results, a Bayesian procedure frequently leads to practically the same conclusions while involving considerable more work.

*Final Remarks* Hypothesis testing is a procedure used to help reach a managerial decision utilizing empirical data and quantifying uncertainty through probabilities. Since it is a probabilistic decision, we cannot guarantee that the decision reached is correct, even when no mistakes were made in the procedure. Instead, our decision is the one with the greatest probability of being the correct one. There are two types of errors when conducting hypothesis testing, both errors are important, though in practice more attention is given to type I error than to type II errors. But type II errors can have serious consequences as seen in this chapter.

In essence, the procedure of any type of hypothesis testing can be represented through the steps discussed in Section 10.3. The way we implement each step will vary a little according to each individual problem. Except how we make our decision about $H_0$ that never changes, if *p*-value $\leq \alpha$ then we reject $H_0$. The preferred tool to make a decision about the null is the *p*-value. Its appeal is perhaps

---

21 Note how $P(H_0$ is true|Data) $\neq P(\text{Data}|H_0$ is true).

Table 10.3 The decision that we make about $H_0$ versus the true statement about the population.

| Attribute | Confirmatory research | Exploratory research |
|---|---|---|
| Goal | Determine if null is rejected | Gain insight from data |
| Setting hypotheses | A priori | Post-hoc |
| Style | Structured and rigorous | Open minded and speculative |
| Advantage | Solidifying knowledge, reproducibility | Innovation and new knowledge |

because a decision about the null is made just looking at that number. This convenience comes at a price. The $p$-value is often misinterpreted in practice, even abused. Furthermore, the $p$-value does not provide direct indication of practical importance. Statisticians recommend to either use confidence intervals or combine hypothesis test results with other information to reach managerial decisions: effect size, sample size, and bayesian methods.

Our coverage of hypothesis testing was generally within the confines of **confirmatory research**: where hypotheses are first established, and data is gathered later. In contrast, in **exploratory research**: data exploration leads to interesting hypotheses to be tested. When you hear of data mining, this refers to exploratory research. Table 10.3 summarizes the main differences between confirmatory and exploratory research. In academia, confirmatory research is overwhelmingly preferred because, by definition, objective null and alternative hypothesis will be established. Yet, exploratory analysis can be insightful, if done right, and it is commonly used by nonacademic data analysts.

## Chapter Problems

10.57 Which statement about the null hypothesis is correct?
a) Statement must always have a strict inequality in it ($>$, $<$, or $\neq$).
b) It must always say something about the population mean.
c) Statement will always have a sign of either ($\geq$, $\leq$, or $=$).
d) The null is always wrong.

10.58 Which of the following is the correct combination:
a) $H_0: \mu > 25; H_1: \mu < 25$.
b) $H_0: \mu > 100; H_1: \mu \leq 150$.
c) $H_0: \mu \leq 75; H_1: \mu < 75$.
d) $H_0: \mu \leq 300; H_1: \mu > 300$.
e) $H_0: \mu \leq 100; H_1: \mu > 125$.

10.59 Choose the appropriate alternative to go with $H_o$: $\pi \le 0.42$.
   a) $H_1$: $\pi > 0.42$.
   b) $H_1$: $\pi \ne 0.42$.
   c) $H_1$: $\pi \ge 0.42$.
   d) Not possible to tell.

10.60 If $H_o$: $\mu \le 12$ is not rejected when the true population mean is 15, determine the best answer below.
   a) No error was incurred.
   b) Type I error was incurred.
   c) Type II error was incurred.
   d) Both Type I and Type II errors were incurred.

10.61 If $H_o$: $\mu \ge 75$ is rejected when the true population mean is 80, determine the best answer below.
   a) No error was incurred.
   b) Type I error was incurred.
   c) Type II error was incurred.
   d) Both Type I and Type II errors were incurred.

10.62 The power of a test is found to be 0.75. What is the meaning of this?

10.63 What is the meaning of the level of significance in hypothesis testing? For example what does 0.05 significance mean?

10.64 If

$$H_o: \mu \le 2.5$$
$$H_1: \mu > 2.5$$

and during a Z-test, $z = 1.44$, which of the following best represents the meaning of the p-value? (You may choose more than one if deemed appropriate.)
   a) $P(Z \le 2.5)$.
   b) Probability that the null is correct.
   c) $P(\mu \le 2.5)$.
   d) $P(Z \le 1.44)$.
   e) $P(Z \ge 1.44 | \mu = 2.5)$.
   f) All of the above.

10.65 Analysts for a firm tested if a new computer design allowed to perform high computation algorithms faster than competitors, $H_o$: $\mu \ge 5$ versus $H_o$: $\mu < 5$ (minutes). The analysts argue that since the power of the test was 1, and the sample size large, the results are practically important. Explain why this is not necessarily true.

10.66 A company has designed a new lithium battery. They conducted a study to see if the new battery lasted over two times longer than the average of competitors ($\mu_o$) and found that it did. What would be the implication for the company of rejecting the null hypothesis in the study, when in fact the null hypothesis was true?

10.67 After much effort, you are convincing your "old school" boss to conduct a hypothesis test as part of the decision-making process for a new service program. Regarding possible errors when deciding about $H_o$, you tell him "when hypothesis test procedures were developed, one of the objectives was to keep $\alpha$ and $\beta$ as small as possible. Since it is not possible to minimize them both simultaneously, $\alpha$ is set to a fixed value." To which he responds "fine then. Set $\alpha = 0$." How would you explain to your boss that it is not reasonable to set $\alpha = 0$?

10.68 While explaining the results of an analysis to superiors, it comes up that a two-sided hypothesis test was used and the null was rejected. But the superiors are really interested in one direction testing and argue that a one-sided test should have been used. In a one-sided test, at same significance and same data, would the null be rejected?

10.69 With some given data, if you perform a two-sided test, you do not reject the null. But if you choose a one-sided test you do. How do you explain this?

10.70 An employee at a water management agency is studying the daily rainfall for last November at region $X$. The goal is to determine if the region's mean daily rainfall from last November was below the 4 mm daily average that historically occurs there for the month, at a 5% significance.
a) Write down the null and alternative hypothesis.
b) Rainfall was collected at four randomly chosen locations in region $X$ in five random days last November (a sample size of 20), which resulted in a sample mean of 3.2 smm with a standard deviation of 1 mm. It is well known that the distribution of daily rainfall for November in region $X$ is right asymmetric and highly skewed. Can the employee use a $t$-test to conduct hypothesis testing here? Why or why not?

10.71 A company is aware that the mean cost of repairs on production machines is at least $625 a week. After a program to train employees on using the machines more effectively, the next 25 weeks are sampled and a mean of $600 and a standard deviation of $34.37 is obtained. Test if the training program worked at a significance level of 0.01, if the costs follow a normal distribution.

10.72 A researcher is claiming that from 2008 to 2012, the per capita income of Chicago residents was $26 000 or less. Socioeconomic indicators during this period for the city of Chicago are available.[22] Use the data in "Census_Data_Selected_socioeconomic_indicators_in_Chicago_2008_2012" to test the researchers claim at 5% significance.

10.73 After many years working at a beauty counter at a department store, Lisa decides to open her own store. She needs to set up a business plan proposal to get a loan from a bank. Her preliminary research found that women in the country spend a mean of $59 on fragrance products every three months. If she can prove that the mean amount spent in her town is higher, then she is more likely to get a loan. Lisa takes a random sample of 25 women and finds a sample mean of $68 and a sample standard deviation of $15.
a) Write down the null and alternative hypothesis.
b) What is the value of the test statistic?
c) According to part (10.73b), which of the following appears to be the $p$-value?
(i) 0.15; (ii) 0.003; (iii) 0.006; (iv) 1
d) For a significance level of $\alpha = 0.01$, do you reject the null hypothesis?
e) Write a statistical conclusion.
f) Any assumptions being taken to perform the hypothesis test?

10.74 Los Angeles authorities are aware that monthly mean residential water usage at the 90007 zip code was 45.14 (measured in hundred cubic feet). Residents in one neighborhood claimed they had always taken measures to save water. A random sample of water usage during that period at 20 residences found a mean of 33.21 and a standard deviation of 3.76. Test the residents' claim at 1% significance.

10.75 For a one-sided hypothesis on the population mean, the population is known to follow a normal distribution. While using a small sample size, an analyst chooses a $Z$ statistic as a test statistic by mistake, instead of a $t$ statistic.
a) What is the effect of this error on the calculation of the $p$-value?
b) How can the decision on $H_o$ be affected by the issue with the $p$-value?

---

22 Source: data.cityofchicago.org (accessed July 26, 2019).

10.76 An administrator believes that the mean proportion of people who car-pooled to work in Baltimore from 2008 to 2012 was less than 0.15. Use the data (carpool12) in the file[23] "Sustainability__2010-2012_" to per-form a hypothesis test at 5% significance.
*Hint:* This is a population mean problem, not a proportion problem. Also, you will have to reformat the carpool12 variable and pay attention to the meaning of last row of data.

10.77 Officials in Chattanooga, Tennessee are aware that in the past, 12% of their residents own a business. One of the 2016 community survey[24] questions was "do you own a business in Chattanooga?" Out of 1910 respondents, 217 answered yes.
a) Check the conditions of the normal approximation if $\pi_o = 0.12$.
b) Test at 5% significance if in 2016 the proportion of residents that owned a business was 12%.

10.78 Suppose San Francisco airport (SFO) administrators have made deci-sions based on the belief that no more than half of airport visitors make restaurant purchases. According to a 2011 survey, they found that 501 out of 938 respondents made a purchase at a restaurant.
a) Test the administrators' belief at 10% significance.
b) Build an appropriate one-sided confidence interval. Based on this confidence interval, do you think the inference results are practically important?

10.79 Reproduce with Minitab the results seen in Example 10.11 by using the entire data[25] "Sustainability__2010-2012_." The variable is called "wlksc11." Do you think the normal distribution assumption is justified?
*Hint:* Remove last row.

## Further Reading

Agresti, Alan and Franklin, Christine. (2016). *Statistics: The Art and Science of Learning from Data*, 4th edition. Pearson.

Chang, Mark. (2013). *Paradoxes in Scientific Inference*. CRC Press.

Harris, Richard. (2017). *Rigor Mortis: How Sloppy Science Creates Worthless Cures, Crushes Hope, and Wastes Billion*. Basic Books.

Wasserstein, R. L., Schirm, A. L., and Larzar, N. A. (2019). Moving to a world beyond "p < 0.05." The American Statistician, *73*:sup1, pages 1–19.

---

23 Source: data.baltimorecity.gov (accessed July 26, 2019).
24 data.chattlibrary.org (accessed July 26, 2019).
25 data.baltimorecity.gov (accessed July 26, 2019).

# 11

# Statistical Inference to Compare Parameters from Two Populations

## 11.1 Introduction

The need for inference does not only apply regarding one population. Another common need is to compare some numerical trait of two populations. Let's return to the example about the gasoline additive at the beginning of Chapter 10, but with a slight modification. Now, in the experiment, some cars are driven with the gasoline additive and others are driven without the additive. To determine if the gasoline additive is any good, the mean miles per gallon of the cars using it and of the cars driving without it will be compared. As another example, suppose a professor has two groups taking business statistics. Both groups have similar qualifications for the course and simply take the same class at different times of the day. The professor gives each group a different version of an exam. Are mean exam scores of the two groups different? If so, this would be a problem to the professor, it would not be fair to give exams of differing difficulty to these two groups.

Our main tools for statistical inference will still be confidence intervals and hypothesis testing. Although it may seem that statistical inference is only slightly more complicated, there are several things that require consideration.

- Are the samples independent or dependent?
- Are the standard deviations known? If not, are the standard deviations unknown but can be assumed to be equal or are they unknown and cannot be assumed to be equal?

---

**Less Math Coming Up!**
At this point in the book, we start backing off from the mathematical details and start emphasizing more a general understanding of the concepts, the required assumptions, and the interpretation of the results. Besides comparing population means, comparing proportions and variances from two populations will also be discussed.

---

*Principles of Managerial Statistics and Data Science*, First Edition. Roberto Rivera.
© 2020 John Wiley & Sons, Inc. Published 2020 by John Wiley & Sons, Inc.
Companion website: www.wiley.com/go/principlesmanagerialstatisticsdatascience

## 11.2 Inference on Two Population Means

If the population means are equal, then the sample averages should be close to each other. To illustrate, consider two scenarios when there are two independent samples taken and their sample means computed:

**Scenario 1**: Both samples have the same population mean, $\mu$.

**Scenario 2**: The first sample comes from a population with a much smaller population mean ($\mu$) than the second population.

Figure 11.1 displays the sampling distributions in both scenarios when both sample means are normally distributed but differ in standard errors. The resulting sample means have also been added. In Scenario 1, both sample means are statistically close to $\mu$. Under Scenario 2, $\bar{x}_1$ is close to $\mu$, but $\bar{x}_2$ is too far above $\mu$. $\bar{x}_2$ is very unlikely under the sampling distribution with mean $\mu$ (black curve).

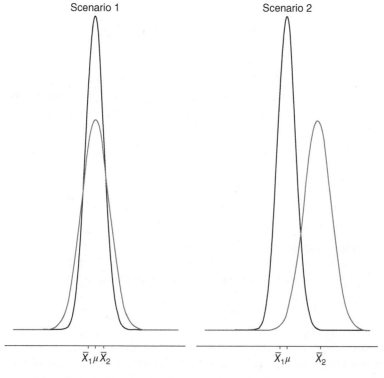

**Figure 11.1** The sampling distributions of two sample means under two scenarios. Scenario 1: they all have the same population mean. Scenario 2: population mean for $\bar{x}_2$ is bigger.

Define $\mu_1$ as the population mean of the first population and $\mu_2$ as the population mean of the second population. We have previously seen it is convenient to use a pivotal quantity to draw inference. Each population mean can be estimated by a sample mean, $\bar{x}_1$ and $\bar{x}_2$, respectively. When comparing two population means, there are two ways to determine how different they are, a ratio of the two statistics and their difference. A value close to 1 of the ratio would indicate similarity of the two sample means. A value close to 0 of the difference would indicate similarity of the two sample means. Which one to use, the ratio or the difference of the sample means depends on the properties of these statistics. In more simple terms, which statistic is more convenient to build a pivotal quantity. It turns out that the sample mean differences is the most convenient to work with. In Chapter 8, we implied that linear combinations of normally distributed random variables are still normally distributed. Therefore, the difference $\bar{x}_1 - \bar{x}_2$ will lead to a useful pivotal quantity. A ratio type of statistic will be used later on to compare estimates of variance.

## 11.3 Inference on Two Population Means – Independent Samples, Variances Known

Let's start conducting inference on two population means using confidence intervals. The set up is that two independent samples are drawn: one with population mean $\mu_1$ and known standard deviation $\sigma_1$, the other sample from a population with mean $\mu_2$ and known standard deviation $\sigma_2$. Each sample has its own sample size, $n_1$ and $n_2$, respectively. $\bar{X}_1$ and $\bar{X}_2$ will each be approximately normally distributed (see Figure 8.7) and therefore, $\bar{X}_1 - \bar{X}_2$ will also be normally distributed with

$$E(\bar{X}_1 - \bar{X}_2) = \mu_1 - \mu_2$$

and

$$Var(\bar{X}_1 - \bar{X}_2) = \frac{\sigma_1^2}{n_1} + \frac{\sigma_2^2}{n_2}$$

Recall that the intuition of the confidence interval is to take a point estimate of the needed value and add/subtract a margin of error. Hence, for our current situation, using properties of the normal distribution, it can be shown that the two-sided confidence interval of confidence level $1 - \alpha$ is

$$\bar{x}_1 - \bar{x}_2 \pm Z_{\alpha/2} \sqrt{\frac{\sigma_1^2}{n_1} + \frac{\sigma_2^2}{n_2}} \tag{11.1}$$

---

**Slight Tweak in Interpretation of Confidence Interval**
Note that the confidence interval is on the difference on the population. So our interpretation is that we are $1 - \alpha\%$ confident that the difference between the population means is between the values obtained from the equation above. Just like for one parameter inference, one-sided confidence intervals are also possible where only one finite bound is attained for the confidence interval.

---

**Why Not Compare Two Separate Confidence Intervals?**
When comparing population means one possibility that springs to mind is to construct confidence intervals separately for each population mean, a tactic sometimes implemented in practice.

- If these confidence intervals overlap, the lower limit of one population mean is lower than the upper limit of the other population mean, then this indicates that the data does not support a difference between the population means.
- If the confidence intervals do not overlap, the opposite interpretation is reached.

Yet, when constructing two separate confidence intervals, we are not inferring directly on a measure of how the means compare. Due to different sample sizes and population variances, $\overline{X}_1$ and $\overline{X}_2$ tend to have different sampling distributions[1] and as a consequence, comparing two separate confidence intervals for each population mean can result in misleading results.

---

**Example 11.1** *Imagine circumstances where $\overline{x}_1 = 125$, with $\sigma_1/\sqrt{n_1} = 10$, and $\overline{x}_2 = 150$, with $\sigma_2/\sqrt{n_2} = 5$. If two separate 95% confidence intervals are constructed, they would be (105.4, 144.6) and (140.2, 159.8), respectively. Since the confidence intervals overlap, it implies no difference between population means. However, a 95% confidence interval for the population mean differences is (−47.36, −2.64). With both limits of this other interval being strictly below zero, it indicates that the second population mean is larger than the first population mean. The discrepancy of both approaches can be explained by the sampling distributions differing on expected value and standard error.*

Generally, when separate confidence intervals for the population means of two independent samples do not overlap, there will indeed be a statistically significant difference between the means. However, the opposite is not necessarily true: two separate confidence intervals may overlap, yet there may be a statistically significant difference between the means. In statistical lingo, the

---

1 Their population mean may differ as may their standard errors. Or their population means differ but not their standard errors and so on so forth.

confidence interval of mean differences has more power than checking for overlapping confidence intervals.

**Example 11.2** *Through a 90% confidence interval, a business statistics professor wants to answer the following question: "Do quizzes help Business undergraduate students get a better grade in their introductory statistics course?" To perform the comparison of average exam scores, students from two past semesters were selected, one with quizzes, the other without. The number of students that dropped the course or stopped attending class were similar for each semester (Why do you think it is necessary to recognize this?). The 28 students in the group that took quizzes had an average test score of 79.95, while the 27 students that had no quizzes obtained an average test score of 75.54. Historically, test scores follow a normal distribution with a standard deviation of 12.14 and there is no evidence that the standard deviation of the two groups is different.*

Define

$\mu_1$ = true mean exam scores of the students who took quizzes.
$\mu_2$ = true mean exam scores of the students who did not take quizzes.

$\sigma_1 = \sigma_2 = 12.14$. For a 90% confidence level, $Z_{0.1/2} = 1.645$. Using Eq. (11.1)

$$79.95 - 75.54 \pm 1.645\sqrt{\frac{12.14^2}{28} + \frac{12.14^2}{27}}$$
$$4.41 \pm 5.39$$

With 90% confidence, it is concluded that the true difference in mean exam scores from both groups is between $(-0.98, 9.80)$. Since this confidence interval includes 0, there is no evidence that the quizzes helped improve the exam grades of students. This example was introduced in Chapter 1, where the reader was warned that the conclusion does not necessarily extend to quizzes in other subjects and it does not even necessarily extend to all introductory business statistics courses.

---

**Choosing What is $\mu_1$; $\mu_2$**
Assigning what is called $\mu_1$ and $\mu_2$ is arbitrary. If we reverse our definitions in the example above,

$\mu_2$ = true mean exam scores of the students who took quizzes.
$\mu_1$ = true mean exam scores of the students who did not take quizzes,

then the 90% confidence interval becomes $(-9.8, 0.98)$. The conclusion remains intact, no evidence of difference in mean exam scores among the groups.

**Inference on Two Population Means Using Hypothesis Testing – Independent Samples**
**Variances Known**  Returning to the example of two groups taking different versions of an exam in the introduction, the null and alternative can be written as

$$H_o: \mu_1 = \mu_2$$
$$H_1: \mu_1 \neq \mu_2$$

A pivotal quantity is required and it has already been determined that the comparison of population means will be done through differences. Therefore, the null and alternative may be written in the equivalent form,

$$H_o: \mu_1 - \mu_2 = 0$$
$$H_1: \mu_1 - \mu_2 \neq 0$$

More generally, hypothesis testing can be done to determine any kind of difference, $d_o$, between the population means:

$$H_o: \mu_1 - \mu_2 = d_o$$
$$H_1: \mu_1 - \mu_2 \neq d_o$$

Lower tail tests hypotheses are

$$H_o: \mu_1 - \mu_2 \geq d_o$$
$$H_1: \mu_1 - \mu_2 < d_o$$

and upper tail tests hypotheses are

$$H_o: \mu_1 - \mu_2 \leq d_o$$
$$H_1: \mu_1 - \mu_2 > d_o$$

Moreover, the null can always have an equality sign as explained in Chapter 10.

Using the properties of expectation and variance from Section 6.3, it can be shown that $E(\overline{X}_1 - \overline{X}_2) = \mu_1 - \mu_2$ and $Var(\overline{X}_1 - \overline{X}_2) = \frac{\sigma_1^2}{n_1} + \frac{\sigma_2^2}{n_2}$. We will assume that each sample mean follows a normal distribution (see Figure 9.7). Thus, knowing that our pivotal quantity

- should use an adequate estimator of the population mean difference;
- should not depend on the value of the population difference under the null and;
- if possible, should not depend on any nuisance parameter,

the following test statistic is attained:

$$Z = \frac{\overline{X}_1 - \overline{X}_2 - d_o}{\sqrt{\frac{\sigma_1^2}{n_1} + \frac{\sigma_2^2}{n_2}}} \tag{11.2}$$

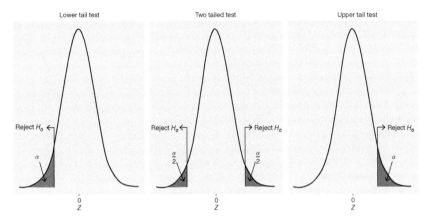

Figure 11.2 If $H_o$ is true, in each of the three types of tests seen here, the observed test statistic should not be too extreme. What is considered too extreme from the expected value of $Z$ depends on the alternative and is the region presented in red.

Therefore, when drawing inference on two population means using hypothesis testing with independent samples where the variances are known, a $Z$ statistic is used. Figure 11.2 sketches the probability distribution of $Z$ when $H_o$ is true along with decision rules for the three types of hypothesis tests. Specifically, to reject the null in a lower tail test, the value of $Z$ must be "extremely" below 0. In contrast, to reject the null in a upper tail test, the value of $Z$ must be "extremely" above 0, and to reject the null in a two-tailed test, the value of $Z$ must be either "extremely" below or "extremely" above 0.

**Example 11.3** *We revisit Example 11.2, now to test at 10% significance if quizzes increase the mean exam scores of students. The summary of the random samples is* $n_1 = 28$, $\bar{x}_1 = 79.95$, $n_2 = 27$, $\bar{x}_2 = 75.54$, $\sigma_1 = \sigma_2 = 12.14$, *and the exam scores are normally distributed.*

**Step 1: State null and alternative** Since $\mu_1$ = true mean exam scores of the students who took quizzes, $\mu_2$ = true mean exam scores of the students who did not take quizzes,

$$H_o: \mu_1 - \mu_2 \leq 0$$
$$H_1: \mu_1 - \mu_2 > 0$$

**Step 2: Choose and calculate the test statistic to be used** From Eq. (11.2), we have

$$Z = \frac{79.95 - 75.54}{\sqrt{\frac{12.14^2}{28} + \frac{12.14^2}{27}}} = 1.34$$

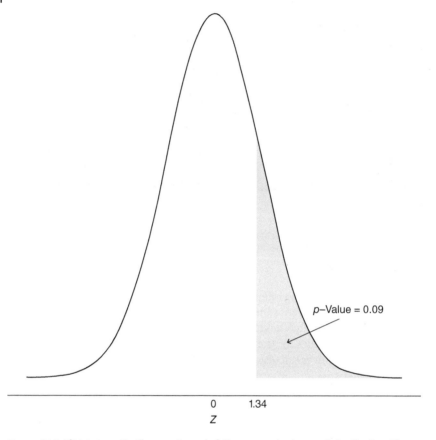

$p$–Value = 0.09

0      1.34

$Z$

Figure 11.3  If $H_o$ is true, $Z$ will approximately follow a standard normal distribution. The $p$-value is shaded in blue.

**Step 3: Find the value of the decision rule, the p-value**  To reject the null, the $Z$ value must be "extreme" enough, which in this example, according to $H_1$, means that it must be far enough above 0 (Figure 11.3).

$$p\text{-Value} = P(Z \geq 1.34) = 1 - P(Z \leq 1.34) = 0.09$$

**Step 4: Make a decision about $H_o$**  Since $0.09 < 0.10$, the null is rejected.

**Step 5: Statistical conclusion**  At 10% significance, it is concluded that the mean exams scores of students that take quizzes is higher than the mean exam scores of students that do not takes quizzes.

> **One Side versus Two-Sided Tests**
> Now wait a minute, did you notice that? In Example 11.2, no difference in mean exam scores was found, while in the test above there was a difference; what gives? In Section 10.4, we saw that the two-sided test is a more conservative test than the one-sided test. In this one-sided test, the null may have been rejected, but the 90% lower bound for the one-sided confidence interval for the difference in mean scores, 0.22, indicates that practically speaking, the mean score of those who take quizzes is not much greater than the mean score of students who did not take quizzes.[2]

## Practice Problems

11.1   Find the standard deviation of $\overline{X}_1 - \overline{X}_2$ when
   a) $n_1 = 50$, $\sigma_1^2 = 25$, $n_2 = 65$, $\sigma_2^2 = 25$.
   b) $n_1 = 200$, $\sigma_1^2 = 100$, $n_2 = 150$, $\sigma_2^2 = 47$.

11.2   Find the standard deviation of $\overline{X}_1 - \overline{X}_2$ when
   a) $n_1 = 115$, $\sigma_1 = 37$, $n_2 = 88$, $\sigma_2 = 52$.
   b) $n_1 = 33$, $\sigma_1 = 61$, $n_2 = 94$, $\sigma_2 = 59$.

11.3   If $\overline{X}_1 = 199.61$, $n_1 = 75$, $\sigma_1^2 = 25$, $\overline{X}_2 = 203.27$, $n_2 = 75$, and $\sigma_2^2 = 25$.
   a) Construct a two-sided 90% confidence interval for $\mu_1 - \mu_2$.
   b) Interpret the confidence interval.

11.4   If $\overline{X}_1 = 17.50$, $n_1 = 100$, $\sigma_1 = 2.07$, $\overline{X}_2 = 10.15$, $n_2 = 100$, and $\sigma_2 = 1.25$.
   a) Construct a two-sided 99% confidence interval for $\mu_1 - \mu_2$.
   b) Interpret the confidence interval.

11.5   In Problem 11.3, what would happen to the confidence interval if everything is the same except the confidence level was 95%?

11.6   In Problem 11.3, what would happen to the confidence interval if everything is the same except $n_1 = n_2 = 100$?

11.7   A retail company operates two stores in different locations. The manager notices that products that sell well in one store do not always sell well in the other and thus believes this may be due to different customer demographics. The manager would like to investigate the difference in mean age of customers of these two stores. A random sample of

---

2 Also, keep in mind the statements made in Chapter 1 regarding careful interpretation of these results.

36 customers in the first store gave a mean age of 40, while a random sample of 49 customers in the second store gave a mean of 35. Based on previous data, $\sigma_1 = 9$ years and $\sigma_2 = 10$ years.

a) Find a 95% confidence interval estimate of the difference between the mean age of customers at each store.

b) Interpret the confidence interval.

c) What assumption is made to get the confidence interval in Problem 11.6?

11.8  In Problem 11.7, how would you argue that the results are not necessarily practically important?

11.9  A one-sided 90% confidence interval for the difference in mean monthly cost of supplies for a printing service shop from two stores gave an upper bound of $714.

a) Interpret the result of this confidence interval.

b) Would the equivalent hypothesis test be an upper tail test or a lower tail test?

11.10  A one-sided 99% confidence interval for the difference in mean number of visits to a theme park in July and August gave a lower bound of 11 321.44.

a) Interpret the result of this confidence interval.

b) Would the equivalent hypothesis test be an upper tail test or a lower tail test?

11.11  If $H_o: \mu_1 - \mu_2 \geq 0; H_1: \mu_1 - \mu_2 < 0$, which of the following is an equivalent statement? (More than one choice is possible.)

a) $H_o: \mu_1 - \mu_2 < 0; H_1: \mu_1 - \mu_2 \geq 0$

b) $H_o: \mu_2 - \mu_1 \leq 0; H_1: \mu_2 - \mu_1 > 0$

c) $H_o: \mu_1 - \mu_2 = 0; H_1: \mu_1 - \mu_2 < 0$

d) $H_o: \mu_2 - \mu_1 = 0; H_1: \mu_2 - \mu_1 < 0$

11.12  If $H_o: \mu_1 - \mu_2 \leq 0; H_1: \mu_1 - \mu_2 > 0$, which of the following is an equivalent statement? (More than one choice is possible.)

a) $H_o: \mu_1 - \mu_2 < 0; H_1: \mu_1 - \mu_2 \geq 0$

b) $H_o: \mu_2 - \mu_1 \geq 0; H_1: \mu_2 - \mu_1 < 0$

c) $H_o: \mu_1 - \mu_2 \geq 0; H_1: \mu_1 - \mu_2 < 0$

d) $H_o: \mu_2 - \mu_1 = 0; H_1: \mu_2 - \mu_1 < 0$

11.13  For a new car model from Company $M$, engineers proposed a slight tweak in the engine design that could lead to better fuel efficiency. To test this, 10 cars were randomly assigned the originally designed motor, and their average miles per gallon was 27.90. Another 10 cars were randomly assigned the motor with the tweaked design, and their average miles per gallon was 25.62. Miles per gallon are normally distributed for both motor designs and assume a variance of 0.97 miles per gallon for both engines. If $\mu_1$ = mean miles per gallon of cars with the original motor, and $\mu_2$ = mean miles per gallon of cars with the tweaked motor, at 5% significance,

a) write the null and alternative hypothesis.
b) calculate the value of the $Z$ statistic.
c) determine what the $p$-value is.
d) is $H_o$ rejected?
e) write a statistical conclusion.

11.14 Numbers that linger in our mind can influence subjective numerical estimates we make later on. In a 2010 study,[3] students had to subjectively estimate the number of calories in a cheeseburger. The goal was to determine if the subjective mean calorie estimate from a group presented with a high-calorie meal is lower than the subjective mean calorie estimate from a group presented with a low calorie meal. Sixty-five students were randomly chosen to do their subjective estimate after thinking about a high calorie cheesecake (group 1), and 70 students were randomly chosen to do their subjective estimate after thinking about an organic fruit salad (group 2). The first group's estimated mean was 780 calories, while the second group estimated a mean of 1041 calories. For our purposes, we assume that $\sigma_1$ = 112 and $\sigma_2$ = 120. If $\mu_1$ = group 1 subjective estimate of mean number of calories, and $\mu_2$ = group 2 subjective estimate of mean number of calories, at 1% significance,

a) write the null and alternative hypothesis.
b) calculate the value of the $Z$ statistic.
c) determine what the $p$-value is.
d) is $H_o$ rejected?
e) write a statistical conclusion.
f) why can we use the $Z$-test in this problem?

11.15 Construct an equivalent one bound confidence interval for Problem 11.14 (*Hint:* See how $Z$ is modified in Table 10.2).

11.16 Construct an equivalent one bound confidence interval for Problem 11.13 (*Hint:* See how $Z$ is modified in Table 10.2).

---

3 Chernev, Alexander. (2011). Semantic anchoring in sequential evaluations of vices and virtues, *Journal of Consumer Research*, 37, pages 761–774.

## 11.4 Inference on Two Population Means When Two Independent Samples are Used – Unknown Variances

When the population variances are not known, the pivotal quantity $Z$ no longer comes into play. In Chapter 10, a work-around was to use a statistic based on the sample standard deviation to perform hypothesis testing. The same line of thought can be followed here but a bit more care is needed. Like before, hypothesis tests or confidence intervals can be used, but from now on, less emphasis is given on the math, and we focus more on the implementation through software and the required assumptions.

*Unknown Variances But They Are Assumed Equal*   Assuming,

- the two random samples are independent.
- $\sigma_1 = \sigma_2$.
- both random samples come from approximately a normal distribution.

Then, hypothesis tests and confidence interval procedures can be derived through the $t$-distribution. The first assumption tends to hold when the data comes from a designed randomized experiment or from random sampling. The second and third assumptions can be defended using historical data or the context of the process.

**Example 11.4**   *The Canadian Dairy Information Centre compiles information from several companies on the cheeses made from cow, goat, sheep, and buffalo milk.*[4] *Nine hundred and thirty-two nonorganic cheeses resulted in a mean fat percentage of 26.54, with a standard deviation of 7.28, while 92 organic ones resulted in a mean fat percentage of 27.09 and a standard deviation of 7.53. Is the mean fat percentage of nonorganic cheeses smaller than mean fat percentage of organic cheeses? Inference will be made at 5% significance.*

*Step 1: State null and alternative*   Let $\mu_1$ = true mean fat percentage of nonorganic cheeses, $\mu_2$ = true mean fat percentage of organic cheeses,

$$H_o: \mu_1 - \mu_2 \geq 0$$
$$H_1: \mu_1 - \mu_2 < 0$$

*Step 2: Choose and calculate the test statistic to be used*   We are provided with two independent samples and the population standard deviations are unknown. Suppose that according to historical evidence, the standard deviations of fat percentage are known to be equal. Both sample sizes are large enough that the central limit theorem can be invoked, and hence a two sample $t$-test can be used. Figure 11.4 shows the necessary commands in Minitab to perform this

---

4  Data source: open.canada.ca (accessed July 26, 2019).

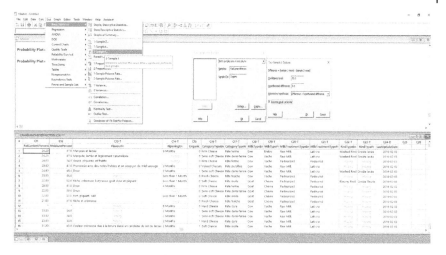

**Figure 11.4** Minitab steps to perform a two population mean *t*-test assuming equal variances.

test: `Stat > 2-Sample t` … Since fat percentage is in one column and organic classification in another, we choose `Both samples are in one column` in the "Two Sample *t* for the Mean" window followed by entering the variables. We must also press the `Options` button to select in the new window the proper sign for the alternative and check `Assume equal variances`.

**Step 3: Find the value of the decision rule, the p-value** Figure 11.5 summarizes the results of the test.

To reject the null, the *t* value must be "extreme" enough, which in this example, according to $H_1$, means that it must be far enough below 0 (Figure 11.6).

$$p\text{-Value} = P(t \leq -0.71) = 0.239$$

**Two-Sample T-Test and CI: FatContentPercent, Organic**

```
Two-sample T for FatContentPercent

Organic   N   Mean  StDev  SE Mean
0        932  26.54  7.28    0.24
1         98  27.09  7.53    0.76


Difference = μ (0) - μ (1)
Estimate for difference: -0.550
95% upper bound for difference: 0.727
T-Test of difference = 0 (vs <): T-Value = -0.71 P-Value = 0.239 DF = 1028
Both use Pooled StDev = 7.3061
```

**Figure 11.5** Minitab output. "0" stands for nonorganic cheese, "1" for organic.

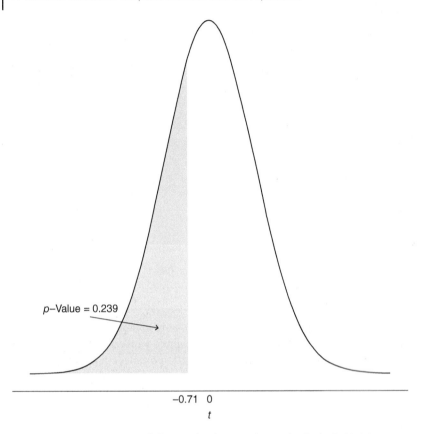

$p$–Value = 0.239

−0.71  0

$t$

Figure 11.6  If $H_o$ is true, $t$ will follow a $t$-distribution. The $p$-value is shaded in blue.

***Step 4: Make a decision about $H_o$***   Since 0.239 $\not<$ 0.05, the null is not rejected.

***Step 5: Statistical conclusion***   At 5% significance, we find no evidence that the fat percentage of organic and nonorganic cheeses in Canada is different. Minitab also provides a one-sided 95% confidence interval with a 0.727 upper bound. This confidence interval includes zero in it.

***Unknown Variances, They Are Not Assumed Equal***   When there is no evidence that the unknown variances are equal, which is usually a case, then an approximate method is required. The test statistic will still be a pivotal quantity, but the denominator will be different from the case when the population variances are assumed equal. Also, this test statistic will now follow approximately a $t$-distribution with degrees of freedom that may be smaller than the degrees of freedom in the case when the population variances are assumed equal.

**Two-Sample T-Test and CI: FatContentPercent, Organic**

```
Two-sample T for FatContentPercent

Organic   N   Mean  StDev  SE Mean
0       932  26.54   7.28    0.24
1        98  27.09   7.53    0.76


Difference = μ (0) - μ (1)
Estimate for difference: -0.550
95% upper bound for difference: 0.772
T-Test of difference = 0 (vs <): T-Value = -0.69 P-Value = 0.246 DF = 116
```

Figure 11.7 Minitab output.

**Example 11.5**   *We revisit Example 11.4. The idea is to consider whether not assuming the population variances are equal would have any effect on the conclusion.*

The steps to perform this test in Minitab are just like Figure 11.4, except that now in the `Options` button the `Assume equal variances` is left blank (the default in Minitab). Figure 11.7 summarizes the results of the test.

The estimated degrees of freedom are much smaller than in Example 11.4, yet this plays not role in the result: $p$-value $\not< 0.05$ and the null is not rejected.

---

**Why Assume $\sigma_1 = \sigma_2$?**
If it is possible to assume the test statistic follows approximately a $t$ distribution when the population variances are unknown without having to assume that they are equal, then why bother with the equal variance version? The degrees of freedom of the $t$ test that does not assume that the variances are equal will always be smaller than the degrees of freedom of the $t$ test under the equal variance assumption unless $s_1 = s_2$ and $n_1 = n_2$. A consequence of this is that it becomes harder to reject the null hypothesis. Thus, if we just rely on the approximation, a less powerful test may be applied when the more powerful unknown but equal variances test is applicable. The distinction between both types of test is generally not important when the degrees of freedom of the approximate version of the test are greater than 30. Also, when $s_1 \approx s_2$ as well as $n_1 \approx n_2$, the equal variance and not equal variance tests will lead to identical conclusions because their degrees of freedom will be close. Although the equal variance $t$ test version is generally robust to a violation of the equal variance assumption, in practice it is avoided when one sample standard deviation is at least double the other sample standard deviation.

### 11.4.1 A Quick Look at Data Science: Suicide Rates Among Asian Men and Women in New York City

Data for leading causes of death in New York City can be retrieved from the internet.[5] The death rates per 100 000 people are organized by gender, race, and over 15 causes of death. Our aim is to determine if suicide rates among Asian or Pacific Islanders men and women in New York City are different. Suicide rates for the groups were obtained by processing the leading cause of death data (Figure 11.8). The 2013 suicide rate for Asian women was not available, but no major issues with the data are detected (Table 11.1).

We define

$\mu_1$ = mean suicide rate a year of Asian men in New York City.
$\mu_2$ = mean suicide rate a year of Asian women in the city.

A significance level of 10% will be used. The sample standard deviations are 1.10 for men and 0.95 for men and women, respectively. Inference will be performed first through a 90% confidence interval on the difference of means. Without prior knowledge on the matter, no assumption of equal variances is made. Figure 11.9 illustrates the required steps to perform a two population mean $t$ test using Minitab, while Figure 11.10 summarizes the results.

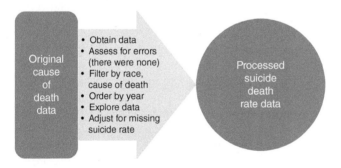

**Figure 11.8** Steps required to filter out suicide rates for Asians in New York City from cause of death data.

**Table 11.1** Suicide rates for Asian men and women in New York City from 2007 until 2014.

| | Year | | | | | | | |
|---|---|---|---|---|---|---|---|---|
| Gender | 2007 | 2008 | 2009 | 2010 | 2011 | 2012 | 2013 | 2014 |
| Male | 7.4 | 6.0 | 6.9 | 8.8 | 9.3 | 7.4 | 7.4 | 8.7 |
| Female | 4.3 | 4.9 | 3.1 | 3.0 | 3.7 | 5.6 | — | 4.5 |

---

5 Source: data.cityofnewyork.us (accessed July 26, 2019).

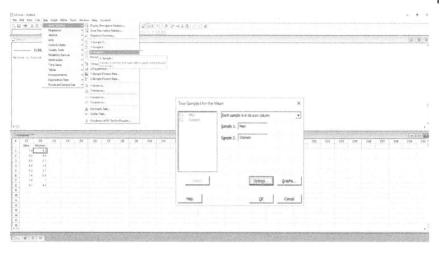

Figure 11.9 Minitab steps to perform two population mean *t*-test without assuming equal variances.

**Two-Sample T-Test and CI: Men, Women**

```
Two-sample T for Men vs Women

          N    Mean   StDev   SE Mean
Men       8    7.74   1.11    0.39
Women     7    4.157  0.952   0.36


Difference = μ (Men) - μ (Women)
Estimate for difference: 3.580
90% CI for difference: (2.633, 4.527)
T-Test of difference = 0 (vs ≠): T-Value = 6.74 P-Value = 0.000 DF = 12
```

Figure 11.10 Minitab output.

Thus, with 90% confidence, it is concluded that the difference in mean suicide rates among Asians in New York City is between (2.42, 4.74). Since the confidence interval does not include zero, it provides evidence of their being a significant difference between the mean suicide rates of the two genders.

Hypothesis testing of

$$H_o: \mu_1 - \mu_2 = 0$$
$$H_1: \mu_1 - \mu_2 \neq 0$$

leads to the same conclusions as the confidence interval since the *p*-value is 0 and the null is rejected. However, the confidence interval provides a better idea of how different the death rates are.

The presented inference methods assume that the populations are normally distributed. The small sample sizes make it difficult to assess this assumption.

Also, the analysis here would be less meaningful if they'd be any trend in time in the suicide rates. From Table 11.1, no trend is visible in the suicide rates.

## Practice Problems

11.17  In the 2016 General Social Survey, one question was "What is the ideal number of children for a family to have?" Of the 738 men who answered,[6] the mean response was 2.52 with a standard deviation of 0.84. In contrast, of the 867 women who answered, the mean response was 2.63 with a standard deviation of 0.89. A 99% confidence interval for the difference in mean ideal number of children is (−0.22, 0.0048).

a) The interval is estimating $\mu_1 - \mu_2$. Which population was chosen as number 1?

b) What would happen if we were to get a 99% confidence interval for $\mu_2 - \mu_1$?

c) How meaningful is the true difference in mean ideal number of children?

d) Do you think the distribution of answers to the questions are normally distributed?

11.18  A realtor wants to compare asking prices per square foot for land in neighborhood 1 and neighborhood 2. He takes a random sample of 16 listings from neighborhood 1. From neighborhood 2, a random sample of 20 is collected. Below, the Minitab output of the hypothesis test is provided.

**Two-Sample T-Test and CI**

```
Sample   N   Mean   StDev   SE Mean
1        16  75.10  5.10    1.3
2        20  73.30  8.20    1.8


Difference = mu (1) - mu (2)
Estimate for difference: 1.80
95% CI for defference: (-2.75, 6.35)
T-Test of defference: = 0 (vs ≠): T-value = 0.81 P-Value = 0.426 DF = 32
```

a) What is the sample mean from neighborhood 1?

b) What is the sample standard deviation from neighborhood 2?

c) Show that the estimated standard error of the sample mean from neighborhood 1 is 1.3.

---

6  This question had answers, 0, ... , 6 along with "7 or more" and "Don't Know." Statistics come from omitting responses "7 or more" and "Don't Know" responses.

d) Write the hypothesis to test if there is a difference in mean asking price per square foot for land in neighborhood 1 and neighborhood 2.
e) Obtain the test statistic (from the Minitab output).
f) Based on the Minitab output, make a decision about the null hypothesis at 5% signficance.
g) Write a statistical conclusion.

11.19 Enrollment data of schools in Connecticut for the 2013–2014 school year are available[7] in the file "School_Enrollment_Connecticut" $\alpha = 0.10$. Note that the last row is for all of Connecticut and thus must be deleted for your analysis.

a) Do you think that there is a need to perform statistical inference to determine if there is a difference between the mean number of male and mean number of female students for the 2013–2014 school year?
b) Based on the data, test if there is a difference between the mean number of male and female students for the 2014–2015 school year at 10% significance.
c) Construct a two-sided 90% confidence interval.
d) What are the assumptions being made to perform the hypothesis test above?
e) Any limitations in the usefulness of the inference performed here?
f) Perform a hypothesis test to determine if the mean number of females at Bridgeport school district is different than the mean number of females at the Fairfield School District.

11.20 A new machine was purchased to can tuna. With the old machine, a random sample of 16 cans revealed an average weight of 6.83 oz and a standard deviation of 0.64 oz. In an independent random sample with the new machine, the average weight was 6.24 oz, with a standard deviation of 0.75 oz.

a) Test whether the old machine has higher mean weight at $\alpha = 0.025$, assuming the unknown population variances are not equal. Interpret decision about null.
b) Test whether the old machine has higher mean weight at $\alpha = 0.025$, assuming the unknown population variances are equal. Interpret decision about the null.

11.21 A company's data analyst performed a two sample $t$-test to compare the mean time to pack some jewelry using two separate machines. Curious, a manager asks the analyst if he used an equal variance $t$-test. His answer was no, that the degrees of freedom of the approximate $t$-test that he

---

7 data.ct.gov (accessed July 26, 2019).

used was 60 and therefore the results of either method should be similar. Comment on the analyst's logic.

11.22 You work for an insurance firm and you must choose between two companies to handle your employer's investment portfolio. Although both companies have claims about the mean 1-year return for a $10 000 investment, you prefer to take independent samples from company A and company B. All you care about is whether there is a difference in the mean 1-year return (measured in dollars) of each of the two companies. The Minitab summary of the analysis is given below.

**Two-Sample T-Test and CI**

```
                        SE
Sample   N   Mean  StDev   Mean
1        15  2013   175    45
2        20  2200   425    95


Difference = μ (1) - μ (2)
Estimate for difference: -187
95% CI for difference: (-403, 29)
T-Test of defference: = 0 (vs ≠): T-value = 1.78 P-Value = 0.087 DF = 26
```

a) Write the null and alternative hypothesis.
b) Based on the Minitab output, make a decision about the null hypothesis at 5% signficance.
c) Write a statistical conclusion.

## 11.5 Inference on Two Means Using Two Dependent Samples

Dependent samples are those where each unit has two observations, one forms part of the first sample, the other forms part of the second sample.

**Example 11.6** *A firm determines to have its employees participate in a training program to reduce the daily cost of production of a popular item. One avenue is to measure the daily cost before the training and for each employee match the daily production cost after the training. The goal is to determine if there is an improvement in mean daily production cost after the training program.*

In the case of dependent samples, the variance of the difference between sample means is harder to calculate and the test statistics previously presented are no longer accurate. To overcome this, we may build a testing procedure based on the average of each individual difference. This commonly leads to a reduction in standard errors of the difference statistic.

**The Paired *t*-Test**

Instead of taking the averages of each sample first and then estimating the difference, the difference is determined first followed by a calculation of the average difference. The idea is that if the null is true, the sample mean difference should be close to zero. From the sample mean difference a pivotal quantity can be obtained, analogous to the *t* test in Chapter 10. This type of inferential method is called a **paired *t*-test**. This test commonly leads to a reduction in standard errors of the difference statistic. For the dependent samples test, similar to one sample tests, it is assumed that

- the differences are a random sample from the population,
- that the differences follow approximately a normal distribution.

Violation of the second assumption is mainly of concern when the sample size is small, differences are highly skewed, and a one-sided test is being performed.

**Example 11.7**   *Returning to Example 11.6, if the difference is defined as, the production cost before the training minus the cost after the training, then the alternative hypothesis is that the true mean difference is greater than zero. The hypothesis test will be performed at 1% significance. A random sample of 11 employees resulted in a mean difference of $22.27 and the standard deviation of the difference was $333.*

These statistics can be enteredinto Minitab to get results seen in Figure 11.11. Since the *p*-value $= 0.414 < 0.01$, the null is not rejected. At 1% significance, we do not have evidence that the training was able to reduce daily mean cost.

**The Paired *t*-Test for Independent Samples?**

How about applying the paired *t*-test for independent samples? That would not be best. There is a price to pay when using the paired *t*-test: degrees of freedom are smaller. As result, for independent samples, it has unnecessary lower power.

**Paired T-Test and CI**

```
              N  Mean  StDev  SE Mean
Difference   11    22    333      100


99% lower bound for mean difference: -255
T-Test of mean difference = 0 (vs > 0): T-Value = 0.22 P-Value = 0.414
```

Figure 11.11  Paired *t*-test Minitab output.

Table 11.2 All cases of testing the difference for two population means.

| Case | Comments |
| --- | --- |
| 1 | Variances known, independent samples are large or populations are normal. |
| 2 | Variances unknown, but equal. Independent samples and populations are normal. |
| 3 | Variances unknown, not assumed equal. Independent samples and populations are normal. |
| 4 | Dependent (matched) samples and populations are normal. |

Table 11.2 summarizes all four cases of drawing inference on the difference of two population means from two samples. With large sample sizes, by the central limit theorem, sample means will be normally distributed. Yet if the populations are highly skewed, population means may not be adequate measures of centrality.

## Practice Problems

11.23 In their 2017 book, *Teaching Statistics: A Bag of Tricks*, Andrew Gelman and Deborah Nolan provide a series of ideas for active student participation. In one of the activities, while taking an exam, students get to guess what score they will get. If their guess is close enough to their actual exam score, they get bonus points. This activity presents an opportunity to assess the following question: "Are students typically overconfident of how they will perform in a test?" The activity was implemented during an exam for the first semester of a business statistics course. Significance level is 1%. Below is a screen grab of the Minitab output.

**Paired T-Test and CI: Guessed score, Actual Score**

```
Paired T for Guessed score - Actual Score

               N   Mean  StDev  SE Mean
Guessed score  31  79.19  12.19    2.19
Actual Score   31  75.39  14.10    2.53
Difference     31   3.81  12.26    2.20

99% lower bound for mean difference: -1.61
T-Test of mean difference = 0 (vs > 0): T-Value = 1.73 P-Value = 0.047
```

   a) How many students were part of the sample?
   b) Write the null and alternative hypothesis.
   c) What is the estimated mean difference?
   d) At 1% significance, is the null rejected?
   e) Write a statistical conclusion.
   f) Why does it make sense to use a paired $t$-test?

**11.24** A company has been receiving complaints that the machines they sell to can tuna constantly can too much of the product. Engineers designed a new machine and assigned 12 machine operators to use the new machine and the older model to manufacture 5 oz cans of tuna. If the difference is defined as the new model can mean weight minus the old machine mean can weight, use the output below to answer the following questions. Use 5% significance.

**Paired T-Test and CI**

```
             N     Mean    StDev  SE Mean
Difference   12  -0.4514  0.2409   0.0695

95% upper bound for mean difference: -0.3265
T-Test of mean difference = 0 (vs < 0): T-Value = -6.49   P-Value = 0.000
```

a) Write the null and alternative hypothesis.
b) What is the estimated mean difference?
c) At 5% significance, is the null rejected?
d) Write a statistical conclusion.

**11.25** Ten people participated in a taste experiment to determine if there is a difference in the mean score of two cereals (where 100 is best).

**Paired T-Test and CI**

```
             N    Mean   StDev  SE Mean
Difference   10   5.03    9.37     2.96

95% CI for mean difference: (-1.67, 11.73)
T-Test of mean difference = 0 (vs ≠ 0): T-Value = 1.70   P-Value = 0.124
```

a) Write the null and alternative hypothesis.
b) What is the estimated mean difference?
c) At 5% significance, is the null rejected?
d) Write a statistical conclusion.

## 11.6 Inference on Two Population Proportions

It should come as no surprise that on occasion, the comparison of two population proportions is needed. Also, just like when comparing population means, it is convenient to compare two proportions by their difference.[8] The set up is that two independent samples are drawn: one with population proportion $\pi_1$ and the other sample from a population with proportion $\pi_2$. Each sample has its own sample size $n_1$ and $n_2$, respectively. Recall that $\hat{p}$ is a convenient estimator of $\pi$.

---

8 However, in some fields, the comparison of two proportions is performed by their ratio, a method that we do not pursue here.

Using the properties of expectation and the central limit theorem, one can find expressions to construct confidence intervals or test statistics to infer on the difference of the two population proportions using the difference between the two sample proportions $\hat{p}_1 - \hat{p}_2$ as the estimator. For example, for a two-sided confidence interval of $\pi_1 - \pi_2$ at $1 - \alpha$ confidence,

$$\hat{p}_1 - \hat{p}_2 \pm z_{\alpha/2} \sqrt{\frac{\hat{p}_1(1 - \hat{p}_1)}{n_1} + \frac{\hat{p}_2(1 - \hat{p}_2)}{n_2}}$$

These normal distribution-based confidence intervals are satisfactory as long as $n_1 p_1$, $n_1 p_2$, $n_1(1 - p_1)$, and $n_2(1 - p_2)$ are all at least 15.

**Example 11.8**   *According to the 2016 police report for Austin, Texas, that year out of 104 928 motor vehicle stops, 9569 times searches were deemed necessary. In contrast, in 2015 out of 120 056 motor vehicle stops, 9253 times searches were deemed necessary.[9] Construct a 99% confidence interval for the true difference in proportion of moving vehicle stops that involved searches.*

Define

$\pi_1$ = proportion of 2016 motor vehicle stops that involved searches.

$\pi_2$ = proportion of 2015 motor vehicle stops that involved searches.

To get the limits of the confidence interval, a margin of error will be added and subtracted to the point estimator of the true difference. As we know well at this stage, the margin of error is a function of the confidence level and the standard error of the point estimate. Analogous to the one proportion confidence interval, the normal approximation condition can be assessed for both samples; it does hold. Using statistical software, it can be found that the 99% confidence interval is (0.011, 0.017). Thus, we are 99% confident that the true difference in 2016 and 2015 proportions of traffic stops that involved a search was between 0.011 and 0.017.

Hypothesis testing to compare the proportion of two populations follows the principles applied before in other contexts. The variance of $\hat{p}_1 - \hat{p}_2$ is a function of $\pi_1$ and $\pi_2$. Thus, when applying the $Z$ statistic for testing on $\pi_1 - \pi_2$, there are two situations to consider: when in the null hypothesis, the difference is equal to zero, and when in the null hypothesis, the difference is equal to some value different from zero. If $H_o: \pi_1 - \pi_2 = 0$, a pooled estimate using information from both samples is applied to estimate the variance of $\hat{p}_1 - \hat{p}_2$. Then, $\hat{p}_1 - \hat{p}_2$ is divided by its standard error to get and observed $Z$. When $\pi_1 - \pi_2 = d_o$ and $d_o$ is not zero, it is no longer reasonable to use a pooled estimate of the standard error of $\hat{p}_1 - \hat{p}_2$ and instead a function of the variance of each sample proportion is applied.

---

9 Source: www.austintexas.gov (accessed July 26, 2019).

**Example 11.9** *According to Philadelphia city records, in 2015 out of 1251 shootings, 227 were fatal; while in 2016 out of 1294 shootings, 212 were fatal. Test if there was a difference in the proportion of 2015 and 2016 fatal shootings at 1% significance*

$\pi_1$ = proportion of 2015 fatal shootings in Philadelphia.

$\pi_2$ = proportion of 2016 fatal shootings in Philadelphia.

The hypotheses are

$$H_o: \pi_1 - \pi_2 = 0$$
$$H_1: \pi_1 - \pi_2 \neq 0$$

Figure 11.12 illustrates the required steps to perform a two population mean $t$ test using Minitab, while Figure 11.13 summarizes the results.

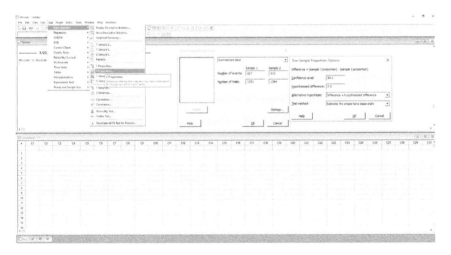

Figure 11.12  Minitab steps to perform two population proportion test.

**Two and CI for Two Proportions**

```
Sample    X     N   Sample p
1        227  1251   0.181455
2        212  1294   0.163833


Difference = p (1) - p (2)
Estimate for difference: 0.0176218
99% CI for difference: (-0.0209809, 0.0562244)
Test of difference = 0 (vs ≠ 0): Z = 1.18 P-Value = 0.240

Fisher's exact test: P-Value = 0.249
```

Figure 11.13  Minitab output.

The $p$-value is greater than 0.01 and therefore the null is not rejected. At 1% significance, we find no evidence that the proportion of fatal shootings in Philadelphia for 2015 and 2016 are different.

## Practice Problems

11.26 Have you ever wondered if the proportion of high school students that smoked cigarettes in the past 30 days is different for Hispanics and Whites? Perform a hypothesis test to check if this was the case in California for the 2015–2016 school year using $\alpha = 0.05$ and the data[10] file "Ca_proportion_smoke."

11.27 Enrollment data of schools in Connecticut for the 2013–2014 school year are available[11] in the file "School_Enrollment_Connecticut." Note that the last row is for all of Connecticut and thus must be deleted for your analysis. Perform a hypothesis test with $\alpha = 0.10$ to determine if the proportion of male and females enrolled in schools is different.

11.28 According to the Centers for Disease Control and Prevention (CDC), in 2015, of 208 595 kids younger than 6 years old tested for lead in the blood in Massachusetts, 5889 had levels between 5 and 9 µg/dl. In contrast, in 2014, 212 014 kids younger than 6 years old were tested in the state and 6429 had levels[12] between 5 and 9 µg/dl. Calculate a 95% confidence interval of the true difference in 2015 and 2014 proportions of children with levels of lead between 5 and 9 µg/dl. If the difference appears statistically important, do you think it is practically important?

11.29 From the New York 2013 youth behavioral risk survey, we can be 95% confident that the proportion of youth who are at least moderate drinkers[13] was between[14] 0.098 and 0.118. Also, we can be 95% confident that the proportion of youth who are at least moderate drinkers in 2011 was between 0.117 and 0.139. Build a confidence interval for the true difference in proportion of youth who are at least moderate drinkers in 2013 and 2011. *Hint:* Get the standard errors from each margin of error first.

---

10 Source: data.chhs.ca.gov (accessed July 26, 2019).
11 data.ct.gov (accessed July 26, 2019).
12 Source: www.cdc.gov (accessed July 26, 2019).
13 Meaning they answered affirmatively when asked if they drank five or more alcoholic drinks in the past 30 days.
14 Source: data.cityofnewyork.us (accessed July 26, 2019).

# Chapter Problems

11.30 Perform the hypothesis test for Example 11.4 at same significance, but using a $Z$-test.
a) What is your conclusion?
b) How can you explain how the conclusion of this test compares to the conclusion of Example 11.4?

11.31 Inference is needed on the difference between the mean math SAT score of high school students who went to a 10-minute presentation ($\mu_1$, group A) and the mean math SAT score of high school students who did not go to the 10-minute presentation ($\mu_2$, group B). A 95% confidence interval on the difference of two means gave (25.15, 38.69). What would be the impact of reversing the population mean definitions: defining $\mu_1$ as for group B and $\mu_2$ as for group A?

11.32 A professor has two possible routes to get to work from home. The distances are not the same but for her it is all about getting to work faster. She has tried both routes, but finds it hard to tell which one on average is best, though she suspects that for route 2 there is a higher variance in the time to get to work. For a whole month, she randomly chooses one of the routes to go to work. Of 10 times that she took route 1, it took her an average of 42 minutes to get to work with a standard deviation of 5 minutes. Of 12 times that she took route 2, it took her an average of 35 minutes to get to work with a standard deviation of 9 minutes. Using statistical software, determine if there is a difference in the mean times of both routes at 5% significance. What assumptions are required to perform the test?

11.33 Data on daily units and legal sales (in $) since July 2014 of several marijuana products in Washington are available in the file[15] "Marijuana_Sales_w_Weight_Daily" Products include marijuana extract for inhalation. A researcher suspects the August mean daily sales per marijuana extract for inhalation unit is decreasing. The August 2016 sales per unit averaged 21.62 with a standard deviation of 0.30, while the August 2017 sales per unit averaged 18.18 with a standard deviation of 0.17. Perform the hypothesis test at a 1% significance.

11.34 New York City chronic lower respiratory disease death rates for White non-Hispanic men and women are available below,[16]

---

15 data.lcb.wa.gov (accessed July 26, 2019).
16 Source: data.cityofnewyork.us (accessed July 26, 2019).

| | Year | | | | | | | |
|---|---|---|---|---|---|---|---|---|
| Gender | 2007 | 2008 | 2009 | 2010 | 2011 | 2012 | 2013 | 2014 |
| Male | 25.8 | 25.9 | 27.6 | 28.9 | 33.3 | 26.3 | 32.5 | 29.7 |
| Female | 32.9 | 36.1 | 31.9 | 35.0 | 36.0 | 35.7 | 36.6 | 36.3 |

a) Calculate the sample standard deviations of both populations.
b) Is it reasonable to assume that the population standard deviations are equal?
c) Using Minitab, construct a 90% two-sided confidence interval on the difference mean chronic lower respiratory disease death rate in New York City.
d) Interpret the confidence interval above.
e) What assumptions were needed to construct the confidence interval?

11.35 Who is more violent, superheroes or villains? A *CNN article*[17] reported on a research presentation on the matter. The research considered (a random sample) of 10 movies and evaluated the number of violent acts committed by superheroes and villains in each movie. Which type of analysis makes most sense: an independent two sample $t$-test or a paired $t$-test? Why?

11.36 Returning to Problem 11.35, if the difference is defined as superhero violent events per hour minus villain violent events per hour, use the Minitab output below to answer the following questions at 1% significance.

**Paired T-Test and CI**

```
            N   Mean   StDev   SE Mean
Difference  10  5.200  3.000   0.949


99% CI for mean difference: (2.117, 8.283)
T-Test of mean difference = 0 (vs ≠ 0): T-Value = 5.48 p-Value = 0.000
```

a) What was the average difference in violent events per hour?
b) If we are testing if their is a difference in mean violent events per hour, is the null rejected?
c) Interpret the confidence interval.

---

17 Superheroes or villains: Who's more violent may surprise you. November 2, 2018.

11.37 From the file[18] "Baltimore_Arts" get the number of art-related businesses per 1000 residents for Baltimore in 2011 and 2012 (artbus11 and artbus12 respectively).

*Hint:* Look carefully at the meaning of the last row.

a) Perform a paired $t$-test at 5% significance to determine if there is any difference between the 2011 and 2012 rates.

b) Construct a histogram of the differences. Does it seem that they follow a normal distribution?

11.38 Download Baltimore education data[19]: "BaltimoreEducation_and_Youth __2010-2012_"

*Hint:* Look carefully at the meaning of the last row.

a) Perform a two mean test at 5% significance to determine if the expected proportion of students receiving free or reduced meals in 2010 (farms10) is higher than in 2012 (farms2012).

b) Perform a two mean test at 5% significance to determine if the expected proportion of students suspended in 2011 (susp11) is lower than in 2012 (susp12).

11.39 An analyst at the firm you work in was comparing the proportions from two independent samples. He was able to use a normal approximation and build a confidence interval for each proportion. Since they overlapped, he determined that the true proportions are not different. What is the main concern about this conclusion?

11.40 In Problem 11.20, the two sample $t$-test assumed equal variances, and the $t$-test not assuming equal variances led to a different decision about the null.

a) Perform a two variance hypothesis test to determine whether they are equal at $\alpha = 0.025$.

b) Construct a 97.5% confidence interval of the ratio of population variances. Interpret results.

c) Based on 11.40 or 11.40, which $t$-test in Problem 11.20 is more reasonable? Explain.

11.41 In September 20, 2017 Puerto Rico was struck by Hurricane Maria, a powerful storm with 155 mph sustained winds. In May 2018, multiple media outlets reported the findings of a new study[20] in which researchers used September to December 2017 data to estimate that

18 data.baltimorecity.gov (accessed July 26, 2019).
19 data.baltimorecity.gov (accessed July 26, 2019).
20 Kishore et al. (2018).

4645 people had died in Puerto Rico, directly or indirectly, due to Hurricane Maria. The news shocked many, since the local government had stated that 64 people had died, directly or indirectly, due to the storm.

a) The study authors were 95% confident that between 793 and 8498 died due to the Hurricane Maria aftermath. Explain how the confidence interval makes the government's official count of 64 doubtful.

b) Explain how the confidence interval provided above makes reporting a 4465 death toll highly misleading.

11.42 In Problem 11.41, one limitation of the study is that it estimated the number of deaths due to Hurricane Maria and relied on survey data from a few neighborhoods to extrapolate the death toll across the entire island. In February 2018, a study[21] used death records before the storm (September 1–19, 2017) and after the storm (September 20 to October 31, 2017) to estimate the death toll from Hurricane Maria in Puerto Rico. It found a 95% confidence interval of (605, 1039).

a) Interpret this confidence interval.

b) Explain why the conclusion from this study are compatible with the conclusion from the study referred to in Problem 11.41.

c) Explain why the confidence interval of this analysis is different from the confidence interval described in Problem 11.41.

## References

Kishore et al. (2018). Mortality in Puerto Rico after Hurricane Maria, *New England Journal of Medicine*, 379:2, pages 162–170.

Rivera, Roberto and Rolke, Wolfgang (2018). Estimating the death toll of Hurricane Maria. *Significance Magazine*. February.

## Further Reading

Doane, D. P. and Seward L. E. (2011). *Applied Statistics in Business and Economics*, 3rd Edition. McGraw-Hill.

Gelman, A. and Nolan, D. (2017). *Teaching Statistics: A Bag of Tricks*. Oxford.

---

21 Rivera and Rolke (2018). 'Estimating the death toll of Hurricane Maria'. Signicance Magazine.

# 12

## Analysis of Variance (ANOVA)

## 12.1 Introduction

Let's suppose that we want to determine if there is a difference in the mean miles per gallon of cars that have three different types of motors. How do we answer this question through hypothesis testing? More generally, how do we conduct hypothesis testing when we want to compare the mean of 5 populations? Or to compare the mean of 10 populations?

In this chapter, we will study comparing means of three or more populations. There are various points to consider that make the process distinct to hypothesis testing of one mean. The extension to subdividing populations more than one way will also be discussed, as well as some other features.

**Example 12.1**  *A company is designing a gas additive to increase the miles per gallon of cars. Management would like to see if methanol content in the additive has an impact. Their options are 50% methanol, 60% methanol, and 75% methanol. In an experiment, 10 cars are randomly assigned to each methanol content and after a predetermined distance, they measure the miles per gallon of the cars.*

***Hypothesis Testing When Comparing Three or More Populations, Intuition***  To conduct hypothesis testing, when comparing three or more populations, it is useful to think of the procedure when we compare two populations. For example, we had established that if we wish to analyze the following null hypothesis and alternative hypothesis:

$$H_o: \mu_1 = \mu_2$$
$$H_1: \mu_1 \neq \mu_2$$

*Principles of Managerial Statistics and Data Science*, First Edition. Roberto Rivera.
© 2020 John Wiley & Sons, Inc. Published 2020 by John Wiley & Sons, Inc.
Companion website: www.wiley.com/go/principlesmanagerialstatisticsdatascience

We may write equivalently:

$$H_o: \mu_1 - \mu_2 = 0$$
$$H_1: \mu_1 - \mu_2 \neq 0$$

Let's go back to the example of the mean miles per gallon at the beginning of this chapter, where we focus on $H_o$:

$$H_o: \mu_1 = \mu_2 = \mu_3$$

Maybe we could say that

$$H_o: \mu_1 - \mu_2 - \mu_3 = 0$$

Sadly, this null is not equivalent to $H_o: \mu_1 = \mu_2 = \mu_3$ since if, say, $\mu_1 = 3, \mu_2 = 2, \mu_3 = 1$, then $\mu_1 - \mu_2 - \mu_3 = 0$; but obviously the three means are not equal. There are infinite values of the means that will lead to a difference of zero. More importantly, when the means are equal, then $\mu_1 - \mu_2 - \mu_3$ will never be equal to zero. The equivalence of $H_o: \mu_1 = \mu_2$ to $H_o: \mu_1 - \mu_2 = 0$ only applies when we are comparing two population means.

The arguments stated until now can bring us to another alternative. We can perform various hypotheses tests of two population means. Going back to the example of the mean miles per gallon, we can conduct the following hypothesis tests:

$$H_o: \mu_1 = \mu_2$$
$$H_1: \mu_1 \neq \mu_2$$

$$H_o: \mu_1 = \mu_3$$
$$H_1: \mu_1 \neq \mu_3$$

$$H_o: \mu_2 = \mu_3$$
$$H_1: \mu_2 \neq \mu_3$$

This option to infer on whether the means are equal seems reasonable. But it has a more subtle problem. Recall that hypothesis testing leads us to a conclusion that has the highest probability of being correct; but for every hypothesis test we conduct, we can incur in type I error or type II error. Assuming $\alpha = 0.05$, it can be demonstrated that the probability of at least one type I error when conducting 3 hypothesis tests on all pairs of three means, is approximately 0.14, which is rather high. With 5 means, the probability of at least one type I error jumps to 0.4 when conducting hypothesis tests for all 10 possible mean pairs. The probability of at least one type I error when realizing $q$ hypothesis tests of two population means will increase as $q$ increases.

**ANOVA Terminology** The procedure to generalize hypothesis testing to determine if the means of two or more populations are equal is known as **Analysis of Variance: ANOVA** for short. First, a series of concepts need to be defined.

**Response variable** is the random variable of interest in the study. Typically represented by $Y$. In practice, $Y$ is also known as a[1] dependent variable.

**Experimental unit** is the object on which you are measuring the response variable.

**Factor** is a qualitative variable which may potentially have an impact on the response variable $Y$.

**Level or treatment** is each possible value or category that a factor can have.

The aim of ANOVA is to determine if one or more factors influence the mean of $Y$. In Example 12.1, the response variable is the miles per gallon, the experimental units are the vehicles, the factor is the amount of methanol, and the levels are the 3 amounts of methanol used. When only one factor is available, the procedure is called **one-way ANOVA** or one-factor ANOVA.

---

**ANOVA Assumptions**

To perform ANOVA, three assumptions are required:

- Observations of the response variable are independent.
- The variances of each population are equal, $\sigma^2$.
- Response variable observations come from a normal distribution.

---

Note that we are considering the potential impact that factor levels may have on the mean of the response variable. This means that under the null hypothesis, all the sample means for each one of the factor levels have the same population mean and the same population standard deviation. That is, the sample averages obtained from each factor have the exact same probability distribution. To illustrate, consider two scenarios when there is one factor with three levels, and sample means are calculated at each level:

Scenario 1: All three levels of the factor have the same population mean, $\mu$.

Scenario 2: The population means at level 1 and 3 are equal, but at level 2 it is much smaller.

Figure 12.1 displays the sampling distributions in both scenarios when all sample means are normally distributed and have the same standard error. The resulting sample means have also been added. In Scenario 1, all sample means are statistically close to $\mu$. Under Scenario 2, $\bar{x}_1, \bar{x}_3$ are close to $\mu$, but $\bar{x}_2$ is too far below $\mu$.

We will be focusing on the case of fixed effects ANOVA. The **fixed effects model** applies when we have collected data for all the possible levels of the factor. When only a sample of the possible levels of the factor have been collected, a different type of ANOVA is needed.

---

1 We prefer to call $Y$ the response variable here, since we have a specific meaning for "dependence" in this book.

Scenario 1                                    Scenario 2

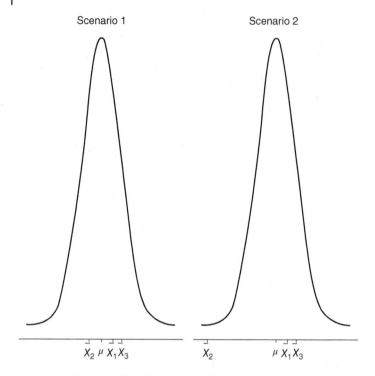

Figure 12.1 The sampling distributions of three sample means under two scenarios. Scenario 1: they all have the same population mean. Scenario 2: population mean for $\bar{x}_2$ is much smaller.

## Practice Problems

12.1 Every year, administrators of the San Francisco airport conduct a survey to analyze guest satisfaction. Suppose management wants to compare the restaurant quality scores for four terminals. Restaurant scoring ranges from 1 to 5, with 1 = Unacceptable, 2 = Below Average, 3 = Average, 4 = Good, and 5 = Outstanding. Identify the following:
   a) Experimental unit
   b) Response variable
   c) Factor
   d) Levels of the factor

12.2 A study is proposed to evaluate three systems to reduce stress among police officers. Police officers are randomly assigned to each system, and a stress score is calculated from a questionnaire.
   a) Experimental unit
   b) Response variable

c) Factor
d) Levels of the factor

12.3 State what are the required assumptions to implement ANOVA.

## 12.2  ANOVA for One Factor

Ronald Fisher recognized the problem of how total probability of incurring in a type I error increases with the number, $q$, of two mean hypothesis tests. He determined that, instead of performing $q$ hypothesis tests of two population means, we should verify if the population means are equal simultaneously with just one test. If there are $a$ levels to the factor, then one-way ANOVA tests

$H_o: \mu_1 = \mu_2 = \ldots = \mu_a.$
$H_1:$ at least two of the population means are not equal.

Note that rejecting $H_o$ does not mean that all means are different (Figure 12.1).

*Partitioning the Sum of Squares*   One can deduce that if the mean of the response variable is not equal for all levels of the factor, then there is an association between the response and the level of the factor. Another way to see the goal of determining the impact of a factor on a response variable is as a statistical modeling problem. That is, the values of the response variable are treated as an additive function of the overall mean $\mu$, a treatment parameter that shifts the mean of $Y$ based on the level of the factor, and a random error term. The error term accounts for random fluctuations around the mean and has an expected value of zero. If the factor has no effect on the response, then the treatment parameter for each factor level should be equal to zero. If $Y_j$ is the response variable at factor level $j$ then,

$$E(Y_j) = \mu + \tau_j$$

where $\tau_j$ is the shift effect of the $j$th level of the factor on $E(Y_j)$. If the null is true, all levels of the factor have the same effect. If the null is false, at least one level of the factor is associated with a higher or lower response expected value than other levels. We proceed to estimate $\mu$ with $\bar{y}$ and then focus on deviations between the response variable values and the sample mean to answer the following question: Do the levels of the factor explain part of the variability in $Y_j$? In Chapter 4 we learned that a measure of variability of a random variable is the variance,

$$s^2 = \frac{1}{n-1} \sum_{i=1}^{n} (Y_i - \bar{Y})^2$$

But for ANOVA, it is convenient to express the response variable according to the factor level and observation taken at that level. We write $Y_{ij}$ where

$j = 1, \ldots, a$ is an index representing the factor level and $i = 1, \ldots, n_j$ is an index of the observation at level $j$. Thus, $Y_{11}$ is the first observation at level 1 of the factor and $Y_{53}$ is the fifth observation at level 3 of the factor. The notation allows us to have a different number of observations for each factor level. We can express the total sample variance of $Y$ as

$$s^2 = \frac{1}{n-1} \sum_{j=1}^{a} \sum_{i=1}^{n_j} (Y_{ij} - \overline{Y})^2$$

where $\overline{Y}$ is the overall average (over all observations without considering factor level at which $Y$ was measured). The numerator of this equation is known as the **total sum of squares, TSS**. ANOVA is based on the partition of $TSS$ into two statistically independent components: a sum of squares based on treatments and a sum of squares component according to the random error. It is due to this partition of the TSS, and the variance estimators based on them that the procedure is called ANOVA. The partition is done with a simple mathematical trick. If $\hat{Y}_j$ is the mean of the response for each level of the factor,

$$(Y_{ij} - \overline{Y}) = (\hat{Y}_j - \overline{Y}) + (Y_{ij} - \hat{Y}_j)$$

Squaring, plus some additional math[2] gets

$$TSS = SSA + SSE$$

The first term on the right of this equation is the **sum of squares due to treatments**, which for notational convenience we call **SSA**, and the second term on the right, **SSE**, is the **sum of squares due to random error**.

- $SSA$ accounts for deviations of the average of the response variable according to each level of the factor from the overall sample mean.
- $SSE$ accounts for deviations of each observation from the average of the variable response according to each level of the factor.

Another way of seeing it is that $SSA$ quantifies the deviations in $Y$ from different levels of the treatment (between samples), while $SSE$ quantifies deviations within the treatments (within samples).

---

**Sum of Squares Partitioning: Intuition**

If the null is true, then the variability from different treatments (between treatments or groups) is no different than variability within treatments, which is random variability around the population mean. Thus, when the null hypothesis is correct, we can get an unbiased estimator of $\sigma^2$ using either $SSA$ or $SSE$.

---

2 Squaring the right side will give us three terms, but due to independence of the main terms, the third term becomes unimportant.

If we define these two estimators as, Mean Square due to factor A treatments (*MSA*),

$$MSA = \frac{SSA}{a - 1}$$

and, Mean Square due to Error (*MSE*),

$$MSE = \frac{SSE}{n - a}$$

then both are unbiased estimators of $\sigma^2$ when $H_o$ is correct. A convenient way of testing this is that $\frac{MSA}{MSE}$ should be statistically close to 1. When $H_o$ is incorrect, then the expected value of *MSA* will be greater than $\sigma^2$, since it captures more variability of the response variable of what should occur randomly. As a result, $\frac{MSA}{MSE}$ will be statistically greater than 1. This fraction is known as the *F* statistic,

$$F = \frac{MSA}{MSE}$$

and it follows an **F-distribution** when the null hypothesis is true (Figure 12.2). The *F* distribution has two parameters, $df_1$ and $df_2$, which in ANOVA correspond to the degrees of freedom of the numerator and denominator of the *F*-statistic, respectively. Specifically,

$$df_1 = a - 1$$
$$df_2 = n - a$$

Thus, $df_1 + df_2 = n - 1$; the degrees of freedom of the TSS.
The analysis is usually summarized through an ANOVA table (Table 12.1),

**Example 12.2** *A mailing company must choose between three different routes to go from point A to point B. If one route takes consistently less mean time to complete deliveries, then this can result in considerable cost savings. Members of the analytics team randomly select 18 truck drivers and randomly assign 6 of them to each route. Then the hours from point A to point B are measured. Is there a difference in mean trip time among the routes at 5% significance?*

The experimental unit in this example is the shipment trips, the response variable is the time to complete the trip, the factor is the route and it has three levels: route 1, route 2, and route 3.
Defining

$\mu_1 = $ population mean trip time for route 1.
$\mu_2 = $ population mean trip time for route 2.
$\mu_3 = $ population mean trip time for route 3.

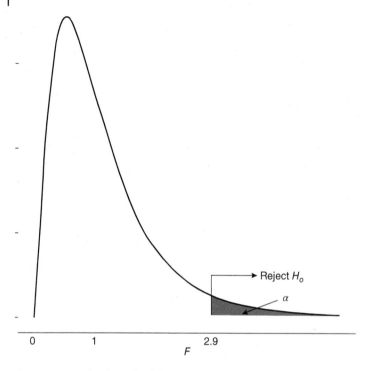

Figure 12.2 Under the null, $F$ follows an $F$-distribution. Here, $df_1 = 5$ and $df_2 = 15$. If $H_o$ is true, the observed test statistic should not be too extreme. What is considered too extreme from the expected value of $F$ is the region presented in red.

Table 12.1 One-factor ANOVA table.

| Source | DF | SS | MS | F | P-value |
|--------|------|------|------|------|---------|
| Factor | $a - 1$ | SSA | $MSA = \frac{SSA}{a-1}$ | $\frac{MSA}{MSE}$ | Value |
| Error | $n - a$ | SSE | $MSE = \frac{SSE}{n-a}$ | | |
| Total | $n - 1$ | TSS | | | |

$H_o: \mu_1 = \mu_2 = \mu_3$

$H_1:$ at least two of the population means are not equal.

Table 12.2 displays the times to complete the trips through each route.

To perform ANOVA using Minitab, go to Stat > ANOVA > One Way ... Figure 12.3 illustrates the steps for the example above. (Since the responses are entered by factor level, the first box in the one-way ANOVA window must

Table 12.2  Time from point *A* to point *B* through each route.

| Route 1 | Route 2 | Route 3 |
| --- | --- | --- |
| 5.48 | 6.60 | 5.58 |
| 5.80 | 6.13 | 5.58 |
| 5.87 | 5.33 | 5.57 |
| 5.76 | 6.03 | 5.05 |
| 6.16 | 5.99 | 5.98 |
| 6.12 | 6.14 | 6.04 |

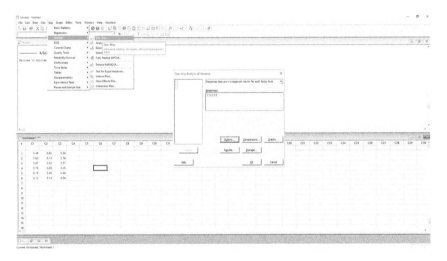

Figure 12.3  Performing one-factor ANOVA for this Example using ANOVA

be changed to `Response data are in a separate column for each factor level.`)

The ANOVA table (Figure 12.4) demonstrates how the degrees of freedom for *SSA* are $df_1 = 3 - 1 = 2$, that $df_2 = 15$, and that TSS degrees of freedom are $18 - 1 = 17$. The estimate of $\sigma^2$ based on between sample variation is 0.4916, while based on within sample variation is 1.7894. Most importantly, the $p$-value is 0.16 (Figure 12.4). Since the $p$-value $\not< 0.05$, the null is not rejected. The conclusion is that at 5% significance, we find no evidence that there is a difference in true mean trip time of each route. Our conclusion was drawn while performing just one hypothesis test.

```
One-way ANOVA: C1, C2, C3

Method

Null hypothesis        All means are equal
Alternative hypothesis At least one mean is different
Singificance level     α = 0.05

Equal variances were assumed for the analysis.


Factor Information

Factor Levels Values
Factor       3 C1, C2, C3


Analysis of Variance

Source DF  Adj ss  Adj MS  F-Value  P-Value
Factor  2  0.4916  0.2458    2.06    0.162
Error  15  1.7894  0.1193
Toral  17  2.2810
```

Figure 12.4 Minitab ANOVA summary.

**Example 12.3** *A company's website design team has come up with four configurations for the new interface of a popular website. Unsure of which one to choose, decision makers determine that they would get feedback from future end users. They randomly select 20 future end users and randomly assign five of them to each configuration. Through a battery of questions, participants give the configurations an overall score in a scale of 0 and 100, where 0 is worst and 100 best. Is there a difference in website mean score among the configurations at 1% significance?*

The experimental unit in this example is the website interface, the response variable is the score, the factor is the configuration and it has four levels: configuration 1, configuration 2, configuration 3, and configuration 4.

Defining,

$\mu_1$ = population mean score for configuration 1.
$\mu_2$ = population mean score for configuration 2.
$\mu_3$ = population mean score for configuration 3.
$\mu_4$ = population mean score for configuration 4.

$H_o: \mu_1 = \mu_2 = \mu_3 = \mu_4$

$H_1$: at least two of the population means are not equal.

Table 12.3 displays the scores given by participants by configurations.

The ANOVA table (Figure 12.5) indicates a $p$-value of 0. Since the $p$-value $< 0.01$, the null is rejected. The conclusion is that at 1% significance, we find evidence that there is a difference in true mean score of the website by

Table 12.3  Software scores by each experimental configuration.

| Configuration 1 | Configuration 2 | Configuration 3 | Configuration 4 |
|---|---|---|---|
| 69 | 78 | 66 | 38 |
| 74 | 85 | 68 | 51 |
| 75 | 78 | 61 | 52 |
| 73 | 67 | 61 | 56 |
| 79 | 77 | 60 | 46 |

```
One-way ANOVA: C1, C2, C3, C4

Method

Null hypothesis          All means are equal
Alternative hypothesis   At least one mean is different
Significance level       α = 0.01

Equal variances were assumed for the analysis.

Factor Information

Factor Levels Values
Factor      4 C1, C2, C3, C4

Analysis of Variance

Source DF  Adj ss  Adj MS  F-Value  P-Value
Factor  3   2476.2  825.40   28.71    0.000
Error  16    460.0   28.75
Toral  19   2936.2

Model Summary

      S   R-sq  R-sq(adj) R-sq(pred)
5.36190 84.33%    81.40%     75.52%

Means

Factor N   Mean StDev     99% CI
C1      5  74.00  3.61 (67.00, 81.00)
C2      5  77.00  6.44 (70.00, 84.00)
C3      5  63.20  3.56 (56.20, 70.20)
C4      5  48.60  6.91 (41.60, 55.60)

Pooled StDev = 5.36190
```

Figure 12.5  Minitab ANOVA summary.

configuration. Table 12.3 demonstrates that configuration 2 has the highest mean score. But, is this significantly different than configuration 1 or 3? The answer to this question requires more testing.

The two examples just presented have the same number of observations per level of the factor, but this is not a requirement to carry out ANOVA. In summary, ANOVA allows us to conduct a hypothesis test of equality for more than two population means in an efficient way. We may conclude that all means are equal through a single test instead of many tests. However, if we reject the null hypothesis, more tests may be needed to establish where the differences occur. But at least ANOVA indicates that only if we reject the null hypothesis, we must do more hypotheses testing to determine specific mean differences. These additional tests are the topic of Section 12.3.

## Practice Problems

12.4  A sample of 25 observations gave a sample mean of 32.6 and a sample variance of 5.11. What is the TSS?

12.5  Fill the blanks in the ANOVA table below:

| Source | DF | SS | MS | F | P-value |
|--------|----|----|-----|---|---------|
| Factor $A$ | 4 | 113.12 | — | — | 0.188 |
| Error | — | — | — | | |
| Total | 40 | 737.69 | | | |

12.6  Refer to the partial ANOVA table below.

| Source | DF | SS | MS | F | P-value |
|--------|----|----|-----|---|---------|
| Factor $A$ | — | 504.77 | — | — | 0.395 |
| Error | 25 | — | — | | |
| Total | 28 | 4581.61 | | | |

a) Fill in the missing values.
b) How many levels does the factor have?
c) How many observations were used?
d) What is the estimated variance of the response variable using all the data?
e) What is the null and alternative hypothesis?
f) Do you reject the null at 10% significance?

12.7  Refer to the partial ANOVA table below, obtained from an experiment where a total of 16 observations were made.

| Source | DF | SS | MS | F | P-value |
|--------|----|----|-----|---|---------|
| Factor $A$ | 3 | — | 772 | — | 0.0038 |
| Error | — | — | — | | |
| Total | — | 2713.93 | | | |

a) Fill in the missing values.
b) How many levels does the factor have?
c) What is the estimated variance of the response variable using all the data?
d) What is the null and alternative hypothesis?
e) Do you reject the null at 10% significance?

12.8 Returning to Problem 12.1, the table below summarizes ANOVA.

| Source | DF | SS | MS | F | P-value |
| --- | --- | --- | --- | --- | --- |
| Factor $A$ | 3 | 110.50 | 36.82 | 19.37 | 0.000 |
| Error | 3868 | 7352.40 | 1.90 | | |
| Total | 3872 | 7462.90 | | | |

a) What is the estimate of $\sigma^2$ based on between sample variation?
b) What is the value of the $F$ statistic?
c) Do you reject the null at 5% significance?

12.9 Course grades in business statistics are obtained from five randomly chosen students from five different programs. Use Minitab to determine if the data shows significant difference in means at 5% significance.

| Accounting | Finance | Human resources | Management | Marketing |
| --- | --- | --- | --- | --- |
| 70.74 | 66.80 | 73.12 | 66.55 | 69.19 |
| 78.84 | 79.87 | 61.90 | 66.84 | 67.82 |
| 68.64 | 82.38 | 51.79 | 76.44 | 60.74 |
| 92.95 | 80.76 | 35.85 | 75.21 | 40.11 |
| 80.29 | 71.95 | 69.25 | 72.94 | 66.20 |

## 12.3  Multiple Comparisons

When performing ANOVA, if the null is not rejected, then no further hypothesis testing is needed. But when the null hypothesis is rejected, the possibility that all means are equal is ruled out without exactly determining which population means are different. From some treatments $i, j$

$$H_o: \mu_i - \mu_j = 0$$
$$H_1: \mu_i - \mu_j \neq 0$$

Sometimes, where these differences lie are obvious, but not always. When we need to find out where such differences are, we must continue performing hypothesis tests.

> **The Challenge with Many Hypothesis Tests**
> The more hypothesis tests we do, the more likely we incur in an error in at least one of these hypothesis tests. Therefore, comparing many pairs of factor level means using $t$-tests will not be optimal. To be more precise, if $\alpha$ is the probability of incurring in type I error when conducting hypothesis tests on two population means, the probability of incurring in type I error in at least one of those $m$ hypothesis tests is $1 - (1 - \alpha)^m$.

Hence, if $\alpha = 0.05$ and five mean population difference hypothesis tests are performed, the probability of incurring in type I error in at least one of those five hypothesis tests is 0.23, much bigger than $\alpha$. Figure 12.6 displays the probability of incurring in at least one type I error when performing different amounts of hypothesis testing. Clearly, the more pairwise hypothesis tests, the higher the probability of incurring in at least one type I error. Also, the higher the $\alpha$ for each individual tests, the quicker the probability of incurring in at least one type I error tends to 1. For $m = 25$, when $\alpha = 0.01$ the probability of incurring in at least one type I error is 0.22, when $\alpha = 0.10$ the probability of incurring in at least one type I error is 0.93.

Performing simultaneous hypothesis tests in an optimal way AFTER rejecting the null hypothesis in an ANOVA is known as **multiple comparisons**. The

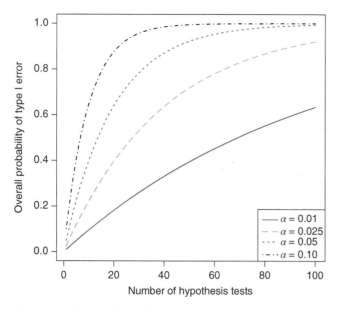

**Figure 12.6** The probability of incurring in at least one type I error when performing $m$ hypothesis tests.

claim of "optimal" raises some questions: How do we perform these additional hypothesis tests efficiently? Of course, we will want to keep type I error to a minimum. But, do we focus on keeping the type I error over all hypothesis tests to a minimum or type I error per hypothesis tests to a minimum? Do we perform all possible pairwise comparisons or just some? Is the intention to consider only differences among two population means or other type of comparisons are also of interest? And which test has the most power? What is considered the optimal procedure will depend on how these questions are answered which in turn will depend on the application at hand. Only a very superficial introduction is given to some multiple comparison testing methods.

**Tukey's Test** This test compares all factor level means to each other. When used to construct confidence intervals, it ensures confidence levels of at least $1 - \alpha$. For example, if a 95% confidence level is chosen, the method develops confidence intervals such that there is a 0.95 chance that all confidence intervals are simultaneously correct. A consequence of this is that confidence intervals using the Tukey method will be wider than confidence intervals constructed separately for each factor level. Note that the "at least" means that in some cases, Tukey's test results may be rather conservative: smaller type I error rates than other methods, but less power.

**LSD Test** Fisher's least significance difference (LSD) works by comparing observed factor level mean differences with a threshold, the LSD. If the absolute value of the difference exceeds the LSD, the population level means are concluded to be significantly different at a significance level $\alpha$. The LSD procedure has a valid $\alpha$ if implemented for preplanned mean comparisons or if samples are independent. It is the least conservative multiple comparison method (easiest to incorrectly determine that a pair of means is different when indeed they are not).

**Duncan's Test** This is another preferred multiple comparison method when comparing all pairs of means. It is an effective method to detect real differences in treatment means, but this is so at the expense of a slightly higher chance of type I error.

**Dunnet's Test** This method is a favorite when one treatment is the control, a benchmark to which other $a - 1$ factor levels will be compared. Suppose a new version of a drug for bad cholesterol is being developed by a company. Researchers aim to compare mean reduction in bad cholesterol after two months of use of the drug with different amounts of an ingredient relative the amount of the old version of the medicine. This is one example where Dunnet's method is a good candidate to use for the comparisons.

```
Tukey Pairwise Comparisons

Grouping Information Using the Tukey Method and 95% Confidence

Factor N Mean Grouping
C2      5 77.00 A
C1      5 74.00 A
C3      5 63.00   B
C4      5 48.00     C

Means that do not share a letter are significantly different.
```

Figure 12.7  A multiple comparison using the Tukey method.

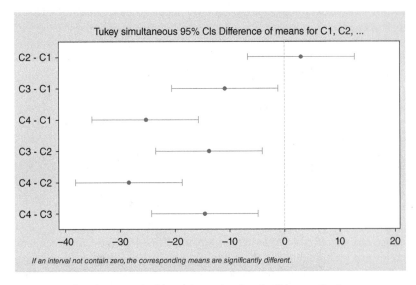

Figure 12.8  Simultaneous confidence intervals using the Tukey method.

**Example 12.4**  *Returning to Example 12.3, the null was rejected. Clearly configuration 4 had lower scores but what about the other 3? This will be addressed using the Tukey method at 5% significance. The Minitab ANOVA output (Figure 12.5) provided some individual confidence intervals but these do not control for family-wise type I error.*

The Tukey method leads us to conclude that only configurations 1 and 2 have equal true mean scores, the others have different means (Figure 12.7).

Simultaneous confidence intervals of the difference in mean scores (Figure 12.8) show that only the confidence interval for the difference in mean score of configuration 1 and 2 includes 0. With these two configurations having the highest mean score, one of them can be selected considering some other criteria such as cost or ease of maintenance.

# Practice Problems

**12.10** In Example 12.2, why it is not necessary to perform multiple comparisons?

**12.11** Returning to Problem 12.8. Below are simultaneous 95% confidence intervals.

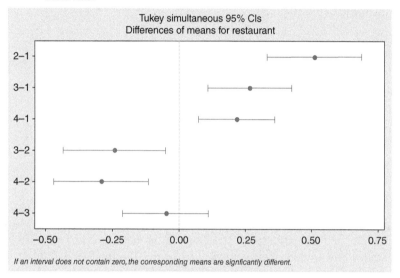

Tukey simultaneous 95% CIs
Differences of means for restaurant

If an interval does not contain zero, the corresponding means are signficantly different.

a) Interpret the 95% confidence level.
b) Write the null and alternative to compare restaurant mean scores from terminals 3 and 4 using Tukey's method.
c) Is the null above rejected?

**12.12** Returning to Problem 12.11, use the Tukey simultaneous 95% confidence intervals to compare restaurant mean scores from terminals 1 and 2.

## 12.4 Diagnostics of ANOVA Assumptions

Three assumptions are made to conduct ANOVA. If these assumptions are badly violated, then the conclusions of the analysis cannot be trusted.

To run diagnostics on the validity of the assumptions recall that one-factor ANOVA can be thought of as building a statistical model to estimate the expected value of the response variable based on the levels of the factor.

$$E(Y_j) = \mu + \tau_j$$

If no levels of the factor affect the expected value of $Y$, then the $\tau_j$ are all zero. Otherwise, essentially some factor levels increase the mean of $Y$ and others decrease it. The difference between the observed value of the response for a given level of the factor and the value of the response variable based on the model (for the same level of the factor) is known as a **residual**. Think of the residual as an estimate of the model error. The residual will have an expected value of zero, making residual-based diagnostics generally applicable, regardless of the mean of the response.

---

**Checking the Residuals**

The residuals should

- be normally distributed,
- have a constant variance for every level of the factor, and
- be independent.

---

*Checking for Constant Variance* $\hat{Y}$ stands for the estimated mean of the response variable for each level of the factor, also called the ANOVA model fit. Residual variances holding constant for all levels of the factor imply the residuals by treatment should have the same dispersion. This can be visually assessed through a plot of residuals against $\hat{Y}$, a function of factor levels. Figure 12.9 demonstrates how the residual versus fit plot works. On the left, we see that residuals have about the same spread per level of the factor; no evidence against the constant variance assumption. On the right, the spread of the residuals for

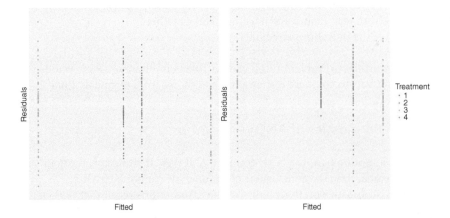

**Figure 12.9** Residuals versus fitted value plots. Case on the left shows no concern with the constant variance assumption. Case on the right suggests much higher variance of level 4 than level 1.

level 4 of the factor is much larger than the spread of the residuals for level 1; constant variance may not hold.

*Normal Probability Plot*   Violations of the normal distribution are typically verified with what is called a **normal plot**, where the ordered residuals are plotted against theoretical points under the normal distribution. If the normal distribution assumption is correct, then the points would lay on a straight line. In practice, however, points with a shape of an outstretched "s" is considered an indication that the normal distribution has been satisfied. The normal plot also helps pinpoint possible unusual observations. Sometimes, a histogram is also used to check the normal distribution, but histograms are dependent on the number of observations and how the binning was done, which can hinder its usefulness in checking normality. Figure 12.10 presents examples of normal plots.

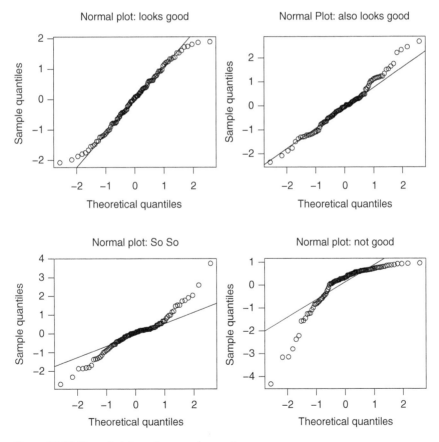

Figure 12.10 Normal plots under several scenarios.

- The upper left panel shows the expected stretched out "s" shape, suggesting the normality assumption is valid.
- Although the upper right panel normal plot does not present a symmetric curve, the shape does not deviate considerably from what is expected under the normal distribution.
- The normal plot in the lower left panel is symmetric, but the points on the tails of the curve deviate considerably from the line. This indicates some concern with the assumption of normality.
- The normal plot in the lower right panel shows that the lower tail is away from the line. The shape of the curve is skewed to the left, normal assumption here is not appropriate.

The normality assumption can also be formally checked through hypothesis testing. There are multiple tests to determine if the null hypothesis of a normal distribution should be rejected or not. In practice, hypothesis testing is used when residual plots rise some suspicions about any of the model assumptions.

*Validity of the Independence of Errors Assumption*    An option to check the independence assumption is to plot the residuals versus the order of the observations. Although some software provide this plot, it can be difficult to detect from it when independence is invalid. Sequence of positive and negative residuals may occur just by coincidence. The sheer number of observations can make it difficult to see anything at all in this plot. Furthermore, there are different ways that independence in the observations can occur: it could be due to temporal dependence (i.e. monthly sales) or due to repeated measurements on participants (i.e. monthly carbohydrate consumption for different participants). Other alternative plots are possible, but will not be discussed here. The way how the data was obtained could give us an indication on whether we should be concerned about the feasibility of the independence assumption. When observations are made at different points in time or space, residuals may no longer be independent. This dependence must then be taken into account. Chapter 17 introduces some models that deal with temporal dependence. On the other hand, data from certain designed experiments are already independent by design.

---

**ANOVA and Assumption Violations**

- ANOVA is robust to even moderate violation to the normality assumption. The $F$ statistic will still follow approximately an $F$ distribution, especially for reasonably large sample sizes.
- Moderate violation to the equal population variances assumption is not of concern either. In fact, if the sample sizes from each factor level are equal, ANOVA will be able to handle even severe violations to this assumption.

> If the largest sample standard deviation is no more than two times the smallest sample standard deviation, then ANOVA will still work quite well.
>
> Because of these reasons, the ANOVA assumptions are generally assessed visually.

Figure 12.11 presents residual plots for Example 12.2. The normal probability plot does not indicate any clear indication that the normality assumption has been violated. The constant variance is a bit harder to check here by eye (right upper panel), but the standard deviations within each level differ by less than a ratio of 2.

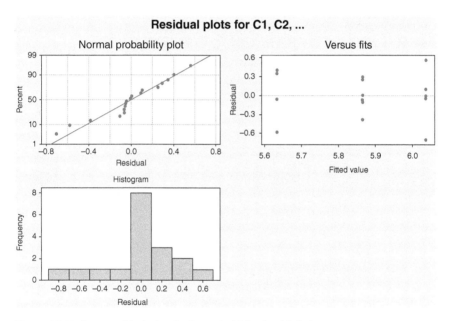

Figure 12.11 Some residual plots for Example 12.2 using Minitab.

### 12.4.1 A Quick Look at Data Science: Emergency Response Time for Cardiac Arrest in New York City

A city administrator wants to determine if emergency dispatch response time (in seconds) for cardiac arrest is associated to the time of day: (0,6], (6,12], (12,18], and (18,24]. Data from the computer-aided Emergency Management System (EMS) for New York City is available online.[3] The raw data includes almost five million records for incidents from January 1, 2013 until June 30,

———
3 data.cityofnewyork.us (accessed July 26, 2019).

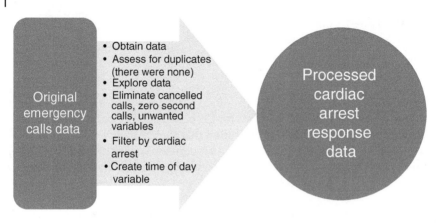

Figure 12.12 Steps required to process New York City emergency call data.

Table 12.4 Number of observations, sample average dispatch response time (in seconds), and sample standard deviation per time of day.

| Time of day | Sample size | Mean | Std. dev. |
| --- | --- | --- | --- |
| (0,6] | 10,052 | 26.09 | 13.92 |
| (6,12] | 18,701 | 27.39 | 16.26 |
| (12,18] | 17,978 | 28.89 | 19.30 |
| (18,24] | 12,517 | 27.20 | 17.47 |

2016. After preprocessing the data, 59 248 incidents are left (Figure 12.12). ANOVA will be performed at 5% significance.

In this application, the response variable is emergency dispatch response time. The factor is the time of day, which has four levels: (0,6], (6,12], (12,18], and (18,24]. From Table 12.4, means per group are not widely different. The biggest estimated difference in dispatch response time was between the 12–6 p.m. and 12–6 a.m. times of day, 2.8 seconds. Moreover, standard deviations are similar, while incidents occur least frequently early on the day. In the boxplots of dispatch response times, Figure 12.13, it is reaffirmed that there is no obvious difference between the median dispatch response times by time of day. Secondly, for each time of day, the dispatch response times are highly skewed; although generally the responses occur within just a few seconds, on occasion it can take a long time. For each time of day, the number of outlying observations are substantial (in red), making the normality assumption doubtful.

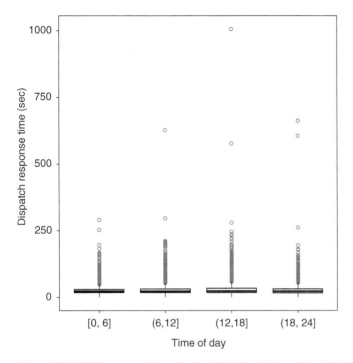

Figure 12.13 EMS response times for New York City by time of day. Outliers appear in red.

Defining

$\mu_1$ = population mean dispatch response time from just past midnight to 6 a.m.
$\mu_2$ = population mean dispatch response time just past 6 a.m. to 12 p.m.
$\mu_3$ = population mean dispatch response time just past 12 p.m. to 6 p.m.
$\mu_4$ = population mean dispatch response time just past 6 p.m. to midnight.

$H_o$: $\mu_1 = \mu_2 = \mu_3 = \mu_4$
$H_1$: at least two of the population means are not equal.

The ANOVA table (below) indicates a $p$-value of zero. Since the $p$-value <0.05, the null is rejected. The conclusion is that at 5% significance, we determine that the true mean dispatch response time is different for at least two times of the day.

| Source | DF | SS | MS | F | P-value |
|---|---|---|---|---|---|
| Time of day | 3 | 55 612 | 18 537 | 63.11 | 0.000 |
| Error | 59 244 | 17 402 504 | 294 | | |
| Total | 59 247 | 17 458 116 | | | |

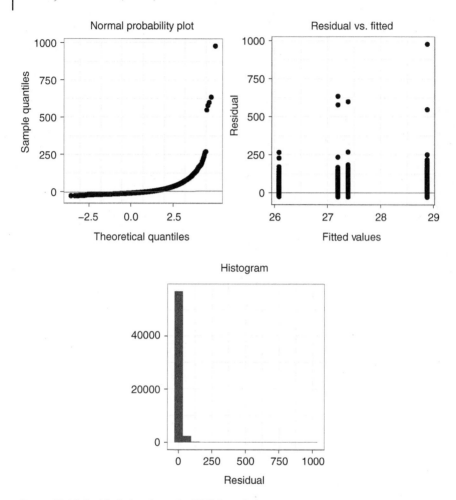

Figure 12.14 Residual plots from the ANOVA analysis.

Figure 12.14 displays some residual plots. The normal probability plot reaffirms that the normality assumption is violated, as does the histogram. The residual versus fitted plot does not indicate that the variances are different, and the largest standard deviation is not two times larger than the smallest standard deviation.

Going back to Table 12.4, the difference between sample averages was largest between the 12–6 p.m. and 12–6 a.m. times of day. Thus, at least this difference is statistically significant. The second largest difference is 1.5 seconds. Is this difference also statistically significant? To find out, the Tukey multiple comparison test may be performed. But, although a statistically significant difference in at least two mean dispatch response time was found, this result is not surprising since so many observations were used. Also, we are in no position to argue if the difference are practical other than to say that, if so, the differences are not very large. One last important final argument to make is that considering the distribution is so skewed, perhaps it is more valuable to draw inference on the population medians. In Chapter 4, we saw that the mean does not represent well the "typical values" of a variable when the distribution is highly skewed.

Statistical software sometimes provides confidence intervals for each population mean. But as indicated in Section 11.2, the temptation of assessing if confidence intervals overlap to determine if the means are different should be avoided. Although ANOVA assumes that the variance of the response is constant over all levels of the factor, the sample sizes may be different; leading to different standard errors and making the "confidence interval overlap evaluation" potentially misleading. Instead, assess confidence intervals of the difference in means.

## Practice Problems

12.13 Download the data file "EMS_Incident_Dispatch_Data_PRE PROCESSED." Instead of dispatch response time as Section 12.4.1, our variable of interest is now incident response time,[4] and the aim is to determine if it is affected by time of day.
   a) Perform ANOVA at 5% significance.
   b) If the null is rejected in the ANOVA test, perform a Tukey test on all pairs.
   c) Check the residual plots and determine if there are any concerns.

12.14 Data on daily units and sales (in $) of several marijuana products in Washington are available[5]: "Marijuana_Sales_w_Weight_Daily." Using 2016 and 2017 data, a researcher aims to test if the mean August sales per unit is associated to the product (Liquid Marijuana Infused Edible, Marijuana Extract for Inhalation, Marijuana Infused Topicals, Marijuana Mix Infused, Marijuana Mix Packaged, or Solid Marijuana Infused Edible).
   a) Construct a boxplot of sales per unit for each product. Interpret.
   b) Perform ANOVA at 5% significance.
   c) If the null is rejected in the ANOVA test, perform a Tukey test.
   d) Check the residual plots and determine if there are any concerns.

---

4 The time elapsed in seconds between the incident time and the first on scene time.
5 data.lcb.wa.gov (accessed July 26, 2019).

## 12.5 ANOVA with Two Factors

**Example 12.5** *A company is studying the impact of a type of gas additive and tires on miles per gallon of cars. Specifically, management would like to see if methanol content in the additive and tire size have an impact on miles per gallon. Their options are 50% methanol, 60% methanol, and 75% methanol and 14, 15, 17, and 21 inch tires. In an experiment, 36 cars are randomly assigned to each methanol content and tire size and after a predetermined distance, they measure the miles per gallon of the cars.*

In this scenario, two factors may affect miles per gallon, $Y$: methanol content and tire size.

---

**Run two one-way ANOVA?**

It might come to mind that when there are two factors, one can simply run one-way ANOVA twice. But, it is better not to do inference this way.

- For one, it is generally cheaper to perform an experiment considering two factors, instead of running two experiments, each testing one factor.
- Also, a two-factor ANOVA test will have more power and, most importantly, it will be more informative.

As a model, the intuitive representation is that the values of the response are an additive function of the overall mean $\mu$, a treatment parameter for factor $A$, another treatment parameter for factor $B$, and a random term. However, this model implies that the impact of factor $A$ on the response is the same for all levels of factor $B$ and vice versa. It is possible that the effect of factor $A$ on $Y$ may depend on the level of factor $B$ (and vice versa). When this occurs, we say that there is an **interaction effect**.

---

Let's suppose that there are two factors. One has levels of high or low (e.g. concentration of a chemical), and the other has levels yes or no (e.g. whether an item was produced using some technique). When both factors have only two levels, it is quite simple to visually appraise if an interaction effect is present. The **interaction plot** shows the response means of all factor level combinations. Response means of one factor are connected with lines.

For our two-factor scenarios:

- The $x$-axis will distinguish between the levels of factor $A$.
- The $y$-axis will indicate the value of the response means.
- Different lines will be drawn by the level of factor $B$.

Figure 12.15 shows interaction plots for four different scenarios.

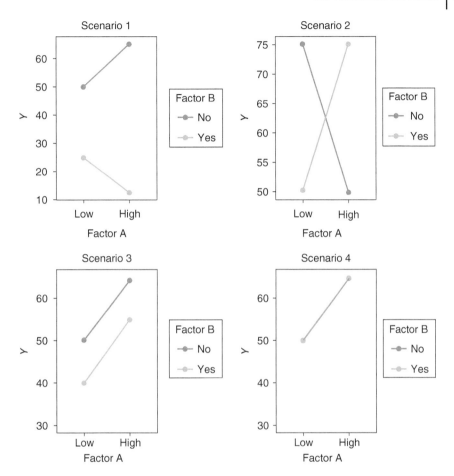

Figure 12.15 Interaction plots for four scenarios of two-factor ANOVA where each factor has two levels.

- Scenario 1 indicates there is an interaction effect. Specifically, when factor $B$ is at the "No" level, the mean response is higher for the "High" factor $A$ level than for the "Low" factor $A$ level (red line). On the other hand, when factor $B$ is at the "Yes" level, the mean response is lower for the "high" factor $A$ level than for the "low" factor $A$ level (blue line). Thus, the impact of the factor $A$ level was reversed when the level of factor $B$ was changed.
- Scenario 2 indicates that there is an interaction effect as well. Note that in both scenarios, the slope of the lines has changed significantly with the level of factor $B$.
- In scenario 3, factor $A$ and factor $B$ have an impact on the mean response. The slope of the two lines is about the same, meaning that the type of increase

in response when moving from "Low" level to "High" level of factor $A$ is not affected by the level of factor $B$.

- Scenario 4 has no interaction effect, and only factor $A$ appears to have an impact on the response variable.

For more than two levels per factor, interactions plots are still helpful, but harder to interpret. Boxplots can also be useful in detecting factor effects, including interactions. Yet visualization is not enough to determine the impact of any factor on the response variable, making statistical inference necessary. To test for an interaction effect, the hypotheses are as follows:

$H_o$: No interaction effect among factor $A$ and factor $B$.

$H_1$: There is an interaction effect among factor $A$ and factor $B$.

For two-factor ANOVA, the possibility of an interaction effect is tested first.[6] Why? Because if there is an interaction effect, then there is no need to test the main factors separately. If the null is not rejected, then the main factors may be tested.

$H_o$: Mean response is equal for all levels of factor $A$ at each fixed level of factor $B$;

and

$H_o$: Mean response is equal for all levels of factor $B$ at each fixed level of factor $A$.

**Example 12.6** *Online courses are becoming increasingly common. There is debate on whether there is a difference in student learning when using online or traditional approaches. The answer to this question may depend on the subject of the class and gender of students. A study was conducted to determine if type of course is online or traditional or gender has any effect on student learning. Twenty people who enrolled in a continued education accounting class were randomly assigned to the two types of course. One professor taught each type of class a different day of the week, for 10 weeks. The students' final grade was recorded as the response variable. Tests will be performed at 10% significance.*

Table 12.5 displays the grades of participants by type of course and gender.

From the interaction plot (Figure 12.16), it appears the type of course affects the students' grades. However, there is no strong indication of an interaction or of gender being important.

The Minitab windows and commands to run a two-way ANOVA with an interaction are illustrated in Figure 12.17. Observe how in the "Model" button

---

6 To test for interaction effects, at least one replication is needed for every combination of factor levels. Statistical software will generally give you an error if the data violates this condition.

Table 12.5 Grades of participants by type of course and gender.

| Gender | Type of course | |
| --- | --- | --- |
| | Online | Traditional |
| Female | 70.81 86.06 71.27 84.19 71.43 | 89.56 86.78 77.08 78.58 87.04 |
| Male | 78.38 75.04 83.10 69.68 77.57 | 83.35 77.77 91.09 83.58 91.70 |

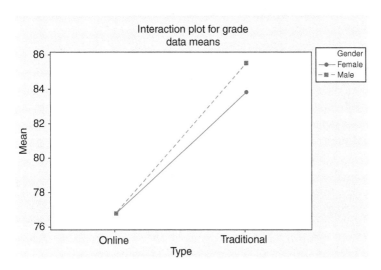

Figure 12.16 Interaction plot for the student learning example.

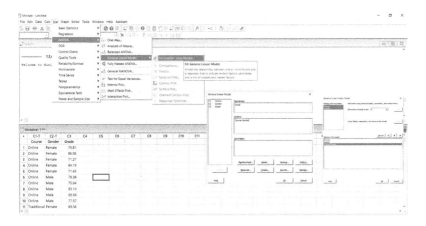

Figure 12.17 Minitab windows to two-way ANOVA with an interaction.

```
Analysis of Variance

Source      DF  Adj SS  Adj MS F-Value  P-Value
  Type       1  312.037 312.037   8.39    0.011
  Gevnder    1    3.589   3.589    0.10    0.760
  Type*Gender 1   3.547   3.547    0.10    0.761
Error       16  595.050  37.191
Total       19  914.224
```

Figure 12.18 Minitab ANOVA output for the student learning example.

of the General Linear Model window one must choose the factors and add them to the model.

The Minitab ANOVA output is shown in Figure 12.18. The interaction effect appears as "Type*Gender." Results affirm that only the type of course is significant. See Problem 12.20 for more.

When at least one factor is significant and there is no interaction effect, multiple comparisons methods can be used to compare treatment means. However, with type of course having only two levels, a simple two population inferential method would suffice.

**Example 12.7** *Business undergrads are generally not fans of the statistics course. There is a rating scale that uses several questions to measure statistics anxiety. Scores go from 1 (No Anxiety) to 5 (Very High Anxiety). Forty students majoring in Accounting, Finance, Management, and Marketing are sampled. Do mean anxiety scores depend on major or gender of the student? Test is performed at 5% significance.*

Table 12.6 displays the statistics anxiety scores of participants by major and gender.[7]

Table 12.6 Statistics anxiety scores of participants by major and gender.

| Gender | Business major | | | |
| | Accounting | Finance | Management | Marketing |
| --- | --- | --- | --- | --- |
| Female | 4.62 4.02 4.42 | 4.32 4.24 4.06 | 3.28 3.37 3.46 | 2.87 2.95 2.89 |
| | 4.29 4.42 | 4.26 4.10 | 3.38 3.52 | 3.09 3.48 |
| Male | 3.00 3.15 3.60 | 2.58 3.21 3.00 | 4.36 4.56 4.00 | 4.13 4.44 4.29 |
| | 3.39 3.10 | 3.15 3.17 | 4.32 4.50 | 4.12 4.37 |

---

7 Statistics anxiety has been a topic of research for a while now. However, the numbers provided in this example are simulated.

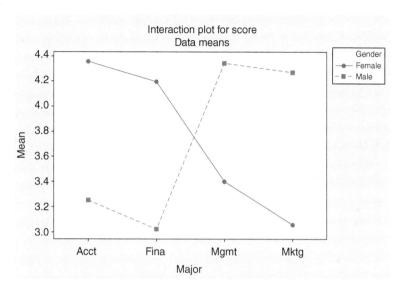

Figure 12.19 Interaction plot for the statistics anxiety example.

```
Analysis of Variance

Source         DF    Adj SS   Adj MS    F-Value   P-Value
   Gender       1    0.0100   0.01000      0.24     0.626
   Major        3    0.4487   0.14956      3.63     0.023
   Gender*Major 3   12.4360   4.14534    100.71     0.000
Error          32    1.3171   0.04116
Total          39   14.2118
```

Figure 12.20 Minitab ANOVA output for the student learning example.

From the interaction plot (Figure 12.19), it appears that mean statistics anxiety score depends on business major. Moreover, an interaction effect appears to occur, and the association between score and major depends on gender.

The Minitab ANOVA output is shown in Figure 12.20, showing that there is an interaction effect.

When there is a significant interaction effect, multiple comparisons for the levels of one factor must be compared for each level of the other factor separately.

## Practice Problems

12.15 Using the ANOVA table below at 5% significance

| Source | DF | SS | MS | F | P-value |
|--------|-----|--------|--------|-------|---------|
| Factor $A$ | 1 | 6.13 | 6.13 | 1.94 | 0.17 |
| Factor $B$ | 1 | 267.43 | 267.43 | 84.44 | 0 |
| Interaction | 1 | 17.01 | 17.01 | 5.37 | 0.026 |
| Error | 36 | 114.01 | 3.17 | — | — |

a) Is the interaction statistically significant?
b) Is factor $B$ relevant?
c) Is factor $A$ relevant?

12.16 Using the ANOVA table below at 5% significance

| Source | DF | SS | MS | F | P-value |
|--------|-----|---------|-------|------|---------|
| Factor $A$ | 1 | 6.97 | 6.97 | 0.25 | 0.62 |
| Factor $B$ | 1 | 78.80 | 78.80 | 2.81 | 0.10 |
| Interaction | 1 | 40.99 | 40.99 | 1.46 | 0.23 |
| Error | 36 | 1008.50 | 28.01 | — | — |

a) Is the interaction statistically significant?
b) Is factor $B$ relevant?
c) Is factor $A$ relevant?

12.17 Data on daily units and sales (in \$) of several marijuana products in Washington are available online.[8] A researcher aims to test if the mean August sales per unit is associated to the product (Liquid Marijuana Infused Edible, Marijuana Extract for Inhalation, Marijuana Infused Topicals, Marijuana Mix Infused, Marijuana Mix Packaged, or Solid Marijuana Infused Edible) or the year (2016, 2017).

| Source | DF | SS | MS | F | P-value |
|--------|-----|---------|---------|-----------|---------|
| Product | 5 | 9070.90 | 1814.18 | 12 608.70 | 0 |
| Year | 1 | 302.48 | 302.48 | 2102.23 | 0 |
| Interaction | 5 | 100.50 | 20.10 | 139.67 | 0 |
| Error | 360 | 51.80 | 0.14 | | |

---

8 data.lcb.wa.gov (accessed July 26, 2019).

a) Is the interaction statistically significant?
b) Is the year factor relevant?
c) Is the product type factor relevant?
d) Does it make sense to do multiple comparisons for product types without accounting for year?

12.18 From the interaction plot below, do you think that factor $A$ has an impact on response? Factor $B$? Do you think there is an interaction effect?

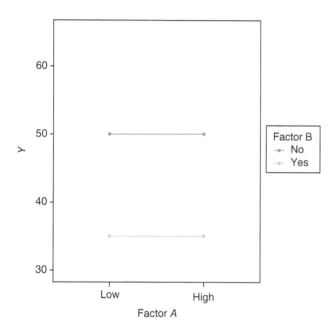

12.19 Assume the same amount of observations were collected for each combination of levels of two factors. From the interaction plot below, answer the following
a) How many levels does factor $A$ have?
b) How many levels does factor $B$ have?
c) Do you think there is an interaction effect?
d) Do you think that factor $A$ has an impact on response? (*Hint:* Based on the given assumption, you can guess the mean response by eye per level of factor $A$.)

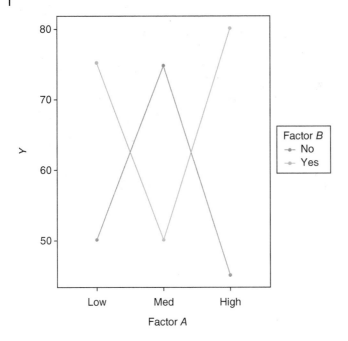

12.20 In Example 12.6, it was found that the type of course was significant.

a) Use Minitab to build a two-sided 95% confidence interval on the difference of the mean grades.

b) The results imply that the type of course, online or traditional face to face, has a significant impact in students' grades. Do the results imply that this is true overall?

12.21 One of the best success stories in big data is the use of statistical algorithms to translate text from one language to another. In an attempt to tap this market, a company will assess three statistical approaches in translating four languages into English: German, Mandarin, Spanish, and French. Thirty-six third-grade level books written in the different languages are randomly assigned to each translation algorithm. A panel of experts then scores (0 lowest, 100 highest) each translation. Data is provided below. With the assistance of statistical software[9], and at 5% significance,

---

9 If using Minitab, Go to `Stat > ANOVA > General Linear Model > Fit General Linear Model` ... In the General Linear Model window, after selecting the response and factors, press the Model button, click on the two factors, and then press the first Add button, followed by the OK button.

| | Language | | | |
|---|---|---|---|---|
| Algorithm | German | Mandarin | Spanish | French |
| 1 | 33 29 | 25 17 | 26 28 | 32 27 |
| | 18 | 22 | 29 | 24 |
| 2 | 31 35 | 35 29 | 48 44 | 53 43 |
| | 22 | 30 | 50 | 42 |
| 3 | 45 41 | 39 38 | 48 59 | 39 47 |
| | 30 | 31 | 51 | 40 |

a) Is the interaction statistically significant?
b) Is factor Language relevant?
c) Is factor Algorithm relevant?

## 12.6 Extensions to ANOVA

What has been covered in this chapter only scratches the surface of ANOVA. In this section, a few additional extensions are presented.

*Randomized Block Design ANOVA*  ANOVA can be applied to experimental or observational studies. In **experimental studies**, experimental units are randomly assigned to treatments. In an **observational study**, there is no attempt to control potential influential variables, all variables are simply recorded from the units. Experimental studies are better than observational studies to argue **causal associations** because the groups are equal, except that treatments were randomly assigned. However, costs, ethical considerations, and resources can make it difficult to perform an experimental study. The simplest type of experimental study is the **completely randomized design**.

**Example 12.8**  *In a completely randomized design, a group of computers will run a task with the company's processor, and another group of computers will run a task with a competitor's processor. The aim is to determine if there is a difference in mean running time of the task of the two groups. In this example, the response variable is the running time of the task, and the factor that may affect running time is the type of processor. Computer processors are randomly assigned to each group. That way, the two groups are identical with the exception that one group has the company's processor installed and the other has the competitor's processor installed. If a difference is found between the mean running times, the team feels more comfortable in arguing that the difference is due to the type of processor, since random assignment makes it likely that the two groups are homogeneous in terms of other traits.*

**Confounding variables** are those variables that are not being controlled in the experiment, yet they may influence the difference in response means. Another experimental design is the **randomized block design**, where there is one factor of interest and another factor is known as a **blocking variable**. When there is a confounding variable, the *MSE* can become large and consequently the *F* statistic becomes small. In a randomized block design, the variation due to confounding variables is controlled for, leading to better capability of the hypothesis test procedure to detect differences among treatment means.

**Example 12.9** *As part of their restructuring plan, a company wants to know if mean delivery time of a product component is affected by order size: small (less than 50 pounds), medium (between 50 pounds and 75 pounds), or large (over 75 pounds). Analysts are aware that one of the confounding variables was supplier, and the company uses four suppliers to provide the component. Twelve orders were randomly created for the study, and three from each supplier. The three orders were created under different sizes. Table 12.7 provides the data. Let $\alpha = 5\%$.*

To analyze this randomized block design using Minitab, apply the commands illustrated in Figure 12.17, but without including an interaction term. The results are presented in Figure 12.21. Since $p$-value $= 0.002 < 0.05$, we reject the null. At 5% significance, we conclude that the mean delivery times of orders are associated to size order.

Table 12.7 A randomized block design to perform a test on order size.

| Supplier | Order size | | |
| --- | --- | --- | --- |
| | Small | Medium | Large |
| 1 | 90.7 | 95.8 | 97.9 |
| 2 | 78.9 | 87.0 | 87.4 |
| 3 | 87.4 | 94.7 | 93.9 |
| 4 | 85.6 | 90.6 | 98.2 |

```
Analysis of Variance

Source   DF  Adj SS   Adj MS  F-Valuer  P-Value
  size    2  162.31   81.157    20.57    0.002
  block   3  175.07   58.356    14.79    0.004
Error     6   23.68    3.946
Total    11  361.68
```

Figure 12.21 Minitab randomized block design ANOVA output.

Had the data been incorrectly treated as coming from a completely randomized design, then the block effects would not have been separated from random variation, leading to a smaller $F$ statistic (see Problem 12.29).

*ANOVA with More than Two Factors*   There is no theoretical limitation to the number of factors in ANOVA. However, conceptually the analysis becomes more difficult. For example, with three factors, $A, B$, and $C$, there are three type of two-way interactions possible, between

- $A$ and $B$
- $A$ and $C$
- $C$ and $B$

Furthermore, a three-way ANOVA allows for a three-factor interaction between $A, B$, and $C$. With four factors or more, additional interactions are possible. These growing number of interactions raise concerns regarding interpretation, and increasing minimum sample size requirements, which brings us to design of experiments. **Factorial experiments** are experiments that permit to properly study the association between multiple factors and a response variable, while considering aspects such as sample size and number of interactions. **Full factorial designs** are when the design collects data for all factor level combinations. To save on cost, **fractional factorial designs** collect data for only a fraction of factor level combinations.

*Random Factors and Mixed Models*   Recall that the focus of this chapter has been fixed effects ANOVA, when the analysis considers all the possible levels of the factor. Sometimes we are interested in a factor that has a large amount of levels, but it is not possible to collect data for all the possible levels. We can however use a randomly selected amount of levels to draw inference about the entire population of factor levels. When this is the case, we say we are working with a **random effects model**. These type of models still perform a partitioning of $TSS$, but now hypothesis testing is done on the variance parameter of the treatment effect (since it is now a random variable), and not on the mean per treatment. When the levels of some factors are fixed and the levels of others are random, a **mixed effects model** is required.

**Example 12.10**   *Recall that in Example 12.1, management wanted to see if methanol content in a gas additive had an impact in miles per gallon. If researchers were to choose a random set of possible levels, say 45% methanol, 65% methanol, and 70% methanol, a random effects analysis would be needed.*

*Final Remarks*   In this chapter, we have presented the principles of ANOVA. We have intentionally kept equations to a bare minimum. There are a lot of important details that must be taken into account when considering implementing ANOVA: study design, constraints required to be able to estimate effects, use of fixed or random effects, verifying assumptions, etc. Although these details are important, including them would obscure the main principles of ANOVA. The basics discussed here should help clarify the intuition, basic application and interpretation of results. For a complete implementation of ANOVA methods, the expertise of a statistician will often be needed.

## Chapter Problems

12.22 State what are the main assumptions required to implement one-factor ANOVA.

12.23 A company is going through some tough times. Some executives have recommended closing down one of its four customer centers. The four centers have equal expenses, so management decides to assess mean customer satisfaction at each center measured on a scale of 0–10 (which higher scores representing higher satisfaction) and the center with the lowest mean score will be closed. For each center 153, 85, 104, and 72 scores respectively were available.
   a) Identify the experimental unit, response variable, and factor and its levels.
   b) Define parameter notation and specify the null and alternative to be tested.
   c) The boxplot below summarizes the scores per center. Does it indicate that any center has a different mean score?
   d) Does the boxplot indicate any concern with the equal variances assumption?

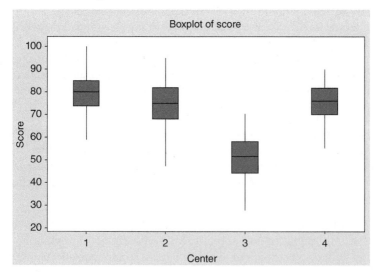

**12.24** Returning to Problem 12.23, from the Minitab output below:
  a) What is the value of the *F*-statistic.
  b) Why is the null rejected at 10% significance?

```
One-way ANOVA: score versus center

Method

Null hypothesis          All means are equal
Alternative hypothesis   At least one mean is different
Singificance level       α = 0.1

Equal variances were assumed for the analysis.


Factor Information

Factor Levels Values
center      4 1, 2, 3, 4


Analysis of Variance

Source  DF Adj ss  Adj MS F-Value P-Value
center   3  55742  18580.7 233.12   0.000
Error  410  32678    79.7
Toral  413  88420


Model Summary

      S   R-sq  R-sq(adj) R-sq(pred)
8.92764 63.04%    62.77%     62.31%


Means

center   N    Mean   StDev      90%  CI
1      153  79.810   8.414 (78.621, 81.000)
2       85  74.68   10.08 ( 73.09,  76.28)
3      104  51.067   9.310 (49.624, 52.511)
4       72  75.431   7.917 (73.696, 77.165)

Pooled StDev = 8.92764
```

Figure 12.22 **Minitab results.**

**12.25** Returning to Problem 12.23, using Tukey's method (below):
  a) Write the null and alternative when comparing the mean score of centers 1 and 3 and determine from the Minitab output if it should be rejected at 10% significance.
  b) According to these results, which center has the lowest mean score?

**Tukey Pairwise Comparisons**

```
Grouping Information Using the Tukey Method and 90% Confidence

center    N     Mean  Grouping
1        153  79.810  A
4         72  75.431    B
2         85   74.68    B
3        104  51.067      C

Means that do not share a letter are significantly different.
```

**12.26** Returning to Problem 12.23:

a) From the Minitab output, do the sample standard deviations (Figure 12.22) indicate any concern about the assumption of equal variances?

b) Interpret the residual plots below:

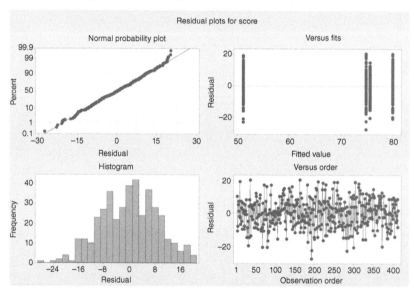

**12.27** You run a two-way ANOVA at 5% significance and find an interaction effect between Factor $A$ and Factor $B$. While presenting your results to management, it turns out that your competitive cubicle neighbor ran an analysis with the same data, same significance. However, he ran two one-way ANOVA–one on factor $A$ and one on factor $B$–and found that factor $B$ was not significant. Explain how the differing results are possible?

**12.28** One thing about ANOVA is that the larger the sample size, the greater the power. Does this mean that if $n$ is large enough, ANOVA will detect differences that are false? Explain.

12.29 Suppose that by mistake, an analyst does not realize that the data in Example 12.9 comes from a randomized block design and runs a one-way ANOVA. The results are presented below. At 5% significance, is there a difference between the mean delivery times?

```
Analysis of Variance

Source   DF   Adj SS   Adj MS   F-Value   P-Value
  size    2    162.3    81.16      3.68     0.068
Error     9    198.7    22.08
Total    11    361.1
```

12.30 Use the data from Example 12.9 to reproduce the incorrect ANOVA table in the problem above.

12.31 Returning to Example 12.9, run a randomized block analysis. Assess the standardized residuals to answer the following questions:
   a) Does the normal assumption seem adequate?
   b) Is there any evidence against the constant variance assumption?
   c) Do the standard residuals indicate any major outliers?

## Further Reading

Goos, Peter and Jones, Bradley. (2011). *Optimal Design of Experiments*. Wiley.
Montgomery, Douglas, C. (2009). *Design of Analysis of Experiments*, 7th edition. Wiley.

# 13

# Simple Linear Regression

## 13.1 Introduction

There is an immense amount of situations where one variable depends on another. For a given amount of money, the new balance in a bank account after a year is a function of the bank's interest rate. Similarly, the revenue of selling shirts at $12 a shirt is a function of the amount of shirts sold. Suppose the cost of a service is a linear function of usage of the service. For example, the monthly electricity bill may be

$$C = 5 + 0.15K$$

where $C$ is the amount to be paid, $K$ is the kilowatt-hour for the month. 0.15 ¢ represents the cost per kilowatt-hour. The equation tells us there is a basic monthly charge of $5. Hence, if a client is on vacation (and has unplugged everything) and consumed a total of 0 kWh, they must still pay $5 to the power company. If a client uses 100 kWh, their bill for the month should be $C = 5 + 0.15 * 100 = 20$. Figure 13.1 presents the association between power bill amount and kilowatt-hour usage.

This example shows us that it is very convenient to express a variable as a function of another. If we know the kilowatt-hours, cost per kilowatt-hour and basic charge, we will be able to determine the total bill for electricity service with certainty.

***Maybe Variables Are Associated?*** We do not always encounter situations were we can determine one variable exactly from another. Often, there is doubt on whether there is an association at all between the two variables. Also, due to randomness, the association may occur in an approximate way. In these cases, statistical methods must be used. The need to consider the association between variables is ubiquitous. Not a day passes by that we hear about a new study that found "a link" between variable $Y$ and variable $X$. For example, the link between alcohol consumption and some measure of health (heart disease, cholesterol

*Principles of Managerial Statistics and Data Science*, First Edition. Roberto Rivera.
© 2020 John Wiley & Sons, Inc. Published 2020 by John Wiley & Sons, Inc.
Companion website: www.wiley.com/go/principlesmanagerialstatisticsdatascience

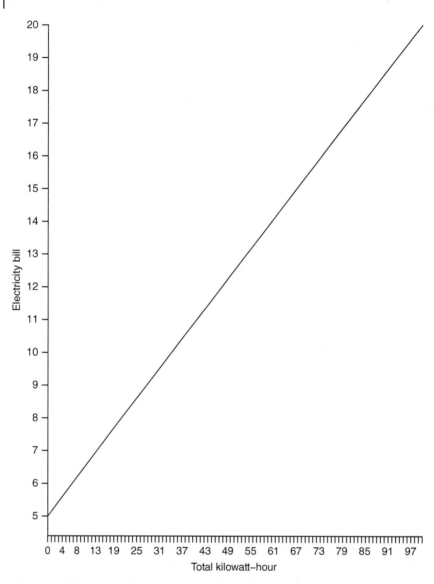

Figure 13.1 Monthly cost of electricity as a function of total kilowatt-hour use for a given month.

levels, blood pressure, etc.). It is equally important to study links or associations between variables in business. Doing so helps us either

- understand a process of interest or
- predict a process of interest.

In this chapter, we cover the fundamental principles of linear regression analysis in the easiest situation, determining the association between one numerical variable with another. We will move on to the association of one numerical variable with many others in Chapter 14.

## 13.2    Basics of Simple Linear Regression

We want to see if there is an association between two numerical variables, $Y$ and $X$. Our main variable of interest is $Y$, and we want to see if it is associated with $X$. In practice, $Y$ is known as the **response or dependent variable**. Recall that we prefer to call $Y$ the response variable here, since we have a specific meaning for "dependence" in this book. On the other hand, $X$ is known in practice as the **predictor, explanatory variable, covariate, or independent variable**. Again, to not confuse "independent variable" with our probabilistic definition of independence, we avoid the use of this terminology and refer to $X$ as the explanatory variable or predictor, mainly because it is easy to remember the meaning of the terminology.

The true relationship between $Y, X$ can be generally expressed as $Y = g(X)$, where $g(\cdot)$ is any function. Hence, $Y$ may

- be a linear function of $X$,
- be a nonlinear function of $X$, and
- have no association with $X$ at all (in which case $g(\cdot)$ is a constant for all values of $X$).

But now we will work with situations were the true relationship is unknown, all we have is data on $X, Y$. Since we do not know the true relationship between $Y$ and $X$, a more realistic expression is $Y = g(X) + \epsilon$, where $\epsilon$ accounts for

- not knowing the true relationship between the variables,
- leaving out other important variables, and
- measurement error.

Another simple way of putting it is that $\epsilon$ accounts for variability in $Y$ that cannot be accounted for by $g(X)$. Only when $\epsilon$ is always zero will we be able to find the exact value of $Y$ using $X$. Although possible to estimate the type of general function $g(\cdot)$ linking these two variables, it is easier and often useful to assume that the association between the two variables is approximately linear. We still want to express the relationship quite generally, and we do so the following way:

$$Y = \beta_o + \beta_1 X + \epsilon \tag{13.1}$$

$\beta_o$ and $\beta_1$ are parameters of the model, while $\epsilon$ is an unobservable random variable known as an **error term**. $\epsilon$ is what makes the pairs $(x_i, y_i)$ not fall exactly along a line, and as a random variable, for given $x_i$, we may have different values $y_i$. As parameters, $\beta_o$ and $\beta_1$ are constant. This equation is known as the **simple linear regression equation**. We suppose that $E(\epsilon) = 0$ and $Var(\epsilon) = \sigma^2$. Therefore,

$$E(Y|X) = \beta_o + \beta_1 x \tag{13.2}$$

and $Var(Y) = \sigma^2$. More precisely, this expectation is conditional on values of $X$, $E(Y|X = x)$. Hence, for a given value $x_i$, this conditional expectation gives us the average value of $Y_i$. Another way of viewing this is that the average value of $Y$ is modeled as a function of $X$. It is useful to clarify that the expectation is conditional on $X$ for the sake of interpretation, but also it helps us remember an important thing. The expectation assumes that the values of $X$ are known, that is, $X$ is not a random variable. We will return to this point later.

In simple linear regression, we **fit (estimate)** the response variable values based on a linear relationship with the values of the predictor. That is,

$$\hat{Y} = b_o + b_1 X \tag{13.3}$$

where $\hat{Y}$ ("y-hat") is the value of $Y$ according to the simple linear regression model – an estimate of $E(Y)$ given values of $X$ – and $b_o, b_1$ are statistics estimating $\beta_o, \beta_1$, respectively.

---

**Interpretation of the Linear Model Parameters**

You may recognize the expected value of the simple linear regression equation (13.2) from your math classes. It is the equation of a line. The parameters of the model have interpretations analogous to line equations. $\beta_o$ is the **intercept** of the model. It gives the numerical value at which the $y$ axis is intercepted when $X = 0$. More importantly, $\beta_1$ is known as the **slope of the model**. It measures the change in the expected value of $Y$ as $X$ changes.

- $\beta_1 > 0$ implies that as $X$ increases so does $E(Y|X)$.
- $\beta_1 < 0$ implies that as $X$ increases $E(Y|X)$ decreases.
- $\beta_1 = 0$ there is no linear association between $E(Y|X)$ and $X$.

---

Figure 13.2 displays how the simple linear equation works. On the left panel, we see a line with a negative slope. Then as $X$ increases, on average, the response variable decreases. In contrast, when the slope is positive, as $X$ increases, on average, the response variable increases as well (middle panel). Finally, if as $X$ increases there is no visible pattern in $E(Y|X)$, the slope is zero (right panel).

For example, if $\beta = -5$, then a one unit increase in $X$ leads to an expected change in $Y$ of $-5$ units. In contrast, if $\beta = 7$, then a one unit increase in $X$

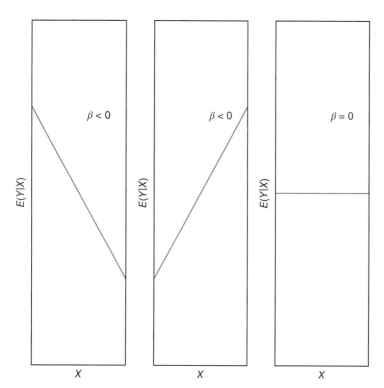

Figure 13.2 Lines with a negative slope (left panel), positive slope (middle panel), and zero slope (right panel).

leads to an expected change in $Y$ of 7 units. In our quest to make decisions based on empirical evidence, linear regression plays an important role.

## Practice Problems

13.1 If $Y = 10 - 43X$, is the association between $Y, X$ positive, negative, or neither?

13.2 If $Y = -245 + 71.4X$, is the association between $Y, X$ positive, negative, or neither?

13.3 When $\beta_1 = 12$, what happens to the response variable as $X$ increases?

13.4 When $\beta_1 = -105.41$, what happens to the response variable as $X$ increases?

## 13.3 Fitting the Simple Linear Regression Parameters

As in other inferential problems we have encountered so far, we can estimate the linear regression parameters based on statistics obtained from data, Table 13.1,

Assuming screening of the data has occurred, the next step is to visually explore the data. A scatterplot will indicate whether there is an approximate association between $X$ and $Y$, the type of association, and the strength of this association. We may also obtain summary statistics such as the correlation coefficient. Recall from Chapter 4 that the correlation coefficient ($r$) is always between $-1$ and 1. Figure 13.3 demonstrates how the correlation coefficient works by partitioning scatterplots into quadrants.

- The quadrants are established by lines at $\bar{x}$ and $\bar{y}$.
- Points in quadrant I pertain to $x_i$ greater than $\bar{x}$ and $y_i$ greater than $\bar{y}$.

Table 13.1 *n* paired observations $(x_1, y_1), \ldots, (x_n, y_n)$ of the predictor and response variable.

| X | Y |
| --- | --- |
| $x_1$ | $y_1$ |
| $x_2$ | $y_2$ |
| $\vdots$ | $\vdots$ |
| $x_n$ | $y_n$ |

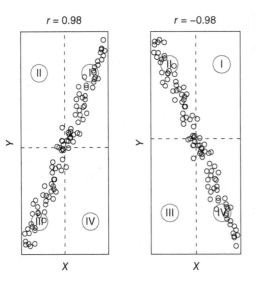

$r = 0.98$    $r = -0.98$

Figure 13.3 Scatterplots partitioned into four quadrants to better appreciate the interpretation of *r*.

Figure 13.4 Scatterplots partitioned into four quadrants to better appreciate interpretation of *r*.

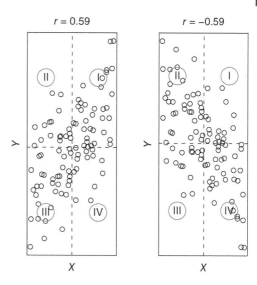

- Points in quadrant II pertain to $x_i$ less than $\bar{x}$ and $y_i$ greater than $\bar{y}$.
- Positive correlation indicates that most points are in quadrant I and quadrant III, when both values of $x$ and $y$ are mostly greater or smaller than the respective means (left panel of Figure 13.3).
- In contrast, negative correlation is indicated in the right panel of Figure 13.3.

Figure 13.4 illustrates moderate correlations.

---

**$r$ Versus $b_1$**

$r = 1$ occurs when there is a perfect positive linear association between $X$ and $Y$. When $r = 1$, it is possible to figure out the value of $Y$ exactly if you know $X$. $r = 0$, or $b_1$ close to zero indicates no linear association between the two variables. However, there may be an association between the two variables that is not approximated well with a line, as Figure 13.5 shows. That is why it is always best not to just rely on $r$ or $b_1$ to determine association, drawing a scatterplot is important too.

Importantly, correlation is not affected by unit of measurement. If there is a correlation between sales and ad budget of 0.75 when these variables are measured in dollars, the correlation will be the same if we convert the measurements to thousands of dollars. In contrast, the value of $b_1$ cannot be used to describe the strength of linear association because it is scale dependent. Thus, if $b_1 = 4.75$ was estimated measuring $y$ in pounds, changing $y$ to kilograms would change the slope to $4.75/2.2 = 2.16$. Changing $y$ to grams would make $b_1 = 2160$. The value of the slope alone is not indicative of strength of association.

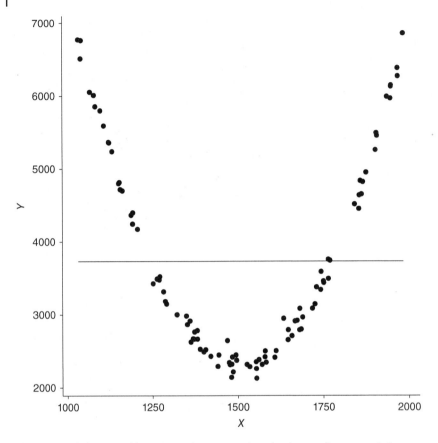

Figure 13.5 A data set with an *r* very close to zero, but clearly a nonlinear association between *X* and *Y*.

One criteria to determine the best estimators $b_o, b_1$ of $\beta_o, \beta_1$ is based on an ordinary least squares criteria,

$$\sum_{i=1}^{n} (y_i - (b_o + b_1 x_i))^2$$

where we wish to find $b_o$ and $b_1$, such that this summation of squared differences is at its minimum value. This is an optimization problem that is solved using differential calculus. In the case of simple linear regression, we obtain convenient formulas for the $b_o, b_1$ estimators. We will refrain from performing these calculations by hand and hence do not present the formulas here. For inference, we will need to estimate $\sigma^2$ as well.

Now, $b_o$ and $b_1$ in Eq. (13.3) minimize the sum of squared error between the response variable and the regression model, which is nice. But do they have other properties such that we can trust their results? The quick answer is yes, least squares estimators have some other good properties. We summarize these properties below.

---

**Least Square Estimator Properties**

- They are unbiased: $E(b_o) = \beta_o$ and $E(b_1) = \beta_1$.
- Assuming errors have expectation of zero, are uncorrelated, and have equal variances, least squares estimators have the smallest variance among all unbiased estimators.
- Assuming errors are normally distributed, $b_o$ and $b_1$ are also normally distributed.

---

Accordingly, although we do not know $\beta_o$ and $\beta_1$, we expect their least square estimators will do a good job in estimating them, giving us the opportunity to draw inference on the unknown parameters. The second and third properties of the ordinary least squares are reasons why model assumptions are important.

**Example 13.1**  *Using ordinary least squares with some data, estimates* $b_o = 5, b_1 = 2.2$ *were obtained. Therefore, the fitted regression model is*

$$\hat{Y} = 5 + 2.2X$$

The slope of 2.2 means that it is estimated that every unit increase in $X$ increases the expected value of $Y$ by 2.2 units. Also if $X = 2$, the estimated mean of the response variable becomes

$$\hat{Y} = 5 + 2.2X = 5 + 2.2 * 2 = 9.4$$

## Practice Problems

13.5  If $\hat{Y} = -0.5 + 11.6X$,
    a) interpret the value of the slope.
    b) estimate the mean of the response variable when $x = 0.48$.

13.6  If $\hat{Y} = 3.5 - 7.34X$,
    a) interpret the value of the slope.
    b) estimate the mean of the response variable when $x = 10.25$.

13.7  In the figures below, determine if the estimated slope of a simple linear regression line will be negative, positive, or close to zero.

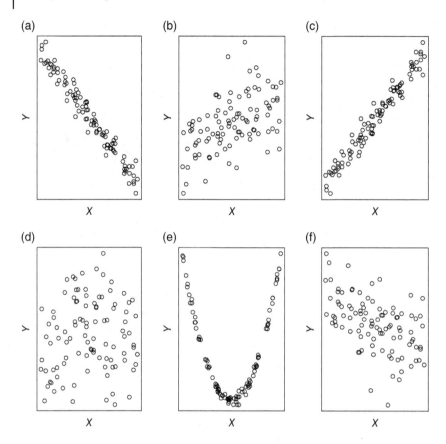

**13.8** A new analyst has been hired and you have been asked to take him under your wing. To get things started, you ask him to check if there is an association between two variables $X$ and $Y$. After a few minutes, the new analyst excitedly returns to your office because he completed his first task. He answers that there is virtually no association whatsoever because the sample correlation was only 0.05. Is the analyst's reasoning correct?

**13.9** A popular way to measure the volatility of a financial asset, such as a stock, is its beta value. If $Y$ = daily return of an asset, and $X$ = daily return of the S&P 500 index fund (the market), then we may write

$$Y = \alpha + \beta X + \epsilon$$

An asset with $\beta = 1$ has the same expected return as the market; with $\beta = 1.5$, the expected return is 50% higher than the market; with $\beta = 0.75$, the expected return is 25% smaller than the market; with $\beta = 0$, the asset is uncorrelated to the market; and with $\beta < 0$, the asset moves in the opposite direction as the market. The following table provides beta values for some company stocks.

| | |
|---|---|
| Company A | 1.44 |
| Company B | 0.29 |
| Company C | <0.17 |

a) Interpret the beta value of company A.
b) Interpret the beta value of company B.
c) Interpret the beta value of company C.
d) If an investor wanted to include in their portfolio an asset that hedges against downturns in the market, which of the stocks presented should they use?

## 13.4  Inference for Simple Linear Regression

In Example 13.1, the estimated slope was 2.2. Recall that $b_1$ is a random variable, a statistic, and as so, we must account for its uncertainty to determine from the estimate if $\beta_1 = 0$. Simply looking at how far from 0 the statistic $b_1$ is will not be enough to know if the true slope is zero or not. For example, a fitted regression model is

$$\hat{Y} = 1000 - 20X$$

$b_1 = -20$ may be well below 0, but from a statistical point of view, it may not be enough. How far is enough depends on the scale of the measurements of $x, y$ that lead to the regression model and the number of observations. Thus, if $x, y$ are on the scale of $10^9$, that $-20$ will likely not be far enough from zero to conclude there is a linear association between the two variables. By just knowing that $b_1 = -20$, it is not possible to determine if the linear association is strong, weak, or even significant. The process to conduct inference must transform the information in such a way that whether $\beta_1 = 0$ or not can be determined probabilistically.

### Simple Linear Regression Model Assumptions
Regression analysis allows us to make conclusions about association between variables by simplifying the real world. This simplification is done by treating the association as approximately linear, ignoring the impact of other variables

(intentionally or unintentionally) and through model assumptions. The four main assumptions required for inference are as follows:

1) The linear model is reasonable.
2) $\epsilon$ are independent.
3) Variance of $\epsilon$ is fixed for all values of $X$.
4) $\epsilon$ are normally distributed.

The first assumption can easily be assessed through a scatterplot. If a nonlinear association is apparent in the scatterplot, then drawing inference on $\beta_1$ may be misleading about how strong the association is between the response variable and the predictor. When the errors are independent, then the error in the first observation does not tell us anything about the error in the second observation or any other error. The third assumption means that the variance does not become larger or smaller as $X$ changes. We should remark that if the regression errors are independent, then so are the values of the response. Similarly, if errors are normally distributed, then so is the response and

$$Var(Y) = Var(\epsilon) = \sigma^2$$

For inference on $\beta_1$ to be appropriate, we need the four assumptions above to be approximately true (as opposed to strictly true). When the regression model assumptions are very inadequate, our conclusion from the inferential analysis may be unreliable.

Technically, there is another assumption that we make: there is no measurement error when obtaining $X$. This assumption is important if we are trying to explain the association between $Y$ and $X$. It is beyond the scope of this book to cover how to deal with measurement error in the predictor, but if our goal is to perform prediction, this assumption is not as important.

***Testing for Linear Association Between Y and X*** Now, we must determine if the linear association between $Y$ and $X$ is "statistically important" or as typically referred to in practice, **statistically significant**. This can be answered by making inference on the linear correlation parameter, $\rho$, or the slope $\beta_1$. In this book, we focus on hypothesis testing based on $\beta_1$.

In Example 13.1, the slope was found to be 2.2. Is this value close enough to 0? Not only $b_1$ will depend on the units of measurement of the response and the predictor, but also $b_1$ is a statistic estimating $\beta_1$, another set of $n$ observations is likely to provide a different estimate of $\beta_1$. However, $b_1$ is an unbiased estimator of $\beta_1$. Moreover, among unbiased estimators, $b_1$ has the smallest variance and is consistent. Therefore, we should be confident inferring about $\beta_1$ through $b_1$ when the regression assumptions hold. The aim is to test

$$H_o: \beta_1 = c_o$$
$$H_1: \beta_1 \neq c_o$$

where $c_o$ is any constant. However, virtually always $c_o = 0$,

$$H_o: \beta_1 = 0$$
$$H_1: \beta_1 \neq 0$$

**A t-Test** Analogous to other hypothesis testing problems we have seen, a convenient standardized statistic is based on the difference between the estimated coefficient $b_1$ and its value under the null hypothesis divided by the standard deviation of the estimator,[1] a $t$ statistic. Under the four linear regression assumptions, the test statistic will follow a $t$ distribution when the null hypothesis is true. A $p$-value can be calculated from this statistic.

---

**When to Reject $H_o$**

If the $p$-value $\leq \alpha$, $H_o$ is rejected. We may also construct a confidence interval for $\beta_1$ using a $t$ distribution. Interval estimates are an attractive option because they can then be used to quantify how practical the statistical significance is. See Section 13.8 for an example.

---

**An F-Test** The $t$-test above addresses whether the $\beta_1$ is zero or not to determine linear association. Another possibility is to try to determine if the simple linear regression is overall useful in "explaining $Y$." Since the simple linear regression model only depends on one explanatory variable, if we find that the model is useful in "explaining $Y$," then there must be an important linear association between $Y$ and $X$. To determine this, we must quantify what we mean by "explaining $Y$." A natural way is through the variability of $Y$. Which leads us to using sum of squares, analogous to Section 12.2. This is what an $F$-test does. Figure 13.6 illustrates the geometry of this approach. The idea is to express the difference between each $Y_i$ and $\overline{Y}$ as two components:

- the difference between $\hat{Y}_i$ and $\overline{Y}$; the regression component.
- the difference between $Y_i$ and $\hat{Y}$; the error component.

The first component can be used to measure how much better the model estimates values of the response over the sample mean. The second component can be used to measure the error of the model.

Specifically, the sum of squares of the deviation between the response observation and response mean is expressed in terms of two independent components: the sum of squared deviations of the regression output from the response

---

[1] Since we will not perform these computations by hand, we leave out this equation.

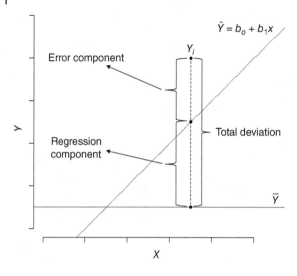

Figure 13.6 Geometry of an $F$-test. Total deviation = $Y_i - \bar{Y}$; Regression component = $Y_i - \bar{Y}$; and Error component = $Y_i - \hat{Y}_i$.

mean, and the sum of squared deviations of the data and the regression model (random error).

$$TSS = SSR + SSE \tag{13.4}$$

where $SSR$ is the sum of squares explained by the regression, and $SSE$ is the sum of squares of the errors unexplained by the model. Analogous to Section 12.2, if there is no association between $Y$ and $X$, then the estimators of $\sigma^2$ based on $SSR$ and $SSE$ should perform similarly. The Mean Square due to Regression (MSR) is the estimator of $\sigma^2$ as a function of $SSR$. The Mean Square Error (MSE) is the other estimator of $\sigma^2$, a function of $SSE$. When $\beta_1 \neq 0$, $MSR$ is no longer an unbiased estimator of $\sigma^2$, in fact it will over estimate it. Hence, the $F$ statistic

$$F = \frac{MSR}{MSE}$$

is a ratio of these two estimators of $\sigma^2$ that can test for significance of the simple linear regression model. If $X$ explains a significant amount of variability of $Y$, then the $F$ statistic should be statistically higher than 1. The results of the $F$-test tend to be arranged in an analysis of variance table (see Table 13.2).

---

**$t$-Test or $F$-Test?**

In the case of simple linear regression, the results of a $t$-test and an $F$-test are equivalent. Specifically, it can be shown that

$$F = t^2$$

These two tests will not be equivalent when using multiple predictors as we will see in Chapter 14.

Table 13.2 ANOVA table for simple linear regression.

| Source | DF | SS | MS | F | P-Value |
|---|---|---|---|---|---|
| Regression | 1 | $SSR$ | $MSR = \frac{SSR}{1}$ | $\frac{MSR}{MSE}$ | Value |
| Error | $n-2$ | $SSE$ | $MSE = \frac{SSE}{n-2}$ | | |
| Total | $n-1$ | $TSS$ | | | |

**Example 13.2** *Many cities have programs to remove graffiti. For 67 weeks, Chicago officials recorded the average number of days to complete graffiti removal and the total completed requests. Assuming weekly observations are independent, is average number of days to complete graffiti removal (response) linearly associated to the total completed requests (predictor)? Inference will be made at 5% significance.*

The scatterplot of the two variables (Figure 13.7, left panel) does not provide strong support for a linear association. However, it also reveals that in one week, the amount of total completed requests was much higher than any other week, 10 220 requests. It is unclear if this is a data entry error or events for that week resulted in much higher requests. Since 10 220 appears to not be typical, this

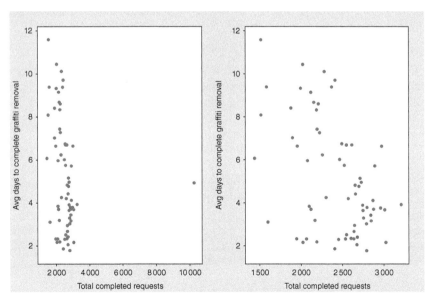

Figure 13.7 Scatterplots of total completed requests versus average number of days to complete graffiti removal. In the right panel, one extreme observation has been removed.

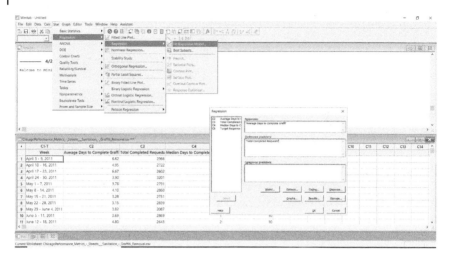

Figure 13.8 Minitab windows to perform regression.

```
Coefficients

Term                              Coef    SE Coef   T-Value   P-Value   VIF
Constant                         12.41       1.70      7.28     0.000
Total Completed Requests     -0.003031   0.000698     -4.34     0.000   1.00


Regression Equation

Average Days to Complete Graffi = 12.41 - 0.003031 Total Completed Requests
```

Figure 13.9 Summary of a *t*-test for the simple linear regression model.

row was removed before performing the analysis. The scatterplot without the outlier (Figure 13.7, right panel) suggests a weak negative linear association. The menus to perform linear regression in Minitab are presented in Figure 13.8.

Our hypotheses are

$$H_o: \beta_1 = 0$$
$$H_1: \beta_1 \neq 0$$

The *t*-test indicates (Figure 13.9) the predictor is statistically significant because the Total completed requests *p*-value $< 0.05$. We also reject $H_o$ with the *F*-test (Figure 13.10).

It is estimated that for every unit increase in total completed requests, the response variable decreases by $-0.003\,031$. The standard error of the coefficient is $0.000\,698$.

**Goodness of Fit**   It would be useful to have a measure to let us know how close to the data the results of the model are. This property is called the **goodness of fit**.

**Regression Analysis: Average Days to Complete Graffi versus Total Completed Requests**

```
Method

Rows unused   1

Analysis of Variance

Source                      DF  Adj SS   Adj MS  F-Value  P-Value
Regression                   1   100.8  100.792    18.86    0.000
  Total Completed Requests   1   100.8  100.792    18.86    0.000
Error                       64   342.0    5.343
Total                       65   442.8


Model Summary

       S    R-sq  R-sq(adj)  R-sq(pred)
 2.31154  22.76%     21.56%      17.70%
```

Figure 13.10 Summary of an *F*-test for the simple linear regression model.

There are many measures of goodness of fit, and one measure is known as the **coefficient of determination**. Returning to Eq. (13.4), the magnitude of each sum of squares is dependent on the unit of measurement (feet, inches, etc.). We can assess goodness of fit of the model with a unit-free statistic, which is derived from Eq. (13.4).

---

$r^2$

The statistic

$$r^2 = \frac{SSR}{TSS} \tag{13.5}$$

is the coefficient of determination.

- Since it is a ratio of two positive numbers, and the numerator is a component of the denominator, $0 \le r^2 \le 1$.
- The closer $r^2$ is to 1, the better the fit of the model.
- If $r^2 = 1$, then $SSE = 0$, which means we have attained a perfect fit, and the model predicts exactly the observed response values. That is, all points fall perfectly on the regression line.
- When $X$ does not help determine $Y$ at all, then $r^2 = 0$.
- Equations (13.4) and (13.5) drive a convenient interpretation of $r^2$ as the percent of variability in $Y$ explained by $X$.

---

Using Eq. (13.5) with the computer output (Figure 13.10) from Example 13.2, we have

$$r^2 = \frac{100.8}{442.8} = 0.2276$$

Thus, total completed requests explains 22.76% of the variability in average number of days to complete graffiti removal. The remaining 77.24% is explained by the random variation component of the model.

---

**$r^2$, Part 2**

- As convenient as the $r^2$ interpretation above is, we remark that by "explain" we do not mean there is a causation link.
- Another interpretation of $r^2$ is as a measure of how much better it is to use the model to predict $Y$ over using $\bar{y}$. Specifically, if $r^2 = 0.2276$, the error when using the model to predict $Y$ is 22.76% smaller than using $\bar{y}$ to predict $Y$.
- Also, the notation of $r^2$ is used for a reason. When there is an intercept in the model (which is always the case for us), the coefficient of determination is the square of the correlation coefficient $r$. Hence, we can get coefficient of determination from the correlation coefficient and vice versa. Specifically,

$$r = (\text{sign of } b_1)\sqrt{r^2}$$

---

*Inference on the Intercept* Note that we have not attempted to perform inference on the model intercept $\beta_o$, nor we have attempted to interpret the intercept within a given problem. This can certainly be done. But we need to be a bit careful when drawing inference or interpreting $\beta_o$. In many applications, it is not possible for the predictor to have a value of zero, making the interpretation of the intercept unnecessary. Also, even when the intercept is not statistically significant or makes no sense, it is often best to leave it in the model to ensure adequate properties of the model and model parameters.

**Example 13.3** *In Example 3.4, we started exploring 2014 housing market data from 37 ZIP Codes in Austin, Texas. Inference will be performed at 5% significance on whether median income (predictor) is linearly associated with median home value (response).*

Our hypotheses are

$$H_o: \beta_1 = 0$$
$$H_1: \beta_1 \neq 0$$

In Figure 13.11, we see the scatterplot of ZIP Code median household income and median home value. A moderate to weak linear association is apparent.

The Minitab output shows the $t$-test and $F$-test rejecting $H_o$ because $p$-value $< 0.05$ (Figure 13.12). The coefficient of determination can be interpreted the following ways:

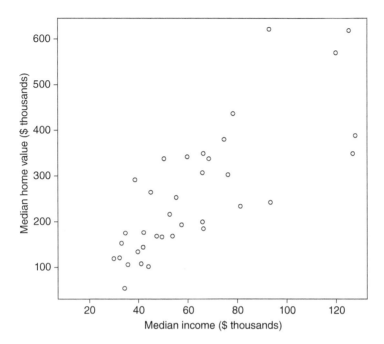

**Figure 13.11** Scatterplot showing different association of housing market variables from 37 Austin, Texas, ZIP codes.

**Regression Analysis: Median home value versus Median household income**

```
Analysis of Variance

Source                   DF       Adj SS      Adj MS  F-Value  P-Value
Regression                1  3.88110E+11  3.88110E+11    42.05    0.000
  Median household income  1  3.88110E+11  3.88110E+11    42.05    0.000
Error                    35  3.23018E+11   9229072263
Total                    36  7.11127E+11


Model Summary

      S   R-sq  R-sq(adj)  R-sq(pred)
96068.1  54.58%     53.28%      46.29%


Coefficients

Term                      Coef  SE Coef  T-Value  P-Value      VIF
Constant                 40292    37370     1.08    0.288
Median household income  3.604    0.556     6.48    0.000     1.00


Regression Equation

Median home value = 40292 + 3.604 Median household income
```

**Figure 13.12** Minitab output for a linear regression model.

- Median household income explains 54.58% of the variability in median home value.
- The error when using the model to predict median home value is 54.58% smaller than using the response mean.

Finally, observe that the estimated intercept is $40 292. Thus, if the median household income of a ZIP code is 0, the estimated median home value would be $40 292. It does not make sense to interpret the intercept in this case.

## Practice Problems

13.10 If you were given the regression equation $\hat{Y} = 4 + 500x$ and told that since the slope is 500 there is a strong linear association between the two variables, what would you say?

13.11 A new model of home selling price has been developed by a realty company. One way to assess if the coefficient estimates have been properly recorded is to construct a simple regression model of the observed selling prices versus the new model predicted prices and check how close the model prices and observed prices are.

a) Let $E(sellingprice) = \beta_o + \beta_1 modelprice$. Close to what value should the estimate of $\beta_1$ be to determine that there is no error in the recording of the model coefficients?

b) If $sellingprice = -0.05 + 0.98 * modelprice$, interpret the meaning of the slope.

c) The 95% confidence interval for the slope was (0.93,1.03). How do you think the new model performed based on this confidence interval.

d) If a house sold at $250 000, what is the predicted selling price of the new model?

13.12 Using data from 55 communities in Baltimore on affordability index for mortgages (affordm15), and for rent (affordr15), a regression model was constructed to fit affordm15 based on affordr15. Below is the Minitab output. Any inference is performed at 1% significance.

a) Write the null and alternative to assess if there is a significant linear association between the two variables.

b) According to the $F$-test, is the null above rejected?

c) According to the $t$-test, is the null above rejected?

d) What is the estimated value of the slope?

e) Interpret the value of the coefficient of determination.

f) Estimate the coefficient of correlation.

## Regression Analysis: affordm15 versus affordr15

```
Analysis of Variance

Source        DF   Adj SS   Adj MS   F-Value   P-Value
Regression     1    528.7   528.69    10.08     0.002
  affordr15    1    528.7   528.69    10.08     0.002
Error         53   2779.7    52.45
Total         54   3308.3


Model Summary

      S    R-sq  R-sq(adj)  R-sq(pred)
7.24199  15.98%     14.40%      10.71%


Coefficients

Term         Coef  SE Coef  T-Value  P-Value   VIF
Constant    18.69     6.11     3.06    0.003
affordr15   0.369    0.116     3.17    0.002  1.00


Regression Equation

affordm15 = 18.69 + 0.369 affordr15
```

**13.13** Based on 2014 housing market data from 37 ZIP Codes in Austin, Texas, use the output below to determine whether average monthly transportation cost (predictor) is linearly associated with median home value (response) at 5% significance.

## Regression Analysis: Median home value versus Average monthly transportation

```
Analysis of Variance

Source                             DF         Adj SS          Adj MS   F-Value   P-Value
Regression                          1        5478856         5478856      0.00     0.987
  Average monthly transportation    1        5478856         5478856      0.00     0.987
Error                              35    7.11122E+11     20317760836
  Lack-of-Fit                       8    3.00082E+11     37510218285      2.46     0.038
  Pure Error                       27    4.11040E+11     15223699369
Total                              36    7.11127E+11


Model Summary

      S   R-sq  R-sq(adj)  R-sq(pred)
142540  0.00%      0.00%       0.00%


Coefficients

Term                              Coef  SE Coef  T-Value  P-Value   VIF
Constant                        256959   182100     1.41    0.167
Average monthly transportation       4      264     0.02    0.987  1.00


Regression Equation

Median home value = 256959 + 4 Average monthly transportation
```

a) Write the null and alternative to assess if there is a significant linear association between the two variables.
b) According to the $F$-test, is the null above rejected?
c) According to the $t$-test, is the null above rejected?
d) What is the estimated value of the slope?
e) Interpret the value of the coefficient of determination.
f) Estimate the coefficient of correlation.

13.14 A confidence interval for the slope coefficient can be obtained using

$$b_1 \pm t_{\alpha,n-2} s_{b_1}$$

$s_{b_1}$ is the standard error of $b_1$. Get a 95% confidence interval for the slope coefficient in Example 13.2. Interpret.

13.15 Get a 99% confidence interval for the slope coefficient in Problem 13.12. Interpret.
*Hint:* See Problem 13.14.

13.16 Using the 2014 housing market data ("2014_Housing_Market_Analysis_Data_by_Zip_Code") from 37 ZIP Codes in Austin, Texas,
a) construct a scatterplot of average monthly transportation (predictor) and median rent (response). Interpret.
b) write the null and alternative hypotheses testing a linear association between average monthly transportation (predictor) and median rent (response).
c) use statistical software to determine if the $H_o$ rejected at 1% significance.

13.17 Returning to Problem 13.16,
a) interpret $r^2$.
b) calculate $r$. Interpret.

13.18 Show that the $F$ statistic in Figure 13.12 is equal to the $t$ statistic calculated from the data, squared.

13.19 Download "Minneapolis_Property_Sales_20102013", a data set[2] of property sales in Minneapolis.
a) Using statistical software, extract data for transactions such that "*PropType = Residential*." For Gross Sale Price (response) and Adjusted Sale Price (predictor), remove any values $\leq 10\,000$.
b) Build a regression model for Gross Sale Price and interpret the coefficient of determination.

---

2 Source: opendata.minneapolismn.gov (accessed July 26, 2019).

**13.20** Is there an association between starting post-college salary and hours a week playing video games? The table below summarizes the results from a sample of 100 randomly surveyed participants. Calculate the coefficient of determination.

| Source | DF | SS | MS | F | p-Value |
|---|---|---|---|---|---|
| Regression | 1 | 131 798 255 | 131 798 255 | 23.83 | 0 |
| Error | 98 | 542 108 218 | 5 531 717 | | |
| Total | 99 | | | | |

  a) Calculate the coefficient of determination.
  b) It can be shown that $r = b_1 \frac{s_x}{s_y}$, where $s_y$ is the standard deviation of the response variable and $s_x$ is the standard deviation of the predictor. If the sign of the slope was negative, $s_x = 5.57$, and $s_y = 2609.05$, find $b_1$.

## 13.5  Estimating and Predicting the Response Variable

Statisticians distinguish between estimation of the response and prediction. There is a subtle difference.

***Confidence and Prediction Intervals***  Here is a question: for what should an interval estimate be constructed, to infer about $E(Y|X)$ or to infer about $Y$? When it comes to constructing intervals for the response variable, we must distinguish between prediction and estimation because their uncertainty is different. In simple linear regression, prediction is when we want to determine the value of $Y$, a random variable, using values of $X$. Estimation is when we want to infer on $E(Y|X)$, a parameter, using values of $X$. We stated the difference between these goals in Section 13.1.

- Prediction intervals are used to infer about individual values of $Y$.
- Confidence intervals are used to infer about where its mean $E(Y|X)$ falls.

If we define

$$\hat{\epsilon}_p = Y - \hat{Y}$$

and

$$\hat{\epsilon}_m = E(Y|X) - \hat{Y}$$

we see that $\hat{\epsilon}_p$ depends on two random variables, while $\hat{\epsilon}_m$ depends on just one. As a result, prediction intervals will always be wider than confidence intervals.
  In Example 13.2, the regression model was

$$\hat{y} = 12.41 - 0.003\,031x$$

where $\hat{y}$ is the estimated average days to complete graffiti removal requests and $x$ is total completed requests. Figure 13.13 presents the Minitab commands necessary to perform two predictions: when $x = 2500$ and $x = 10\,000$. The estimates and 95% intervals are shown in Figure 13.14. When $x = 2500$, the predicted average days for removal is

$$\hat{y} = 12.41 - 0.003\,031(2500) = 4.83$$

From the output, we determine with 95% confidence that the mean response is between (4.25, 5.41) when $x = 2500$, and that for a randomly chosen week, $y$ will fall within (0.18, 9.48).

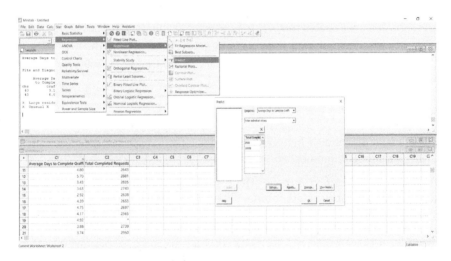

Figure 13.13 Minitab menus to perform a prediction.

```
Average Days to Complete Graffi = 12.41 - 0.003031 Total Completed Requests


Variable                       Setting
Total Completed Requests         2500


    Fit     SE Fit         95% CI                    95% CI
4.82989   0.291826  (4.24690, 5.41288)     (0.175408, 9.48437)


Variable                       Setting
Total Completed Requests        10000


     Fit     SE Fit         95% CI                    95% CI
-17.9039   5.30680  (-28.5055, -7.30234)   (-29.4675, -6.34028)    XX

XX denotes an extremely unusual point relative to predictor levels used to fit the model.
```

Figure 13.14 Minitab prediction; 95% confidence interval and 95% prediction intervals for two predictor values.

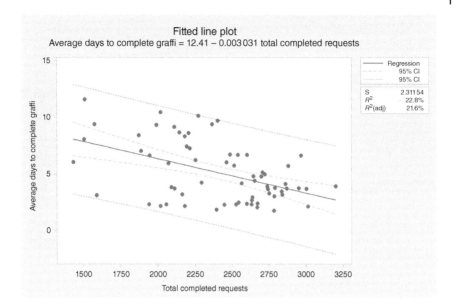

**Figure 13.15** Minitab 95% confidence intervals and 95% prediction intervals.

---

**Going Beyond Given Values of $X$**

**Extrapolation** is when we use the regression model to determine values of $Y$ from values of $X$ outside the observed range. When $x = 10\,000$, the model above predicted unreasonable average days for removal. The reason for this is because 10 000 is way outside the range of values for the predictor used to construct the regression model. Outside the range of values of $x$ used to construct the model, we can no longer assume the relationship between $y$ and $x$ is still approximately linear.

---

Figure 13.15 displays the 95% confidence (green-dashed lines) and prediction intervals (red-dashed lines). The prediction bounds are always further away from the regression line than the confidence bounds are.

---

**A Discrete $X$**

So far, it has been implicitly assumed that $X$ is approximately continuous. This allowed explaining concepts from mathematical and geometrical perspectives. If $X$ were to be discrete with a small range of values, the procedure and math behind simple linear regression do not change, albeit the geometric aspect changes slightly. When $X$ is discrete with a small range of values, then there would be some gaps between the $x, y$ pairs because of the discreteness of $X$. Thus, the pairs are less likely to fall on the regression line, even when the linear association is significant. This does not take away from the interpretation of the results. That is, if as the values of the discrete $X$ increase, $E(Y|X)$ decreases, then it is said that $Y$ and $X$ have a negative linear association, just as with continuous predictors.

## Practice Problems

13.21 In Problem 13.12, affordr15 was a predictor found to have a signifi-cant association with affordm15. Below is the Minitab output when affordm15 is predicted for affordr15 = 53.8.

**Prediction for affordm15**

```
Regression Equation

affordm15 = 18.69 + 0.369 affordr15


Variable    Setting
affordr15     53.8


  Fit    SE Fit        95% CI              95% PI
38.5633   1.00368   (36.5501, 40.5764)   (23.8988, 53.2277)
```

a) What is the estimated value of affordm15 when affordr15 = 53.8?
b) Interpret the 95% confidence interval of $E(Y|X)$.
c) Interpret the 95% prediction interval of $Y$.
d) The predictor has a minimum of 30.26 and a maximum of 69.88. Why do you think the prediction interval was so wide?
   *Hint:* How good do you think the model is?

13.22 Continuing Problem 13.21, below is the Minitab output when affordm15 is predicted for affordr15 = 10.4.
a) What is the estimated value of affordm15 when affordr15 = 10.4?
b) Interpret the 95% confidence interval of $E(Y|X)$.
c) Interpret the 95% prediction interval of $Y$.
d) The predictor has a minimum of 30.26 and a maximum of 69.88. Any concerns about the results are summarized here?

## Prediction for affordm15

```
Regression Equation

affordm15 = 18.69 + 0.369 affordr15


Variable    Setting
affordr15     10.4


   Fit    SE Fit        95% CI              95% PI
22.5299   4.91591   (12.6699, 32.3900)   (4.97390, 40.0860)   xx

xx denotes an extremely unusual point relative to predictor
levels used to fit the model.
```

13.23 Download the file "Sanitation_2010Baltimore," which has 2010 Baltimore data[3] for rate of dirty streets and alleys (dirtyst10) and rate of clogged drain reports per 1000 residents (drains10). Is drains10 a good predictor of dirtyst10?
  a) Construct a scatterplot. Interpret.
  b) Show that the slope of drains10 is statistically different to zero at 1% significance.
  c) Get the 99% confidence and prediction intervals for $E(Y|X)$ and $Y$ respectively when *drains*10 = 8.

13.24 Returning to Problem 13.16, which uses "2014_Housing_Market_Analysis_Data_by_Zip_Code" from 37 ZIP Codes in Austin, Texas,
  a) construct a 95% confidence interval. Interpret.
  b) construct a 95% prediction interval. Interpret.

13.25 A random sample of 100 college grads reported their starting salary (response) and their estimated hours playing video games a week during their last year of college (a discrete predictor). Use the data file "vgames_salary":
  a) To construct a scatterplot. Interpret.
  b) Is your linear model slope statistically different to zero at 5% significance.
  c) To get the 95% confidence and prediction intervals for $E(Y|X)$ and $Y$, respectively, when the predictor is 9.

---

3 Source: https://data.baltimorecity.gov

## 13.6   A Binary $X$

Now, suppose we wish to examine the association between grade (values from 0 to 100) and gender. The simple linear model representation requires a numerical $X$. A solution is to code the categories into numbers. **Binary variables** have only two possible numerical values. Sometimes they are also called **indicator variables**, since the absence or presence of a condition is represented using[4] 0, 1. For example, we can define 0 as a female and 1 as a male. Instead of ambiguously calling the gender variable $X$, we may call it male: 1 represents Yes, 0 represents No, making the meaning of the variable coding obvious. The choice of which gender is coded as 1 is arbitrary and will not affect the conclusions from the model.

---

**Interpreting the Impact of a Binary $X$**

To shed light on the interpretation of the effect of a binary predictor $X$, recall that

$$E(Y|X = x) = \beta_o + \beta_1 x$$

Therefore, if $X = 1$, we have

$$E(Y|X = 1) = \beta_o + \beta_1$$

In contrast, if $X = 0$, then

$$E(Y|X = 0) = \beta_o$$

---

Thus, with a binary predictor,

- Under the null hypothesis, $Y$ has mean $\beta_o$.
- If the binary predictor is statistically significant, then there is a difference in mean of $Y$ depending on the category of $X$, and the mean gets shifted upward or downward (depending on sign of $\beta_1$) from $\beta_o$ when $X = 1$.

---

**Comparing Two Population Means**

Let $\mu_1 = \beta_o + \beta_1$ and $\mu_o = \beta_o$. From this we get,

$$\mu_1 - \mu_o = \beta_1$$

This helps us see that when $X$ is a binary predictor, simple linear regression is equivalent to hypothesis testing for the comparison of two population means using independent samples.

---

4 Technically, any numerical coding can be used. In designed experiments, it can be convenient to use $-1, 1$ coding when categorizing involves quantitative variables. We focus strictly in a 0, 1 coding here.

Figure 13.16 Modeling cheese percentage according to whether cheese is organic or not.

**Regression Analysis: FatContentPercent versus Organic**

```
Method

Rows unused 410


Analysis of Variance

Source        DF   Adj SS   Adj MS  F-Value  P-Value
Regression     1     26.9    26.86     0.50    0.478
  Organic      1     26.9    26.86     0.50    0.478
Error       1028  54873.5    53.38
Total       1029  54900.3


Model Summary

      S   R-sq  R-sq(adj)  R-sq(pred)
7.30608  0.05%      0.00%       0.00%


Coefficients

Term         Coef  SE Coef  T-Value  P-Value   VIF
Constant   26.535    0.239   110.88    0.000
Organic     0.550    0.776     0.71    0.478  1.00


Regression Equation

FatContentPercent = 26.535 + 0.550 Organic
```

The discussion of categorical variables with more than two categories is deferred until Section 14.7.

**Example 13.4** *Instead of the one-sided hypothesis test performed in Example 11.4, a simple linear regression model can be constructed where the predictor is a binary variable: whether the cheese is organic or nonorganic.*

Results are summarized in Figure 13.16. Whether the cheese is organic or nonorganic is not found to be associated with its fat percentage. The estimated shift in fat percentage is 0.55 when the cheese is organic, but this is not statistically significant.

## Practice Problems

13.26 The table below summarizes suicide rates for Asian men and women in New York City from 2007 until 2014. With the assistance of statistical software, test if gender predicts suicide rate. Use $\alpha = 0.10$.

| | Year | | | | | | | |
|--------|------|------|------|------|------|------|------|------|
| Gender | 2007 | 2008 | 2009 | 2010 | 2011 | 2012 | 2013 | 2014 |
| Male   | 7.4  | 6.0  | 6.9  | 8.8  | 9.3  | 7.4  | 7.4  | 8.7  |
| Female | 4.3  | 4.9  | 3.1  | 3.0  | 3.7  | 5.6  |      | 4.5  |

**13.27** Why is the *p*-value in Example 13.4 different than in Example 11.4? *Hint:* What are the hypotheses in each example?

## 13.7 Model Diagnostics (Residual Analysis)

For the model conclusions to be reliable, the simple linear regression assumptions must be approximately true. An intuitive way to check the adequacy of the model and detect unusual observations is through functions of how much the model fit differs from the actual measurement of the response variable. This step of the procedure is known as **model diagnostics or residual analysis**. In Chapter 12, the residual was defined as the difference between the observed response $y_i$ and the regression model fit, $\hat{y}_i$,

$$\hat{\epsilon}_i = y_i - \hat{y}_i$$

We can think of the residuals as estimates of the error term $\epsilon$ in the simple linear regression equation.

---

**Residual analysis**, also referred to as **model diagnostics**, is performed to check two things:

- Adequacy of model assumptions,
- Influence of individual observations, including outliers.

---

Recall from Section 13.4, that to conduct statistical inference using the simple linear regression results, we must make four model assumptions. When these assumptions are not approximately true, our inference conclusions become unreliable. It is worth discussing the consequences of violating these assumptions.

- Linearity assumption – if the association between $Y$ and $X$ is nonlinear, then the predictions from the model will be way off, in part because of bias in the estimated slope coefficient. Also, $H_o$: $\beta_1 = 0$ may not be rejected, but there are some nonlinear associations that cannot be approximated by a line.
- Independent errors assumption – a violation of the independent errors biases the estimate of the variance. Most of the time, the variance will be underestimated, resulting in inflated *t*-statistics and confidence intervals that are too narrow.
- Constant variance assumption – when the constant variance assumption does not hold, then $\sigma^2$ will depend on $X$. This makes our constant $\sigma^2$ estimate unreliable. This has an impact on statistical tests, and confidence intervals may become too wide or too narrow.

- Normally distributed errors assumption – the normality assumption is required for significance tests and interval estimates. It is not required to implement the least square estimation procedure. A highly skewed error distribution will produce confidence intervals that are too narrow or too wide. Highly unusual observations can have a strong influence in parameter estimation affecting statistical inference. If interest is mainly on prediction and your model is correct (not many important predictors have been left out), then violating the normality assumption is of less concern.

*Standardized Residuals*   A second objective of residual analysis is screening for observations that might be unduly influential in the results of our statistical analysis. Residual analysis to check model assumptions evaluates residuals as a group. But when screening for unduly influential observations, we are looking for individual observations that may be causing problems. Traditional residuals make it more challenging to perform this task. For example, a residual of 500 or $-150$ is not necessarily too high or too low, and it will depend on the unit of measurement. Different types of residuals can be used to assess the influence of individual observations. We mainly focus on **standardized residuals**. To get the standardized residuals, each raw residual is divided by the standard deviation of all the raw residuals.

$$\hat{r}_i = \frac{\hat{e}_i}{s_{\hat{e}}}$$

Statistical softwares sometimes use other methods to scale the raw residuals. For example, each raw residual may be divided by their corresponding standard deviation (known as a studentized residual), or each raw residual may be divided by their corresponding standard deviation when the $i$th observation is deleted from the model (known as a jackknife residual). Although these alternative residuals may indeed be preferred to standardized residuals, we stick to standardized residuals for ease of presentation.

*Visually Checking Model Assumptions*   The most common way to evaluate simple linear regression assumptions is through different plots of residuals. In what follows, these figures are described. Any regression analysis on this textbook is based on standardized residuals $\hat{r}_i$.

   **Residuals against $x$ plot:** This is a graph that displays each $\hat{r}_i$ with its corresponding $x_i$. Plot typically represents $x_i, \hat{r}_i$ pairs, with $x_i$ values displayed on the horizontal axis and $\hat{r}_i$ values on the vertical axis. This plot is used to verify the constant variance assumption and the linearity assumption.

   **Residuals against $\hat{y}$ plot:** Another chart used to check the constant variance and linearity assumptions (technically also check whether expected value of the errors is zero). We focus on using this plot since it is more general than the residuals against $x$ plot, as will be seen in Chapter 14.

- If constant variance and linearity assumptions are true, the plot shows uniform random scatter of the residuals around zero, with no trend in mean or variability of the residuals.
- If a trend in mean or variability is apparent as $\hat{y}$ changes, then these assumptions do not hold.
- Typically a trend in local $\hat{r}_i$ values indicates that the model is misspecified.
- By using standardized residuals, one can also use this plot to evaluate the impact of individual observations.

---

**Extreme Standardized Residuals**

In principle, standardized residuals should be mostly between −2 and 2, and if $|\hat{r}_i| \geq 3$ this may be indicative of an outlier. Many standardized residuals outside this range or one $\hat{r}_i$ that is far outside this range (e.g. a standardized residual of −75) imply observations that need a closer look.

---

Conceptually, there is a difference between outliers, influential points, and leverage points, something that is briefly discussed later in Section "Unusual Observations."

Figure 13.17 illustrates the residuals against $\hat{y}$ plot under several different scenarios.

- In the upper left panel, no pattern in $\hat{r}_i$ is detected as $\hat{y}$ changes. Variance stays about uniform as one moves through the horizontal axis.
- The same can be said in the scenario shown in the upper right panel. Although some $\hat{r}_i$ are outside the −2 and 2 range, it is not by much, and no serious concern is apparent.
- The lower panel illustrations are a different story. On the left, it can be clearly seen that as $\hat{y}$ increases, the variability in $\hat{r}_i$ increases. Thus, the constant variance assumption is not appropriate in this case. This outward funnel pattern turns out to be fairly common in practice.
- In the lower right panel, a parabolic trend is shown, indicative of the linear assumption not being appropriate.

Another look at the scatterplot (Figure 13.18) reveals a slight curvature, suggesting a more complex regression model should be fit.

In general, nonlinearity can be detected using scatterplots, a plot of observations versus $\hat{y}$, or a plot of residuals versus $\hat{y}$. In a plot of observations versus $\hat{y}$, the points should be scattered symmetrically around a diagonal line while in a plot of residuals versus $\hat{y}$ the points should be scattered symmetrically around a horizontal line with a variance that is roughly constant.

Formal hypothesis tests for nonconstant variance are also possible, although visual diagnostics are the norm. There are multiple paths to follow to resolve nonconstant variance, but two approaches are used most often.

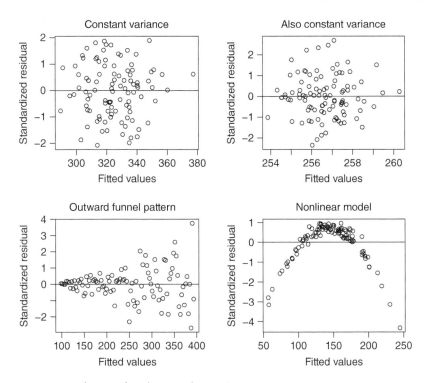

Figure 13.17 $\hat{r}_i$ versus $\hat{y}$ under several scenarios.

- The first one is transforming variables (the response, the predictor, or both) by building the model on some function of $Y$ or $X$. For example, in finance taking the logarithm of the returns of financial assets helps make the variance more constant. The square root is another preferred transformation.
- The second favorite method is to use a version of the least squares known as weighted least squares, where observations are weighted based on nonconstant variances.

*Normal Probability Plot* Recall that the normal plot shows the ordered residuals against theoretical points under the normal distribution. Points with a shape of a outstretched "s" is considered as an indication that the normal distribution has been satisfied. Figure 12.10 presents the normal plots for the same data sets used for Figure 13.17.

*Validity of the Independence of Errors Assumption* An option to check the independence assumption is to plot the residuals versus the order of the observations. Although some software provide this plot, it can be difficult to detect from it when independence is invalid. Sequence of positive and negative residuals may

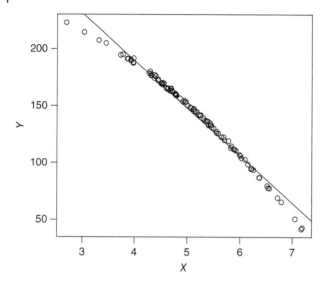

Figure 13.18 Scatterplot for the data used for the residual plot in the lower right panel of Figure 13.17. There is curvature around the included fitted line for the highest and lowest $x_i, y_i$ pairs indicating a nonlinear relationship.

occur just by coincidence. The sheer number of observations can make it difficult to see anything at all in this plot. Furthermore, there are different ways independence in the observations can occur: it could be due to temporal dependence (i.e. monthly sales) or due to repeated measurements on participants (i.e. monthly carbohydrate consumption for different participants). Other alternative plots are possible, but will not be discussed here. The way in which the data was obtained could give us an indication on whether we should be concerned about the feasibility of the independence assumption. When observations are made at different points in time or space, residuals may no longer be independent. This dependence must then be taken into account. Chapter 17 introduces some models that deal with this dependence. On the other hand, data from certain designed experiments are already independent by design.

*Unusual Observations*  Recall that we must be aware of the impact that specific observations might have on the results. Residual analysis helps pinpointing outliers. On their own, outliers are not evidence of a problematic observation. Observations associated to predictor values that deviate considerably from the mean value of the predictor are called **leverage points**. Most importantly, an observation that has a dramatic impact on the regression fit is known as an **influential point**. Figure 13.19 illustrates the effect of influential points.

- On the left, data has 30 observations, one observation has a $y$ value much lower than most $y$ values. As a result, the fitted regression line when this

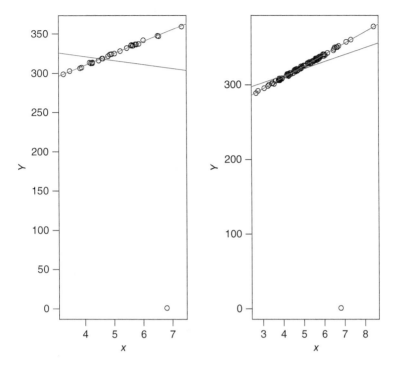

Figure 13.19 Left panel shows how an influential point leads to a negative slope for the regression line (red line). Without the influential point, the regression line has a positive slope (blue line). At the right, an influential point again affects the slope estimate (red line), although this time not as much.

observation is part of the data has a negative slope. But if that observation were to be removed, the slope of the line is positive and follows most of the data much more closely.

● The right panel data has 100 observations. An influential point again distorts the estimated slope, but this time not as much, in part because of the larger size for the data set.

---

**Observation Sensitivity**
One simple way to double check if an outlier is an influential point is to rerun the regression fit excluding the outlier.

● If the results do not change significantly, then the outlier is of no concern.
● If the regression fit has a substantial change, then the outlier requires further inspection.

> Of course, the refitting strategy only works if there is only a handful of outliers present to assess. Influential points are not always as easy to detect as in the figure above. On occasion, there can be multiple unusual observations which may be grouped themselves. There is an array of statistics that can be used to help detect these problematic observations, but this topic is beyond the scope of this textbook.

For our purposes, it is enough to know that outliers can have a dramatic impact on the regression results. Sometimes, assessing the data collection process may help determine if the outliers were due to errors, or that the observations have a nonnormal distribution.

## Practice Problems

13.28 A simple linear regression model is constructed to see how well average American College Testing (ACT) scores in Chicago help predict college enrollment rate. Below, the residuals of the model are plotted against the fitted values. What does the figure imply to be wrong about the model?

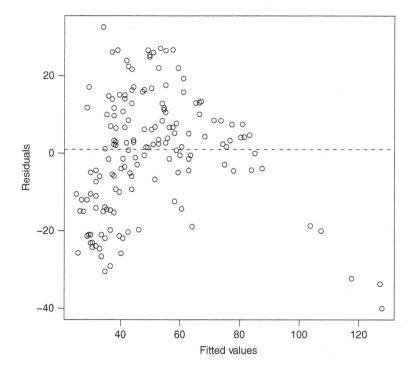

**13.29** Below are residual plots for the simple linear regression model fit in Problem 13.12. Interpret the plots.

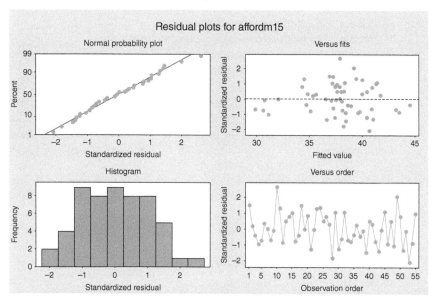

**13.30** In Example 13.2, an extreme observation was removed. Below are the Minitab results if the observation were to be left in.

**Regression Analysis: Average Days to Complete Graffi versus Total Completed Requests**

```
Analysis of Variance

Source                         DF    Adj ss   Adj MS   F-Value   P-Value
Regression                      1     16.96   16.960      2.59     0.112
  Total Completed Requests      1     16.96   16.960      2.59     0.112
Error                          65    425.83    6.551
Total                          66    442.79


Model Summary

      S    R-sq   R-sq(adj)   R-sq(pred)
2.55955   3.83%       2.35%        0.00%


Coefficients

Term                           Coef    SE Coef   T-Value   P-Value    VIF
Constant                      6.341      0.827      7.66     0.000
Total Completed Requests  -0.000488   0.000304     -1.61     0.112   1.00


Regression Equation

Average Days to Complete Graffi = 6.341 - 0.000488 Total Completed Requests
```

a) Write the null and alternative to assess if there is a significant linear association between the two variables.
b) According to the *t*-test, is the null above rejected?
c) What is the estimated value of the slope?
d) Interpret the value of the coefficient of determination.
e) Estimate the coefficient of correlation.
f) Visually explore the model standardized residuals. Interpret.

13.31 Returning to Problem 13.23 with rate of dirty streets and alleys (dirtyst10) and rate of clogged drain reports per 1000 residents (drains10) for Baltimore.[5] Use residual plots to assess all regression assumptions.

13.32 While revising a report you received from an analyst, you come across a residual versus fit plot. It shows that all residuals were negative. How is this indicative of an issue?

## 13.8 What Correlation Doesn't Mean

When the null hypothesis of $\beta = 0$ is rejected at a given significance level, there is evidence of a linear association between $Y$ and $X$, correlation. But care is needed in the interpretation of this. Three misconceptions of what rejecting the null means commonly occur that it implies that $X$ causes $Y$, or an approximately linear association in general between the variables, or that statistical significance means practical significance.

*Correlation Does Not Imply Causation*   There are many ways to argue that correlation between $Y$ and $X$ does not imply causation. Based on math, we can state that

$$Y = \beta_o + \beta_1 X + \epsilon$$

can be expressed as

$$X = \frac{1}{\beta_1}(Y - \beta_o - \epsilon)$$

Hence, a nonzero $\beta_1$ does not imply that $X$ causes $Y$ or the other way around.

Based on logic, correlation among $X$ and $Y$ may be caused by a third unobserved variable $Z$. This argument can be complemented with a third type of argument, a counterexample. A favorite example found online is the negative correlation among number of pirates around the world and the global average temperatures. Naturally, it does not make sense to think

---

5  Source: data.baltimorecity.gov (accessed July 26, 2019).

that the decrease in number of pirates around the world has led to global warming.

***Correlation Does Not Imply a Linear Association in General***  Recall that in Section 13.2, we emphasized below Eq. (13.2) that the expectation of $Y$ is conditional on the values of $X$. This implies that the values of $X$ are known. Since a line is defined by two points, any value of $X$ within the observed range can be treated as known. But outside the range of observed values, we will be less certain what the association between $Y$ and the unobserved $X$ will be. When the values of $X$ are not known, we should expect more variability in the results of the model.

---

**Going Beyond Given Values of $X$ (Again)**
Extrapolation must be done with care (Figure 13.20). Unless we have further support of the linear association extending beyond the range of observed $X$ values, predictions at these values of $X$ should be avoided. Problem 3.24 showed us school college enrollment rate versus average ACT scores using 2016–2017 Chicago school data[6] (see Figure 13.21). But let's suppose that some of the data is not available, specifically ACT scores are available only up to 17.5 and college rate only up to 60%. As we can see from the scatterplot, this version of the data indicates a fairly strong linear association between the two variables. Only higher values of these variables help us identify that their association is nonlinear. Attempting to predict college rates with a simple linear regression model and an ACT score over 20 would lead us to overestimate the college rate. Hence, extrapolation should always be done with care.

---

***Statistical Significance Does Not Imply Practical Importance***  When the null $H_o: \beta_1 = 0$ is rejected, even when done correctly, this does not mean $X$ is important in explaining or predicting $Y$. Recall from hypothesis testing that as sample size increases, even small effects will become statistically significant.

An alternative is to use one-sided hypothesis testing instead of two-sided hypothesis testing. More precisely, we may choose our alternative as $H_1: \beta_1 > \Delta$ as opposed to $H_1: \beta_1 \neq 0$, where $\Delta$ is a constant. The choice of $\Delta$ requires careful consideration. A value that is too conservative could dismiss a potentially important predictor as insignificant. Also, one must feel confident about the type of relationship (positive or negative) between $Y$ and $X$. Another option is to conduct inference based on a confidence interval and then compare data-based results with our understanding of the problem at hand. For example, management may state that for practical significance they would need more than a 5 unit change in $Y$ for every unit change in $X$. A

---

6 Source of data: data.cityofchicago.org (accessed July 26, 2019).

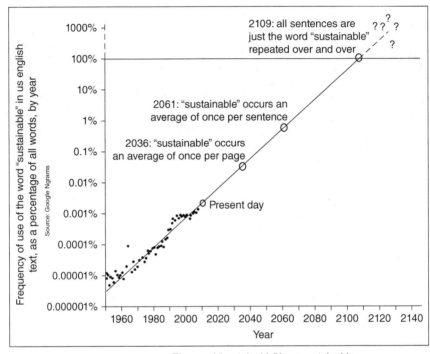

The word "sustainable" is unsustainable

Figure 13.20 Extrapolation gone wrong. *Source:* https://www.xkcd.com/1007.

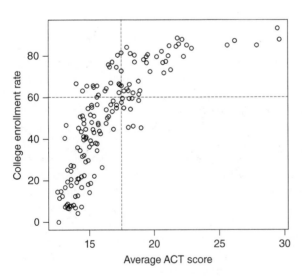

Figure 13.21 ACT mean score versus college entrance rate for Chicago schools. Assume any ACT score above the red-dashed line or any college rate to the right of the red-dashed line is not available.

one-sided 95% confidence interval for $\beta_1$ of $(12, \infty)$ is strictly above 5, implying statistical and practical importance.

### 13.8.1   A Quick Look at Data Science: Can Rate of College Educated People Help Predict the Rate of Narcotic Problems in Baltimore?

As part of an internship for a nonprofit, a student is challenged to predict crime using some measurement of education. In her search of open data, she came across Baltimore's 2015 vital signs data which are available[7] from several sources and are indicators of the health of Baltimore's neighborhoods. Specifically, for 55 community statistical areas, data are provided on education, crime, demographics, and other topics. After a quick look over the indicators, the intern decides to build a simple linear regression model that predicts "Number of Narcotics Calls for Service per 1000 Residents," abbreviated narc15, using "Percent Population (25 years and over) with a Bachelor's Degree or Above," abbreviated bahigher15. The null and alternative are

$$H_o: \beta_1 = 0$$
$$H_1: \beta_1 \neq 0$$

and $\alpha = 0.05$ will be used. Table 13.3 shows data for 2 of the 55 communities. The first community shown has 51.7 narcotic calls for service per 1000 residents, while it is estimated that 11% of residents have bachelor degrees or above. Limited data wrangling is needed for this data set (Figure 13.22).

Descriptive summaries (Figure 13.23) show reasonable minimum and maximum values for both variables; no further data clean up is needed. Moreover, the mean of the response variable is two times bigger than the median, hinting at a distribution that is skewed to the right and in violation of the normality assumption. This should not be too surprising, we would expect that most communities have relatively low narcotic calls for service per 1000 residents.

The scatterplot of the two variables (Figure 13.24) gives some support to an association between narc15 and bahigher15, albeit a weak negative association. Indeed, the relationship between the two variables appears to be nonlinear.

Table 13.3  Indicators for 2 of the 55 community statistical areas in Baltimore.

| Community | narc15 | bahigher15 |
| --- | --- | --- |
| Allendale/Irvington/S. Hilton | 51.7 | 11.0 |
| Beechfield/Ten Hills/West Hills | 21.7 | 24.6 |
| ⋮ | | |

---

7  data.baltimorecity.gov (accessed July 26, 2019).

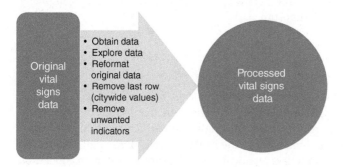

Figure 13.22 Steps to set up the data to perform simple linear regression analysis.

### Descriptive Statistics: narc15, bahigher15

```
Variable      Mean  StDev Minimum      Q1 Median      Q3 Maximum
narc15       68.41  71.00    0.90   20.60  34.40  105.80  286.40
bahigher15   28.42  21.58    5.50   11.00  19.50   41.70   80.40
```

Figure 13.23 Minitab descriptive statistics of the two variables.

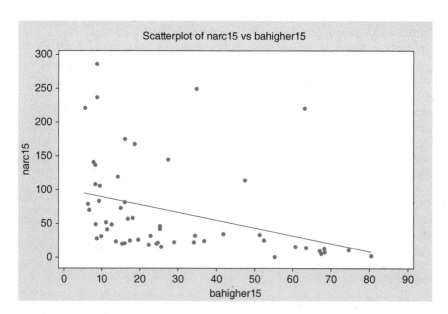

Figure 13.24 A scatterplot of narc15 versus bahigher15. A regression line has been added as a reference.

**Figure 13.25** Minitab regression output.

**Regression Analysis: narc15 versus bahigher15**

```
Analysis of Variance

Source          DF   Adj SS   Adj MS   F-Value   P-Value
Regression       1    33482    33482      7.43     0.009
  bahigher15     1    33482    33482      7.43     0.009
Error           53   238697     4504
  Lack-of-Fit   48   198828     4142      0.52     0.892
  Pure Error     5    39869     7974
Total           54   272180


Model Summary

       S    R-sq   R-sq(adj)   R-sq(pred)
 67.1098   12.30%     10.65%        5.94%


Coefficients

Term           Coef   SE Coef   T-Value   P-Value    VIF
Constant      101.2      15.1      6.72     0.000
bahigher15   -1.154     0.423     -2.73     0.009   1.00


Regression Equation

narc15 = 101.2 - 1.154 bahigher15
```

Since $p$-value $= 0.009 < 0.05$, the intern determines that the null should be rejected (Figure 13.25). However, $R^2 = 0.12$, bahigher15 explains only 12% of the variability in narc15. The model is not expected to predict well the response variable.

The normal probability plot of the standardized residuals do not support the normal distribution assumption (Figure 13.26). Moreover, the residuals versus $\hat{y}$ plot gives some indication that the variance of the model errors is not constant. Already Figure 13.24 made the intern doubt the adequacy of the linearity assumption.

The intern shows the preliminary results to a college friend, who recommends applying a natural logarithm transformation to the response variable. Calculating this transformation is very easy (Figure 13.27).

The intern proceeds to construct a linear regression model on $log(narc15)$. The regression on the transformed response variable is still significant (Figure 13.28), it now has a more respectable $R^2 = 0.37$, and the standardized residual plots show less concern with the modeling assumptions (Figure 13.29).

If the intention is to predict $log(narc15)$ when bahigher15 $= 11.3$, the model gives 4.25 (Figure 13.30). The statistical software also shows 95% confidence and prediction intervals. Thus, when bahigher15 $= 11.3$, we can be 95% confident that the predicted $log(narc15)$ is between (2.35, 6.14). To get values for $narc15$, we transform backwards; an exponential function is applied to the $log(narc15)$ values: when bahigher15 $= 11.3$, the model predicts

$$narc15 = e^{4.25} = 70.10$$

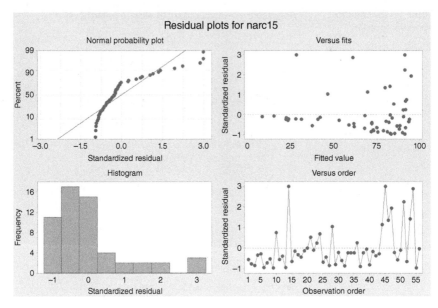

Figure 13.26 Minitab standardized residual plots.

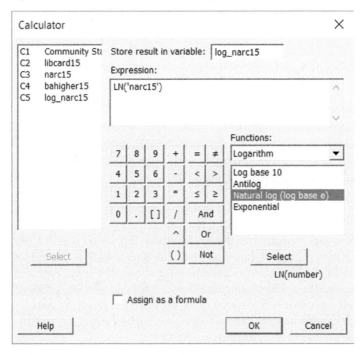

Figure 13.27 To transform narc15 in Minitab, go to `Calc > Calculator`.

## Regression Analysis: log_narc15 versus bahigher15

```
Analysis of Variance

Source        DF   Adj SS   Adj MS   F-Value   P-Value
Regression     1   27.363   27.3631    31.44     0.000
  bahigher15   1   27.363   27.3631    31.44     0.000
Error         54   47.001    0.8704
  Lack-of-Fit 49   42.701    0.8714     1.01     0.565
  Pure Error   5    4.301    0.8602
Total         55   74.365


Model Summary

       S    R-sq   R-sq(adj)   R-sq(pred)
0.932951   36.80%     35.63%       31.27%


Coefficients

Term          Coef   SE Coef   T-Value   P-Value   VIF
Constant     4.619     0.209     22.15     0.000
bahigher15  -0.03299   0.00588   -5.61     0.000   1.00


Regression Equation

log_narc15 = 4.619 - 0.03299 bahigher15
```

Figure 13.28  Minitab regression output.

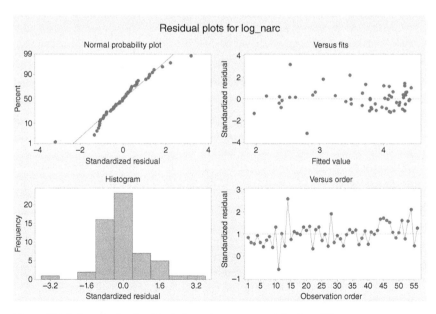

Figure 13.29  Standardized residual plots when response is *log*(*narc*15).

Figure 13.30 To predict *log(narc15)* in Minitab go to `Stat > Regression > Regression > Y Predict` .... Then use the settings shown in the left panel.

and we can be 95% confident that the predicted *narc15* is between

$$(e^{2.35}, e^{6.14}) = (10.48, 464.05)$$

This is a wide interval, which in part is due to the model explaining only a moderate amount of variation of the response variable.

## Chapter Problems

13.33 Using Eq. (13.1), show that Eq. (13.2) holds.
*Hint:* See Section 6.3.

13.34 Using Eq. (13.1), show that $Var(Y) = \sigma^2$.

13.35 According to simple linear regression notation, explain what is the difference between $Y$ and $\hat{Y}$.

13.36 State what are the four main assumptions required for inference during simple linear regression analysis.

13.37 Technically, in the simple linear regression method, $Y|X = x \sim N(\beta_o + \beta_1 x, \sigma)$. If $\beta_o = 11.45$ and $\beta_1 = -3.21$, what would be the expected value of $Y$ given $X = 1.79$?

13.38 A multinational corporation completed an analysis to model weekly sales, in dollars, using average production work hours per week. To present their work to pundits in Germany, they must convert the currency to the euro: €1 equaled $1.20 at the time of the analysis.
   a) Explain the impact, if any, the change of currency will have on $b_1$.
   b) Explain the impact, if any, the change of currency will have on $r$.
   c) Why can we say that the change of currency will not affect the results of a $t$-test.

13.39 Returning to Problem 4.39, suppose that a simple linear regression model is fit to evaluate actual score using guessed score. Below is a screen grab of the Minitab output.

**Regression Analysis: Actual Score_ versus Guessed Score_**

Analysis of Variance

| Source | DF | Adj SS | Adj MS | F-Value | P-Value |
|---|---|---|---|---|---|
| Regression | 1 | 6086.4 | 6086.35 | 72.60 | 0.000 |
| Guessed Score_ | 1 | 6086.4 | 6086.35 | 72.60 | 0.000 |
| Error | 23 | 1928.3 | 83.84 | | |
| Lack-of-Fit | 18 | 1151.0 | 63.95 | 0.41 | 0.925 |
| Pure Error | 5 | 777.3 | 155.45 | | |
| Total | 24 | 8014.6 | | | |

Model Summary

| S | R-sq | R-sq(adj) | R-sq(pred) |
|---|---|---|---|
| 9.15634 | 75.94% | 74.89% | 71.31% |

Coefficients

| Term | Coef | SE Coef | T-Value | P-Value | VIF |
|---|---|---|---|---|---|
| Constant | 18.63 | 6.51 | 2.86 | 0.009 | |
| Guessed Score_ | 0.7629 | 0.0895 | 8.52 | 0.000 | 1.00 |

Regression Equation

ctual Score_ = 18.63 + 0.7629 Guessed Score_

Fits and Diagnostics for Unusual Observations

| Obs | Actual Score_ | Fit | Resid | Std Resid | |
|---|---|---|---|---|---|
| 6 | 25.00 | 33.89 | -8.89 | -1.14 | X |

X Unusual X

a) What are the null and alternative hypotheses?
b) At 5% significance, is the null rejected?
c) Write your regression equation.
d) Use the regression equation to estimate the actual exam score when a student's guessed score was 85.
e) Interpret the coefficient of determination.
f) Find the coefficient of correlation. Interpret.

13.40 The normal probability plot from the previous problem is shown below. Interpret.

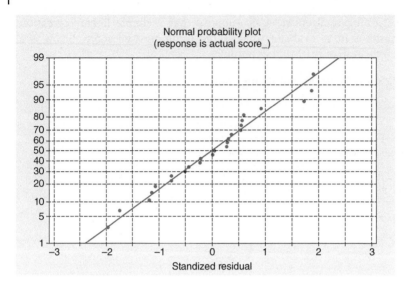

Normal probability plot
(response is actual score_)

13.41 A teacher wants to check if there was a linear association between the final grade and the number of absences during the semester. Treating his class of 27 students as a random sample, Minitab returned the following output:

**Regression Analysis: grade versus absences**

```
Analysis of Variance

Source          DF   Adj ss   Adj MS   F-Value   P-Value
Regression       1    60.98    60.98      1.53     0.227
   absences       1    60.98    90.98      1.53     0.227
Error           25   995.92    39.84
   Lack-of-Fit    2   143.24    71.62      1.93     0.168
   Pure Error    23   852.67    37.07
Total           26  1056.90


Model Summary
       S    R-sq   R-sq(adj)  R-sq(pred)
 6.31164   5.77%      2.00%       0.00%


Coefficients

Term       Coef   SE Coef   T-Value   P-Value   VIF
Constant   75.50    4.07      18.57     0.000
absences   -1.83    1.48      -1.24     0.227   1.00


Regression Equation

grade = 75.50 - 1.83 absences
```

a) What are the null and alternative hypotheses?
b) At 1% significance, is the null rejected?
c) Interpret the coefficient of determination.
d) Find the coefficient of correlation. Interpret.

13.42 The Canadian Dairy Information Centre compiles information on the cheeses made from cow, goat, sheep, and buffalo milk.[8] Simple linear regression was used to determine if cheese fat percentage is associated to moisture percentage. Output of the analysis is seen below:

**Regression Analysis: FatContentPercent versus MoisturePercent**

```
Method

Rows unsed 424


Analysis of Variance

Source              DF   Adj SS   Adj MS   F-Value   P-Value
Regression           1    24319  24319.0    910.93     0.000
  MoisturePercent    1    24319  24319.0    910.93     0.000
Error             1014    27071     26.7
  Lack-of-Fit       65     5106     78.6      3.39     0.000
  Pure Error       949    21964     23.1
Total             1015    51390


Model Summary

      S    R-sq   R-sq(adj)   R-sq(pred)
5.16690  47.32%      47.27%       47.05%


Coefficients

Term               Coef   SE Coef   T-Value   P-Value    VIF
Constant         50.618     0.811     62.42     0.000
MoisturePercent -0.5091    0.0169    -30.18     0.000   1.00


Regression Equation

FatContentPercent = 50.618 - 0.5091 MoisturePercent
```

a) At 1% significance, is the null rejected?
b) Interpret the slope.
c) Interpret the coefficient of determination.
d) Write your regression equation.
e) Use the model to estimate fat percentage, when moisture percentage is 47.
f) If moisture percentage is statistical significant, does it mean that it causes an impact on fat percentage?

---

8 Source: open.canada.ca (accessed July 26, 2019).

**13.43** Some years ago, *NBC News* published an article titled "Praise the Lard? Religion Linked to Obesity in Young Adults." It revealed how a study found that young adults that attended religious activities were 50% more likely to become obese by middle age than those who did not attend religious studies.[9] Given that these findings are scientifically valid, does this mean that religion or going to church causes obesity among young adults? Explain.

**13.44** Based on 2014 housing market data from 37 ZIP Codes in Austin, Texas, a model for median rent was fit using average monthly transportation. The output below shows the model.

```
Regression Equation

Median rent = 711 + 0.463 Average monthly transportation
```

Explain why the model should not be used to determine the median rent when the average monthly transportation cost is zero.

**13.45** Returning to Problem 13.25, a researcher wants to use the regression model to estimate the mean starting post-college salary for those who played video games for 30 hours. Why should the researcher be careful with such a prediction?
*Hint:* Perform summaries of the variables in "vgames_salary."

**13.46** In Problem 13.26, it is found that the suicide rate among Asian men in New York City is higher than among Asian women. Discuss the practical significance of the results.

**13.47** Using the file "VitalSignsBaltimore15wdictionarytoprows" that includes an affordability index for mortgages, affordm15, and for rent, affordr15, for 55 communities in Baltimore,
a) reproduce the inferential results presented in Problem 13.12.
b) reproduce the standardized residual plots presented in Problem 13.29.
*Hint:* The last row indicates overall measurements for the city and therefore should not be considered for computations.

**13.48** Download the file "Census_Data_Selected_socioeconomic_indicators_in_Chicago_2008_2012" with data[10] from 77 communities in Chicago. Using statistical software, we will assess if percent of Housing Crowded (PCNT_H_C) helps predict per capita income (INC). Statistical

---

9  Source: www.nbcnews.com (accessed July 26, 2019).
10  Source: data.cityofchicago.org(accessed July 26, 2019).

inference will be conducted at 1% significance. Note that last row in the data set is for all of Chicago and thus must be deleted for the analysis.
a) Construct a scatterplot. Interpret.
b) Perform inference on significance of the model. Interpret.
c) Interpret $R^2$.
d) Find the correlation coefficient and interpret.
e) Construct histogram of significant model residuals. Discuss what it checks and how it looks.
f) Construct a plot of per capita income versus its model-predicted value. Discuss what it checks and how it looks.

13.49 Download "Chicago_Public_Schools_-_Progress_Report_Cards__2011-2012_," Chicago open data.[11] We will assess the relationship between rate of misconduct per 100 students (response) and teachers' score. Inference will be conducted at 5% significance.
a) Construct a scatterplot. Interpret.
*Hint:* Data clean up is required.
b) Perform inference on significance of the model. Interpret.
c) Interpret $R^2$.
d) Find the correlation coefficient and interpret.
e) Predict misconduct per 100 students if teachers' score is 75. How reliable do you think this prediction is?
f) Construct histogram of the residuals from the significant model. Discuss what it checks and how it looks.
g) Construct a plot of rate of misconduct per 100 students versus its model-predicted value. Discuss what it checks and how it looks.

13.50 Using the data from Problem 13.49
a) Get a 95% confidence and prediction interval for (mean) misconduct per 100 students if teachers' score is 43.
b) Get a 95% confidence and prediction interval for (mean) misconduct per 100 students if teachers' score is 1.25 and explain why the results for this predictor value should be used with caution.

13.51 Download the file "Sustainability__2010-2013_Baltimore", which has Baltimore data.[12] Is dirtyst12 (rate of dirty streets and alleys Reports per 1000 residents) a good predictor of voted12 (% of population over 18 who voted)?
a) Construct a scatterplot. Interpret.
b) Perform inference at 5% significance.
c) Check residual plots to assess assumptions required for inference. Interpret.

---

11  Source: data.cityofchicago.org (accessed July 26, 2019).
12  Source: data.baltimorecity.gov (accessed July 26, 2019).

**13.52** In Problem 13.23, it was found that dirtyst10 is linearly associated to drains10. Moreover, in Problem 13.31, it was seen that the constant variance assumption appears to be violated. Download the data[13] file "Sanitation_2010Baltimore." Construct a scatterplot with 95% intervals in Minitab by going to Stat > Regression > Fitted Line Plot .... Then in the window that pops up, press the "Options" and check "Display confidence interval" and "Display prediction interval." What is the impact of the violation of the constant variance assumption?

**13.53** Recall that under the right circumstances, transforming one or both of the regression variables will alleviate assumption violations. Using data from Problem 13.49, build a regression model to see if Teaching Score can model $log(rateofmisconductper100students + 1)$. Also check the residuals. Does this model look better?

**13.54** Download the "VitalSignsBaltimore15SUBSETforSLRdatascience" data and reproduce the $log(narc15)$ model found in Section 13.8.1.
*Hint:* Review the data wrangling steps taken in Section 13.8.1 to reproduce the results and to guide you on the rest of this problem.
a) Interpret $R^2$ for the $log(narc15)$ model.
b) Using the $log(narc15)$ model, predict $narc15$ when $bahigher = 28.42$.
c) Based on the $log(narc15)$ model, what is the 95% prediction interval for $narc15$ when $bahigher = 28.42$.
d) Based on the $log(narc15)$ model, what is the 95% confidence interval for $narc15$ when $bahigher = 28.42$.
e) Explain why the interval found in 13.54c is wider than the one found in 13.54d.
f) Explain why one should be extra careful in using the $log(narc15)$ model to predict $narc15$ when $bahigher = 97.75$.

## Further Reading

Ruppert, David. (2004). *Statistics and Finance.* Springer.
Suarez, Erick, Perez, Cynthia M., Rivera, Roberto, and Martinez, Melissa N. (2017). *Applications of Regression Models in Epidemiology.* Wiley.

---

13 Source: data.baltimorecity.gov (accessed July 26, 2019).

# 14

## Multiple Linear Regression

### 14.1 Introduction

Simple linear regression is used when our aim is to determine the association between a response variable and only one predictor. Of course, there are situations when we observe several predictors along with the response variable.

---

**What If Important Predictors Are Not Accounted For?**
Leaving important predictors out of the model leads to a biased slope estimate in the simple linear model. Also, by including important predictors, better estimates of $E(Y)$ are obtained from the model.

---

In some instances, all predictors are of interest. In other instances, only some are of real interest, but we must account for other variables. Multiple linear regression is the extremely useful procedure performed when more than one predictor is used in the linear model. It may seem that this is a trivial extension to simple linear regression, but there are complications that arise. Although statistical software will perform multiple linear regression with a click of a button, there are many ways to conduct this type of analysis and one must know how to select the apt methods. We will cover the fundamental aspects of multiple linear regression, but because of these complications, we highly recommend that this type of analysis be done by a professional statistician.

---

**Goal of the Analysis**
Generally, regression methods have two main objectives:

- explaining a variable or
- predicting a variable.

---

*Principles of Managerial Statistics and Data Science*, First Edition. Roberto Rivera.
© 2020 John Wiley & Sons, Inc. Published 2020 by John Wiley & Sons, Inc.
Companion website: www.wiley.com/go/principlesmanagerialstatisticsdatascience

For the first objective, the idea is to use measurements from several variables to determine if they can explain the response variable, and if so, which predictors are useful. For the second objective, the idea is to use measurements from several variables to determine if they can predict the response variable. The adequate principles of properly constructing a regression model will depend on the goal of the analysis.

## 14.2  The Multiple Linear Regression Model

We now have $k$ predictors, $X_1, X_2, \ldots, X_k$, that we suspect might be linearly associated with the response variable $Y$. The true linear relationship between $Y$ and all predictors is written as

$$Y = \beta_o + \beta_1 X_1 + \beta_2 X_2 + \cdots + \beta_k X_k + \epsilon$$

As before, $\epsilon$ is an unobservable random error term and it accounts for

- not knowing the true relationship between the variables,
- disregarding other important variables, or
- measurement error.

$\beta_o, \beta_1, \ldots, \beta_k$ are the main parameters of interest. Each $\beta_i$, where $i = 1, \ldots, k$, corresponds to the change in the expected value of $Y$ for a change of one unit in $X_i$ when holding all other predictors fixed.

**Example 14.1**  *A company that performs local deliveries in Austin, Texas, wants to estimate gas consumption of its trucks, which leave the warehouse on a full tank and only refill when they return. It is believed the gas consumption (in gallons) is related to the number of deliveries and the distance traveled (in miles).*

Instead of using $X$ and $Y$ notation, the true linear relationship can be expressed as

$$gas = \beta_o + \beta_1 delivery + \beta_2 miles + \epsilon$$

Since all $\beta_i$ are unknown, we must estimate them. Observations are available for the response variable and all predictors.[1]

A random sample of 96 trips provided data to determine (through least squares estimation) the regression equation, Figure 14.1.

---

1 Be aware that equivalently to this equation, one may define $X_1$ as a variable of "1"s and hence omit $\beta_o$ in the equation above. Also, software for statistical analysis often automatically includes an intercept by default.

| ↓ | C1 | C2 | C3 |
|---|---|---|---|
| | gas | del | mil |
| 1 | 14.9559 | 11 | 105.057 |
| 2 | 22.1234 | 21 | 111.054 |
| 3 | 18.9058 | 21 | 91.786 |
| 4 | 16.8227 | 14 | 88.955 |
| 5 | 18.7403 | 15 | 108.479 |
| 6 | 21.6785 | 16 | 124.118 |
| 7 | 17.8929 | 19 | 96.171 |
| 8 | 18.3342 | 22 | 104.648 |
| 9 | 16.7319 | 21 | 94.561 |
| 10 | 15.9529 | 15 | 108.997 |

Figure 14.1 First 10 gas consumption observations (gas) along with number of deliveries (del), and distance traveled in miles (mil).

***Fitting the Multiple Linear Regression Parameters*** Since now we have several predictors, we will need multiple scatterplots to explore the association between $y$ and each $x_i$. What's more, it is important to consider the possibility of association between the predictors.

Before we get into all that, analogous to simple linear regression, the regression equation is

$$\hat{y} = b_o + b_1 x_1 + \cdots + b_k x_k$$

where $b_i, i = 0, \ldots, k$ are estimates of $\beta_i$ respectively.

---

**Where Are the Formulas for $b_i$?**
Formulas to compute $b_o, b_1, \ldots, b_k$, by minimizing the least squares, require linear algebra and are not presented here. Instead, we emphasize the use of computers to obtain these estimates and how the computer output should be interpreted.

In multiple linear regression, we must interpret each coefficient $b_i$.

---

**Interpretation of $b_i$**

$b_i$ is the change in expected value of the response variable for any one unit change in $x_i$ when all other predictors are held fixed.

---

Returning to Example 14.1, the estimated regression equation is

$$\widehat{gas} = 0.978 + 0.350 delivery + 0.117 miles \tag{14.1}$$

Thus, 0.350 is the estimated expected increase in gas consumption associated to an increase of one delivery, while miles are kept fixed. Meanwhile, 0.117 is the estimated expected increase in gas consumption associated to an increase of one mile while deliveries are kept fixed. Importance of predictors should not be judged by comparing the magnitude of coefficient estimates. For example, that $0.350 > 0.117$ does not mean that *delivery* is more important in predicting *gas* than *miles* because the coefficient estimates are dependent of the scale of predictor measurements. Now, if 15 deliveries are performed while driving 100 miles,

$$\widehat{gas} = 0.978 + 0.350(15) + 0.117(100)$$
$$= 17.93$$

So the model predicts that 17.93 gallons of gasoline will be consumed by the delivery truck.

---

**Goodness of Fit**

In multiple linear regression, the performance of the model (i.e. goodness of fit) can also be assessed with the coefficient of determination, $R^2$. Similar to Section 13.4 (box $r^2$ under section "Goodness of Fit") $R^2$ is interpreted as the percent of variability in $Y$ explained by the model. Notice the slight change in interpretation, since we now have many predictors, it is the percent of variability explained by multiple predictors, not just one.

---

**Example 14.2** *Later on we will see that for Example 14.1, $R^2 = 0.3285$. That is, the model explains 32.85% of the variability in gas consumption.*

Not all of the predictors in a model are guaranteed to be useful, and given this interpretation of $R^2$, it is tempting to use it to determine which predictors should be kept in the model. We will see later on that $R^2$ is not a good statistic to choose predictors.

## Practice Problems

14.1 In an effort to determine if advertisement expenditures lead to greater revenue from online sales of surfboards, a firm examined 32 weeks of data. The regression model was $Y$ = revenue (in thousand of dollars), $X_1$ = total expenditures in web advertising (in thousand of dollars), $X_2$ = total expenditures in print advertising (in thousand of dollars), $X_3$ = average surfboard volume for the week (liters). The table below presents the estimated coefficients.

| Predictor | Coefficient |
|---|---|
| Intercept | 6.79 |
| $X_1$ | 3.97 |
| $X_2$ | 1.48 |
| $X_3$ | −2.85 |

a) Write the fitted regression model.
b) Interpret the web ad expenditures coefficient.
c) Interpret the surfboard volume coefficient.

14.2 At the admissions office of a university, they have been using past data to study how high school GPA (*HSgpa*) and combined math and critical reading SAT scores (*SAT*) are associated to college freshmen GPA (*Fgpa*). Their fit is

$$\widehat{Fgpa} = -1.29 + 0.79HSgpa + 0.001\ 03SAT$$

a) Interpret the *HSgpa* coefficient.
b) Interpret the *SAT* coefficient.

14.3 Download the file "gasConsumption_deliverytrucks."
a) Use statistical software to get the regression equation (14.1).
b) Construct a scatterplot of gas consumption and weight of the cargo, *cargoweight*. Interpret.
c) Also get the regression equation for gas consumption when *cargoweight* is included in the model with the predictors, *delivery* and *miles*. Comment on its estimated coefficient.
d) Interpret the coefficient of determination of the model that adds *cargoweight*.
e) Predict gas consumption when *delivery* = 15, *miles* = 15, *cargoweight* = 200.

**14.4** In Example 14.1, the estimated intercept was 45.63. Why do you think this estimate is so far above zero?

**14.5** In Problem 14.2, the coefficient of determination was 0.158.
a) Interpret the coefficient of determination.
b) Universities tend to have admission indexes based on high school GPA and combined math and critical reading SAT scores. Based on the interpretation in 14.5, how reliable are indexes in predicting freshmen GPA?

**14.6** In Problem 14.1, $R^2 = 0.434$. Interpret this statistic.

**14.7** In Problem 14.2, Does the magnitude of the coefficients in the equation mean that *HSgpa* is more important than *SAT*?

**14.8** In simple linear regression, the squared root of the coefficient of determination is known as the coefficient of correlation ($r$) measuring linear association between $Y$ and $X$. But in multiple linear regression, there are several predictors, and the squared root of the coefficient of determination, called the multiple correlation ($R$) that measures the correlation between $Y$ and $\hat{Y}$. The higher $R$ is, the better $\hat{Y}$ predicts $Y$. Find the multiple correlation for Example 14.2.

**14.9** In simple linear regression, $r$ falls between $-1$ and $1$. From the definition of $R$ Problem 14.8, what values do you think $R$ falls within?

## 14.3 Inference for Multiple Linear Regression

In the Section 14.2, it was argued that coefficient estimation for multiple linear regression models is very similar to simple linear regression models. But how useful is the model and which predictors are truly statistically significant? To answer these questions, we can assess the regression model by considering the following six criteria:

- **Objective** for obtaining regression model. Generally, we state that a model has two objectives: we either want to predict $Y$ or we want to explain $Y$.
- **Logic** in choice of coefficients. Does background knowledge about the problem tells us something about the type of association $Y$ should have with the predictors?
- **Fit** of model. How close is the model output to the response variable observations?
- **Stability** of coefficients. Do coefficient estimates vary considerably if another predictor is present/absent in the model?

- **Simplicity** of model (Parsimony). Model should only include predictors that are useful to predict or explain $Y$ and not more. When important predictors are left out, the model provides biased estimates of the model coefficients. In contrast, when too many unimportant predictors are included, the model suffers from high variability.
- **Cost** of incorporating predictors. In the case of prediction, how costly is it to measure the predictor? It might not be worth to include a predictor when it is too costly to measure and it appears to be of little importance.

Least square estimators in multiple linear regression are unbiased estimators, just as in simple linear regression. Moreover, among unbiased estimators, they have the smallest variance and are consistent. Therefore, we should be confident inferring about multiple linear regression coefficients through least square estimators when the regression assumptions hold. Multiple linear regression assumptions are similar to those for simple linear regression but slightly modified:

1) Linear model between response variable and all predictors.
2) Model errors are independent.
3) Variance of errors are constant for all values of each predictor.
4) Model errors are normally distributed.

Just as with simple linear regression, we can conduct hypothesis testing for inference. But there are some differences with simple linear regression and some issues with hypothesis testing we must address. Two tests are commonly used for inference in multiple linear regression.

***An Overall F-Test*** Analogous to ANOVA, for multiple regression, the $F$-test checks if the model is worthwhile looking into more carefully. If the null is rejected, at least one covariate is statistically significant. The ANOVA null and alternative have to be slightly tweaked,

$$H_o: \beta_1 = \beta_2 = \cdots = \beta_k = 0$$
$$H_1: \text{at least one coefficient is not zero}$$

Although the alternative implies that at least one predictor is statistically significant, it does not tell us which ones. Often, at least some of the covariates are reasonably expected to be associated to the response variable and therefore, the null is rejected.

***A t-Test*** If we do not reject the null hypothesis of the $F$-test, no further tests are needed. However, if we reject the null hypothesis of the $F$-test, then further analysis may be needed to check which of the predictors are significant.

One alternative is to individually conduct hypothesis testing in each coefficient, through a $t$-test. For $i = 1, \ldots, k$, we have

$$H_o: \beta_i = 0$$
$$H_1: \beta_i \neq 0$$

The $t$-test uses a standardized statistic based on the difference between each estimated coefficient $b_i$ and its value under the null hypothesis (zero), divided by the standard deviation of the estimator.[2] Under the multiple linear regression assumptions, the test statistic will follow a $t$ distribution when the null hypothesis is true. A $p$-value can be calculated from this statistic. If the $p$-value $\leq \alpha$, the null hypothesis is rejected. We may also construct confidence intervals for each $\beta_i$, using a $t$ distribution.

---

**Overfitting**

It is important to keep only the predictors that we need in the model. When a model has too many predictors, it will actually start to fit the error component in the regression equation, something called **overfitting**, something that can lead to misleading conclusions and in fact hinder the ability of the model to make predictions.

---

The overall $F$-test and $t$-test are no longer equivalent as in simple linear regression. If we are not sure about the relationship between $Y$ and all predictors, we should conduct an overall $F$-test first. This comes from the same arguments for the ANOVA procedure:

- We first want to see if we have evidence that at least one predictor is significant.
- If so, we then proceed to see which predictors are significant, which is what the $t$-test does.

**Example 14.3** *Using 2008–2012 data[3] from 77 communities in Chicago, we will assess the relationship between per capita income and three predictors summarized in Table 14.1. All tests will be conducted at 5% significance level.*

First, we would expect negative associations, if any, between the per capita income and the predictors. That is, the higher the percent of residents of a community without a high school diploma, the lower, on average, the per capita income, and similarly with the other variables. Table 14.1 also includes some

---

2 Since we will not perform these computations by hand, this equation is not included here.
3 Data source: data.cityofchicago.org (accessed July 26, 2019).

Table 14.1 Summary of variables for Example 14.3.

| Abbreviation | Description | Minimum | Mean | Maximum |
|---|---|---|---|---|
| INC | Per capita income | 8 201 | 25 563 | 88 669 |
| PCNT_H_C | % of housing crowded (more than one person per room) | 0.30 | 4.92 | 15.80 |
| PCNT_B_POV | % below poverty | 3.30 | 21.77 | 56.50 |
| NO_HS | % with no high school diploma | 2.50 | 20.34 | 54.80 |

summaries that allow us to quickly scan for any obvious errors[4] in the data.[5] All predictors are percentages, and thus they must be between 0 and 100. Let us visually assess the association between the response variable and the predictors (Figure 14.2). Associations do indeed appear to be negative, and association is

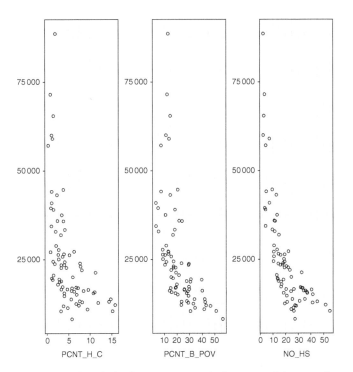

Figure 14.2 Association between per capita income and three predictors.

---

4 Note that last row in the data set is for all of Chicago and thus must be deleted for our analysis.
5 Socio demographic knowledge from Chicago could also be taken into consideration. This may indicate that some communities have percentages that are too low or too high. However, this type of assessment was not performed here.

strongest between per capita income and percent of people with no high school diploma. Also, it appears associations are nonlinear.

The null and alternative hypothesis are

$$H_o: \beta_1 = \beta_2 = \beta_3 = 0$$

$H_1$: at least one coefficient is not zero

and according to the ANOVA table (Table 14.2), at least one of the predictors is statistically significant. To find out which predictors are statistically significant, $t$-tests are performed (Table 14.3). It is seen that PCNT_H_C is not statistically significant ($p$-value > 0.05). The insignificant predictor is removed and the model is refit with the remaining predictors. Why refit the model? Table 14.4 shows that the coefficient estimates have changed with this new model. We will explain why this happens later in the chapter. The two remaining predictors are significant. Thus, our simplest model is:

$$\widehat{INC} = 49\,977 - 429.80 PCNT\_B\_POV - 740.30 NO\_HS$$

Table 14.2 ANOVA summary for Example 14.3.

| Source | DF | SS | MS | F | p-Value |
|---|---|---|---|---|---|
| Regression | 3 | 10 793 566 014 | 3 597 855 338 | 37.62 | 0.000 |
| Error | 73 | 6 981 226 917 | 95 633 245 | | |
| Total | 76 | 17 774 792 931 | | | |

Table 14.3 *t*-Test results for Example 14.3.

| Predictor | Coef | SE Coef | T | P |
|---|---|---|---|---|
| Constant | 50 382.50 | 2 659.90 | 18.94 | 0.000 |
| PCNT_H_C | 1141.90 | 636.00 | 1.80 | 0.077 |
| PCNT_B_POV | −406.60 | 108.20 | −3.76 | 0.000 |
| NO_HS | −1061.40 | 207.30 | −5.12 | 0.000 |

Table 14.4 *t*-Test results when PCNT_H_C is removed.

| Predictor | Coef | SE Coef | T | P |
|---|---|---|---|---|
| Constant | 49 977.00 | 2 689.80 | 18.58 | 0.000 |
| PCNT_B_POV | −429.80 | 109.00 | −3.94 | 0.000 |
| NO_HS | −740.30 | 106.30 | −6.96 | 0.000 |

Computer output indicates that the coefficient of determination of the simplest model is 59%. That is, the model explains 59% of the variability in per capita income. Also, according to the model, the predicted income per capita when the percent below poverty is 27.8% and percent aged 25 without high school diploma is 14.5% is $27 294.21.

---

**Multiple Regression Inference Tips**

- Statistical software will often give you analysis results with asterisks or notations stating that some predictors are significant at, say, 0.01 significance, while others are significant at 0.05, 0.1, and so forth. This is done in the statistical software for "convenience." However, the analysts must choose one (and only one) significance level before conducting the analysis.
- Technically, estimation is reliable as long as $n > k$, where $n$ is the number of observations and $k$ is the number of predictors. But, careful with using too many predictors in a multiple linear regression model. When $n/k$ is small, the power of the tests become low and $r^2$ becomes untrustworthy. Roughly, the data set should have at least 10 observations per predictor in the model to conduct inference. For example, to use five predictors, there should be at least 50 observations available. The higher $n/k$, the better.

---

## Practice Problems

14.10  Refer to the ANOVA table below.

| Source | DF | SS | MS | F | p-Value |
|---|---|---|---|---|---|
| Regression | 4 | 113.12 | 28.28 | 1.63 | 0.188 |
| Error | 36 | 624.57 | 17.35 | | |
| Total | 40 | 737.69 | | | |

a) How many predictors were tested?
b) Write the null and alternative hypothesis.
c) Is the null rejected at 5% significance?

14.11  Refer to the ANOVA table below.

| Source | DF | SS | MS | F | p-Value |
|---|---|---|---|---|---|
| Regression | 3 | 341 642 | 113 881 | 241 732.92 | 0.000 |
| Error | 46 | 22 | 0.48 | | |
| Total | 49 | 341 664 | | | |

a) How many predictors were tested?
b) Write the null and alternative hypothesis.
c) Is the null rejected at 5% significance?

14.12 At the admissions office of a university, past data has been used to study how high school GPA (*HSgpa*) and combined math and critical reading SAT scores (*SAT*) are associated to college freshmen GPA (*Fgpa*). Refer to the ANOVA table below.

| Source | DF | SS | MS | F | p-Value |
|---|---|---|---|---|---|
| Regression | 2 | 161.16 | 80.58 | 171.96 | 0.000 |
| Error | 1833 | 858.94 | 0.47 | | |
| Total | 1835 | 1020.09 | | | |

a) Write the null and alternative hypothesis.
b) Is the null rejected at 5% significance?

14.13 Download the file "gasConsumption_deliverytrucks."
a) Perform an *F*-test at 5% significance on a regression model for gas consumption when *cargoweight*, *delivery*, and *miles* are the predictors.
b) Do *t*-tests indicate that any of the predictors should be removed from the model at 5% significance?

14.14 In an effort to determine if advertisement expenditures lead to greater revenue from online sales of surfboards, a firm considers 32 weeks of data. The regression model was $Y$ = revenue (in thousand of dollars), $X_1$ = total expenditures in web advertising (in thousand of dollars), $X_2$ = total expenditures in print advertising (in thousand of dollars), and $X_3$ = average surfboard volume for the week (liters). The ANOVA table is presented below.

| Source | df | SS | MS | F | p-Value |
|---|---|---|---|---|---|
| Regression | 3 | 167.05 | 55.68 | 39.92 | 0 |
| Error | 28 | 39.06 | 1.40 | | |
| Total | 31 | 206.11 | | | |

a) Write the null and alternative for the *F*-test.
b) Is the null rejected at $\alpha = 0.10$?

14.15 Continuing the Problem 14.4, the table below presents the estimated coefficients and *t*-test *p*-values. Which coefficients differ significantly from zero at 10% significance?

| Predictor | Coefficient | p-Value |
|-----------|-------------|---------|
| Intercept | 6.79 | 0 |
| $X_1$ | 3.97 | 0.005 |
| $X_2$ | 1.48 | 0.001 |
| $X_3$ | −2.85 | 0.73 |

14.16 Continuing Problem 14.14, if $R^2 = 0.81$, interpret the meaning of this.

14.17 Using the simplest model found in Example 14.3, predict per capita income when
a) the percent below poverty is 11.6%, and percent aged 25 without high school diploma is 19.3%. Interpret.
b) the percent below poverty is 20.5%, and percent aged 25 without high school diploma is 41.6%. Interpret.
c) the percent below poverty is 30.7%, and percent aged 25 without high school diploma is 54.8%. Interpret.

14.18 Universities use indexes to measure the eligibility of admission candidates. These indexes vary, but they are generally a function of SAT scores[6] and GPA. For University Free for All (UFfA), the admission index (*ai*) is obtained through

$$ai = 50gpa + \frac{0.5}{3}(tsat - 400)$$

where *tsat* is the sum of SAT Critical Reading and Math Scores. Download the file "Admissions13" which has 2013 data for 1850 admissions to UFfA. *dai* is the recorded admission index.
a) Construct a simple linear regression equation to see if *gpa* explains *dai* at 5% significance.
b) Construct a 95% confidence interval for the slope of *gpa*.

14.19 In Problem 14.18, the confidence interval suggests a slope for *gpa* that is much larger than 50, the expected value according to the *ai* equation in Problem 14.18. One explanation for this is that by leaving out *tsat* while constructing the simple linear equation, an important predictor, the estimate of the *gpa* coefficient in the simple regression model is biased.

---

6 They are also versions that are a function of ACT scores but this is omitted here for simplicity.

a) Show that by constructing a regression model for *dai* using *gpa* and *tsat*, the estimates are close to the values from the *ai* equation in Problem 14.18.

b) When associations are exactly linear, estimated coefficients have standard errors of zero. In 14.19a, the standard errors of the coefficients are small but not small enough. Find out why.
*Hint:* Use the *ai* equation in Problem 14.18 to get *ai* for all *gpa* and *tsat*. Then draw a scatterplot of *ai* versus *dai*.

14.20 Using 30 observations and 15 predictors, a firm has fit a multiple regression model for the salary of managers. Explain why inference is not reliable in this situation.

14.21 In a move to make more data-driven decisions, the CEO of a technology company aims to build a regression model for customer experience using 25 predictors. Why should the company use a data set of at least 250 observations?

## 14.4 Multicollinearity and Other Modeling Aspects

So far, we have not addressed the possibility of dependence among predictors. It is quite common to have some form of dependence among predictors and this dependence brings some challenges.

---

**Dependence Among Predictors**
**Multicollinearity** occurs when two or more predictors are highly correlated. With multicollinearity,

- it becomes difficult to identify the individual contribution of predictors.
- coefficient estimates become unstable and dependent on whether other predictors are present or absent from the model.
- the $F$-test could be significant and all $t$-tests nonsignificant.

When data is not obtained through a designed experiment, it is not likely that all predictors are completely uncorrelated. However, the concern is high correlation, moderate to low correlation has less of an impact on the results.

---

Multicollinearity can be detected in several ways. Scatterplots are screened not only for association of the predictors to the response, but for association among predictors as well. A matrix of scatterplots, simultaneous scatterplots to explore association among all variables, is known as a **scatterplot matrix**. Sample correlations between all variables are also useful.

Figure 14.3 shows the scatterplot matrix for all variables of Example 14.3. The first column presents the association between the response variable and all predictors. The other columns express the association among predictors. Does it appear that there is an association among predictors?

Table 14.5 summarizes sample correlations among all variables. The predictors percent with no high school diploma and percent of housing crowded are highly correlated with $r = 0.88$.

As a sensitivity test of multicollinearity, one can drop a predictor from the regression model to see what happens to the fitted coefficients in the reduced model. If the coefficient estimates or their significance does not change much, then multicollinearity is not a concern. If dropping the predictor causes sharp changes in at least one of the remaining estimated coefficients in the model, then multicollinearity may be causing instability. From Tables 14.3 and 14.4,

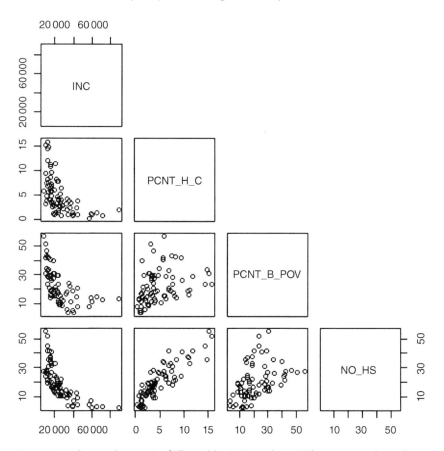

Figure 14.3 Scatterplot matrix of all variables in Example 14.3. There is some dependence among predictors.

Table 14.5 Sample correlations for Example 14.3.

| | INC | PCNT_H_C | PCNT_B_POV |
|---|---|---|---|
| PCNT_H_C | −0.54 | | |
| PCNT_B_POV | −0.57 | 0.32 | |
| NO_HS | −0.71 | 0.88 | 0.42 |

it can be seen that the coefficient estimate for NO_HS changed by about 25% when PCNT_H_C was removed. Since high multicollinearity causes instability in the coefficients, their estimates will have "inflated variances." A statistic known as the **variance inflation factor** can be computed.

---

**Variance Inflation Factor (VIF)**

Although sample correlations and scatterplots are obtained easily, they only present dependence between pairs of predictors. The VIF for the $i$th predictor is

$$VIF_i = \frac{1}{1 - R_i^2}$$

where $R_i^2$ is the coefficient of determination when predictor $i$ is fit against all other predictors (not $Y$). The higher $R_i^2$, the higher $VIF_i$. A rule of thumb is that $VIF_i \geq 10$ is sign of concern. However, a fairly large value, say $VIF_i = 10$, does not automatically mean that there is instability in the least squares coefficient estimates.

---

*Hypothesis Testing in Practice*  Practitioners need to be careful when applying hypothesis testing for multiple linear regression. There are three main issues.

- $t$-Tests are affected by multicollinearity. The stronger the relationship between predictors, the bigger the impact in $t$-tests. Only when all predictors are independent, the $t$-tests for individual predictors are unaffected.
- The more hypothesis tests we do, the more likely we will incur in a type I error. In an age of Big Data, erroneously rejecting a null hypothesis is something that can easily happen, leading to spurious relationships. We saw examples of spurious relationships in Section 1.1.
- $F$-Tests for model selection as discussed here require that models be nested. Two models are nested when one has a subset of variables from the other model. Hence, $Y = \beta_o + \beta_1 X_1 + \epsilon$ is nested with $Y = \beta_o + \beta_1 X_1 + \beta_2 X_2 + \epsilon$ but $Y = \beta_o + \beta_1 X_1 + \epsilon$ is NOT nested with $Y = \beta_o + \beta_1 X_3 + \beta_2 X_4 + \epsilon$.

Statisticians have been warning practitioners for years about relying too much on hypothesis testing when fitting multiple linear regression models.

***Optional: Other F-Tests*** If the null is rejected during a overall $F$-test, a **partial F -test** can be used to determine a subset of predictors that are important. This type of test compares a model with $p$ predictors with a model that uses a subset of those predictors. This is done by further partitioning $SSR$. The partitioning can be performed in different ways. For example, sometimes we already have evidence that some predictors are significant and wish to study if other predictors might help explain the variability of $Y$. In this case, we can compare the original simpler model with the more complicated model, and one way of doing this is through partial $F$-tests. A rejection of the null hypothesis during a partial $F$-test would imply that at least one of the new variables being added is significant. Once again, this test will not tell us which of the new variables added are significant that requires further analysis. The exception is if we are adding or removing only one predictor. Some semiautomatic algorithms implement these partial $F$-tests in different ways. For example, one version adds variables one by one until no more predictors are significant. Another version does the opposite. There is also an algorithm that adds and removes predictors along the way.

***Optional: Other Model Selection Criteria*** Since hypothesis testing is not ideal to choose between regression models, we are left in search of alternative methods to choose between predictors. Several possibilities exist, $R^2$, adjusted $R^2$, Akaike Information Criterion (AIC), Bayesian Information Criterion (BIC) and predicted residual sum of squares to mention a few.

$R^2$: Since the coefficient of determination is interpreted as the amount of variability in the response explained by the model, the more variability in $Y$ explained the better the model. $R^2$ can also be used in multiple linear regression. However, it can be shown that increasing the number of predictors in the model will never reduce the $R^2$ of the model no matter how useless the added predictors are. Slight increases in $R^2$ may seem to be adequate, although in reality the model is starting to fit the error component of the model.

**Adjusted $R^2$ ($R_a^2$):** The problem that arises with increasing the number of predictors in the regression model never reducing the $R^2$ of the model can be attributed to $R^2$ not accounting for the number of predictors in the model. This can be remedied through what is known as the adjusted $R^2$, which penalizes a model for the inclusion of useless predictors. $R_a^2$ is always smaller than or equal to $R^2$. In theory, a large discrepancy between $R_a^2$ and $R^2$ would indicate that a model with less predictors will provide just as good as a fit. However, there is no general rule to compare $R_a^2$ and $R^2$. Also one must be aware that $R_a^2$ cannot be interpreted the same way[7] as $R^2$.

---

7 In fact, for a model with predictors that poorly predict $Y$, $R_a^2$ could potentially be negative while $0 \leq R^2 \leq 1$. When $R_a^2 < 0$, many software programs will simply provide an output of zero for $R_a^2$.

$C_p$: This statistic combines two terms: one a measure of goodness of fit, and another that penalizes for the number of parameters. When a model has a lot of bias, $C_p$ will be much greater than the number of parameters; a sign that the model does not have enough predictors.

**Akaike Information Criterion**: Hypothesis testing selects the best model based on how close the model output is to the data on the response variable. We can also choose models based on which does best in estimation of[8] $E(Y)$. $AIC$ examines the performance of the model in estimation of $E(Y)$ while simultaneously penalizing the model if it has too many unnecessary predictors. The strategy is to calculate the $AIC$ for competing models and choose the one with the smallest value.

**Bayesian Information Criterion** (**BIC**): $BIC$ works under the same premise of $AIC$, but has a stronger penalty for having too many unnecessary predictors.

Benefits of $AIC$ and $BIC$ over hypothesis testing is that the order of imputation of predictors into the model does not matter (this impacts hypothesis testing because of multicollinearity). Also, models do not need to be nested. Unfortunately, no model selection criteria is universally better than the other. For example, $AIC$ has been known to choose a model with too many predictors as best. Also, not all models are linear on the predictors. Some models are based on artificial neural networks, splines, and other statistical tools. These type of models are known as nonparametric regression models. See Section 14.9 for a brief introduction to nonparametric regression models. The performance of nonparametric models cannot be compared with the performance of multiple linear regression models using criteria such as $AIC$ and $BIC$. Analysts will often implement a combination of model selection criteria depending on the situation, to determine what the best model is.

Many other regression methods are also available. For example, the **least absolute selection and shrinkage operator (LASSO) method** performs least squares estimation, penalizing the sum of the absolute value of the coefficients for the predictors in the model. It has become popular when there are many predictors available. We do not attempt to cover all model selection methods.

## Practice Problems

14.22 An organization wishes to build a model that estimates a ZIP Codes median home value based on median household income, unemployment, and median rent. Using the "2014_Housing_Market_Analysis_Data_by_Zip_Code" data from 37 ZIP Codes in Austin, Texas,

---

8 Technically, it is a mean conditional on specific predictors. But for simplicity, we skip this notation.

a) construct a scatterplot matrix of median rent, unemployment, median household income, and median home value.
b) do median rent and unemployment seem to be strongly correlated?
c) do median rent and median household income seem to be strongly correlated?
d) do unemployment and median household income seem to be strongly correlated?

14.23 The table that follows presents $R^2$ and $R_a^2$ (percent) values for models with between five and eight predictors.

| Predictors | $R^2$ | $R_a^2$ |
|---|---|---|
| 5 | 63.39 | 62.48 |
| 6 | 65.97 | 64.71 |
| 7 | 66.18 | 63.91 |
| 8 | 66.82 | 63.06 |

a) What are the respective $R^2$ and $R_a^2$ values for the model with 8 predictors?
b) What are the respective $R^2$ and $R_a^2$ values for the model with 5 predictors?
c) Based on the table, which model appears best? Explain.

14.24 Download the file "gasConsumption_deliverytrucks."
a) Construct a scatterplot matrix of *gas*, *cargoweight*, *delivery*, and *miles* and recall that the last three are the predictors. Do the predictors appear to be correlated?
b) Construct a regression model for gas consumption when *delivery* and *miles* are the predictors. Then build another regression model, but with *cargoweight*. How does $R_a^2$ compare with $R^2$?

14.25 Download the file "Sustainability__2010-2013_Baltimore"; which has Baltimore data.[9] We will model clogged13 (2013 rate of clogged storm drain reports per 1000 residents) considering clogged12, clogged11, dirtyst13 (rate of dirty streets and alleys reports per 1000 residents), and carpool13 (2013 percent of population that carpool to work) as predictors.
a) Construct a scatterplot matrix of all the variables. Do the predictors appear to be correlated? Are the findings expected?

---

9 Source: data.baltimorecity.gov (accessed July 26, 2019).

b) According to $t$-tests, are any predictors statistically insignificant? Use $\alpha = 10\%$.

c) Construct a regression model for clogged13 using clogged12 and dirtyst13. Then build another regression model, but adding *carpool*13. Compare $R^2$ and $R_a^2$ from both models.

## 14.5 Variability Around the Regression Line: Residuals and Intervals

The regression line allows a point estimate, $\hat{y}$, of the response variable. The variability around $\hat{y}$ is also important to

- help assess assumptions for inference,
- detect troublesome observations, and
- build confidence or prediction intervals.

---

**Residual Analysis**

Model adequacy and unusual observations are checked by analyzing the residuals. There is little difference from how residual diagnostics are performed in simple linear regression.

- To check the constant variance assumption for all $X$, either the residuals versus the fitted value plot is drawn, or the residuals versus each predictor.
- The linearity assumption is evaluated with scatterplots (for each $x$) and the residual versus fitted value plot.
- Independence could be evaluated by plotting the residuals against the order of observations.
- Normality is assessed with the normal plot of the residuals.

Standardized residuals are one way to pinpoint unusual observations (see Section 13.7), which are evaluated for their possibility of being influential or leverage points through plots and more advanced statistics. Considering that more than one predictor may be in the model, these advanced statistics become more essential.

---

Sometimes, residuals versus a predictor, $x_{k^*}$, not yet included in the model are plotted to explore whether $x_{k^*}$ will improve the model.

Figure 14.4 presents some plots for standardized residuals for the regression model started in Example 14.3. It reaffirms the interpretation of the scatter in Figure 14.2: the linear assumption may be inadequate. The normal assumption does not appear to hold either. Perhaps a better model could be fit? See Problem 14.52.

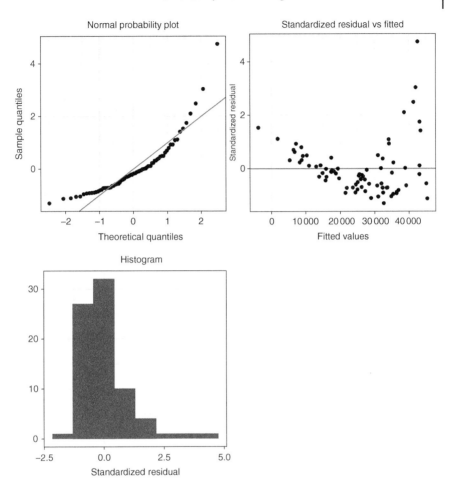

Figure 14.4 Residual plots from the regression model in Example 14.3.

*Intervals for Y and E(Y)*   To estimate[10] $E(Y)$, or to predict a value of the response, we substitute values of the predictors into the final regression model. Prediction intervals are used to infer about individual values of $Y$, while confidence intervals are used to infer about where its mean $E(Y)$ falls. See Section 13.5 for more.

Returning to Example 14.1, recall that if 15 deliveries are performed while driving 100 miles, the model predicts $\widehat{gas} = 17.93$ gallons of gasoline will be consumed by the delivery truck. It can be shown, Figure 14.5, that with 95%

_____

10  Technically, its a mean conditional on specific values of the predictors. But for simplicity, we skip this notation.

**Prediction for gas**

```
Regression Equation

gas = 45.63 + 0.708 delivery + 0.1053 miles

Variable setting
delivery      15
miles        100

     Fit     SE Fit          95% CI                95%  PI
  66.7836   0.557515   (65.6765, 67.8907)    (57.5604,  76.0068)
```

Figure 14.5 Minitab 95% confidence interval and 95% prediction intervals for given predictor values.

confidence, the mean response is between (17.53, 18.26) for the provided predictor values, and that for a randomly chosen trip, gas consumption will fall within (14.85, 20.94).

## Practice Problems

14.26 In Problem 14.25, clogged13 (2013 rate of clogged storm drain reports per 1000 residents) was modeled by some predictors.
   a) Check the normal probability plot of the standardized residuals of the regression model for clogged13 using clogged12 and dirtyst13. Interpret.
   b) Check the standardized residuals versus fitted values plot of the regression model for clogged13 using clogged12 and dirtyst13. Interpret.

14.27 In Problem 14.26, build 95% confidence and prediction intervals when
   a) clogged12 = 5.5 and dirtyst13 = 45. Interpret.
      *Hint:* See Figure 13.13.
   b) clogged12 = 25 and dirtyst13 = 72. Interpret.

## 14.6 Modifying Predictors

When some multiple linear regression assumptions are violated, or sometimes to ease interpretation, predictors may be modified.

*To Ease Interpretation* The interpretation of the regression coefficients is sensitive to the scale of measurement of the predictors, and no coefficient is more affected by this than the intercept. This is why up to this point, no attempt has been made to infer on the intercept. Recall that, mathematically, the estimated intercept is the value of the fitted $Y$ when all predictors are zero. But in many situations, predictors cannot be equal to zero. For example, if $X$ is the height of a product box, $X$ cannot possibly be zero. One strategy often used to be able to draw inference on the intercept is to take each predictor and subtract their mean. This is called **centering**. This way, the centered representation of the predictors can now potentially be zero (when the predictor has a value equal to its mean). When all predictors are centered, then the intercept is interpretable: it is the expected value of $Y$ when all predictors are equal to their means. Another advantage of centering the predictors is that it reduces the impact of multicollinearity and also helps in the interpretation of interactions. Other similar techniques are scaling and standardization. With **scaling**, a predictor is multiplied by some constant. Scaling is done to have more appealing coefficient values (e.g. to avoid coefficient estimates that are too large or too small). **Standardization** consists of centering followed by dividing by the standard deviation of the respective predictor. This technique makes the regression coefficients more comparable.

## Practice Problems

14.28 The table below is analogous to Table 14.4, but the predictors have been centered: $X_1$ = centered PCNT_B_POV, $X_1$ = centered NO_HS.

| Predictor | Coef | SE Coef | T | P |
|-----------|------|---------|---|---|
| Constant | 25 563.20 | 1 131.10 | 22.60 | 0.000 |
| $X_1$ | −429.80 | 109.00 | −3.94 | 0.000 |
| $X_2$ | −740.30 | 106.30 | −6.96 | 0.000 |

   a) Interpret the intercept.
   b) How have the coefficient estimates of the predictors changed?

14.29 With the data set "Census_Data_Selected_socioeconomic_indicators_ in_Chicago_2008_2012," use statistical software to show that the coefficients, their standard errors, and $t$-tests are the ones presented in the table from Problem 14.28. Note that the last row in the data set is for all of Chicago and thus must be deleted for your analysis.

14.30 Download the file "gasConsumption_deliverytrucks."
   a) Create the new predictor $m10 = miles/10$.
   b) Fit a regression model to predict gas consumption using *delivery*, $m10$, and *cargoweight*.
   c) Compare the $m10$ coefficient with the *miles* estimate shown in Eq. (14.1).

14.31 Problem 14.12 showed that high school GPA (*HSgpa*), and combined math and critical reading SAT scores (*SAT*) are associated to college freshmen GPA (*Fgpa*). If an intercept interpretation is desired, the predictors can be standardized.[11] This way, the intercept is interpreted as the value of the response when the predictors are equal to their mean value. Download the file "Admissions13cleanFreshmenGPA" and use the standardized predictors: *HSgpa_s*, *SAT_s* to model *Fgpa*. Interpret the estimated intercept.

14.32 Since the Problem 14.31 has standardized the predictors, we can compare coefficient estimates.
  a) Which predictor appears to be more important in predicting *Fgpa*?
  b) Can we say that the predictor with the highest coefficient value is more likely to have a stronger causal link with the response?

## 14.7   General Linear Model

Multiple linear regression can be extended to cover the following scenarios:

- Categorical predictors or combination of categorical and quantitative predictors.
- Functions of the predictors and/or the response variable, a task known as variable transformation.
- More than one response variable.

The regression model that allows for these possible extensions is known as the **general linear model**. Small adjustments to the multiple linear regression procedure or to the interpretation of the results are required for the general linear model. We discuss very briefly the general linear model when using categorical predictors and with transformed variables.

---
**Categorical Predictors**
With the exception of Section 13.6, all models covered thus far involve quantitative predictors. But there are many situations in which the predictor is qualitative or categorical. When a qualitative variable has $c$ categories, the solution is to define $c - 1$ dummy binary variables.

---

**Example 14.4**   *In Example 14.1, the company owns four different trucks. To incorporate truck as a predictor, we define*

---

11 The predictor mean is subtracted to each value, and this difference is divided by the predictors standard devation.

$x_1 = 1$ *for trips using truck no. 1;* $x_1 = 0$ *for trips using other trucks,*
$x_2 = 1$ *for trips using truck no. 2;* $x_2 = 0$ *for trips using other trucks,*
$x_3 = 1$ *for trips using truck no. 3;* $x_3 = 0$ *for trips using other trucks,*

*Observe that if* $x_1 = x_2 = x_3 = 0$, *truck no. 4 was used.*

If $c$ dummy variables were to be included, then statistical software would give an error since it would not be able to perform necessary computations.

In practice, we can state that a variable is categorical and statistical software will define the dummy binary variables and choose the reference level. One-factor ANOVA can be thought of as building a model to estimate the expected value of the response variable based on the levels of the factor. Thus, we may decompose the mean response per factor level into an overall mean and an additional term,

$$E(Y_i) = \mu_i$$
$$= \mu + \tau_i$$

If the levels of the factor affect the expected value of $Y$, then $\tau_i \neq 0$, and essentially, some factor levels will add to the overall mean ($\tau_i > 0$) and others will subtract to it ($\tau_i < 0$).

If we fit a model with only one categorical variable, our multiple linear regression model hypothesis testing is equivalent to a one-way ANOVA procedure. If the null hypothesis is rejected, then at least one of the categories is significant. In this context, the interpretation is that the categorical variable is important and we leave all categories in the model (it does not make sense to test for significance for each category separately).

***Combining Quantitative and Qualitative Predictors*** Suppose there are two qualitative predictors. Then our regression model can be represented as a two-way ANOVA problem. Similarly, more than two qualitative variables may be analyzed by representing the regression model as a higher-order ANOVA problem. Taking into account the requirement of defining dummy binary variables for the analysis, at this point it is worth remembering the need of using a dictionary of variables that includes description of the variables, definition of coding used for qualitative predictors, units of measurement for quantitative predictors, etc.

It is also possible to have quantitative predictors and categorical predictors. When we are interested in the effect of a categorical variable and control for quantitative predictors, a statistical method sometimes referred to as **Analysis of Covariance (ANCOVA)** is employed, a type of general linear model.

---

**When One Predictor Is Quantitative and the Other Is Binary**

If $X_1$ is binary and $X_2$ numerical, when $X_1 = 1$, we have

$$E(Y|X_1 = 1, X_2 = x_2) = \beta_o + \beta_1 + \beta_2 x_2$$

In contrast if $X_1 = 0$, then

$$E(Y|X_1 = 0, X_2 = x_2) = \beta_o + \beta_2 x_2$$

Thus, when statistically significant, the binary predictor will shift the intercept of regression line upward or downward (depending on sign of $\beta_1$).

---

**Example 14.5** *Returning to Example 14.4, if a regression model for gas is fitted using the predictors miles, delivery, and the dummy variables for trucks, the fit is*

$$\widehat{gas} = 2.51 + 0.347delivery + 0.111miles - 1.558x_1 - 1.949x_2 - 0.061x_3$$

*Each truck shifts the intercept of the model. For example, for truck #no.1m the model is*

$$\widehat{gas} = 2.51 + 0.347delivery + 0.111miles - 1.558(1) - 1.949(0) - 0.061(0)$$
$$= 0.952 + 0.347delivery + 0.111miles$$

*while for truck #no.2 we have*

$$\widehat{gas} = 2.51 + 0.347delivery + 0.111miles - 1.558(0) - 1.949(1) - 0.061(0)$$
$$= 0.561 + 0.347delivery + 0.111miles$$

***Functions of Predictors and Response Variables*** Picture a situation where from a scatterplot, you determine that the association between the response and a predictor is not linear. You determine that a candidate model is

$$Y = \beta_o + \beta_1 x + \beta_2 x^2 + \epsilon$$

How do you fit such a model? Turns out that this type of function of $x$ is still linear in the parameters, $\beta_i$. Other examples are

- $log(Y) = \beta_o + \beta_1 x_1 + \beta_2 x_2 + \epsilon$
- $log(Y) = \beta_o + \beta_1 log(x_1) + \beta_2 x_2 + \epsilon$
- $\sqrt{Y} = \beta_o + \beta_1 x_1 + \beta_2 x_2 + \epsilon$

Specifically, the response variable or the predictors can be transformed and these transformations will not affect the least squares estimation procedure, nor the inferential process. In contrast, the model $Y = \beta_o + e^{\beta_1 x_1} + \beta_2 x_2 + \epsilon$ is not linear in $\beta_1$ and requires inferential procedures beyond this text.

**Example 14.6** *Using 2016–2017 Chicago high school data, we want to determine if college enrollment rate can be predicted by school average ACT score, school graduation rate, and whether the school provides bilingual services at $\alpha = 0.05$.*

From Figure 14.6, it appears there is some association between the predictors and the response variable, but there is some association among the two predictors as well.

*F*-test indicates the model is useful (Table 14.6) while *t*-tests indicate that bilingual services is not significant (Table 14.7).

Figure 14.7 displays the diagnostic plots for the model. The pattern in the standardized residual versus fitted plot is indicative of at least one nonlinear association.

Following the revelation from the scatterplot matrix and residual plots, a polynomial model is fit.

$$Y = \beta_o + \beta_1 x_1 + \beta_2 x_1^2 + \beta_3 x_2 + \beta_4 x_3 + \epsilon$$

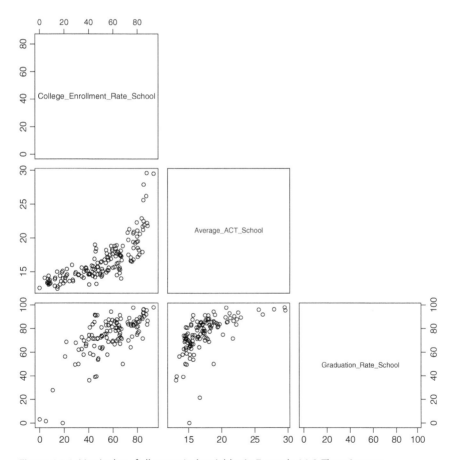

Figure 14.6 Matrixplot of all numerical variables in Example 14.6. There is some dependence among predictors.

Table 14.6 ANOVA summary.

| Source | DF | SS | MS | F | p-Value |
|---|---|---|---|---|---|
| Regression | 3 | 20 041.9 | 6 680.23 | 58.65 | 0.000 |
| Error | 110 | 12 530.6 | 113.9 | | |
| Total | 113 | 32 572.5 | | | |

Table 14.7 *t*-Test results.

| Predictor | Coef | SE Coef | T | P |
|---|---|---|---|---|
| Constant | −15.72 | 6.49 | −2.42 | 0.017 |
| ACT score | 3.48 | 0.42 | 8.22 | 0.00 |
| Graduation rate | 0.22 | 0.10 | 2.14 | 0.035 |
| Bilingual services (yes) | −3.42 | 2.03 | −1.68 | 0.095 |

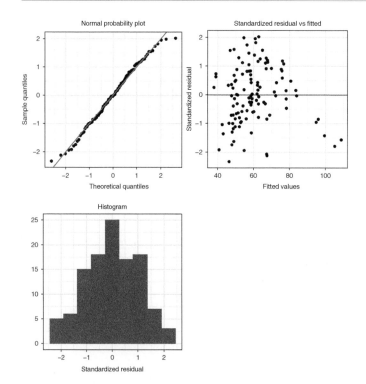

Figure 14.7 Residual plots from regression model to fit college enrollment rate using school average ACT score, school graduation rate, and whether the school provides bilingual services.

where $x_1$ = ACT score, $x_2$ = graduation rate, and $x_3$ = bilingual service. Although school graduation rate has some value in predicting college enrollment rate, the inclusion of school average ACT score in the model and the association between these two predictors lead to school graduation rate being statistically insignificant. The most parsimonious model is

$$\hat{Y} = -128.68 + 16.28x_1 - 0.30x_1^2$$

This model has $R^2 = 0.67$. Also, this model has succeeded in resolving the linearity violation seen in the original model (Figure 14.8).

Another possibility is to transform the response or predictor variables. This is a convenient alternative when the model assumptions of constant

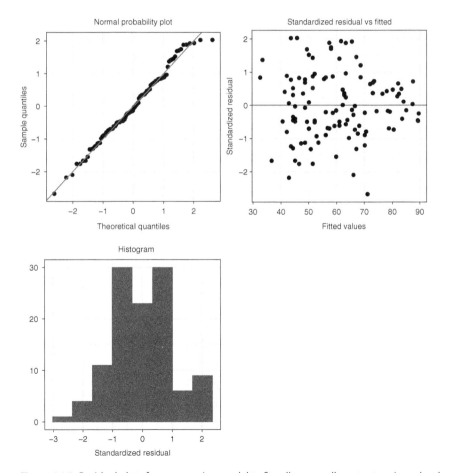

Figure 14.8 Residual plots from regression model to fit college enrollment rate using school average ACT score, and its square.

variance or normality are not valid. Transforming variables can make these assumptions more plausible. Variable transformation is also an option when we have knowledge of a nonlinear association, but wish to express the association linearly in the parameters, so a regression model can be fit. However, care is needed to interpret the regression results in terms of the observation scale. See Section 13.8.1 for an application of a logarithmic transformation and interpretation of results.

In multiple regression modeling, the influence of one predictor on the expected value of the response is constant for all values of other predictors. This condition can be made more flexible by including an interaction term, which implies that the effect of one predictor is dependent on the value of some other predictor. The concept of an interaction was already introduced in the context of two-way ANOVA, but it can be used more generally, even for quantitative variables. Incorporating a possible interaction term in a regression model is rather straightforward, a cross product of the two predictors in question is introduced, and inference is performed to assess if the slope of this cross product is significantly different than zero.

## Practice Problems

14.33 A model to predict prison sentence (in years) for men included the predictors: type of crime (criminal damage, burglary, robbery, drugs, public order, sexual, motoring, possession of weapon, and theft), number of prior felony convictions, whether convict has children ($1 =$ has children), and age.

a) Identify the categorical variables.
b) How many binary variables are required in the model to represent type of crime?

14.34 In an attempt to improve their current text translator, a company will assess three statistical approaches in translating three languages into English: Spanish, French, and Mandarin. Let $Y =$ english translation score $(0-100)$;
$X_1 = 1$ Spanish text; $X_1 = 0$ otherwise,
$X_2 = 1$ French text; $X_2 = 0$ otherwise,
The fitted model was $\hat{Y} = 55.73 + 25.31X_1 + 21.92X_2$.

a) Interpret the estimated effect of translating from Spanish to English.
b) Interpret the estimated effect of translating from French to English.
c) Interpret the estimated effect of translating from Mandarin to English.

14.35 The Canadian Dairy Information Centre compiles information on the cheeses made from cow, goat, sheep, and buffalo milk.[12] Multiple linear regression will be used to determine if cheese fat percentage is associated to moisture percentage, cheese type (Firm, Fresh, Hard, Semisoft, Soft, and Veined), organic status (Yes, No), and type of manufacturing (Artisan, Farmstead, Industrial).

a) Identify the numerical predictors.

b) How many (dummy) binary variables are needed in the model to account for organic status?

c) How many (dummy) binary variables are needed in the model to account for cheese type status?

d) How many (dummy) binary variables are needed in the model to account for type of manufacturing?

14.36 In Example 14.5, write down the regression equation in terms of *miles* and *delivery* for

a) truck no.3

b) truck no.4

14.37 In Example 14.5, predict gas consumption with the model when $delivery = 21, miles = 122$, and $truck = 3$.

14.38 Download "Minneapolis_Property_Sales_20102013," a data set[13] of property sales in Minneapolis. Using statistical software, extract data for transactions such that "*PropType = Residential*." For Gross Sale Price (response), and Adjusted Sale Price (predictor), remove any values $\leq 10\ 000$. Use $\alpha = 0.05$

a) Perform an $F$-test to determine if a regression model for Gross Sale Price using Adjusted Sale Price and InRTOStudy – yes or no depending on whether the sale was used for the Minnesota Department of Revenue Sales Ratio Study, is statistically significant (note that ward number is categorical).

b) Interpret the coefficient for InRTOStudy.

14.39 A teacher wants to check if there was a linear association between the final course grade (from 0 to 100), the number of absences, and gender (sex = 0 for female, sex = 1 for male). Treating his class of 100 students as a random sample, Minitab returned the following output:

## Regression Analysis: grade versus absences, sex

Analysis of Variance

| Source | DF | Adj SS | Adj MS | F-Value | P-Value |
|--------|----|--------|--------|---------|---------|
| Regression | 2 | 1596.2 | 798.12 | 15.31 | 0.000 |
| absences | 1 | 166.7 | 166.68 | 3.20 | 0.077 |
| sex | 1 | 1464.6 | 1464.63 | 28.09 | 0.000 |
| Error | 97 | 5057.0 | 52.13 | | |
| Lack-of-Fit | 5 | 186.0 | 37.20 | 0.70 | 0.623 |
| Pure Error | 92 | 4871.0 | 52.95 | | |
| Total | 99 | 6653.2 | | | |

Model Summary

| S | R-sq | R-sq(adj) | R-sq(pred) |
|---|------|-----------|------------|
| 7.22039 | 23.99% | 22.42% | 19.26% |

Coefficients

| Term | Coef | SE Coef | T-Value | P-Value | VIF |
|------|------|---------|---------|---------|-----|
| Constant | 73.12 | 2.48 | 29.51 | 0.000 | |
| absences | -1.628 | 0.910 | -1.79 | 0.077 | 1.00 |
| sex | 7.66 | 1.45 | 5.30 | 0.000 | 1.00 |

Regression Equation

grade = 73.12 - 1.628 absences + 7.66 sex

a) What is the null and alternative hypothesis for an $F$-test?
b) At 1% significance, is the null rejected?
c) Interpret the coefficient of determination.
d) Find the multiple correlation. Interpret.
e) Write the parsimonious regression equation, removing any unimportant predictors.
f) Use the parsimonious regression equation to estimate the grade when a male misses three classes during the semester.

14.40 Returning to Problem 14.35, multiple linear regression was used to determine if cheese fat percentage is associated to moisture percentage, cheese type (Firm, Fresh, Hard, Semisoft, Soft, and Veined), organic status (Yes = 1, No = 0), and type of manufacturing (Artisan, Farmstead, Industrial). Categorical predictors required dummy variables: *Fresh* = 1 when the cheese type is Fresh, and *Fresh* = 0 when the cheese type is not Fresh. The final model was

$\widehat{fat} = 54.88 - 0.63Moisture + 2.96Fresh - 3.35Hard + 1.09SemiSoft + 4.59Soft + 2.53Veined - 1.06Organic$

a) Which predictors were left out of the model?
b) Interpret the impact on fat percentage of the cheese being Hard.

    c) There are six types of cheese in the data set: Firm, Fresh, Hard, Semisoft, Soft, and Veined. Interpret the impact on fat percentage of the cheese being Firm.

    d) Predict the fat percentage for industrial, soft, nonorganic cheese with moisture percentage of 46.

14.41 In Problem 14.40, no interaction between any of the predictors was considered. Now a multiple linear regression model is developed considering the possibility that effect of moisture percentage on fat percentage will depend on the cheese type. The final model was

$$\widehat{fat} = 65.96 - 0.90 Moisture - 21.35 Fresh - 28.10 Hard - 11.95 SemiSoft + 2.92 Soft - 18.93 Veined - 1.28 Organic + 0.49 Moisture \times Fresh + 0.67 Moisture \times Hard + 0.31 Moisture \times SemiSoft + 0.09 Moisture \times Soft + 0.51 Moisture \times Veined$$

    a) Interpret the interaction effect on fat percentage of the cheese being Hard.

    b) Predict the fat percentage for soft, nonorganic cheese with moisture percentage of 46.

    c) If the coefficient of determination was 57%, interpret its meaning.

## 14.8  Steps to Fit a Multiple Linear Regression Model

Multiple linear regression is an extremely useful tool used to make managerial decisions. With the availability of primary and secondary data, and software for statistical analysis these days, the capacity to perform linear regression analysis is at everyone's fingertips. What follows are summarized steps required for multiple linear regression analysis,

- Determine objectives and get data: see Chapter 2 for tips on proper data collection.
- Data preparation: before using the data through statistical software, we must ensure the level of measurement (nominal, ordinal, etc.) has been properly defined for each variable. Adjustments of unit of measurement are also sometimes necessary and often, data wrangling.
- Screen data.
- Exploratory analysis (scatterplots, tables, and boxplots). Data screening is still needed.
- Estimate parameters and draw inference.
- Choose predictors to keep.
- Residual analysis: check model assumptions and unusual observations. Refit new model if needed.
- Draw conclusions.

Although the tools are available, and some of the principles are easy to grasp, multiple linear regression is not always easy to do. Our objective was to introduce the procedure of regression analysis, so that the reader can interpret the results, understand the potential of the methods, and have a general idea of how it works. For the sake of simplicity, many details of the procedure were left out. If the procedure is implemented without a solid understanding of the steps, inadequate conclusions may be reached. No statistical software will be able to assure you that you have applied multiple linear regression correctly. It is often best to let a statistician or data analyst perform multiple linear regression analysis. Some tips when implementing multiple linear regression are as follows:

- Keep in mind what the goal of the analysis is. Suppose a linear regression model is fitted using 10 predictors. If the intention is prediction and an $F$-test indicates that at least one predictor is statistically significant but with an $R^2$ of just 10%, then you should question the need to continue with the analysis to find the best model. The coefficient of determination is already suggesting that the model explains little variability in the response. Unless some predictors are canceling out the effect of another predictor, or the model has been misspecified, or it is determined that 10% is beneficial, there is little evidence pointing to further analysis. In contrast, if the objective is to explain a response variable, then it is reasonable to extend the analysis, even with an $R^2$ of just 10%.
- Commonly, the $F$-test will reject the null hypothesis. $t$-Tests are a reliable continuation if the predictors are independent, or if you do not have many predictors in the model. Otherwise, other methods should be considered to proceed with the analysis.
- Remember that statistical significance does not imply practical importance.
- Also, if there is evidence that a response is linearly associated to predictors, it does not mean the association is causal.
- As indicated in Chapter 13, we should be careful in using the model for extrapolation, predicting the response for values of the predictor out of the observed range. The association between the variables may no longer be linear.
- If our model currently predicts the response variable well, it does not mean our model will predict the response well in the future, say five years from now. The model could turn slightly less effective in the future or completely useless. The performance of the model must be routinely assessed.
- Multiple linear regression analysis can become complicated fast. In situations when only a handful or predictors are available, or there is no association between the predictors, the analysis is straightforward. But on situations when there are several predictors, there is dependence among predictors, and the analysis is performed frequently, it is best to have an analyst conduct the analysis.

## 14.9 Other Regression Topics

Regression methods have a wide array of applications. For example, when the response variable is continuous, and the predictor is categorical, our choices are as follows:

- hypothesis testing or
- multiple linear regression.

Table 14.8 suggests statistical analysis methods based on the type of response and predictor variables.[14]

---

**Logistic Regression**

The main variable of interest is not always a continuous random variable. It may be a binary random variable, a count, and even a categorical variable. Different statistical models are needed for these scenarios. Linear regression methods construct a model that can be used to develop predictions through a linear function. At first, it is not clear how this can be replicated in the case of a binary variable. How could a linear function produce only 0 or 1s? The trick is that, instead of attempting to model the value of the binary response, $Y$, a function of $P(Y = 1)$ is modeled. This is known as a **logistic regression model**. This type of model requires a different type of procedure for estimation, and the interpretation of the results is a bit harder. But logistic regression modeling is wildly used in practice.

---

Table 14.8 Statistical analysis methods based on type of response and predictor variables.

| Type of response variable | Type of predictors | Analysis method |
|---|---|---|
| Continuous | Categorical | Hypothesis testing, multiple linear regression |
|  | Numerical or categorical | Multiple linear regression |
| Binary | Categorical | Contingency table, logistic regression |
|  | Numerical or categorical | Logistic regression |
| Categorical | Categorical | Contingency tables, generalized linear models |
|  | Numerical or categorical | Generalized linear models |
| Count | Numerical or categorical | Generalized linear models |

Note: multiple linear regression may refer to a general linear model.

---

14 As seen in Section 14.7, multiple linear regression goes under different names depending on the type of predictors: general linear model, analysis of covariance, or ANCOVA are some of the names used. We do not make this distinction to avoid confusion.

If the response variable is binary and the predictor is categorical, another alternative is a contingency table analysis, a topic that will be discussed in Chapter 15. More generally, to model categorical and count responses, analysis relies on the **generalized linear model**. This technique is based on probability distributions that fall within a family of distributions known as the **exponential family**. Probability distributions that fall within the exponential family include the following:

- Normal distribution
- Binomial distribution
- Poisson distribution

In fact, multiple regression and logistic regression are special cases of generalized linear models.

Other regression extensions are needed to incorporate nonlinear associations (see below) or dependent observations. Situations where there is only a response variable with temporal dependence among observations, called time series data, are briefly introduced in Chapter 17.

*Missing Data* An implicit assumption we have made in Chapters 13 and 14 is that all measures on the response and the predictors are available. Often, some measurements are missing. Reasons for missing values may be due to unanswered questions on a survey, issues with the measuring mechanism, or values that were clearly erroneous and had to be removed. The common solution to missing values is to remove rows of the data that is missing values for any of the variables. Other alternatives are possible and which is best will depend on the assumptions we make on why the data is missing.

*Beyond the Linear Regression Model* Quite often, the relationship between $Y$ and the predictors is not approximately linear[15] (Figure 14.9).

Let us revert back to having just one predictor. There are four ways to deal with situations of a nonlinear association between $Y$ and $X$.

- Include high-order terms of predictors: $X^2, X^3, \ldots$
- Choose a specific type of nonlinear function[16] when it is already known that associations will not be linear. Due to the theory of a problem, we may state that $Y = 100e^{-\beta_1 X - \beta_2 X^2/2} + \epsilon$. This type of model has been used in the past in financial engineering.
- Perform variable transformations as presented in Section 14.7.
- Use **nonparametric regression**, which does not make any assumptions about the type of association between $Y$ and $X$.

---

15 Or that the relationship between a function of $Y$ and functions of the predictors is not approximated linearly.

16 In some instances, we can transform the nonlinear model into a linear model. But this sometimes hinders interpretation or leads to inferential problems.

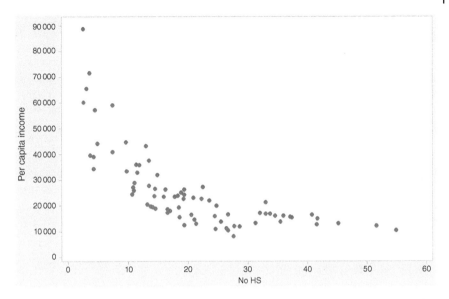

Figure 14.9 Minitab scatterplot of Chicago per capita income and percent of residents of a community without a high school diploma.

---

**Nonparametric Regression**

This regression method lets the data determine what type of nonlinear association there is between $Y$ and $X$. Nonparametric regression has become quite popular. There are many different ways to perform non-parametric regression, each with fancy names: artificial neural networks, splines, support vector machines, and wavelets are just a few types. The distinguishing feature of each of these nonparametric regression methods is the building blocks that they use to fit the model.

---

These days, there are many applications that although compatible with some of the principles of linear regression, require modeling strategies that are far more complex. Computational advances, unstructured data, and availability of massive data sets have led to new fields like big data, data analytics (and its subfield, **predictive analytics**), and data science. Some promising applications include gathering patterns from text data (e.g. how Google translator works), interpreting verbal commands (e.g. how the iPhone's Siri works), or identifying patterns from images (e.g. diagnosing skin cancer). Some have mistakenly thought that the availability of massive amounts of data have diminished the need of statistical literacy and informed skepticism, but as some applications using massive data have shown (e.g. Google Flu Trends), this is far from true.

### 14.9.1  A Quick Look at Data Science: Modeling Taxi Fares in Chicago

Open data portals provide a great opportunity to gather insight, which can serve to the benefit of the community. Data collected from over 113 million taxi rides in Chicago have been made available online.[17] The data includes date of ride, tip amount, fare paid, ride duration, ride distance, and several other variables (Table 14.9). Our aim is to attempt to build a rather elementary regression model to predict fares. Since tip amount is a function of the fare, it will not be used as a predictor.

This case study is considered a big data problem, primarily because most software and computers are unable to open the full data set due to its sheer size. There are multiple ways to address this challenge, but for our purposes, for the analysis we extract a subsample, 10 million rows of data. Substantial data wrangling is required to set up the data set to fit the model (Figure 14.10). For example, fares of $0 are removed, as well as uncommonly high fares (over $200).

Table 14.9  Data for two taxi trips in Chicago.

| Fare | Duration | Distance | Pickup.loc | Dropoff.loc | Tips | Pay.type | Day | Tod |
|------|----------|----------|------------|-------------|------|----------|-----|-----|
| 20.65 | 1440 | 8.7 | Near North Side | Lincoln Square | 0 | Cash | Sunday | (12,18] |
| 35.25 | 2580 | 16.7 | O'Hare | Near North Side | 0 | Credit Card | Wednesday | (12,18] |
| ⋮ | | | | | | | | |

Pay.type stands for payment type, Day for day of the week, and Tod for time of day.

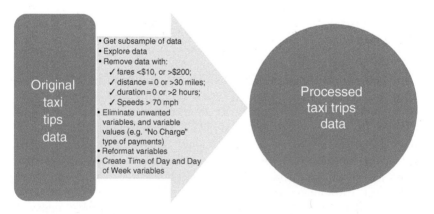

Figure 14.10  Steps taken to process Chicago taxi rides data.

---

17  Source: data.cityofchicago.org (accessed July 26, 2019).

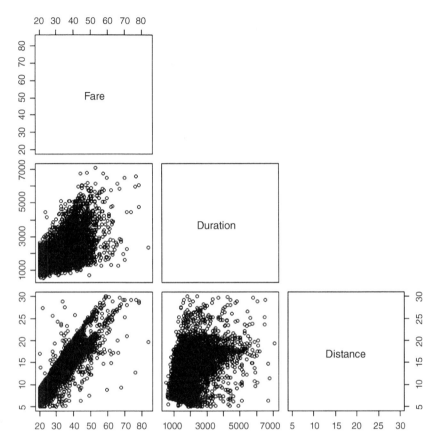

Figure 14.11 Scatterplot matrix of all numerical variables.

Using the ride's time stamp, a day of the week and time of day variables are created. The preprocessed data includes 813 847 taxi rides. Figure 14.11 displays the scatterplot matrix[18] of the numerical predictors and *Fare*. Trip distance is the numerical predictor that appears to have the strongest linear association with the response variable. The strongest correlation between numerical predictors is among *Distance* and *Duration*, 0.35. Table 14.10 presents summaries of taxi fares by time of day. Average fare is lowest in the early hours of the day, highest from 12p.m. until 6p.m.

Further exploration of the data reveals that effect of *Tod* on taxi fare may depend on day of the week. Intuitively, other interactions such as duration based on day of the week or time of day are also reasonable. Also, with Chicago

---

18 For the first 10 000 observations.

Table 14.10 Number of observations, sample average fare, and sample standard deviation per time of day.

| Time of day | Sample size | Mean fare | Std. dev. of fare |
|---|---|---|---|
| (0,6] | 95 368 | 31.70 | 7.87 |
| (6,12] | 217 068 | 34.65 | 7.98 |
| (12,18] | 294 421 | 35.23 | 8.16 |
| (18,24] | 206 990 | 33.17 | 8.12 |

having 77 community areas, over 150 model coefficients require estimation (including dummy variables). Therefore, statistical inference in this problem should be done with caution; with so many observations and predictors, statistical significance is likely to be found, although the association may not be practically important. One way to proceed is to perform lasso estimation first. Once some predictors are determined to be useful, $t$-tests are performed in this subset of predictors. This procedure leads to the model:

$$\widehat{Fare} = 5.42 + 1.59 Distance + 0.00\,256 Duration$$

with each predictor being statistically significant, and $R^2 = 0.8126$. That is, 81.26% of the variation in taxi fare is explained by distance and duration of the ride. Had $t$-tests been conducted without performing the lasso method first, time of day, day of week, pick up and drop-off location, payment type, and interaction effects would have been statistically significant, albeit these predictors would not have any practical effect on predictions.[19]

The standardized residuals do not present any major assumption violations (Figure 14.12), although some observations are outliers. One explanation for these outliers is that unavailable important predictors have been left out. For one, taxis tend to charge by the number of passengers. More careful data cleanup may also help.

The model presented here is for educational purposes. In no way it should be considered the best possible model. Data mining methods could very well result in a better predictive model. Many other interesting insights may be examined. For example, *Fares* between different community areas can be analyzed to determine the effect of taxi company.[20] Alternatively, new predictors can be created and included in the model (e.g. pace or speed).

---

19 *AIC* would have suggested adding time of day and day of week to the model above; but these predictors do not practically contribute to the predictive capabilities of the regression model.
20 This predictor required considerable cleanup. Therefore, it was not used in the analysis presented here.

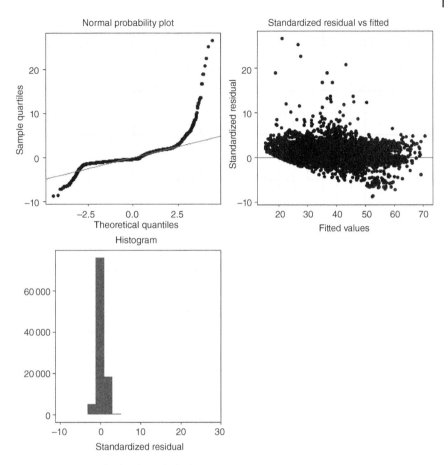

Figure 14.12 Standardized residual plots.

---

**Want to Know More?**
If you want to know more about this case study, in the companion website you will find the R code that reproduces the results presented here. This data set is rich, and you can modify the code to serve your own needs.

---

## Chapter Problems

14.42 State what are the four main assumptions required for inference during multiple linear regression analysis.

14.43 The following regression model is fit to data

$$\hat{Y} = 103.32 + 3.25X_1 - 153.63X_2 - 27.44X_3$$

Why can't we say that $X_2$ is definitely the predictor with the strongest linear association with $Y$?

14.44 Explain how multicollinearity can affect the construction of a regression model.

14.45 Download the file "Admissions13cleanFreshmenGPA" and reproduce the results from Problem 14.12.

14.46 In Problem 14.45, assess the standardized residual plots. Interpret.

14.47 Using Baltimore open data,[21] "Sustainability__2010-2012_," we will assess the relationship between rate of dirty streets (dirtyst12) and the following predictors: rate of dirty streets (dirtyst11), rate of clogged storm drains (clogged12), number of community-managed open spaces (cmos11), percent who carpool to work (carpool12), and number of miles of bike lanes (bkln12). All tests will be conducted at 5% level.

*Hint:* You must remove the last row, which stands for the entire city.

a) Which predictors do you think will be associated to rate of dirty streets? Of those you chose, will the association be positive or negative?

b) Construct a scatterplot matrix. Interpret.

c) Perform an $F$-test. Interpret.

d) Perform $t$-tests to choose which predictors should stay in the model. Write down the statistically significant regression model.

*Hint:* The final model is the one fit with only statistically significant predictors.

e) Interpret $R^2$ for your simplest model.

f) Predict rate of dirty streets with the most parsimonious model if dirtyst11 = 43.32, clogged12 = 5.70, cmos11 = 32, carpool12= 11.27, and bkln12 = 2.25.

g) Construct a histogram of standardized residuals of the most parsimonious model. Discuss what it checks and how it looks.

h) Construct a plot of standardized residuals versus its fitted values. Discuss what it checks and how it looks.

i) Construct a plot of rate of dirty streets versus its predicted value. Discuss what it checks and how it looks.

---

21 Source: data.baltimorecity.gov (accessed July 26, 2019).

**14.48** Download the file "gasConsumption_deliverytrucks."

   a) Perform an $F$-test for a regression model for gas consumption using all other variables in the file as predictors. Note: *truck* and *driver* are coded numerically but are factors.

   b) Perform $t$-tests to choose which predictors should stay in the model. Write down the statistically significant regression model.

   c) Interpret $R^2$.

   d) Predict gas consumption with the most parsimonious model if $delivery = 21, miles = 122, cargoweight = 206, truck = 2,$ and $driver = 1$.

   e) Construct a histogram of standardized residuals of the most parsimonious model. Discuss what it checks and how it looks.

   f) Construct a plot of standardized residuals versus its fitted values. Discuss what it checks and how it looks.

   g) Construct a plot of gas consumption versus its predicted value. Discuss what it checks and how it looks.

**14.49** Referring back to Example 14.3, truth is that more variables are available in the file "Census_Data_Selected_socioeconomic_indicators_in_Chicago_2008_2012." We will assess the relationship between per capita income and five predictors at $\alpha = 0.05$.

| Abbreviation | Description | Minimum | Mean | Maximum |
|---|---|---|---|---|
| INC | Per capita income | 8201 | 25 563 | 88 669 |
| PCNT_H_C | % of housing crowded | 0.30 | 4.92 | 15.80 |
| PCNT_B_POV | % Below poverty | 3.30 | 21.77 | 56.50 |
| UNEMP | % Unemployed | 4.70 | 15.37 | 35.90 |
| NO_HS | % No high school diploma | 2.50 | 20.34 | 54.80 |
| U18O64 | % Aged < 18 or > 64 | 13.50 | 35.75 | 51.50 |

Note that the last row in the data set is for all of Chicago and thus must be deleted for your analysis.

   a) Construct a scatterplot matrix. Interpret.

   b) Write the null and alternative for the $F$-test.

   c) Perform the $F$-test. Interpret.

   d) Perform $t$-tests to choose which covariates should stay in the model. Write down the statistically significant regression model. (*Hint:* If more than one predictor is not significant, remove the one with the highest $p$-value, then refit the model. Continue removing predictors this way until all remaining predictors are statistically significant.)

    e) Compare the best model found in 14.49 with the one from Example 14.3. Explain why the models are different.

    f) Interpret $R^2$.

**14.50** Returning to Problem 14.49,

    a) Predict per capita income when percent aged 25+ without a diploma is 18%, percent of households below poverty is 20%, and percent of housing crowded is 17%.

    b) Do you think it would be appropriate to predict per capita income when percent aged 25+ without a diploma is 85%, percent of households below poverty is 1%, and percent of housing crowded is 22%? (Hint: Refer to Table 14.1).

    c) Construct a histogram of significant model standardized residuals. Discuss what it checks and how it looks.

    d) Construct a plot of standardized residuals versus its fitted values. Discuss what it checks and how it looks.

**14.51** Download "Minneapolis_Property_Sales_20102013," a data set[22] of property sales in Minneapolis. Using statistical software, extract data for transactions such that *"PropType = Residential."* Also, for Gross Sale Price (response), and Adjusted Sale Price (predictor), remove any values $\leq 10\,000$. Use $\alpha = 0.05$.

    a) Perform an $F$-test to determine if a regression model for Gross Sale Price using Adjusted Sale Price and Ward number is statistically significant (note that ward number is categorical).

    b) Interpret $R^2$.

**14.52** Figure 14.4 showed that some regression assumptions were violated in Example 14.3. With the data set "Census_Data_Selected_socioeconomic _indicators_in_Chicago_2008_2012," use statistical software to (Note that the last row in the data set is for all of Chicago and thus must be deleted for your analysis),

    a) create a new response variable, the natural logarithm of per capita income.

    b) Construct a scatterplot matrix for the natural logarithm of per capita income and the three predictors from Example 14.3. Interpret.

    c) Write the null and alternative for the $F$-test on these three predictors.

    d) Perform the $F$-test at 5% significance. Interpret.

    e) Perform $t$-tests at 5% significance to choose which covariates should stay in the model. Write down the statistically significant regression model. (*Hint:* If more than one predictor is not significant, remove the

---

22 Source: opendata.minneapolismn.gov (accessed July 26, 2019).

one with the highest $p$-value, then refit the model. Continue removing predictors this way until all remaining predictors are statistically significant.)

f) Construct histogram of significant model standardized residuals. Discuss what it checks and how it looks.

g) Construct a plot of standardized residuals versus its fitted values. Discuss what it checks and how it looks.

14.53 Recall that in Section 14.9.1, constructing a model with the original predictors, and possible interaction effects, would require estimation of over 100 model coefficients.

a) How could this large number of coefficients cause problems while performing $t$-tests to choose a final model?

b) Including all these predictors would increase the coefficient of determination of the model. Explain why this does not imply all predictors should be used.

## Further Reading

Gelman, Andrew and Hill, Jennifer. (2006). *Data Analysis Using Regression and Multilevel/Hierarchical Models*. Cambridge.

Hastie, Trevor, Tibshirani, Robert, and Friedman, Jerome. (2013). *The Elements of Statistical Learning*, 2nd edition, Springer.

Rivera, R., Marazzi, M., and Torres, P. (2019) Incorporating open data into introductory courses in statistics. *Journal of Statistics Education*. DOI: 10.1080/10691898.2019.1669506

Suarez, Erick, Perez, Cynthia M. Rivera, Roberto, and Martinez, Melissa N. (2017). *Applications of Regression Models in Epidemiology*. Wiley.

Wickam, Hadley and Grolemund, Garrett. (2017). *R for Data Science*. O'Reilly. If possible, please update Rivera et al. (to appear) with published year, volume, and page range.

# 15

## Inference on Association of Categorical Variables

### 15.1 Introduction

In Chapter 3, we saw some evidence suggesting that whether a driver was searched during a traffic stop in San Diego was related to the driver's race (Figure 15.1). Some empirical probabilities also indicated that this was the case, for male and female drivers (Problems 5.44 and 5.45).

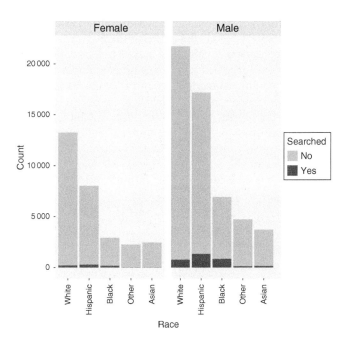

Figure 15.1 Race, gender, and whether a driver was searched during traffic stops in San Diego in 2016.

*Principles of Managerial Statistics and Data Science*, First Edition. Roberto Rivera.
© 2020 John Wiley & Sons, Inc. Published 2020 by John Wiley & Sons, Inc.
Companion website: www.wiley.com/go/principlesmanagerialstatisticsdatascience

Ignoring the limitations of the data, the association between police searches and race may be simply due to chance. Statistical inference allows us to assess further if there is any association between being searched during a traffic stop in San Diego and drivers race.

The regression methods presented in Chapters 13 and 14 rely on the response variable being a numerical variable. These models can be adapted to scenarios when the response variable is categorical. Not only these models with categorical responses will allow us to test for association between categorical variables, but the strength of the association could be assessed as well. However, regression models involving categorical response variables is more technical to implement and a bit harder to interpret than ordinary multiple linear regression. There is a rather simple method to test if there is an association between two categorical variables (without assessing strength), which is the main focus of this chapter.

## 15.2   Association Between Two Categorical Variables

First things first, what is the meaning of two categorical variables being associated? For two numerical variables, we look at what happens, on average, to $Y$ as $X$ increases. In contrast, if $Y$ is numerical and $X$ is categorical, we can use one-factor ANOVA, where the expected value of a response variable can shift based on the level of a factor. We then conclude that the two variables are associated. We say that two categorical variables are associated, or dependent, when the category of one influences the probability, the other variable being in a specific category. This has been presented in the context of events and random variables. Specifically, when two random variables are independent, any value of one random variable, an event, will not affect the probability of the other random variable to have any specific value, another event. One way to assess the association between two categorical variables is through a **contingency table**, for which we present the template in Table 15.1. Variable $A$ has $r$ categories, while variable $B$ has $c$ categories. Cells in the contingency table count the frequency the categories combine. For example,

- $f_{11}$ indicates the times category 1 of variable $A$ occurred together with category 1 of variable $B$.
- $f_{r2}$ indicates the times category $r$ of variable $A$ occurred together with category 2 of variable $B$.

Row and column totals may be presented as the last column and row of the table, respectively. For example, $R_1$ in Table 15.1 counts the frequency of category 1 of variable $A$ overall. The sum of the row totals and sum of column totals add up to $n$, the number of observations.

Table 15.1 General template of a contingency table.

| Variable A | Variable B | | | | |
|---|---|---|---|---|---|
| | 1 | 2 | $\cdots$ | $c$ | Row total |
| 1 | $f_{11}$ | $f_{12}$ | $\cdots$ | $f_{1c}$ | $R_1$ |
| 2 | $f_{21}$ | $f_{22}$ | $\cdots$ | $f_{2c}$ | $R_2$ |
| $\vdots$ | $\vdots$ | $\vdots$ | $\vdots$ | $\vdots$ | $\vdots$ |
| $r$ | $f_{r1}$ | $f_{r2}$ | $\cdots$ | $f_{rc}$ | $R_r$ |
| Column total | $C_1$ | $C_2$ | $\cdots$ | $C_c$ | $n$ |

---

**Null and Alternative to Test Independence**

Hypothesis testing for independence of two categorical variables establishes the following:

$H_o$: The two categorical variables are independent.
$H_1$: The two categorical variables are dependent.

---

It is assumed the null is true and that any association in the data is due to sampling variation. Thus, in theory, if $A$ is any category of the first variable, and $B$ is any category of the second variable,

$$P(A \text{ and } B) = P(A)P(B)$$

But, since a contingency table comes from a sample, the estimated $P(A \text{ and } B)$ should be approximately equal to the estimate of $P(A)P(B)$ if the null is true. Driven by this argument, it can be shown that if the two categorical variables are independent, then the best way to estimate the expected frequency of the cell in row $i$ and column $j$ is by multiplying the total of the row $i$ ($R_i$) by the total of column $j$ ($C_j$), divided by the total number of observations ($n$),

$$e_{ij} = \frac{R_i C_j}{n}$$

Next, a test statistic is used, which is a standardized measure of how much each contingency table cell value deviates from its expected value, assuming independence.

---

**Chi-squared Test**

The test statistic accumulates the squared differences of the frequency of each cell and its expected value, divided by the expected value of each cell. This is known as the **chi-squared test**. If the two categorical variables are independent, then we should expect the numerator of the test statistic to be "close to zero," each observed cell frequency is relatively close to its expected value. Conversely, a large deviation between the observed cell frequency and its expected value would be indicative that the null is not appropriate.

If the null is true, this test statistic will follow approximately[1] a chi-squared distribution, as long as each cell has an expected value greater than or equal to 5.

**Example 15.1** *Administrators of the San Francisco airport conduct a survey every year to analyze guest satisfaction. Among responses is whether guests have checked baggage, and if they have used a restaurant, their rating of it. Responses from 2880 guests are presented in Table 15.2. At $\alpha = 0.10$, test if guests having checked baggage is independent of how they rate restaurants.*

The hypotheses are as follows:

$H_o$: Restaurant rating and Checked baggage status are independent.
$H_1$: Restaurant rating and Checked baggage status are dependent.

How a chi-squared test is performed with Minitab is showed in Figure 15.2.

The results of the chi-squared test are seen in Figure 15.3. Output includes the expected value for each cell (bottom number of each cell). This way, we can check the assumption that all expected cell values are $\geq 5$. For example, the expected value of guests who checked baggage, and rated restaurants as unacceptable, was 26.0. All expected cell values are large enough. The test statistic, 5.886, leads to a $p$-value $= 0.208$, which is not below 0.1 and therefore, the null is not rejected. Thus at 10% significance, it is concluded that the checked baggage status of guests is independent of their rating of the restaurants.

Table 15.2 Checked baggage status and restaurant rating from San Francisco.

| | Restaurant quality | | | | | |
|---|---|---|---|---|---|---|
| Checked baggage | Unacceptable | Below average | Average | Good | Outstanding | Row total |
| Yes | 29 | 143 | 629 | 623 | 236 | 1660 |
| No | 16 | 89 | 435 | 503 | 167 | 1210 |
| Column total | 45 | 232 | 1064 | 1126 | 403 | 2870 |

---

1 The degrees of freedom of this distribution are $(r - 1)(c - 1)$.

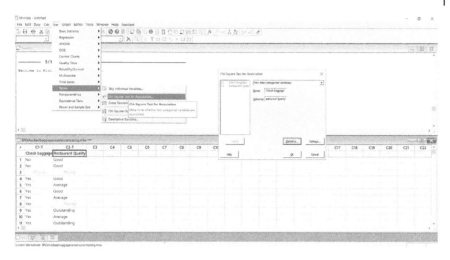

Figure 15.2 Menus to perform a chi-squared test with Minitab.

**Chi-Square Test for Association: Check baggage, Restaurant Quality**

```
Rows : Check baggage Columns : Restaurant Quality

                          Below
             Unacceptable Average  Average   Good  Outstanding Missing    All

Yes                   29     143      629     623          236     471    1660
                    26.0   134.2    615.4   651.3        233.1

No                    16      89      435     503          167     330    1210
                    19.0    97.8    448.6   474.7        169.9

Misssing               0       8       54      80           31      28       *

All                   45     232     1064    1126          403       *    2870


Cell contents :    Count
                   Expected count

Pearson Chi-Square = 5.886, DF = 4, p-Value = 0.208
Likelihood Ratio Chi-Square = 5.901, Df = 4, P-Value = 0.207
```

Figure 15.3 Minitab screenshot of chi-squared test.

---

**Small Frequencies: Option 1**

When at least one cell does not have an expected value greater than or equal to 5, the test statistic will not approximately follow a chi-squared distribution. An alternative is to merge categories of one or both variables. The merger represents a loss of information and is therefore performed on categories of least interest.

Table 15.3 Ride sharing frequency and age category.

| Ride sharing | Age | | | | | | Row total |
|---|---|---|---|---|---|---|---|
| | 18–24 | 25–34 | 35–44 | 45–54 | 55–64 | 65+ | |
| Never tried | 9 | 24 | 48 | 52 | 56 | 66 | 255 |
| Daily | 0 | 8 | 4 | 1 | 1 | 0 | 14 |
| Weekly | 13 | 38 | 28 | 16 | 13 | 4 | 112 |
| Monthly | 15 | 35 | 39 | 18 | 15 | 8 | 130 |
| Rarely | 17 | 42 | 44 | 38 | 36 | 27 | 204 |
| Tried, but don't use it | 2 | 8 | 15 | 10 | 9 | 13 | 57 |
| Column total | 56 | 155 | 178 | 135 | 130 | 118 | 772 |

**Example 15.2** *Does use of ride sharing companies (e.g. Uber, Lyft) depend on age? To answer this question, 772 responses to the San Francisco Travel Decision Survey[2] were used (Table 15.3). Inference will be performed at 5% significance.*

The hypotheses are as follows:

$H_o$: Ride sharing use and Age category are independent.
$H_1$: Ride sharing use and Age category are dependent.

Figure 15.4 summarizes the results of the chi-squared test. A warning is given that for seven cells, the expected value is less than 5. We deal with this issue by

```
Chi-Square Test for Association: Ride Sharing, Age
Rows : Ride Sharing      Columns : Age
                          18-24 25-34  35-44 45-54 55-64    65+    All
Never Tried                   9    24     48    52    56     66    255
                          18.50 51.20  58.80 44.59 42.94 38.98
Daily                         0     8      4     1     1      0     14
                           1.02  2.81   3.23  2.45  2.36  2.14
Weekly                       13    38     28    16    13      4    112
                           8.12 22.49  25.82 19.59 18.86 17.12
Monthly                      15    35     39    18    15      8    130
                           9.43 26.10  29.97 22.73 21.89 19.87
Rarely                       17    42     44    38    36     27    204
                          14.80 40.96  47.04 35.67 34.35 31.18
I've tried it, but I do not use    2     8     15    10     9     13     57
                           4.13 11.44  13.14  9.97  9.60  8.71
All                          56   155    178   135   130    118    772

Cell Contents:     Count
                   Expected count

Pearson Chi-Square = 111.324, DF = 25, P-Value = 0.000
Likelihood Ratio Chi-Square = 116.919, Df = 25, P-Value = 0.000
* NOTE* 7 cells with expected counts less than 5
```

Figure 15.4 Minitab screenshot of chi-squared test.

2 Source: data.sfgov.org (accessed July 26, 2019).

**Chi-Square Test for Association: Ride Sharing, Age**

```
Rows : Ride Sharing    Columns : Age

                18-24   25-34   35-44   45-54   55-64   65+    All

Never Tried        9      24      48      52      56      66    255
                18.50   51.20   58.80   44.59   42.94   38.98

Daily or Weekly   13      46      32      17      14       4    126
                 9.14   25.30   29.05   22.03   21.22   19.26

Monthly           15      35      39      18      15       8    130
                 9.43   26.10   29.97   22.73   21.89   19.87

Rarely            19      50      59      48      45      40    261
                18.93   52.40   60.18   45.64   43.95   39.89

All               56     155     178     135     130     118    772

Cell Contents:     Count
                 Expected count


Pearson Chi-Square = 99.380, DF = 15, P-Value = 0.000
Likelihood Ratio Chi-Square = 104.730, DF = 15, P-Value = 0.000
```

Figure 15.5 Minitab screenshot of a chi-squared test after merging some ride sharing categories.

combining the "Daily" category of ride sharing with the "Weekly" category and the "I've tried it, but don't use it" category with the "Rarely" category.

Figure 15.5 presents the chi-squared test results for the new contingency table. Now all expected values exceed 5. Moreover, the $p$-value $< 0.05$ and the null is rejected. At 5% significance, we conclude that ride sharing use is dependent on age.

---

**Small Frequencies: Option 2, Fisher's Exact Test**

The chi-squared test is a form of large sample test. It is not always acceptable to merge categories of the categorical variables, and sometimes the strategy may not work anyway. Ronald Fisher came up with a solution called **Fisher's exact test**. In a nutshell, this type of test determines a $p$-value without assuming the test statistic follows a chi-squared distribution. Instead, it considers all possible contingency tables with the fixed row and column totals to test independence. The "exact" in the name of the test is a bit of a misnomer, as generally the computation of the $p$-values requires simulations.

Fisher's exact test works under the assumption that the row and column marginal counts are fixed. There are exact tests that discard this assumption at the cost of more complex computations.

---

### 15.2.1 A Quick Look at Data Science: Affordability and Business Environment in Chattanooga

Is the perspective of business owners in Chattanooga, Tennessee, as a place of business, independent of their perspective of its affordability? Let us assess

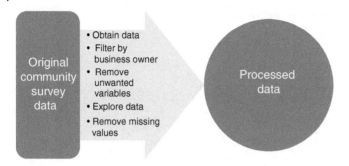

Figure 15.6 Steps required to process Chattanooga Community Survey data.

Table 15.4 Frequencies of the responses pertaining to Chattanooga as a place of business and the location's affordability.

| | Business environment | | | | | |
|---|---|---|---|---|---|---|
| Affordability | Very good | Good | Neutral | Bad | Very bad | Don't know |
| Very satisfied | 15 | 11 | 1 | 1 | 0 | 0 |
| Satisfied | 7 | 28 | 8 | 0 | 0 | 0 |
| Neutral | 6 | 12 | 8 | 3 | 0 | 0 |
| Somewhat dissatisfied | 0 | 1 | 0 | 0 | 0 | 0 |
| Very dissatisfied | 0 | 0 | 0 | 0 | 0 | 0 |
| Don't know | 29 | 58 | 14 | 1 | 3 | 4 |

this question at 1% significance. 2172 responses to 87 variables from the 2016 community survey are available online.[3] Level of education, ethnic background, and cleanliness are some of the variables for which responses are available. Some data wrangling is required (Figure 15.6). For example, responses must be filtered to strictly use people who own a business (one of the questions participants answered) and blank responses must be eliminated. This leaves us with 210 responses.

Table 15.4 summarizes the responses to both questions by business owners:

- More than half of the respondents gave a "Don't Know" response to the affordability question, and the rest rated the city as neutral or better on the topic.
- Regarding Chattanooga as a place of business, the majority of respondents gave the city a rating of good or better.

---

3 data.chattlibrary.org (accessed July 26, 2019).

**Tabulated Statistics: Worksheet rows, Worksheet columns**

Rows: Worksheet rows Columns: Worksheet columns

| | Very Good | Good | Neutral | Bad | Very Bad | Do not Know | All |
|---|---|---|---|---|---|---|---|
| 1 | 15 | 11 | 1 | 1 | 0 | 0 | 28 |
| | 7.600 | 14.667 | 4.133 | 0.667 | 0.400 | 0.533 | |
| 2 | 7 | 28 | 8 | 0 | 0 | 0 | 43 |
| | 11.671 | 22.524 | 6.348 | 1.024 | 1.024 | 1.024 | |
| 3 | 7 | 28 | 8 | 0 | 0 | 0 | 43 |
| | 11.671 | 22.524 | 6.348 | 1.024 | 1.024 | 1.024 | |
| 4 | 7 | 28 | 8 | 0 | 0 | 0 | 43 |
| | 11.671 | 22.524 | 6.348 | 1.024 | 1.024 | 1.024 | |
| 5 | 0 | 0 | 0 | 0 | 0 | 0 | 0 |
| 6 | 7 | 28 | 8 | 0 | 0 | 0 | 103 |
| | 11.671 | 22.524 | 6.348 | 1.024 | 1.024 | 1.024 | |
| All | 57 | 110 | 31 | 5 | 3 | 4 | 210 |

Cell Contents:     Count
                   Expected count

Pearson Chi-Square = 36.030, DF = 20
Likelihood Ratio Chi-Square = 36.009, DF =20

* WARNING * 14 cells with expected counts less than 1
* WARNING * Chi-Square approximation probably invalid

* NOTE * 20 cells with expected counts less than 5

Figure 15.7 Chi-squared test of independence using Minitab. The software indicates expected values are too small for this test.

- Only one person gave Chattanooga a "Somewhat Dissatisfied" rating in affordability, while rating the place as "Good" for business.
- Overall, many business owners gave good or better ratings to both questions.
- Finally, several cells related to lower ratings in Table 15.4 have zero counts, a sign that small expected frequencies may occur.

Minitab output (Figure 15.7) corroborates our concerns, with 20 cells with expected value below 5, and 14 cell below 1. Marginal totals are not fixed for this data and thus Fisher's test is not an applicable test.

To cope with the issue, categories will be merged. After several attempts to comply with the expected frequency condition without losing too much

Table 15.5 Frequencies of the responses pertaining to Chattanooga as a place of business and the location's affordability.

| Affordability | Business environment | | |
| --- | --- | --- | --- |
| | Very good | Good | Neutral or else |
| Very satisfied | 15 | 11 | 2 |
| Satisfied | 7 | 28 | 8 |
| Neutral or else | 6 | 13 | 11 |
| Don't know | 29 | 58 | 18 |

**Tabulated Statistics: Worksheet rows, Worksheet columns**

```
Rows: Worksheet rows   Columns: Worksheet columns

           Very          Neutral
           Good2  Good2   or Else   All

1            15     11        2     28
           7.75  14.95     5.30

2             7     28        8     43
          11.90  22.96     8.14

3             6     13       11     30
           8.30  16.02     5.68

4            29     58       18    105
          29.05  56.07    19.88

All          57    110       39    206

Cell Contents :        Count
                       Expected count


Pearson Chi-Square = 19.448, DF = 6, P-Value = 0.003
Likelihood Ratio Chi-Square = 18.040, DF = 6, P-Value = 0.006
```

Figure 15.8 Chi-squared test of independence after merging some categories. The software indicates expected values are now adequate for this test.

information, we find adequate to merge the categories Neutral, Somewhat Dissatisfied, and Very Dissatisfied on the affordability variable and the categories Neutral, Bad, Very Bad, and Don't Know of the place for business variable (Table 15.5). People that answered strictly "Don't Know" to both questions were removed. Figure 15.8 indicates that now all expected frequencies are above 5. Since the p-value $< 0.01$, independence of the two categorical variables is rejected. At 1% significance, we find that the perspective of respondents on business environment and affordability is not independent.

## Practice Problems

15.1 Does use of ride sharing companies (e.g. Uber, Lyft) depend on gender? To answer this question, 799 responses to the San Francisco Travel Decision Survey[4] were used. Below is a screenshot of Minitab output.

**Chi-Square Test for Association: Recoded Q21A, Recoded Q30**

```
Rows: Recoded Q21A   Columns: Recoded Q30

                              Male    Female   All

Never tried                    143      130    273
                            145.55   127.45

Daily                            7        7     14
                              7.46     6.54

Weekly                          68       45    113
                             60.25    52.75

Monthly                         67       64    131
                             69.84    61.16

Rarely                         108      102    210
                            111.96    98.04

Tried it, but don't use it.     33       25     58
                             30.92    27.08

All                            426      373    799

Cell Contents:        Count
                      Expected count


Pearson chi-Square = 3.142, DF = 5, P-Value = 0.678
Likelihood Ratio chi-Square = 3.162, DF = 5, P-Value = 0.675
```

a) Write the null and alternative hypothesis.
b) Do the expected cell counts indicate any concerns?
c) Do you reject the null at 5% significance?
d) Write your statistical conclusion.

15.2 Section 5.6.1 explored whether reports of a missing person in Boston were more likely in some days than others. Below is a screenshot of Minitab output.

---

4 Source: data.sfgov.org (accessed July 26, 2019).

**Chi-Square Test for Association: Day of week, Worksheet columns**

```
Rows: Day of week    columns : Worksheet columns

                    missing    not missing      All

Sunday                 339          29608      29947
                     370.5        29576.5

Monday                 408          33565      33973
                     420.3        33552.7

Tuesday                416          34009      34425
                     425.9        33999.1

Wednesday              384          34386      34770
                     430.2        34339.8

Thursday               411          34378      34789
                     430.4        34358.6

Friday                 524          35455      35979
                     445.1        35533.9

Saturday               451          32740      33191
                     410.6        32780.4

ALL                   2933         234141     237074


Cell Contents:        Count
                      Expected count


Pearson chi-Square = 27.382, DF = 6, P-Value = 0.000
Likelihood Ratio chi-Square = 26.784, DF = 6, P-Value = 0.000
```

a) Write the null and alternative hypothesis.
b) Do the expected cell counts indicate any concerns?
c) Do you reject the null at 1% significance?
d) Write your statistical conclusion.

15.3 A company classifies quality of the furniture it manufactures based on four categories: Excellent, Good, Acceptable, and Unacceptable. The quality department wants to test if quality rating is independent of the work shift at which it was produced. 927 products were evaluated, leading to the contingency table below. Use 5% significance.

| Shift | Quality | | | | Total |
|-------|-----------|------|------------|--------------|-------|
|       | Excellent | Good | Acceptable | Unacceptable |       |
| 1     | 65        | 63   | 116        | 36           | 280   |
| 2     | 69        | 64   | 121        | 35           | 289   |
| 3     | 86        | 80   | 148        | 44           | 358   |
| Total | 220       | 207  | 385        | 115          | 927   |

15.4 The table below summarizes whether a female driver was searched by San Diego police, in 2016, according to race. Test whether a female driver was searched by police depends on race at 10% significance.

| Searched | | | Race | | |
|---|---|---|---|---|---|
| | Asian | Black | Hispanic | Other | White |
| Yes | 11 | 59 | 120 | 9 | 90 |
| No | 1 983 | 1 805 | 5 694 | 1 920 | 10 730 |

15.5 The table below summarizes alcohol-related deaths in England in 2014, considering classification and gender.[5] Test if the classification is dependent on the gender of the person at 5% significance.

| Category | Gender | |
|---|---|---|
| | Male | Female |
| Mental/behavioral disorders | 339 | 150 |
| Alcoholic cardiomyopathy | 68 | 16 |
| Alcoholic liver disease | 2845 | 1488 |
| Fibrosis and cirrhosis of liver | 911 | 609 |
| Accidental poisoning | 242 | 127 |
| Other | 28 | 8 |

15.6 There is an excellent open data portal for the City of Edmonton, Canada. One data set comes from a survey to gather community insight, including residents, opinion about drones. Two drone aspects participants answered were as follows:
- Unrestricted use will cause an invasion of privacy.
- I have concerns with the noise unmanned aerial vehicle (UAV)/drones could create in residential areas or parks.

Does the opinion of residents about drones invading their privacy depend on their concerns of the noise drones may create? Answer this question using the variables "Q17_DroneandPrivacy" and "Q17_DroneandNoise" in the file[6] "Edmonton_Insight_Community." Use 10% significance. *Hint*: It will be useful to remove the "Don't Know" answers from the data.

5 Source of data: https://data.gov.uk/dataset/87dd943e-54fd-4219-8294-33d27c6c6a11/alcohol-related-deaths-in-the-united-kingdom (accessed July 26, 2019).
6 Source of data: data.edmonton.ca (accessed December 28, 2018).

## Chapter Problems

**15.7** Download the file "RideSharingandGenderSanFrancisco" and reproduce the results seen in Problem 15.1.

**15.8** Delaware has a rebate program for the purchase or lease of alternative fuel vehicles and data is available to the public.[7] Is vehicle type (plug-in hybrid or electric) independent of gender? Download "Delaware_State_Rebates_for_Alternative-Fuel_Vehicles." Construct a contingency table of gender and vehicle type. Test at 1% significance if vehicle type is independent of gender. *Hint*: Subset the data first, discarding any vehicle type other than plug-in hybrid and electric.

**15.9** Using the data set from Problem 15.8, is vehicle type (plug-in hybrid or electric) independent of whether vehicle was leased or not? Test at 5% significance.

## Further Reading

Black, Ken. (2012). *Business Statistics*, 7th edition. Wiley.

Levine, David M., Stephan, David F., and Szabat, Kathryn A. (2017). *Statistics for Managers: Using Microsoft Excel* 8th edition. Pearson.

---

7 Source: data.delaware.gov (accessed July 26, 2019).

# 16

# Nonparametric Testing

## 16.1  Introduction

Most of the hypothesis tests performed in this book use $Z$-tests, $t$-tests, or $F$-tests. These tests rely on either the population being normally distributed or the sample mean following approximately a normal distribution. These assumptions are sometimes far from reality, and researchers are less willing to make such assumptions when the sample size is small. Furthermore, occasionally it is not reasonable to infer on the population mean (i.e. when there is support for the population having a highly skewed distribution), or it might be necessary to test the probability distribution assumption that has been made. $Z$-tests, $t$-tests, and $F$-tests all fall into the category of **parametric tests**, which rely on a parametric probability distribution for the population or main estimator. **Nonparametric tests**, also known as **distribution-free tests**, either drop the probability distribution assumption for the population or estimator in use, or test assumptions made for parametric tests. For example, the presence of outliers in data makes nonparametric tests viable alternatives. Moreover, there are nonparametric tests that also apply to ranked or ordinal data.

In this chapter, nonparametric tests are introduced for several statistical inference needs. You will see some similarities in the essence of the testing procedures, except that nonparametric procedures have some tweaks that allow for less stringent assumptions. As in the last few chapters, mathematical computations and technicals are kept to a minimum, instead emphasizing the intuition and interpretation of the methods.

## 16.2  Sign Tests and Wilcoxon Sign-Rank Tests: One Sample and Matched Pairs Scenarios

The nonparametric procedure in one sample and two paired sample situations is similar. Let's start with the one sample scenario. Suppose $X$ represents

*Principles of Managerial Statistics and Data Science*, First Edition. Roberto Rivera.
© 2020 John Wiley & Sons, Inc. Published 2020 by John Wiley & Sons, Inc.
Companion website: www.wiley.com/go/principlesmanagerialstatisticsdatascience

measurements from some population, such that there are $n$ observations. The objective is to assess the typical value of the measurement. For highly skewed distributions, the population mean is no longer the best measure of a typical value of the random variable, the median ($M$) is better. The null hypothesis will assign a specific value to the population median.[1]

| Lower tail test | Two-tailed test | Upper tail test |
| --- | --- | --- |
| $H_o: M \geq M_o$ | $H_o: M = M_o$ | $H_o: M \leq M_o$ |
| $H_1: M < M_o$ | $H_1: M \neq M_o$ | $H_1: M > M_o$ |

Let $C$ be the number of observations that are above the population median under the null. If $H_o$ is correct, we should expect about half of the observations to be above the population median. The **sign test** is a nonparametric method that can be used to test whether half of the observations are above the population median under the null hypothesis. If $\pi$ is arbitrarily defined as the probability that an observation is above the median, then $H_o: \pi = 0.5$. Under $H_o$, $C$ follows a binomial distribution with $\pi = 0.5$. This premise can be used to find a $p$-value. If $p$-value $\leq \alpha$, $H_o$ is rejected.

---

**The Sign Test Assumption**

- Independent observations.

---

If $n$ is large enough, the test may be performed using a normal approximation to the test statistic.

**Example 16.1** *Administrators for the City of Austin, Texas, randomly send surveys to residents with the aim to gauge their opinion about services. One item of the questionnaire asks to evaluate the city as a place to raise children. Answers are Very Satisfied, Satisfied, Neutral, Dissatisfied, and Very Dissatisfied, and each is recoded to 5, 4, 3, 2, and 1, respectively. We will perform a sign test at 5% significance to assess if the median is 3.*

To perform a sign test with Minitab, go to Stat > Nonparametrics > 1-Sample Sign .... In the 1-Sample Sign window, enter your ordinal data in the first box and leave the default "Alternative" sign to "equal." The Minitab output is displayed in Figure 16.1. It tells us that the median of the data was 4. Since the $p$-value $= 0 < 0.05$, the null is rejected. This way, at 5% significance, we conclude that the median answer is not 3 (Neutral).

---

1 The null can have strictly an equality sign as explained in Chapter 10.

### Sign Test for Median: raise children

```
Sign test of median = 3.000 versus ≠ 3.000

                 N   Below  Equal  Above       P   Median
raise children  5283    468   1118   3697  0.0000    4.000
```

Figure 16.1 Minitab sign test output.

*Wilcoxon Sign-Rank Test*  As its name suggests, the sign test relies on the signs of the data (positive difference between observation and median) to test the hypothesis. In a way, this throws out quite a bit of information available from the data. Specifically, order of the data is useful too. An alternative is to rely on the sign and rank of the data. Another way of putting it is that instead of focusing on the proportion of observations above the population median, we could focus on the expected number of observations above the median, $n/2$. The goal of the **Wilcoxon sign-rank test** is to assess the value of the population median by ranking observations based on the difference between the observation and the median under the null.[2] This gives us a test statistic that can be used to find a $p$-value. If $p$-value $\leq \alpha$, $H_o$ is rejected.

---

**The Wilcoxon Sign-Rank Test Assumptions**

- Independent observations.
- Roughly symmetric population.

---

If $n$ is large enough, the test may be performed using a normal approximation for the test statistic.

*Paired Samples*  Suppose that $X$ represents measurements from some population, and $Y$ from a second population, such that there are $n$ pairs of observations $(X_i, Y_i)$. The objective is to test whether there is an important difference among pairs. Let

$$D_i = X_i - Y_i$$

for matched pair $i$. If there is no real difference between the pairs (the null), then we should expect about half of the $D_i$ to be positive and the other half negative. The **sign test** is a nonparametric method to test whether half of the differences are positive. If $\pi$ is arbitrarily defined as the probability that $X > Y$, then $H_o$: $\pi = 0.5$. The method proceeds by counting the number of positive differences, which under $H_o$ follows a binomial distribution with $\pi = 0.5$.

---

2 There are some intricacies in the rankings such as what is done with zero differences and ties. These details can be found in "Further Reading" section.

---

**Paired Samples Sign Test Assumptions**

- Pairs $(X_i, Y_i)$ are random (and thus, $D_i$ are independent).

---

If $n$ is large enough, the test may be performed using a normal approximation for the test statistic.

**Example 16.2** *A company develops an algorithm to perform English–Mandarin translations. They have updated their algorithm. Researchers randomly assign seven translation tasks and have an expert count translation errors from both versions of the algorithm (see Table 16.1). Using $\alpha = 0.05$, determine whether translation errors decreased substantially.*

To perform this sign test with Minitab, first calculate the differences between the errors for each translator. Then, go to Stat > Nonpara-metrics > 1-Sample Sign .... In the 1-Sample Sign window, enter the differences column in the first box and change the "Alternative" sign to "less than." Figure 16.2 shows the Minitab output. Since $p$-value $= 0.0625 > 0.05$, we do not reject the null. Thus, at 5%, there is no evidence that the updated algorithm has less translation errors.

Table 16.1 Errors counted in versions of the algorithm.

| Translator | Baseline version | New version |
|------------|------------------|-------------|
| 1 | 299 | 292 |
| 2 | 317 | 310 |
| 3 | 301 | 314 |
| 4 | 288 | 287 |
| 5 | 309 | 299 |
| 6 | 299 | 295 |
| 7 | 315 | 309 |

**Sign Test for Median: Difference**

```
Sign test of median  =  3.00000 versus  <  0.00000

              N   N*  Below  Equal  Above     P   Median
Difference    7   1     6      0      1   0.0625   -6.000
```

Figure 16.2 Minitab sign test output.

## Wilcoxon Signed Rank Test: Difference

```
Test of mediam  =  0.000000  versu  medium  <  0.000000

                 N for      Wilcoxon                Estimated
            N    Test       Statistic        P      Median
Difference  7    7              7.0      0.136      -5.250
```

Figure 16.3  Minitab Wilcoxon sign-rank test output.

The Wilcoxon signed-rank test can also be applied to matched paired rank data by ranking each $D_i$.

---

**Paired Wilcoxon Sign-Rank Test Assumptions**

- Pairs $(X_i, Y_i)$ are random (and thus $D_i$ are independent).
- Differences have a roughly symmetric distribution.

---

**Example 16.3**   *Returning to Example 16.2, assuming the distribution of differences is symmetrical, and $\alpha = 0.05$, determine whether translation errors decreased substantially using a Wilcoxon sign-rank test.*

To perform a Wilcoxon sign-rank test with Minitab, first calculate the differences between the errors for each translator. Then, go to Stat > Nonparametrics > 1-Sample Wilcoxon .... In the 1-Sample Wilcoxon window, enter the differences column in the first box and change the default "Alternative" sign to "less than." Figure 16.3 shows the Minitab output. Since $p$-value $= 0.136 > 0.05$, we do not reject the null. Thus, at 5%, there is no evidence that the updated algorithm has less translation errors.

*Sign Test Versus Wilcoxon Signed-Rank Test*   By also using rank information from the data, the Wilcoxon sign-rank test has more power than the sign test when the assumptions hold. A disadvantage of the Wilcoxon sign-rank test, in the case of continuous data, is that the symmetric distribution assumption is only a slightly weaker assumption than the normal population assumption for the $t$-test. Moreover, the $t$-test is rather robust to violations in normality, with the exception of small sample sizes and one-sided tests.

## Practice Problems

16.1  As seen in Problem 4.19, the mass of meteorites often follow a skewed distribution. As a result, many scientists prefer to consider the median as a measure of centrality as opposed to the mean. A researcher believes that

Acapulcoite meteorites have a median mass greater than 50 g. With the help of a computer, use the NASA data set[3] "Meteorite_Landings" to run a sign-rank test at 1% significance.

*Hint:* First subset by Acapulcoite classification.

16.2 Referring to Problem 16.1, run a Wilcoxon signed-rank test at 1% significance to determine if the median mass of Acapulcoite meteorites is greater than 50 g. Why should we question running this test?

16.3 With the help of a computer, use NASA's "Meteorite_Landings" to run a Wilcoxon signed-rank test at 5% significance to determine if the median mass of Pallasite meteorites is equal to 100 g. Why should we question running this test?

*Hint:* First subset by Pallasite classification.

16.4 With the help of a computer, use NASA's "Meteorite_Landings" to run a sign test at 5% significance to determine if the median mass of Pallasite meteorites is equal to 100 g.

*Hint:* First subset by Pallasite classification.

16.5 As part of a marketing study of beer $X$, 10 bars of similar size are chosen across the United States. The sales for each bar are recorded for the week before the ad campaign, and the week immediately after completion of the campaign. Perform a Wilcoxon sign-rank test to determine at 10% significance if the ad campaign increased sales. Also, state what assumptions are required for this test.

| Bar | Before | After |
| --- | --- | --- |
| 1 | 101 | 107 |
| 2 | 103 | 106 |
| 3 | 97 | 101 |
| 4 | 115 | 138 |
| 5 | 100 | 100 |
| 6 | 97 | 108 |
| 7 | 143 | 167 |
| 8 | 102 | 101 |
| 9 | 105 | 117 |
| 10 | 99 | 147 |

16.6 Same as Problem 16.5, but run a sign test.

---

3 Source: data.nasa.gov (accessed July 26, 2019).

## 16.3 Wilcoxon Rank-Sum Test: Two Independent Samples

The **Wilcoxon rank-sum test** is the nonparametric alternative to the two sample $t$-tests for independent samples. It works by combining the observations to rank them, followed by obtaining a sum of the ranks from one of the samples, the test statistic.

---

**The Wilcoxon Rank-Sum Test Assumptions**

- Samples have been selected randomly and independently from their populations.

The null hypothesis is that the distributions are equal. Under an additional assumption, the population distributions of both groups have the same shape. Then, the null is equivalently represented as the medians being equal (in case of numerical data).

---

Under the null, the sum of the ranks from one of the samples cannot be too small or too big (depending on the type of alternative hypothesis). Finally, the rank sum can be used to find a $p$-value. If the $p$-value $\leq \alpha$, $H_o$ is rejected. The **Mann–Whitney test** is another nonparametric test often used for two independent samples and provides equivalent results to the Wilcoxon rank-sum test.

**Example 16.4** *A company develops algorithms to translate the following:*

- *English–Mandarin*
- *Mandarin–English*

*Seven trial runs are performed for both types of translations, and a translator counts the errors of each translation. Using $\alpha = 0.05$, determine whether errors for both types of translations have the same distribution.*

| Translation | English–Mandarin | Mandarin–English |
|---|---|---|
| 1 | 299 | 292 |
| 2 | 317 | 310 |
| 3 | 301 | 314 |
| 4 | 288 | 289 |
| 5 | 309 | 303 |
| 6 | 299 | 295 |
| 7 | 315 | 317 |

**Mann-Whitney Test and CI: E-M, M-E**

```
        N   Median
E-M     7   301.00
M-E     7   303.00


Point estimate  for  η1 - η2  is 1.00
95.9 Percent CI for  η1 - η2  is (-15.00, 14.00)
W = 53.5
Test of  η1 = η2  vs  η1 ≠ η2 is significant at 0.9491
The test  is  significant  at 0.9489  (adjusted for ties)
```

Figure 16.4  Minitab Mann–Whitney test output.

To perform this Mann–Whitney test with Minitab, go to Stat > Nonpara-metrics > Mann-Whitney .... In the Mann–Whitney window, enter each column of data and leave the "Alternative" sign at "equal." Figure 16.4 shows the Minitab output. Since $p$-value = 0.9491 > 0.05, we do not reject the null. Thus, at 5% significance, there is no evidence that the distributions of the errors are different.

### 16.3.1  A Quick Look at Data Science: Austin, Texas, as a Place to Live; Do Men Rate It Higher Than Women?

Many cities implement their own community surveys to gauge the perception of residents on the services provided. We aim to determine if men rate Austin, Texas, higher than women as a place to live. Inference will be performed at 10% significance. A total of 6374 responses and over 100 variables for the 2015–2017 surveys are available online.[4] For this analysis, only 2017 will be considered, and data wrangling (Figure 16.5) left us with 2176 responses. Also, answers were numerical coded to scores: Very Satisfied was turned to 5, Satisfied to 4, Neutral to 3, Dissatisfied to 2, and Very Dissatisfied to 1. Missing responses or those who responded "Don't Know" were removed. Table 16.2 summarizes the data by response to each question.

A chi-squared test could be implemented but such a test would ignore the information provided by the order in the categories answering the "City of Austin as a place to live" question. Also, it wouldn't help us directly determine whether men prefer living in Austin more than women. The hypotheses for a Wilcoxon rank-sum test are as follows:

$H_o$: Men and women do not differ on how much they like living in Austin.
$H_1$: Men like more living in Austin than women.

Using statistical software, it is found that the $p$-value = 0.14. Since 0.14 ≮ 0.10, the null is not rejected. At 10% significance, we find no evidence against

---

4  data.austintexas.gov (accessed July 26, 2019).

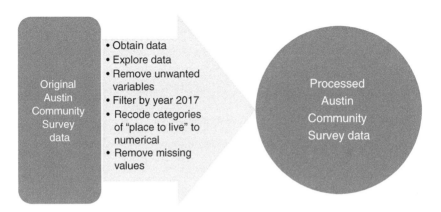

Figure 16.5 Steps required to process Austin Community Survey data.

Table 16.2 Frequencies of the scores pertaining to Austin, Texas, as a place to live, by gender.

| | As place to live | | | | |
|---|---|---|---|---|---|
| Gender | 5 | 4 | 3 | 2 | 1 |
| Male | 319 | 518 | 125 | 80 | 18 |
| Female | 316 | 554 | 126 | 81 | 39 |

men and women equally scoring Austin as a place to live. To determine whether men prefer living in Austin more than women, some analysts could state the statistical inference problem as a comparison of population means. Let $\mu_1$ = true mean score given to Austin as a place to live by men and $\mu_2$ = true mean score given to Austin as a place to live by women. Then,

$$H_o: \mu_1 - \mu_2 \leq 0$$
$$H_1: \mu_1 - \mu_2 > 0$$

and a $t$-test is performed. Figure 16.6 summarizes the results of running this test. We see that since $p$-value $= 0.07 < 0.10$, the null is rejected.

So when a Wilcoxon rank-sum test was performed, the null was not rejected, but when a $t$-test was performed, the null was rejected. What gives? To better understand what happened, take a look at Table 16.3. The median score given to the city by men and women was 4. But, the mean score given by men was slightly higher. Not only is it debatable if the difference in means is of any practical importance, but the use of means for ordinal values that were artificially established is also debatable. In particular, the difference between "Very Satisfied" and "Satisfied" is the same as the difference between "Satisfied" and "Neutral."

**Two-Sample T-Test and CI: place to live_Male, place to live_ Female**

```
Two-sample T for place to live_Male vs place to live_Female

                        N   Mean  StDev  SE Mean
place to live_Male    1060  3.981 0.935   0.029
place to live_Female  1116  3.920 0.998   0.030


Difference = μ (place to live_Male) - μ (place to live_Female)
Estimate for difference: 0.0609
95% lower bound for difference: -0.0073
T-Test of difference = 0(vs >): T-Value = 1.47 p-Value = 0.071 Df = 2173
```

Figure 16.6 Minitab *t*-test output comparing mean score given to Austin as a place to live by gender.

Table 16.3 Summary of Austin scores by gender.

| Gender | *n* | Mean | Median | Std. dev. |
| --- | --- | --- | --- | --- |
| Male | 1060 | 4 | 3.98 | 0.94 |
| Female | 1116 | 4 | 3.92 | 1.00 |

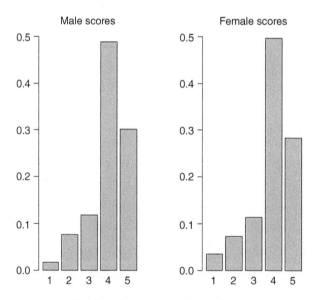

Figure 16.7 Barplots of Austin scores by gender.

Moreover, barplots of the scores show asymmetric distributions, albeit similar shapes (Figure 16.7). Thus, even if the application of means with this ordinal data is defended, the asymmetry puts into doubt whether inference should be done on the population means at all. The Wilcoxon rank-sum test is able to bypass all these concerns that affect a *t*-test.

## Practice Problems

16.7  In Example 15.2, it was found that ride sharing in San Francisco was dependent on age. A problem with chi-squared tests is that they ignore the information provided by age ordering. The question we really want to answer is if younger people use more ride sharing in San Francisco? The response to the ride sharing question, Have you tried any of these new travel options? (e.g. Uber, Lyft), is ordinal: 6 – Daily, 5 – Weekly, 4 – Monthly, 3 – Rarely, 2 – I've tried it, but I do not use it, and 1 – Never tried. We will test if the median answer from the younger group (44 or less) is higher than the median answer from the mature group. Inference is conducted at 1% significance. With the help of statistical software, it is found that the $p$-value $= 0$.

a)  Is the null rejected?

b)  Write a statistical conclusion.

c)  What are the assumptions required for the hypothesis test performed?

16.8  Are drone owners in Edmonton, Canada, less concerned than people who do not own a drone about safety with drones flying over people? In a survey[5] to gather community insight about drones, participants provided feedback on "I have safety concerns regarding drones flying over people." Possible answers were[6] "Strongly Agree, Somewhat Agree, Neutral, Somewhat Disagree, Strongly Disagree." We assume that the shape of the responses from both genders is equal and the inference will be expressed in terms of the medians. $\alpha = 5\%$.

a)  Write the null and alternative hypothesis for a Wilcoxon rank-sum test.

b)  The $p$-value $= 0$. Is the null rejected?

c)  Write a statistical conclusion.

16.9  The city of Chattanooga, Tennessee, implements its own community survey. Similar to the case study in Section 16.3.1, the aim is to determine if Chattanoogan men rate the city higher than Chattanoogan women as a place to live. We assume that the shape of the responses from both genders is equal and the inference will be expressed in terms of the medians. $\alpha = 5\%$.

a)  Write the null and alternative hypothesis for a Wilcoxon rank-sum test.

b)  The $p$-value $= 0.0823$. Is the null rejected?

c)  Write a statistical conclusion.

---

5  Source of data: data.edmonton.ca (accessed December 28, 2018).
6  "Don't Know" answers were removed.

## 16.4    Kruskal–Wallis Test: More Than Two Samples

The nonparametric version of the one-way ANOVA is the **Kruskal–Wallis test**. Assuming there are samples from $k$ populations (the levels of the factor), the test assesses if all samples have the same probability distribution ($H_o$) or not ($H_1$).

$H_o$: Measurements have the same distribution in all groups.
$H_1$: Measurements do not have the same distribution in all groups.

---

**Kruskal–Wallis Test Assumption**
Samples have been selected randomly and independently from their populations. Under an additional assumption,

- The population distributions of both groups have the same shape.

Then, the null is equivalently represented as the medians being equal and the alternative that at least one median differs from the others.

---

The test works by establishing the ranks for the entire sample followed by determining sample mean ranks for each group. The test statistic is a function of the squared differences of the sample mean ranks and average rank over the entire sample. When enough observations are made over the entire sample, the test statistic will approximately follow a chi-squared distribution. The Kruskal–Wallis test can be applied to ordinal data. Analogous to one-way ANOVA, rejection of the null does not specify where the differences are.

**Example 16.5**    *Recall that in section 12.4.1, one-way ANOVA was performed to compare EMS dispatch response times in New York City by time of day. The analysis indicated a highly skewed distribution of the response times making even the idea of inference on population means rather questionable. An alternative is to perform a Kruskal–Wallis test to determine if the distributions are equal. We perform the test at 5% significance.*

The hypotheses are as follows:

$H_o$: Response times have the same distribution in all times of day.
$H_1$: Response times do not have the same distribution in all times of day.

The table below shows the Minitab output. The test statistic is denoted $H$ by Minitab. The $p$-value $< 0.05$. Therefore, at 5% significance, we conclude that the probability distributions of dispatch response times are not equal by time of day. After visually assessing the shape of the distribution of dispatch times for each time of day (Figure 16.8), we determine they are approximately equal, and the

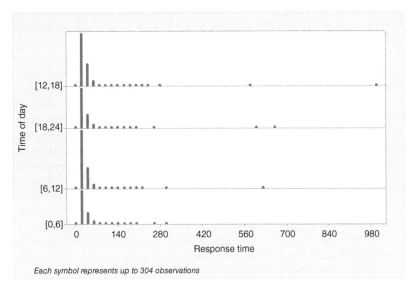

Figure 16.8 Dot plots of dispatch response time by time of day.

procedure is comparing the median response times of each time of day. Note from the Minitab output that the median response times are even closer to each other than the averages when one-way ANOVA was done in Example 12.4.1. Although statistically significant, there is no indication that the difference is large.

Kruskal–Wallis test on dispatch response time

| Time of day | N | Median | Ave rank | Z |
|---|---|---|---|---|
| (12, 18] | 17 978 | 24.00 | 30 792.2 | 10.97 |
| (18, 24] | 12 517 | 23.00 | 28 877.0 | −5.51 |
| (6, 12] | 18 701 | 23.00 | 29 571.5 | −0.51 |
| (0, 6] | 10 052 | 23.00 | 28 565.6 | −6.81 |
| Overall | 59 248 | | 29 624.5 | |
| $H = 146.42$ | $DF = 3$ | $P = 0.000$ | | |
| $H = 146.57$ | $DF = 3$ | $P = 0.000$ | (adjusted for ties) | |

Rejecting the null in a Kruskal–Wallis test can be followed up with a Wilcoxon test to compare pair of groups. Alternatively, nonparametric tests analogous to parametric multiple comparison procedures are available. In the case of randomized block designs, there are nonparametric tests available as well. Which test is best to use depends on the type of block design that is being applied.

## Practice Problems

16.10 Every year, administrators of the San Francisco airport conduct a survey to analyze guest satisfaction. Suppose management wants to compare the restaurant quality scores of four terminals. Restaurant scoring ranges from 1 to 5, with 1 = Unacceptable, 2 = Below Average, 3 = Average, 4 = Good, and 5 = Outstanding. Since it is an ordinal response variable, it is decided a Kruskal–Wallis test will be implemented.

Kruskal–Wallis test on restaurant quality scores

| Terminal | N | Median | Ave rank | Z |
|----------|------|--------|----------|-------|
| 1 | 540 | 3.00 | 727.8 | −8.58 |
| 2 | 277 | 4.00 | 1141.9 | 9.11 |
| 3 | 438 | 4.00 | 983.6 | 4.66 |
| 4 | 514 | 3.00 | 827.6 | −3.02 |
| Overall | 1769 | | 885.0 | |

| | | | |
|---|---|---|---|
| $H = 143.99$ | $DF = 3$ | $P = 0.000$ | |
| $H = 162.36$ | $DF = 3$ | $P = 0.000$ | (adjusted for ties) |

a) Write the null and alternative hypothesis.
b) According to the table above, is the null rejected at 5% significance?
c) Assuming the distribution of scores per terminal have the same shape, does it appear that the difference between medians is large?

16.11 With the help of Minitab, and using the data set "2011_SFO_Customer_ Survey," first subset the rows by whether respondents made a restaurant purchase ("Q4C" = 1). Second, exclude rows where "Q8B" = 0 and "Q8B" = 6. Now reproduce the Kruskall–Wallis test in the table above by using "Q8B" as the response variable, and "TERM" as the terminal variable.

16.12 Is concern about safety with drones, flying over people, related to age? A survey was conducted in Edmonton, Canada, to gather community insight about drones. Participants provided feedback on "I have safety concerns regarding drones flying over people." Possible answers were[7]

---

7 "Don't Know" answers were removed.

"Strongly Agree, Somewhat Agree, Neutral, Somewhat Disagree, Strongly Disagree." Instead of actual age, categories were provided. Kruskal–Wallis test on drone over people concern

| Age | N | Median | Ave rank | Z |
|---|---|---|---|---|
| 18–24 | 48 | 4.00 | 420.9 | −4.26 |
| 25–29 | 119 | 4.00 | 510.3 | −4.16 |
| 30–34 | 224 | 4.00 | 547.3 | −4.34 |
| 35–39 | 162 | 4.00 | 591.6 | −1.97 |
| 40–44 | 121 | 4.00 | 664.5 | 0.59 |
| 45–49 | 124 | 4.00 | 663.8 | 0.58 |
| 50–54 | 113 | 5.00 | 721.7 | 2.28 |
| 55–59 | 144 | 5.00 | 780.7 | 4.62 |
| 60–64 | 99 | 5.00 | 761.2 | 3.22 |
| 65–69 | 70 | 5.00 | 756.5 | 2.56 |
| 70–74 | 48 | 4.00 | 690.5 | 0.85 |
| 75–79 | 18 | 5.00 | 843.9 | 2.28 |
| Overall | 1290 | | 645.5 | |

| | | | |
|---|---|---|---|
| $H = 97.97$ | $DF = 11$ | $P = 0.000$ | |
| $H = 110.82$ | $DF = 11$ | $P = 0.000$ | (adjusted for ties) |

a) Write the null and alternative hypothesis for a Kruskal–Wallis test.
b) According to the table below, is the null rejected at 10% significance?
c) Assuming the distribution of scores per terminal have the same shape, does it appear that the difference between medians is large?

## 16.5 Nonparametric Tests Versus Their Parametric Counterparts

A comparison of features of parametric and nonparametric tests is warranted at this point.

**When the Mean Is Not a Good Measure of Central Tendency** Through simulations, experts have shown that parametric tests are robust to violations to the normality assumption. This provides confidence when inferring about $\mu$, but as we explained in Chapter 4, the population mean is not always the best measure of central tendency. For asymmetric distributions, the population median is better. Thus, when inferring on central tendency of continuous random variables with skewed distributions, nonparametric tests are better choices than parametric tests.

***When Ordinal Data, Ranked Data, or Outliers Are Present*** Parametric tests are designed for continuous random variables. Assuming normality when using ordinal data, when using ranked data, or when outliers are present could be an egregious mistake.[8] Nonparametric tests are distribution-free and nearly as powerful as their parametric counterparts.

For nonparametric methods, confidence intervals are generally constructed using resampling procedures.

## Chapter Problems

16.13 Provide three examples of parametric tests and three examples of nonparametric tests.

16.14 Compare advantages and disadvantages of using a sign test versus a Wilcoxon sign-rank test.

16.15 With the help of Minitab, and using the data set "Edmonton_Insight_Community," first subset the rows by excluding age categories below 18 and above 79. Second, in a new column, recode "Q17_DroneSafetyOver People" with numbers (e.g. "5 Strongly Agree" equals 5, "4 Somewhat Agree" equals 4, etc.). Now reproduce the Kruskall–Wallis results provided in Problem 16.12.

16.16 With the data set "Austin_Community_Survey_15to2017," and the assistance of statistical software, reproduce the results from Example 16.1. Note that the variable of interest is called "the city as a place:1," and you will have to remove "Don't Know" responses.

16.17 Use Minitab to reproduce the results from Example 16.3.

16.18 Using Minitab and the data set "Chattanooga_place:to_live," reproduce the results presented in Problem 16.9. Also, construct barplots to preliminary assess whether it was reasonable to assume that the distribution of responses from both genders was equal.

---

8 For some ordinal data, the concept will not make any sense.

# Further Reading

Agresti, Alan and Franklin, Christine. (2016). *Statistics: The Art and Science of Learning from Data*, 4th edition. Pearson.

Hollander, Myles, Wolfe, Douglas A., and Chicken, Eric. (2014). *Nonparametric Statistical Methods*, 3rd edition. Wiley.

Wackerly, Dennis D., Mendenhall III, William, and Scheaffer, Richard L. (2002). *Mathematical Statistics with Applications*, 6th edition. Duxbury.

# 17

# Forecasting

## 17.1   Introduction

There is a lot of data out there with an index of time. These type of observations are known as a **time series**. For example, monthly sales of winter clothing from a local company for the past 15 years. These data sets have important features over time that are worthwhile taking into consideration. Returning to the winter clothing example, the company may want to determine if the winter clothing sales have been increasing overall. Additionally, the company may wonder if there is a seasonal pattern in the sales of their winter clothing (Do you think they should expect a seasonal pattern?). This type of information can help guide the company's short- and long-term plans.

---

**A Regression Model?**

Some of the regression techniques discussed before may come to mind as a way to model the association between a response variable and time. Although possible, regression techniques may fall short due to differing components of temporal dependency and inefficiency in estimating regression coefficients when observations are not independent. If there is no strong temporal dependence in the data, then regression methods and more generally, predictive analytics can be applied to obtain forecasts.

---

Typically, the time index is at a monthly, quarterly, or yearly scale. It may be discrete and regular (i.e. $t = 1, 2, 3, \ldots$) or discrete and irregular (i.e. $t = 1, 5, 12, \ldots$). The index may be continuous as well and even random. In this chapter, we take a look at some ways to work with time series data with a discrete regular time index.

*Principles of Managerial Statistics and Data Science*, First Edition. Roberto Rivera.
© 2020 John Wiley & Sons, Inc. Published 2020 by John Wiley & Sons, Inc.
Companion website: www.wiley.com/go/principlesmanagerialstatisticsdatascience

## 17.2   Time Series Components

The modeling process for time series data is as follows:

- Define the goal – There are two main goals when analyzing time series data: to understand a process or to forecast future values. When explaining a process, not accounting for dependence leads to biased estimates of variance of a statistic (such as a sample mean), which can lead to inappropriate inference. When forecasting a process, accounting for temporal dependence can lead to better forecasts! Our main focus will be forecasting.
- Establish the **forecast horizon**, which is how far into the future forecasts will be made. In the case of monthly data, it may be 1 month ahead, 3 months ahead, or 12 months ahead.
- Get data.
- Preprocess data – Setting up a time index is one of the requirements. Also, the data is not always available at the desired temporal scale. Data may be available at a hourly scale though our modeling goals call for a daily scale. Moreover, data may be at a scale where missing values are present. Converting to a different scale, say by averaging over the higher resolution scale, can achieve a regular time index.
- Explore time series (reprocess data if necessary) – The aim is to assess each individual observation for the presence of outliers, determine if missing values are (still) an issue, and make a preliminary determination on whether the data has any worthwhile information; and if so, assess the temporal features of the data. This initial assessment could lead to a transformation of the data (to handle outliers) or to converting the time index into a lower resolution time scale.
- Partition time series. The time series should be partitioned in two. The first partition, the one that goes furthest into the past, will be used to construct the time series model. The second partition will be used to assess the forecasting performance of the model. Most often, the majority of the data is used to construct the model, and a smaller fraction is kept to assess its performance.
- Apply modeling method(s) – Taking into consideration the steps above, a model or series of models are constructed.
- Evaluate model performance – The forecasting performance of the model(s) is evaluated. If needed, changes in the model are made.
- Implement model – Once an adequate model is found, the forecasting of the process is made.

---

**A Line Chart to Explore the Data**

One of the first steps when exploring time series data is to construct a line chart. By plotting the observations against time, the analyst can have a first look at the overall behavior of the data. Some questions to consider are as follows:

- What happens to measurements overall as time passes? Do they have about the same average value, or do the average appears to be changing in time?
- Is there a repetitive behavior in the values of the data, a seasonal effect?
- Are there any aberrant observations that require a closer look?

If the average of the data appears to change over time, then we say that a **trend** is present. The line chart can also help determine the appropriate descriptive statistics to use when summarizing the data. Consider all annual corporate taxes assessed[1] by the California Franchise Tax Board from 1950 to 2015 (Figure 17.1). There is a marked increase in tax assessed until 2006, afterwards a drop occurred until 2014. Monthly auto imports[2] reported by the Maryland Port Administration are displayed in Figure 17.2. An increasing

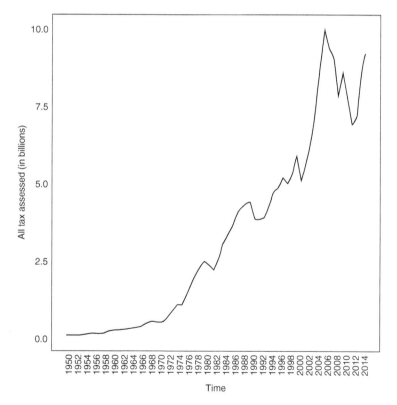

Figure 17.1 Corporate tax assessed by the California Franchise Tax Board from 1950 to 2015.

1 Source: data.ftb.ca.gov (accessed August 20, 2018).
2 Source: opendata.maryland.gov/Transportation/Maryland-Port-Administration-General-Cargo/2ir4-626w (accessed July 26, 2019).

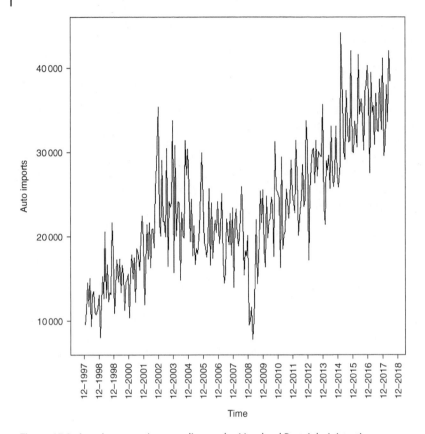

**Figure 17.2** Auto import units according to the Maryland Port Administration.

trend is visible until 2008, when perhaps the financial crisis led to a large drop in imports. Since then, the imports have picked up again.

As another example, Figure 17.3 presents quarterly employment generated in Canada related to tourism activities. As can be seen, employment generation in this sector increases until about 1999, stabilizes until 2009, and has been increasing again in recent years. Furthermore, a consistent variation within every year occurs, a seasonal effect. Specifically, every year more employment is generated in the second and third quarters, when more tourists visit, perhaps because of the warmer weather.

Let $Y_t$ represent the value of the time series at time $t$, such that $t = 1, \ldots, n$, and $n$ is the number of time points. Thus, our time series is $Y_1, \ldots, Y_n$. In previous chapters, each observation was independent. Now, observations are no longer independent. Most generally, with the temporal dependence, the probability distribution of $Y_1, \ldots, Y_n$ has considerably more parameters. To make the

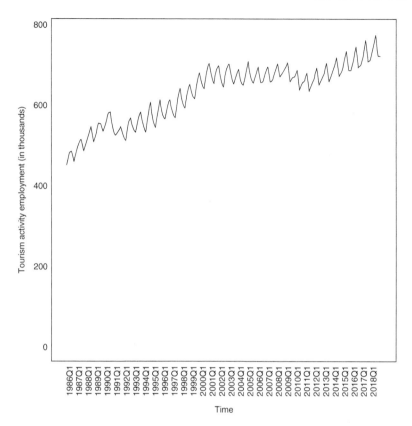

Figure 17.3 Quarterly employment generated by tourism activity in Canada.

modeling more tractable, some simplification of the temporal dependence is needed. For forecasting, $Y_t$ is additively decomposed into three components:

$$Y_t = T_t + S_t + \epsilon_t \tag{17.1}$$

$T_t$ stands for the trend component, $S_t$ for the seasonal component, and $\epsilon_t$ includes short-term temporal dependence or random fluctuations.

- Data may have a constant mean for all $t$ (no trend), or the mean may be some function of $t$.
- For nonseasonal data, $S_t = 0$ for all $t$.
- $\epsilon_t$ are not necessarily independent.

Some temporal processes are expected to also have a cycle component: a predictable upward and downward pattern occurring every few years. But these cycles are hard to model, in part because of their complexity, in part because the amount of data may not be enough to model it. Thus, we disregard any cycle component.

**Multiplicative Model**

Model ((17.1)) is an additive model. A multiplicative model is also possible,

$$Y_t = T_t \times S_t \times \epsilon_t$$

By taking the logarithm on both sides of the equality, this multiplicative model becomes additive.

Figure 17.4 summarizes the steps to explore time series data. First, the line chart is evaluated for a trend. Any visual evidence of a trend requires exploration of the type of trend. This will permit modeling the trend component which can then be "removed" from the observations by subtracting the fitted trend from each observation. Then a similar assessment is done for the other components of temporal dependence in the data. In Sections 17.3 and 17.4 that follow, how to model the components of temporal dependence is briefly addressed.

*Other Ways to Visualize a Time Series* Line charts are not always informative about the temporal dependence of the time series. The two main reasons are too much data or too many aberrant observations (Figure 17.5). When this is the case, it may help to construct line charts for several windows of time through the period covering the data set. Many other visual displays besides line charts are available to explore time series data.

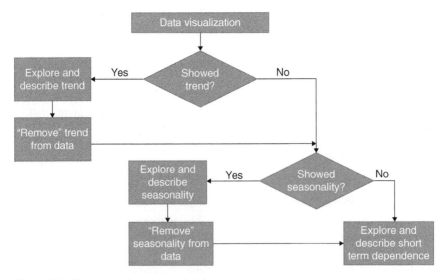

Figure 17.4 Steps to explore time series data.

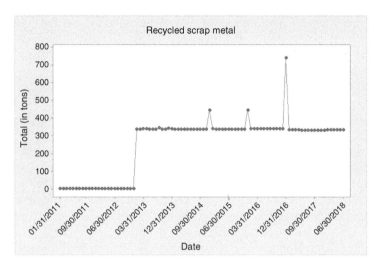

Figure 17.5 Buffalo, New York, scrap metal recycling monthly data. Little variation, with first few months having no recycling followed by some months with much more action than average.

- For example, for monthly data with no trend, a boxplot can be made per month for every year to help detect seasonal effects.[3]
- Histograms can also be of use, sometimes after the data has been processed to evaluate its distribution.
- For some time series, observations that are close together in time are more correlated than observations that are further apart. Specifically, for all values of $t$, $Y_t$ may be more correlated to the previous observation, $Y_{t-1}$, than observations further into the past. The time difference between observations is called the **lag**. Thus, observations that are one unit apart are lag 1. Scatterplots of a time series with no trend or seasonal effect versus some lagged version of itself can help explore this type of short-term temporal dependence.
- Another option are charts that measure temporal correlation over several lags.

## Practice Problems

17.1 Yearly water use per capita in New York City is available[4] ("NYC_Water_ Consumption"). Draw a line chart for water use per capita. Interpret chart.

---

3 In the case of a time series with a trend, it would need to be removed from the data first.
4 Source: data.cityofnewyork.us (accessed July 26, 2019).

17.2 Access the monthly percentage of 311 calls attended within 60 seconds from April 2007 to November 2016[5] in San Francisco ("311_Call_Metrics_by_Month"). Construct a line chart and discuss whether trend or seasonality are visible.
*Hint:* Make sure the rows are chronologically increasing.

17.3 Monthly amount of recycled water (acre feet) delivered in the city of Los Angeles is available[6] ("Recycled_Water_LA").
a) Use Minitab to build a line chart. Does there appear to be a trend? Seasonality?
b) Create boxplots for each month of the amount of recycled water delivered. Any indications of seasonality? (You need to build a new column that defines month factors.)

17.4 The file "canadaemplmntdata" contains quarterly Canada employment data for multiple job sectors. The variable name is VALUE, and it is measured in thousands. With computer software assistance, get a chart similar to Figure 17.3 by subsetting the data by[7] VECTOR = v81673.

17.5 Returning to the data set "canadaemplmntdata" from Problem 17.4, get a line chart of Accommodation jobs by subsetting by VECTOR = v81682.

## 17.3  Simple Forecasting Models

The aim is to forecast values of the process at future times $n + 1, n + 2$, and $n + h$, where $h$ is the forecast horizon, using a model constructed from observations at times $1, ..., n$. First, we consider the scenario where the time series does not show any trend or seasonality. The easiest model for forecasting is the average of the data. Think of it as a benchmark method to assess the strength of the temporal dependence in the data. For example, monthly sales of winter clothing are likely to have a strong seasonal component. The average of the data would not exploit the opportunity to improve on the forecast by accounting for the month of the year, leading to poorer forecasts than models that account for seasonality. Another straightforward model is known as the **naive model**, where forecasts are set to be the last observation.

---

5  Source: data.sfgov.org (accessed July 26, 2019).
6  data.lacity.org (accessed July 26, 2019).
7  A subset corresponding to tourism employment data unadjusted for seasonality.

Table 17.1 Annual chronic lower respiratory mortality rates for White Non-Hispanic men in New York City.

| Year | 2007 | 2008 | 2009 | 2010 | 2011 | 2012 | 2013 | 2014 |
|------|------|------|------|------|------|------|------|------|
| Rate | 20.0 | 18.3 | 16.9 | 20.6 | 21.7 | 21.3 | 18.5 | 20.6 |

**Example 17.1** *Annual New York City chronic lower respiratory disease mortality rates for White Non-Hispanic men are available*[8] *(see Table 17.1). What is the 2015 naive forecast for this variable?*

Because there are only eight time points available, and there is no evidence of a trend over time, applying the naive model is reasonable. Our forecast for 2015 would be the last observation available: 20.6. If forecasts are desired further into the future, the naive model will result in the same forecast for each year.

The implication of the naive model is that the data does not provide much information to forecast the process. Best one can do is use the latest observation, even if one is attempting to forecast a value rather far into the future.

*Moving Average*   When a time series is void of a consistent trend or seasonality, but appears to have a mean that is changing over time, a **moving average** could be applied. The moving average takes windows of $m$ "close by observations," and returns their average as output.

---

**Two Types of Moving Average**
A **trailing moving average** is when the last $m$ observations are used. In contrast, **centered moving average** uses the observations at time points[9] $t - \frac{m-1}{2}, \ldots, t + \frac{m-1}{2}$ to generate an average. Instead of generating a forecast, the centered moving average will generate a smoothed version of the data. On the other hand, the trailing moving average can generate a forecast for time period $n + 1$; beyond that, it cannot be updated.

---

**Example 17.2** *San Francisco monthly percentage of 311 calls attended within 60 seconds*[10] *does not feature consistent trend or seasonality; a perfect time series candidate to illustrate how moving averages work.*

It is easier to visually present moving average results instead of the numbers. Figure 17.6 shows the Minitab windows to obtain a graph of moving averages.

---

8  Source: data.cityofnewyork.us (accessed July 26, 2019).
9  The expression applies when $m$ is odd. When $m$ is even, software will perform a more complicated computation.
10  Source: data.sfgov.org (accessed July 26, 2019).

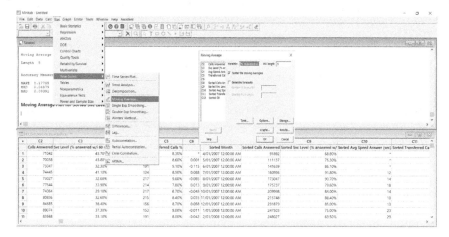

Figure 17.6  Minitab window grabs to generate moving average graphs.

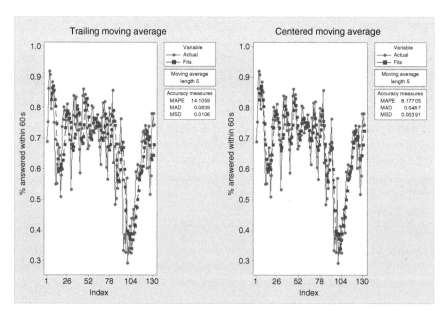

Figure 17.7  Monthly percentage of 311 calls attended within 60 seconds and corresponding moving averages.

Figure 17.7 shows the original time series (blue) and the corresponding trailing moving average at multiple time points (left panel, red) and centered moving averages (right panel) when $m = 5$.

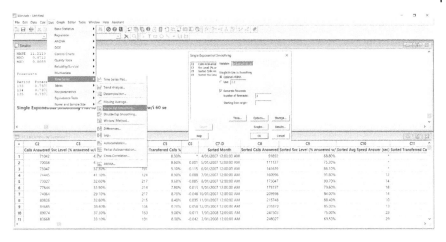

Figure 17.8 Minitab window grabs to generate single exponential smoothing.

**Simple Exponential Smoothing** This method produce forecasts as weighted averages of the observation at time $t$, and the forecast at that time point,

$$F_{t+1} = \alpha^* Y_t + (1 - \alpha^*) F_t$$

where $0 \le \alpha^* \le 1$ is known as the **smoothing parameter**.

---

$\alpha^*$

It should be noted that $\alpha^*$ in this chapter is not the significance level, $\alpha$, used in other chapters. Instead of defining a new Greek symbol for the smoothing parameter, $\alpha^*$ is used for the smoothing parameter for compatibility with the notation used in Minitab.

---

$F_{t+1}$ represents the forecast of the **level**, also known as the **local mean value**. The further into the past from the forecast time point observations are the less weight[11] the observation receives. Simple exponential smoothing produces forecast to horizon $h$ that fall on a flat line (no trend or seasonality assumed).

**Example 17.3** *Continuing with Example 17.2, Figure 17.8 demonstrates how to create simple exponential smooths for San Francisco data on monthly percentage of 311 calls attended within 60 seconds. For this example, we do not alter the default of estimating the smoothing parameter from data. Forecasts at T + 1, T + 2, and T + 3 will be calculated. Figure 17.9 displays the exponential smooth at every t ≤ T (red) along with the three forecasts (green), and 95% prediction intervals (purple). The output also shows us that $\alpha^*$ was estimated to be 0.68.*

---

11 Weights decay exponentially, since it can be shown that $F_t$ is an exponential function of $\alpha^*$.

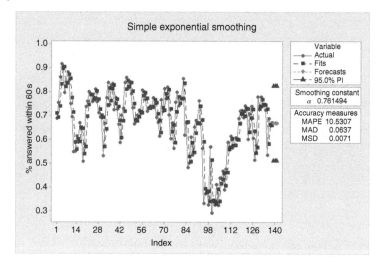

Figure 17.9 Monthly percentage of 311 calls attended within 60 seconds, corresponding exponential smooths, and 95% prediction intervals.

## Practice Problems

17.6 The table below shows the most recent observations of quarterly wine production (in pounds) from a California winery.

| Quarter | Q4-16 | Q1-17 | Q2-17 | Q3-17 | Q4-17 | Q1-18 | Q2-18 | Q3-18 |
|---|---|---|---|---|---|---|---|---|
| Production | 2000.0 | 350.3 | 420.9 | 2100.6 | 2050.7 | 210.3 | 380.5 | 2200.6 |

a) According to the naive method, what is the forecast for Q4-18?
b) According to the naive method, what is the forecast for Q1-19?
c) How good do you think the Q1-19 forecast in part (b) will be? Explain.

17.7 A firm that manufactures snowboards records monthly sales. They have data until August 2018, when sales were $109 303.
a) According to the naive method, what is the forecast for September 2018?
b) According to the naive method, what is the forecast for January 2019?
c) How good do you think the January 2019 forecast in part (b) will be? Explain.

17.8 Using the data from Example 17.2 ("311_Call_Metrics_by_Month"), with $m = 3$ and using statistical software, create a (Note: the original data set will have to be sorted first by the "Month" variable.),
a) centered moving average chart.
b) trailing moving average chart.

17.9 Using the data from Example 17.1, with $m = 2$ and using statistical software, create a
   a) centered moving average chart.
   b) trailing moving average chart.

17.10 Consider annual corporate taxes reported[12] by the California Franchise Tax Board from 1950 to 2015 (file "Ca_Income_Loss_Tax_Assessed"; variable Tax Assessed on Net Income.).
   a) Using statistical software, create a simple exponential smoothing chart.
   *Hint:* Make sure the rows are chronologically increasing.
   b) Do you think simple exponential smoothing would provide reliable forecasts for this data?

17.11 The file from Problem 17.10 also includes the variable Returns w/Net Income. With the help of statistical software, create a simple exponential smoothing chart.

## 17.4 Forecasting When Data Has Trend, Seasonality

The methods from Section 17.3 do not apply when data exhibits trend or seasonality. First, let's consider a time series with trend, but no seasonality.

***The Trend Component*** Trend describes the general movement of the time series over the entire period (Figure 17.10).
   Two methods are commonly used to model the trend component of a time series:

- Curve fitting – The time series is fitted as some function of time. Typically a linear, quadratic, cubic, or exponential function is chosen. In a linear trend, the time series increases or decreases linearly as a function of time. In a polynomial trend, data values change nonlinearly in time; and in an exponential trend, the time series changes (increases or decreases) over time at a rate proportional to its values.
- Filtering – In this method, the trend is modeled as a function of past observations. The further in the past the observation, the less weight the observation is given at each time $t$. There are two types of filtering methods. The first is the **double exponential smoothing** method, also known in practice as **Holt's method**. This method expresses the model in terms of two smoothing parameters: one for the level, and the other for trend. It is used when

---

12 Source: data.ftb.ca.gov (accessed August 20, 2018).

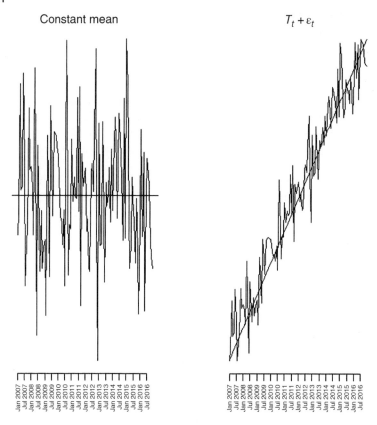

Constant mean $\qquad\qquad\qquad\qquad\qquad\qquad\qquad T_t + \varepsilon_t$

Figure 17.10 The time series on the left has a constant mean, no trend. The one on the right has a linear increasing trend. Neither time series has seasonality.

the time series exhibits a trend but no seasonality. The second type of filtering is **differencing**, used when the trend is not of much interest and instead the aim is to forecast the time series. With this method, a new time series is created from the differences of each observation and the previous one. For nonseasonal data, one or two differences are often sufficient to remove the trend.

The advantages and disadvantages of each method to model trend are summarized in Table 17.2. When a curve is fit through a regression procedure, the output is easily interpretable. However, a linear model may not be flexible enough, while high-order polynomial models tend to overfit the data: model performance will look good, but forecast will be poor. Filtering methods do not have convenient interpretations as curve fitting models do, but are generally efficient in capturing complex trends.

Table 17.2 Pros and cons of methods to model the trend component in a time series.

| Method | Pros | Cons |
|---|---|---|
| Curve fitting | Easy to understand | Limited flexibility |
| | Valuable when trends are of general interest | High-order polynomials can overfit |
| Filtering | Efficient in removing trend | Harder to understand method |
| | Differencing requires | Not useful when trend is of general interest |
| | no parameter estimation | |

**The Seasonal Component** In Eq. (17.1), $S$ models a repetitive upward and downward pattern occurring within every year. The left panel of Figure 17.11 shows a scenario where peak values of the year are seen in July and low values in January. The data for the right panel has the same seasonality and also an increasing linear trend.

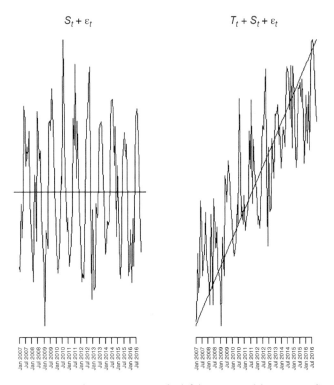

Figure 17.11 The time series on the left has no trend, but seasonality. The one on the right has a linear increasing trend and seasonality.

Slight variations to the techniques to model trend are available to model seasonality. For example, there is the **seasonal naive model**, which can be considered when there is a strong seasonal component to the data. Instead of using the latest observation like the naive model does, the seasonal naive model uses a past observation as the forecast according to the type of seasonality. Thus, if monthly data is seasonal, the forecast for next November would be the observation from last November. Other methods include the following:

- Curve fitting – One way is to set a categorical variable denoting the season of each observation. Computer software then generates dummy binary variables to be included as predictors in the model. To illustrate this, consider monthly hotel registration data from Puerto Rico. Table 17.3 shows how a season category has been created to assess seasonality. $m - 1$ binary variables are required for $m$ seasons (see Section 14.7). Trigonometric functions are another alternative to model seasonality.
- Filtering – The **Holt–Winters exponential smoothing method** is a form of filtering applied in practice to model time series with a trend and seasonality. **Seasonal differencing** is another filtering technique employed to model seasonality. Specifically, for monthly data, a new time series is created from

Table 17.3 Subset of Puerto Rico hotel registration data (in thousands). The month variable has been generated to assess seasonality.

| Date | Season | Registrations |
|---|---|---|
| Jan-04 | Jan | 125.354 |
| Feb-04 | Feb | 131.859 |
| Mar-04 | Mar | 145.920 |
| Apr-04 | Apr | 136.118 |
| May-04 | May | 105.419 |
| Jun-04 | Jun | 113.448 |
| Jul-04 | Jul | 128.527 |
| Aug-04 | Aug | 121.761 |
| Sep-04 | Sep | 74.248 |
| Oct-04 | Oct | 94.348 |
| Nov-04 | Nov | 111.505 |
| Dec-04 | Dec | 123.332 |
| Jan-05 | Jan | 126.046 |
| Feb-05 | Feb | 134.437 |
| Mar-05 | Mar | 157.340 |

the differences of each observation and the observation 12 months in the past.

- Seasonal adjustment – In this method, each value of the time series is adjusted through a seasonal index. The seasonal index helps highlight seasonal differences. How each value is adjusted with a seasonal index depends on whether an additive or a seasonal model is applied.

**Example 17.4** *To assess hotel room demand in Puerto Rico, an island in the Caribbean, we use monthly hotel registration data from January 2004 until September 2012.*

Figure 17.12 shows the line chart of monthly hotel registrations in Puerto Rico. A seasonal pattern is clearly visible. Since no significant trend is present

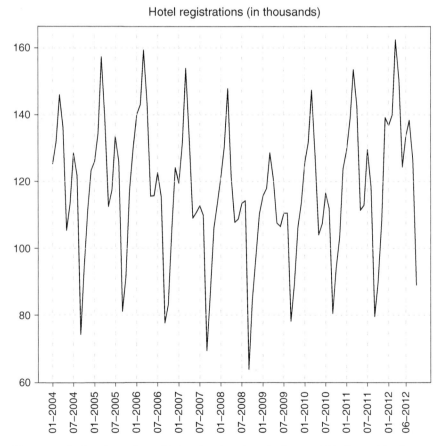

**Hotel registrations (in thousands)**

Figure 17.12 Line chart of monthly hotel registrations (in thousands) in Puerto Rico from January 2004 to September 2012.

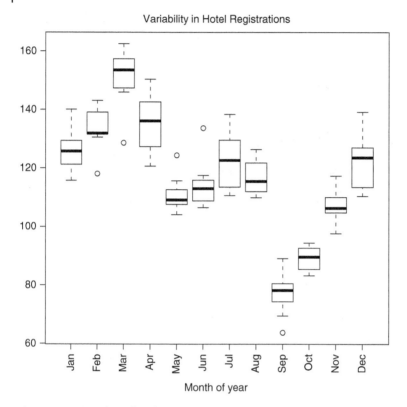

Figure 17.13 Boxplots of total hotel registrations (in thousands) for each month of the year.

in the line chart, the seasonality of the data can be further assessed by building boxplots of hotel registrations for every month of the year (Figure 17.13). Clearly, there is a marked difference in hotel registrations in some months with March having the highest hotel registrations on average, and September experiencing the lowest.

***Modeling the Error Component*** Short-term temporal dependence is captured through $\epsilon$, and the upshot of adequately modeling it are better forecasts. If the time series has no trend or seasonal effect, scatterplots of the time series versus some lagged version of itself can help explore short temporal dependence. Alternatively, sample correlations of the time series and several lagged versions of the time series can be studied visually. Although widely used in practice, these visual aids known as **autocorrelation plot** and **partial autocorrelation plot** are beyond the scope of this book.

Table 17.4 Pros and cons of methods to model a time series with trend and seasonality.

| Method | Pros | Cons |
|---|---|---|
| Curve fitting | Easy to understand | Limited flexibility |
| | Valuable when trends are of general interest | High-order polynomials can overfit |
| Filtering | Efficient in removing trend | Harder to understand method |
| | No parameter estimation needed | Not useful when trend is of general interest |
| Seasonal adjustment | Convenient to decompose time series into components | Limited flexibility |

---

**ARMA Models**

Collectively, methods that model short-term temporal dependence, for time series without trend or seasonality, are known as **Autoregressive Moving Average models, or ARMA models**. These models were popularized by two researchers, George Box and Gwilym Jenkins. Unfortunately, the fitting and selection of ARMA models is not trivial and should be performed by an expert.

---

Table 17.4 provides a quick summary of these methods.

---

**SARIMA Models**

George Box and Gwilym Jenkins extended the ARMA model for time series with trend, seasonality, and short-term dependence through Seasonal Autoregressive Integrated Moving Average Models (SARIMA), also known as Box–Jenkins methods. SARIMA models are indeed a large family of models that provide good flexibility to analysts. They can have one parameter, two parameters, or many more. Theoretically, SARIMA models will outperform any other competing model for time series and deserve consideration when at least 50 observations are available. But, the fitting and selection of SARIMA models demands good knowledge on the topic. Moreover, in practice, simpler models such as the Holt–Winters model can in fact give better forecast than SARIMA models. This is especially true when the temporal dependence in a time series is dominated by the trend and seasonal effects.

---

## Practice Problems

17.12 Returning to Problem 17.2,
    a) using Minitab, fit a linear trend model.

b) using Minitab, fit a quadratic trend model.

c) visually, which of the two fits appears to be best?

17.13 Subset the Buffalo, New York, curb recycling data ("Buffalo_Monthly_Recycling"), by the variable Type = "Curb Recycling," with the help of computer software to fit a

a) linear trend model.

b) quadratic trend model.

c) visually, which of the two fits appears to be best?

17.14 Download the file "Buffalo_Monthly_Recycling." For the variable Type = "Curb Garbage,"

a) construct a line chart and interpret.

   *Hint:* You will need to subset the data first.

b) fit the data using the double exponential smoothing method.

17.15 Figure 17.2 indicates a trend in monthly auto imports[13] reported by Maryland Port Administration. Using the variable Import Auto Units in the file "Maryland_Port_Cargo," forecast imports three months ahead using a quadratic trend model.

17.16 Using the variable Export Auto Units in the file "Maryland_Port_Cargo,"

a) construct a line chart and argue why a quadratic trend would be reasonable.

b) fit the data using the quadratic trend method.

c) forecast exports six months into the future.

17.17 Returning to the data from Problem 17.6, according to the seasonal naive method, what is the forecast for Q4-18?

17.18 Returning to the data from Problem 17.6, according to the seasonal naive method, what is the forecast for Q1-19?

17.19 The file "canadaemplmntdata" contains quarterly Canada employment data for multiple job sectors. The variable name is VALUE, and it is measured in thousands. With computer software assistance, fit a Holt–Winters model after subsetting the data by[14] VECTOR = v81673.

---

13 Source: opendata.maryland.gov/Transportation/Maryland-Port-Administration-General-Cargo/2ir4-626w (accessed July 26, 2019).

14 A subset corresponding to tourism employment data unadjusted for seasonality.

**17.20** Returning to the data set "canadaemplmntdata" from Problem 17.19, fit a Holt–Winters model for Accommodation jobs after subsetting by VECTOR = v81682.

**17.21** Often, it is not enough to assume that seasonality or trend components are present. We must evaluate the need to fit these components. The file "canadaemplmntdata" contains quarterly Canada employment data for multiple job sectors. It provides seasonally adjusted and unadjusted versions of the data.
   a) Construct a line chart for seasonally unadjusted tourism employment data by subsetting the data by VECTOR = v81673.
   b) Construct a line chart for seasonally adjusted tourism employment data by subsetting the data by VECTOR = v81658.
   c) Based on the line charts in parts (a) and (b), does it seem that seasonality is important?

**17.22** Returning to the data set "canadaemplmntdata" from Problem 17.21,
   a) construct a line chart for seasonally unadjusted tourism employment data by subsetting the data by VECTOR = v81675.
   b) construct a line chart for seasonally adjusted tourism employment data by subsetting the data by VECTOR = v81660.
   c) based on the line charts in parts (a) and (b), does it seem that seasonality is important?

**17.23** Another easy way to assess trend or seasonality fits are to plot the residuals, the detrended or deseasonalized data. Consider annual returns with net income[15] reported by the California Franchise Tax Board from 1950 to 2015 (file "Ca_Income_Loss_Tax_Assessed"; variable Returns/w Net Income.).
   a) Fit an exponential trend model, and record the model residuals. *Hint:* If using Minitab, perform a Trend Analysis. Select the "Storage" button in the Trend Analysis window, and place a checkmark on the residuals.
   b) Fit a quadratic trend model, and record the model residuals.
   c) Construct line charts of the residuals in part (a), (b) to visually determine which model appear to have the least systematic error (which has the most nonrandom pattern).

**17.24** Consider annual corporate taxes (file "Ca_Income_Loss_Tax_Assessed"; variable Tax Assessed on Net Income.).
   a) Fit a Holt–Winters model and record the model residuals.

---

15 Source: data.ftb.ca.gov (accessed August 20, 2018).

*Hint:* If using Minitab, perform Winters' Method. Select the "Storage" button in the Winters' Method window, and place a checkmark on the residuals.

b) Construct line charts of the residuals in part (a) to visually determine if the model has any systematic error (if it has a nonrandom pattern).

17.25 Use quarterly criminal domestic violence data[16] from Edmonton, Canada, "EPS_Domestic_Violence."

a) With the help of statistical software, create one lag differences, then chart them.

b) Does the line chart of the differences appear to have a trend?

17.26 With the data on noncriminal domestic violence in "EPS_Domestic_ Violence," repeat the tasks in Problem 17.25.

## 17.5   Assessing Forecasts

> **Inference in Time Series Models**
> Some forecast models were not initially developed using a statistical framework (e.g. Holt–Winters model). Setting up forecast models in a statistical framework allows efficient estimation of parameters, selection of models (say, how many parameters to use in class of model), hypothesis testing, and prediction intervals. Unfortunately, the inference for time series can become rather technical and is therefore not discussed any further.

With several candidate models to choose from to get forecasts, it is wise to compare the forecast accuracy of these candidate models. Many measures that are a function of forecast error, the difference between the true observation and its forecast, are available. If forecast error is

$$e_{t+h} = y_{t+h} - \hat{y}_{t+h}$$

some forecast accuracy measures are as follows:

- **Mean Absolute Error (MAE)** – Average absolute forecast error over all $t$.
- **Mean Absolute Percentage Error (MAPE)** – Average absolute forecast percentage error over all $t$ where each percentage error is $100e_t/y_t$.
- **Root Mean Squared Error (RMSE)** – Square root of the average squared forecast prediction error over all $t$.

---

16 dashboard.edmonton.ca (accessed December 12, 2018).

The forecast performance of several candidate models can be evaluated by comparing these errors from different models. The model that generally has the lowest errors will be best. When any of these error measures are defined considering the observations used to build the model, we call them **in-sample errors**. Problem is, in-sample errors tend to be overoptimistic in terms of estimating forecasting error. Alternatively, the time series can be separated into two sets:

- The first $m$ observations, the training set, will be used to build the models
- The remaining $m + 1, \dots, n$ observations, the test set, will be used to estimate forecast errors.

The error measures from this alternative approach are called **out-of-sample errors**. A model could have a very low in-sample error but perform badly in terms of out-of-sample error. Thus, in-sample error makes us prone to overfitting the data: when your model captures the temporal behavior and the noise. Out-of-sample errors do present a practical challenge in choosing the size of the training and test sets. Enough data is required for fitting the model, and estimating forecasting errors, so both sets should be sizable. But often, the entire data set available is not too big, and using a model that was fit to data far into the past may not capture well current temporal behavior. Thus, when the entire data set is big enough, usually most of the data is used for the training set, and a handful of observations are used for the test set.

**Example 17.5** *Returning to Example 17.4, using monthly hotel registration data from January 2004 until March 2013 for Puerto Rico, an island in the Caribbean, the goal is to select the best forecasting model. Observations from October 2012 until March 2013 will not be used to fit models and instead will help assess the forecasting performance of models.*

Since the data set shows strong evidence of seasonality, while trend is less obvious (Figures 17.12 and 17.13), for simplicity, we will only compare the forecasting performance of a Holt–Winters model with a Seasonal Naive model. Figure 17.14 displays hotel registrations for the last few months of data (black line). For the test set, the data that was left out of the models (red), the Holt–Winters forecast hotel registrations (green), and seasonal naive forecasts (blue) are also presented. Visually, both models perform well in some months, but the seasonal naive model has several instances when the forecast deviates more from the actual hotel registrations than the Holt–Winters forecasts. Table 17.5 summarizes some in-sample errors and out-of-sample errors for the first six months after September 2012. In general, in-sample errors are always smaller than their out-of-sample counterparts. Moreover, all out-of-sample errors indicate that the Holt–Winters model is superior to

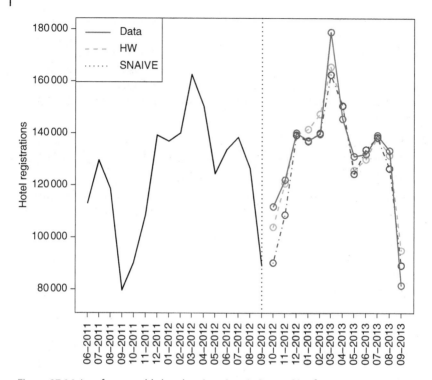

Figure 17.14 Last few monthly hotel registrations in Puerto Rico from January 2004 to September 2012. Additionally test data (red), Holt–Winters model forecast (green), and Seasonal Naive model forecast (blue) have been added.

Table 17.5 Forecast accuracy comparison of models: HW – Holt–Winters model, SNAIVE – seasonal naive model.

| | Forecast accuracy | | | | | |
|---|---|---|---|---|---|---|
| | MAE | | MAPE | | RMSE | |
| Model | In | Out-6 | In | Out-6 | In | Out-6 |
| HW | 4 220.69 | 5 901.93 | 3.80 | 4.14 | 5 422.68 | 7 330.08 |
| SNAIVE | 6 702.11 | 8 822.67 | 5.72 | 6.77 | 8 213.00 | 12 373.1 |

"In" column shows in-sample errors, "Out-6" out-of-sample errors up to six months ahead.

the Seasonal Naive model. This is not to say that the Holt–Winters method is the best model, the meaning is that Holt–Winters performs better than the Seasonal Naive model, and other models may outperform the Holt–Winters model.

---

**Tips to Develop Good Forecasts**

- Team should have at least one well-qualified data analyst.
- Use sufficient data. For the simplest models, not much data is required, but for more flexible models (i.e. SARIMA), at least 50 time points should be used.
- Compare forecast errors of different models. The smaller the error the better, but rely on Occam's razor principle.
- Maintain your forecasting procedure. Just because a model is best now, it does not mean it will be best five years down the line. In the future, a different model may be needed.
- Assess modeling assumptions.
- Apply forecast intervals, not just point forecasts.

---

### 17.5.1  A Quick Look at Data Science: Forecasting Tourism Jobs in Canada

Through its open portal, the Canadian government shares many useful data sets. Among them are quarterly employment data for multiple sectors of the economy, going back to January 1986. In this section, we will forecast tourism employment. Other than filtering by seasonally unadjusted tourism activity data, minor additional data wrangling was necessary (Figure 17.15).

Recall that Figure 17.3 suggested tourism employment in Canada increased until about 1999, stabilized until 2009, and increased again in recent years. A seasonal effect was also apparent; with more employment generated in the second and third quarters, when more tourists visit, perhaps because of the warmer weather. A time series decomposition into temporal components lends further support of a trend and seasonal effect (Figure 17.16).

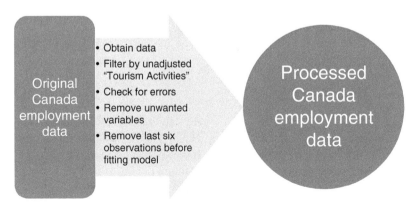

Figure 17.15 Procedure to preprocess Canada's tourism employment data.

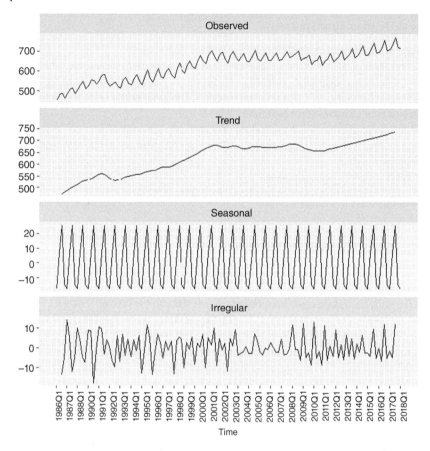

Figure 17.16 Decomposition of Canada tourism employment data.

Based on the exploratory analysis, we will compare three models:

- Holt–Winters.
- SARIMA.
- Seasonal naive.

The last six months of tourism employment data will be left out of the fitting process and used to compare models. Table 17.6 displays in-sample and out-of-sample errors for the three models. The in-sample errors indicate that SARIMA is the superior model; with Holt–Winters having the second smallest errors, and the seasonal naive model having much higher in-sample errors. However, the out-of-sample errors indicate the Holt–Winters model is best, generally being about 15–40% smaller than the SARIMA out-of-sample errors.

Table 17.6 Forecast accuracy comparison of models: HW – Holt–Winters model, SNAIVE – seasonal naive model.

| Model | MAE In | MAE Out-6 | MAPE In | MAPE Out-6 | RMSE In | RMSE Out-6 |
|---|---|---|---|---|---|---|
| | | | **Forecast accuracy** | | | |
| HW | 6.64 | 2.20 | 1.10 | 0.30 | 8.82 | 2.62 |
| SARIMA | 5.78 | 3.50 | 0.94 | 0.49 | 8.05 | 4.44 |
| SNAIVE | 24.30 | 37.11 | 3.91 | 5.15 | 27.60 | 39.59 |

"In" column shows in-sample errors, "Out-6" out-of-sample errors up to six months ahead.

Table 17.7 Employment data removed to assess forecasts, the Holt–Winters forecast, and their prediction interval limits.

| Date | Jobs | Forecast LL | Forecast | Forecast UL |
|---|---|---|---|---|
| 2016 Q4 | 706.80 | 687.70 | 704.98 | 722.27 |
| 2017 Q1 | 711.80 | 689.25 | 712.49 | 735.72 |
| 2017 Q2 | 736.00 | 708.42 | 737.72 | 767.02 |
| 2017 Q3 | 775.10 | 734.65 | 774.20 | 813.75 |
| 2017 Q4 | 721.40 | 672.59 | 718.08 | 763.57 |
| 2016 Q1 | 720.80 | 673.77 | 725.59 | 777.40 |

LL and UL for lower limit, upper limit respectively.

Table 17.7 shows the six observations that were removed from the time series before fitting the models, Holt–Winters point forecasts, and 95% prediction interval[17] limits. Overall, the Holt–Winters model provided good forecasts and actual job data was always within the prediction intervals.

There is no guarantee that the Holt–Winters and SARIMA models will perform best in any given situation. In Section 17.5.2, we compare the performance of the same three methods on a different data set.

### 17.5.2 A Quick Look at Data Science: Forecasting Retail Gross Sales of Marijuana in Denver

To attain our goal, we will use monthly gross sales data for the city from January 2014 until September 2017. Minor data wrangling is required (Figure 17.17). The line chart gives the impression of seasonal and trend effects present (Figure 17.18). When the time series is decomposed into temporal components, trend and seasonal effects are once more implied (Figure 17.19).

---

17 The interpretation of prediction intervals for forecast is the same as in the context of regression. For simplicity, no theoretical discussion of prediction intervals is made here.

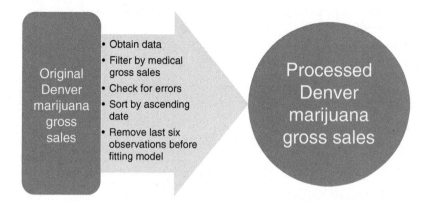

Figure 17.17 Procedure to preprocess Denver's marijuana monthly total retail gross sales data.

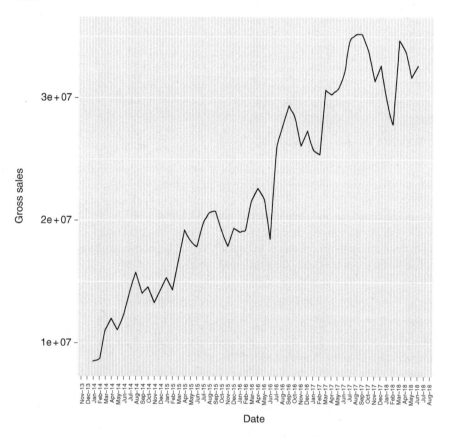

Figure 17.18 Line chart of Denver's marijuana monthly retail gross sales.

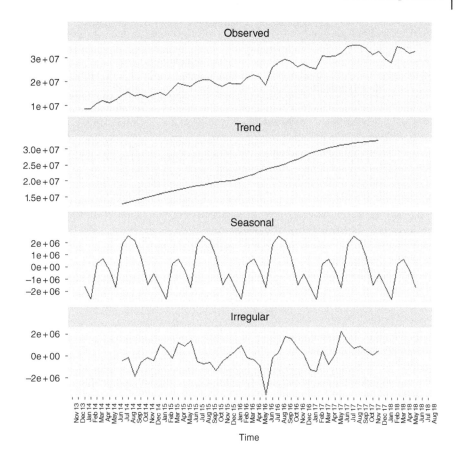

Figure 17.19 Decomposition of marijuana monthly retail gross sales.

Specifically, gross sales have been increasing. Also, during the summer, gross sales are higher than during the winter. However, the peaks and valley are not consistently at the same month every year. For example, in 2014, lowest monthly sales occurred in January, the following year in February, in 2016 it was in June.

Gross sales of the last six months will be left out of the fitting process and used to compare models.

Table 17.8 displays in-sample and out-of-sample errors for the Holt–Winters method, SARIMA, and a seasonal naive model. The in-sample errors indicate that Holt–Winters is the superior model; with SARIMA having the second smallest errors. However, the out-of-sample errors show no benefit of using the more sophisticated Holt–Winters and SARIMA models, and SNAIVE has the smallest forecast error.

Some remarks about these results are warranted:

Table 17.8 Forecast accuracy comparison of models: HW – Holt–Winters model, SNAIVE – seasonal naive model.

| | Forecast accuracy | | | | | |
| | MAE | | MAPE | | RMSE | |
| Model | In | Out-6 | In | Out-6 | In | Out-6 |
| --- | --- | --- | --- | --- | --- | --- |
| HW | 779 109.70 | 2 913 344.70 | 3.82 | 9.25 | 1 063 514 | 3 089 227 |
| SARIMA | 972 370.50 | 2 741 927.30 | 3.91 | 8.74 | 1 605 051 | 3 038 349 |
| SNAIVE | 1 508 552 | 2 067 285 | 7.42 | 6.87 | 2 007 285 | 2 609 489 |

"In" column shows in-sample errors, "Out-6" out-of-sample errors up to six months ahead.

- Holt–Winters and the more advanced SARIMA model can often lead to better forecasts. But this is not always the case.
- Our results do not preclude one of many other methods to forecast time series to perform well.
- The results here in part are explained by the data being dominated by random behavior.

## Chapter Problems

17.27 Consider the data in the file "Ca_Income_Loss_Tax_Assessed"; variable Returns w/Net Income.
a) Using statistical software, create a double exponential smoothing chart.
   *Hint:* Make sure the rows are chronologically increasing.
b) How would you argue that this model is better than simple exponential smoothing to perform forecasts?

17.28 Subset the Buffalo, New York, recycled tires data ("Buffalo_Monthly_Recycling"), by the variable Type = "Recycled Tires." With the help of statistical software to fit a,
a) Simple exponential smoothing.
b) Quadratic trend model.
c) Visually, which of the two fits appears to be best?

17.29 Subset the Buffalo New York recycled tires data ("Buffalo_Monthly_Recycling"), by the variable Type = "E-waste" and exclude data before April 30, 2012. With the help of statistical software to fit a,
a) Simple exponential smoothing.
b) Exponential trend model.
c) Visually, which of the two fits appears to be best?

17.30 The variable name VALUE in the file ("canadaemplmntdata") provides quarterly Canada employment data (in thousands). With statistical software assistance,
    a) fit a Holt–Winters model after subsetting the data by[18] VECTOR = v81676.
    b) for air transportation jobs, fit a simple exponential model.

17.31 Returning to the data set "canadaemplmntdata" from Problem 17.30, for other industries jobs (subsetting by VECTOR = v81687),
    a) fit a Holt–Winters model.
    b) fit a simple exponential model.

17.32 The Denver Colorado marijuana gross sales data file, "CO_mj_gross_sales," includes monthly retail total gross sales.
    *Hint:* Subset data set by gross sales type: retail total gross sales.
    a) Fit a Holt–Winters model.
    b) Fit a double exponential model.

17.33 Using the variable Steel and other metals Break Bulk in the file "Maryland_Port_Cargo,"
    a) construct a line chart and interpret.
    b) fit the data using a simple exponential model and record the model errors.
    c) fit the data using a double exponential model and record the model errors.
    d) fit the data using a Holt–Winters model and record the model errors.
    e) which model appears best according to the model errors?
    f) explain why we should be careful in choosing a forecast model based on the computed model errors above.

## Further Reading

Black, Ken. (2012). *Business Statistics*, 7th edition. Wiley.

Hyndman, Rob J. and Athanasopoulos, George. (2018). *Forecasting: Principles and Practice*, 2nd edition. Otexts.

Rivera, R. (2016). A dynamic linear model to forecast hotel registrations in Puerto Rico using Google Trends data. *Tourism Management*, 57, pages 12–20.

Shmueli, Galit. (2016). *Practical Time Series Forecasting: A Hands-on Guide*, 3rd edition. Axelrod Schnall Publishers.

---

18 A subset corresponding to air transportation data unadjusted for seasonality.

# Appendix A

# Math Notation and Symbols

In this appendix, basic math concepts and notation are reviewed. Also, some Greek and statistical symbols are defined.

## A.1 Summation

Suppose we have a sample of $n$ observations: $X_1, X_2, ..., X_n$. Summation is represented by

$$\sum_{i=1}^{n} x_i = x_1 + x_2 + \cdots + x_n$$

The summation notation is sometimes simplified to $\sum x$ or $\sum_{all} x$.

**Example A.1** *If we have a sample of three observations: 11, 7, and 15, then n = 3, $X_1$ = 11, $X_2$ = 7, and $X_3$ = 15. Therefore,*

$$\sum_{i=1}^{3} x_i = 11 + 7 + 15 = 33$$

## A.2 *p*th Power

$x^p$, means "$x$ to the $p$th power." For example, if $x = 2$, then $x^2 = 4$, while $x^4 = 16$.

*Principles of Managerial Statistics and Data Science*, First Edition. Roberto Rivera.
© 2020 John Wiley & Sons, Inc. Published 2020 by John Wiley & Sons, Inc.
Companion website: www.wiley.com/go/principlesmanagerialstatisticsdatascience

> **Order of Math Operations Matters**
> The rules of math operations learned in elementary school still apply. PEMDAS establishes the order:
>
> - P – (Anything inside) parenthesis first
> - E – Exponents second
> - MD – Multiplication or Division
> - AS – Addition or Subtraction

Specifically,

- $\sum_{i=1}^{n} x_i^2 \neq (\sum_{i=1}^{n} x_i)^2$
- $\sum_{i=1}^{n} \sqrt{x_i} \neq \sqrt{\sum_{i=1}^{n} x_i}$
- $\sum_{i=1}^{n} x_i y_i \neq \sum_{i=1}^{n} x_i \sum_{i=1}^{n} y_i$

However,

- $\sum_{i=1}^{n} (x_i^p + y_i^q) = \sum_{i=1}^{n} x_i^p + \sum_{i=1}^{n} y_i^q$
- $\sum_{i=1}^{n} (x_i^p - y_i^q) = \sum_{i=1}^{n} x_i^p - \sum_{i=1}^{n} y_i^q$

## A.3  Inequalities

For numbers $a, b$, and $c$,

- $a > b$ means $a$ is strictly greater than $b$.
- $a \geq b$ means $a$ is greater than or equal to $b$.
- $a < b$ means $a$ is strictly smaller than $b$.
- $a \leq b$ means $a$ is smaller than or equal to $b$.
- $c \leq a \leq b$ means $a$ is greater than or equal to $c$ and smaller than or equal to $b$.

Be prepared to figure out inequalities from words. For example, "*a* is at least *b*" or "*a* is no less than *b*" means $a \geq b$, while "*a* is at most *b*" or "*a* is no more than *b*" means $a \leq b$.

## A.4  Factorials

In equations, we write $n!$ to read "*n* factorial." It is a useful way to express counting. Specifically,

- $n! = n \times (n-1) \times (n-2) \times \cdots \times 1$
- $0! = 1$

Factorials may be combined with other mathematical operations.

**Example A.2**   *Calculate $\frac{5!}{3!2!}(0.03125)$.*

First, let's turn each factorial into familiar numbers. $5! = 120$, $3! = 6$, and $2! = 2$. Thus,

$$\frac{5!}{3!2!}(0.03125) = \frac{120}{6(2)}(0.03125) = 0.3125$$

Alternatively, the factorials can be simplified,

$$\frac{5!}{3!2!}(0.03125) = \frac{5 \times 4 \times 3!}{3!2!}(0.03125) = \frac{5 \times 4}{2}(0.03125) = 0.3125$$

## A.5   Exponential Function

An exponential function is any function of $x$ expressed as

$$f(x) = a^x$$

where $a$ is any constant. The function arises whenever a quantity increases or decreases at a rate proportional to its current value. We will exclusively use the constant $e$: $e^x$, where $e \approx 2.718$. $e^x$ is very common in statistics.

## A.6   Greek and Statistics Symbols

In statistics, it is imperative that we differentiate between a value that is fixed and possibly unknown and a value that may be random. For the former, we will use some Greek symbols:

- $\mu$, read as "mu."
- $\sigma$, read as "sigma."
- $\pi$, read as "pi."
- $\lambda$, read as "lambda."
- $\alpha$, read as "alpha."
- $\beta$, read as "beta."
- $\epsilon$, read as "epsilon."
- $\tau$, read as "tau."

Two common statistical terms in the book are $\bar{x}$, read as "x bar" and $\hat{p}$, read as "p hat."

# Appendix B

# Standard Normal Cumulative Distribution Function

*Principles of Managerial Statistics and Data Science*, First Edition. Roberto Rivera.
© 2020 John Wiley & Sons, Inc. Published 2020 by John Wiley & Sons, Inc.
Companion website: www.wiley.com/go/principlesmanagerialstatisticsdatascience

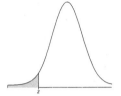

This table shows, for example, $P(Z \leq -1.96) = .025.$

| z | .00 | .01 | .02 | .03 | .04 | .05 | .06 | .07 | .08 | .09 |
|------|------|------|------|------|------|------|------|------|------|------|
| −3.4 | .0003 | .0003 | .0003 | .0003 | .0003 | .0003 | .0003 | .0003 | .0003 | .0002 |
| −3.3 | .0005 | .0005 | .0005 | .0004 | .0004 | .0004 | .0004 | .0004 | .0004 | .0003 |
| −3.2 | .0007 | .0007 | .0006 | .0006 | .0006 | .0006 | .0006 | .0005 | .0005 | .0005 |
| −3.1 | .0010 | .0009 | .0009 | .0009 | .0008 | .0008 | .0008 | .0008 | .0007 | .0007 |
| −3.0 | .0013 | .0013 | .0013 | .0012 | .0012 | .0011 | .0011 | .0011 | .0010 | .0010 |
| −2.9 | .0019 | .0018 | .0018 | .0017 | .0016 | .0016 | .0015 | .0015 | .0014 | .0014 |
| −2.8 | .0026 | .0025 | .0024 | .0023 | .0023 | .0022 | .0021 | .0021 | .0020 | .0019 |
| −2.7 | .0035 | .0034 | .0033 | .0032 | .0031 | .0030 | .0029 | .0028 | .0027 | .0026 |
| −2.6 | .0047 | .0045 | .0044 | .0043 | .0041 | .0040 | .0039 | .0038 | .0037 | .0036 |
| −2.5 | .0062 | .0060 | .0059 | .0057 | .0055 | .0054 | .0052 | .0051 | .0049 | .0048 |
| −2.4 | .0082 | .0080 | .0078 | .0075 | .0073 | .0071 | .0069 | .0068 | .0066 | .0064 |
| −2.3 | .0107 | .0104 | .0102 | .0099 | .0096 | .0094 | .0091 | .0089 | .0087 | .0084 |
| −2.2 | .0139 | .0136 | .0132 | .0129 | .0125 | .0122 | .0119 | .0116 | .0113 | .0110 |
| −2.1 | .0179 | .0174 | .0170 | .0166 | .0162 | .0158 | .0154 | .0150 | .0146 | .0143 |
| −2.0 | .0228 | .0222 | .0217 | .0212 | .0207 | .0202 | .0197 | .0192 | .0188 | .0183 |
| −1.9 | .0287 | .0281 | .0274 | .0268 | .0262 | .0256 | .0250 | .0244 | .0239 | .0233 |
| −1.8 | .0359 | .0351 | .0344 | .0336 | .0329 | .0322 | .0314 | .0307 | .0301 | .0294 |
| −1.7 | .0446 | .0436 | .0427 | .0418 | .0409 | .0401 | .0392 | .0384 | .0375 | .0367 |
| −1.6 | .0548 | .0537 | .0526 | .0516 | .0505 | .0495 | .0485 | .0475 | .0465 | .0455 |
| −1.5 | .0668 | .0655 | .0643 | .0630 | .0618 | .0606 | .0594 | .0582 | .0571 | .0559 |
| −1.4 | .0808 | .0793 | .0778 | .0764 | .0749 | .0735 | .0721 | .0708 | .0694 | .0681 |
| −1.3 | .0968 | .0951 | .0934 | .0918 | .0901 | .0885 | .0869 | .0853 | .0838 | .0823 |
| −1.2 | .1151 | .1131 | .1112 | .1093 | .1075 | .1056 | .1038 | .1020 | .1003 | .0985 |
| −1.1 | .1357 | .1335 | .1314 | .1292 | .1271 | .1251 | .1230 | .1210 | .1190 | .1170 |
| −1.0 | .1587 | .1562 | .1539 | .1515 | .1492 | .1469 | .1446 | .1423 | .1401 | .1379 |
| −0.9 | .1841 | .1814 | .1788 | .1762 | .1736 | .1711 | .1685 | .1660 | .1635 | .1611 |
| −0.8 | .2119 | .2090 | .2061 | .2033 | .2005 | .1977 | .1949 | .1922 | .1894 | .1867 |
| −0.7 | .2420 | .2389 | .2358 | .2327 | .2296 | .2266 | .2236 | .2206 | .2177 | .2148 |
| −0.6 | .2743 | .2709 | .2676 | .2643 | .2611 | .2578 | .2546 | .2514 | .2483 | .2451 |
| −0.5 | .3085 | .3050 | .3015 | .2981 | .2946 | .2912 | .2877 | .2843 | .2810 | .2776 |
| −0.4 | .3446 | .3409 | .3372 | .3336 | .3300 | .3264 | .3228 | .3192 | .3156 | .3121 |
| −0.3 | .3821 | .3783 | .3745 | .3707 | .3669 | .3632 | .3594 | .3557 | .3520 | .3483 |
| −0.2 | .4207 | .4168 | .4129 | .4090 | .4052 | .4013 | .3974 | .3936 | .3897 | .3859 |
| −0.1 | .4602 | .4562 | .4522 | .4483 | .4443 | .4404 | .4364 | .4325 | .4286 | .4247 |
| −0.0 | .5000 | .4960 | .4920 | .4880 | .4840 | .4801 | .4761 | .4721 | .4681 | .4641 |

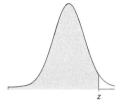

This table shows, for example, $P(Z \leq 1.96) = .975$.

| z | .00 | .01 | .02 | .03 | .04 | .05 | .06 | .07 | .08 | .09 |
|---|-----|-----|-----|-----|-----|-----|-----|-----|-----|-----|
| 0.0 | .5000 | .5040 | .5080 | .5120 | .5160 | .5199 | .5239 | .5279 | .5319 | .5359 |
| 0.1 | .5398 | .5438 | .5478 | .5517 | .5557 | .5596 | .5636 | .5675 | .5714 | .5753 |
| 0.2 | .5793 | .5832 | .5871 | .5910 | .5948 | .5987 | .6026 | .6064 | .6103 | .6141 |
| 0.3 | .6179 | .6217 | .6255 | .6293 | .6331 | .6368 | .6406 | .6443 | .6480 | .6517 |
| 0.4 | .6554 | .6591 | .6628 | .6664 | .6700 | .6736 | .6772 | .6808 | .6844 | .6879 |
| 0.5 | .6915 | .6950 | .6985 | .7019 | .7054 | .7088 | .7123 | .7157 | .7190 | .7224 |
| 0.6 | .7257 | .7291 | .7324 | .7357 | .7389 | .7422 | .7454 | .7486 | .7517 | .7549 |
| 0.7 | .7580 | .7611 | .7642 | .7673 | .7704 | .7734 | .7764 | .7794 | .7823 | .7852 |
| 0.8 | .7881 | .7910 | .7939 | .7967 | .7995 | .8023 | .8051 | .8078 | .8106 | .8133 |
| 0.9 | .8159 | .8186 | .8212 | .8238 | .8264 | .8289 | .8315 | .8340 | .8365 | .8389 |
| 1.0 | .8413 | .8438 | .8461 | .8485 | .8508 | .8531 | .8554 | .8577 | .8599 | .8621 |
| 1.1 | .8643 | .8665 | .8686 | .8708 | .8729 | .8749 | .8770 | .8790 | .8810 | .8830 |
| 1.2 | .8849 | .8869 | .8888 | .8907 | .8925 | .8944 | .8962 | .8980 | .8997 | .9015 |
| 1.3 | .9032 | .9049 | .9066 | .9082 | .9099 | .9115 | .9131 | .9147 | .9162 | .9177 |
| 1.4 | .9192 | .9207 | .9222 | .9236 | .9251 | .9265 | .9279 | .9292 | .9306 | .9319 |
| 1.5 | .9332 | .9345 | .9357 | .9370 | .9382 | .9394 | .9406 | .9418 | .9429 | .9441 |
| 1.6 | .9452 | .9463 | .9474 | .9484 | .9495 | .9505 | .9515 | .9525 | .9535 | .9545 |
| 1.7 | .9554 | .9564 | .9573 | .9582 | .9591 | .9599 | .9608 | .9616 | .9625 | .9633 |
| 1.8 | .9641 | .9649 | .9656 | .9664 | .9671 | .9678 | .9686 | .9693 | .9699 | .9706 |
| 1.9 | .9713 | .9719 | .9726 | .9732 | .9738 | .9744 | .9750 | .9756 | .9761 | .9767 |
| 2.0 | .9772 | .9778 | .9783 | .9788 | .9793 | .9798 | .9803 | .9808 | .9812 | .9817 |
| 2.1 | .9821 | .9826 | .9830 | .9834 | .9838 | .9842 | .9846 | .9850 | .9854 | .9857 |
| 2.2 | .9861 | .9864 | .9868 | .9871 | .9875 | .9878 | .9881 | .9884 | .9887 | .9890 |
| 2.3 | .9893 | .9896 | .9898 | .9901 | .9904 | .9906 | .9909 | .9911 | .9913 | .9916 |
| 2.4 | .9918 | .9920 | .9922 | .9925 | .9927 | .9929 | .9931 | .9932 | .9934 | .9936 |
| 2.5 | .9938 | .9940 | .9941 | .9943 | .9945 | .9946 | .9948 | .9949 | .9951 | .9952 |
| 2.6 | .9953 | .9955 | .9956 | .9957 | .9959 | .9960 | .9961 | .9962 | .9963 | .9964 |
| 2.7 | .9965 | .9966 | .9967 | .9968 | .9969 | .9970 | .9971 | .9972 | .9973 | .9974 |
| 2.8 | .9974 | .9975 | .9976 | .9977 | .9977 | .9978 | .9979 | .9979 | .9980 | .9981 |
| 2.9 | .9981 | .9982 | .9982 | .9983 | .9984 | .9984 | .9985 | .9985 | .9986 | .9986 |
| 3.0 | .9987 | .9987 | .9987 | .9988 | .9988 | .9989 | .9989 | .9989 | .9990 | .9990 |
| 3.1 | .9990 | .9991 | .9991 | .9991 | .9992 | .9992 | .9992 | .9992 | .9993 | .9993 |
| 3.2 | .9993 | .9993 | .9994 | .9994 | .9994 | .9994 | .9994 | .9995 | .9995 | .9995 |
| 3.3 | .9995 | .9995 | .9995 | .9996 | .9996 | .9996 | .9996 | .9996 | .9996 | .9997 |
| 3.4 | .9997 | .9997 | .9997 | .9997 | .9997 | .9997 | .9997 | .9997 | .9997 | .9998 |

# Appendix C

## *t* Distribution Critical Values

*Principles of Managerial Statistics and Data Science*, First Edition. Roberto Rivera.
© 2020 John Wiley & Sons, Inc. Published 2020 by John Wiley & Sons, Inc.
Companion website: www.wiley.com/go/principlesmanagerialstatisticsdatascience

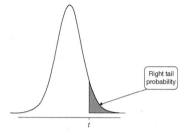

Right tail probability

t

Table shows that, for example, a right-tailed test, degrees of freedom (*df*) 15, and $\alpha = 0.05$, $t = 1.753$.

| | Two-sided confidence level | | | | | | Two-sided confidence level | | | | |
|---|---|---|---|---|---|---|---|---|---|---|---|
| | 0.80 | 0.90 | 0.95 | 0.98 | 0.99 | | 0.80 | 0.90 | 0.95 | 0.98 | 0.99 |
| | $\alpha$ for a right-tailed test | | | | | | $\alpha$ for a right-tailed test | | | | |
| *df* | 0.10 | 0.05 | 0.025 | 0.01 | 0.005 | *df* | 0.10 | 0.05 | 0.025 | 0.01 | 0.005 |
| 1 | 3.078 | 6.314 | 12.706 | 31.821 | 63.657 | 31 | 1.309 | 1.696 | 2.040 | 2.453 | 2.744 |
| 2 | 1.886 | 2.920 | 4.303 | 6.965 | 9.925 | 32 | 1.309 | 1.694 | 2.037 | 2.449 | 2.738 |
| 3 | 1.638 | 2.353 | 3.182 | 4.541 | 5.841 | 33 | 1.308 | 1.692 | 2.035 | 2.445 | 2.733 |
| 4 | 1.533 | 2.132 | 2.776 | 3.747 | 4.604 | 34 | 1.307 | 1.691 | 2.032 | 2.441 | 2.728 |
| 5 | 1.476 | 2.015 | 2.571 | 3.365 | 4.032 | 35 | 1.306 | 1.690 | 2.030 | 2.438 | 2.724 |
| 6 | 1.440 | 1.943 | 2.447 | 3.143 | 3.707 | 36 | 1.306 | 1.688 | 2.028 | 2.434 | 2.719 |
| 7 | 1.415 | 1.895 | 2.365 | 2.998 | 3.499 | 37 | 1.305 | 1.687 | 2.026 | 2.431 | 2.715 |
| 8 | 1.397 | 1.860 | 2.306 | 2.896 | 3.355 | 38 | 1.304 | 1.686 | 2.024 | 2.429 | 2.712 |
| 9 | 1.383 | 1.833 | 2.262 | 2.821 | 3.250 | 39 | 1.304 | 1.685 | 2.023 | 2.426 | 2.708 |
| 10 | 1.372 | 1.812 | 2.228 | 2.764 | 3.169 | 40 | 1.303 | 1.684 | 2.021 | 2.423 | 2.704 |
| 11 | 1.363 | 1.796 | 2.201 | 2.718 | 3.106 | 41 | 1.303 | 1.683 | 2.020 | 2.421 | 2.701 |
| 12 | 1.356 | 1.782 | 2.179 | 2.681 | 3.055 | 42 | 1.302 | 1.682 | 2.018 | 2.418 | 2.698 |
| 13 | 1.350 | 1.771 | 2.160 | 2.650 | 3.012 | 43 | 1.302 | 1.681 | 2.017 | 2.416 | 2.695 |
| 14 | 1.345 | 1.761 | 2.145 | 2.624 | 2.977 | 44 | 1.301 | 1.680 | 2.015 | 2.414 | 2.692 |
| 15 | 1.341 | 1.753 | 2.131 | 2.602 | 2.947 | 45 | 1.301 | 1.679 | 2.014 | 2.412 | 2.690 |
| 16 | 1.337 | 1.746 | 2.120 | 2.583 | 2.921 | 46 | 1.300 | 1.679 | 2.013 | 2.410 | 2.687 |
| 17 | 1.333 | 1.740 | 2.110 | 2.567 | 2.898 | 47 | 1.300 | 1.678 | 2.012 | 2.408 | 2.685 |
| 18 | 1.330 | 1.734 | 2.101 | 2.552 | 2.878 | 48 | 1.299 | 1.677 | 2.011 | 2.407 | 2.682 |
| 19 | 1.328 | 1.729 | 2.093 | 2.539 | 2.861 | 49 | 1.299 | 1.677 | 2.010 | 2.405 | 2.680 |
| 20 | 1.325 | 1.725 | 2.086 | 2.528 | 2.845 | 50 | 1.299 | 1.676 | 2.009 | 2.403 | 2.678 |
| 21 | 1.323 | 1.721 | 2.080 | 2.518 | 2.831 | 55 | 1.297 | 1.673 | 2.004 | 2.396 | 2.668 |
| 22 | 1.321 | 1.717 | 2.074 | 2.508 | 2.819 | 60 | 1.296 | 1.671 | 2.000 | 2.390 | 2.660 |
| 23 | 1.319 | 1.714 | 2.069 | 2.500 | 2.807 | 65 | 1.295 | 1.669 | 1.997 | 2.385 | 2.654 |
| 24 | 1.318 | 1.711 | 2.064 | 2.492 | 2.797 | 70 | 1.294 | 1.667 | 1.994 | 2.381 | 2.648 |
| 25 | 1.316 | 1.708 | 2.060 | 2.485 | 2.787 | 75 | 1.293 | 1.665 | 1.992 | 2.377 | 2.643 |
| 26 | 1.315 | 1.706 | 2.056 | 2.479 | 2.779 | 80 | 1.292 | 1.664 | 1.990 | 2.374 | 2.639 |
| 27 | 1.314 | 1.703 | 2.052 | 2.473 | 2.771 | 85 | 1.292 | 1.663 | 1.988 | 2.371 | 2.635 |
| 28 | 1.313 | 1.701 | 2.048 | 2.467 | 2.763 | 90 | 1.291 | 1.662 | 1.987 | 2.368 | 2.632 |
| 29 | 1.311 | 1.699 | 2.045 | 2.462 | 2.756 | 100 | 1.290 | 1.660 | 1.984 | 2.364 | 2.626 |
| 30 | 1.310 | 1.697 | 2.042 | 2.457 | 2.750 | ∞ | 1.282 | 1.645 | 1.960 | 2.326 | 2.576 |

# Appendix D

# Solutions to Odd-Numbered Problems

## Chapter 2

2.1   (a) Legal sales (in $). (b) 1857. (c) $14 983.

2.3   (b) They may not be representative of the population of interest.

2.7   (a) Unstructured. (b) Unstructured. (c) Structured.

2.9   (a) Ordinal. (b) Nominal. (c) Ordinal. (d) Nominal.

2.11  (a) Discrete. (b) Continuous. (c) Continuous.

2.13  Categorical and the level of measurement ordinal, but can also be interpreted as coming from a count of runners getting to finish line. However, some mathematical operations (e.g. the mean place) do not make sense.

2.15  (a) Ratio. (b) Interval. (c) Ratio. (d) Interval. (e) Ratio.

2.17  (a) A professional directory of all medical doctors. (b) Proportion of doctors who have been involved in a malpractice lawsuit. (c) 100 doctors who have been selected at random from a professional directory. (d) Proportion of doctors who have been involved in a malpractice lawsuit, in the sample of 100.

2.19  Parameter, because it is the true measure of interest, and a statistic is a random estimate of the parameter.

2.21  (a) Primary. (b) Primary. (c) Secondary.

2.23  (a) Experimental. (b) Observational. (c) Observational. (d) Experimental.

*Principles of Managerial Statistics and Data Science*, First Edition. Roberto Rivera.
© 2020 John Wiley & Sons, Inc. Published 2020 by John Wiley & Sons, Inc.
Companion website: www.wiley.com/go/principlesmanagerialstatisticsdatascience

2.25 A discrete variable has countable number of distinct values, whereas a continuous variable can have any value in a given range or interval. Examples: number of cars (discrete), car speed (continuous).

2.27 (a) Secondary. (b) Primary.

2.29 (a) No. (b) No.

2.31 (a) No. (b) Yes. (c) Yes.

2.33 (a) Is there an increasing trend in the ratio of yearly student tuition over median income? (b) What factors influence the quality of restaurants?

## Chapter 3

3.1 (a) 0.22. (b) 128.

3.3 (a) Yes. (b) Because we use sample data, and relative frequencies are estimates of probabilities.

3.5

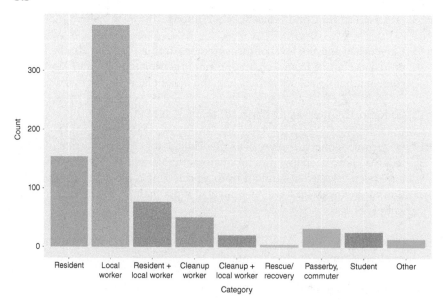

3.7 (a) About 55%. (b) About 7%

3.9 (b)

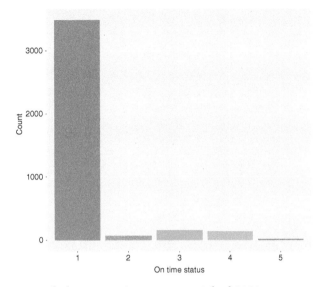

Flight was on time, category 1, had 3490 responses. In contrast, flight was 91 minutes late or canceled, category 5, had 18 responses.

3.11 (a) Minimum = 1, maximum = 2245. (b)

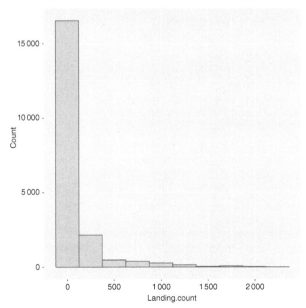

(c) Typical number of landings is 1. (d) Right skewed.

3.13 (a) Right skewed. May look like a bar chart in some software since cost is highly discretized.

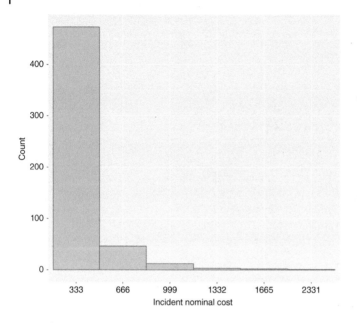

(b) Dog incidents are more common.

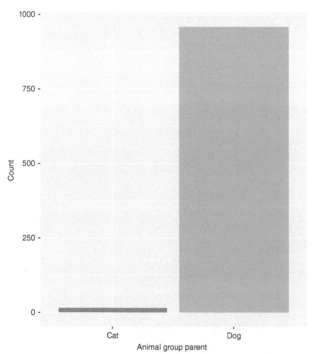

3.17 The trend is decreasing.

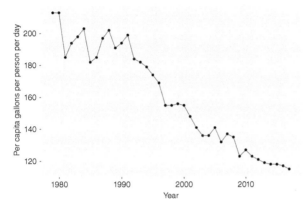

3.19 (a) True. (b) False.

3.21 According to San Diego Census data for the period, Whites were the most prevalent race, followed by Hispanics.

3.23 Commercial has the most variability in consumption, followed by residential.

3.25 (a) Decreases. (b) There is no linear association. There is an outlier. (c) No, particularly number of people exposed to industrial noise is asymmetric.

3.27

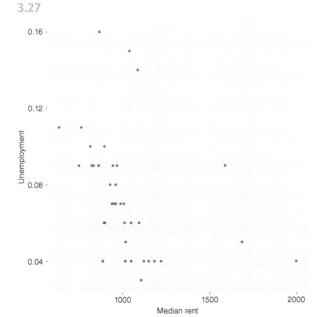

There is correlation between the variables, some outliers.

**3.29** Consider less countries, or use a stacked bar chart.

**3.31** (a) 73%. (b) 16%. (c) Czech Republic.

**3.33** (a)

| Rebate amount $ | Frequency |
|---|---|
| 1000 | 64 |
| 1100 | 7 |
| 1350 | 3 |
| 1500 | 259 |
| 2000 | 1 |
| 2200 | 292 |
| 3500 | 72 |
| Total | 698 |

The two most frequent rebate amounts are $1500 and $2200. The least frequent rebate amount is $2000. (b)

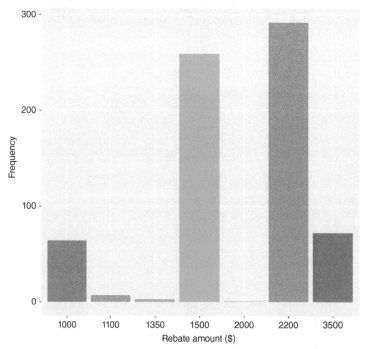

3.35 (a) Transgender, though Gay men were just below. (b) Bisexual women.
(c) 78.

3.37 (a)

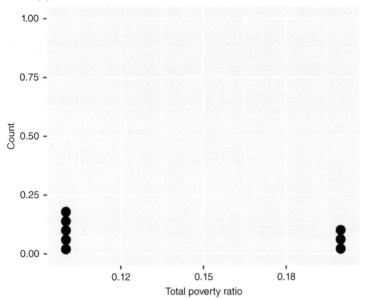

(b)

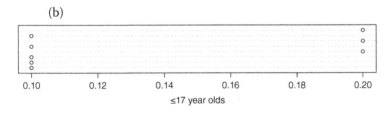

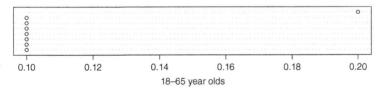

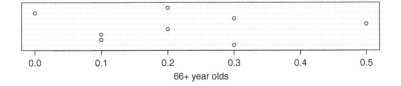

3.39 There is a seasonal component, the peaks are in summer, and in 2013 and 2014 the peaks were higher than other years.

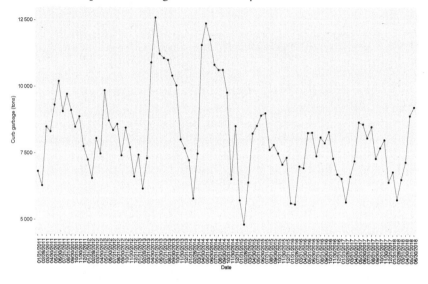

3.41 Not summarizing categorical variable frequencies, slices do not add up to 100%, and too many slices.

3.43 There is good association, the association between cheese fat percentage and moisture percentage in both, organic and nonorganic, is negative. Thus, if there is more percent of moisture, then the fat percentage is less in both, organic and nonorganic.

3.45 (a)

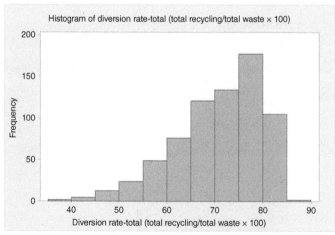

3.47 (a)

| Q1a\Q1c | Don't Know | Very bad | Bad | Neutral | Good | Very good | Total |
|---|---|---|---|---|---|---|---|
| Don't know | 1 | 0 | 1 | 0 | 4 | 0 | 6 |
| Very bad | 1 | 6 | 3 | 2 | 1 | 1 | 14 |
| Bad | 2 | 5 | 19 | 27 | 9 | 1 | 63 |
| Neutral | 10 | 11 | 31 | 100 | 53 | 9 | 214 |
| Good | 41 | 4 | 34 | 201 | 597 | 75 | 952 |
| Very good | 49 | 2 | 10 | 63 | 285 | 408 | 817 |
| Total | 104 | 28 | 98 | 393 | 949 | 494 | 2066 |

(b)

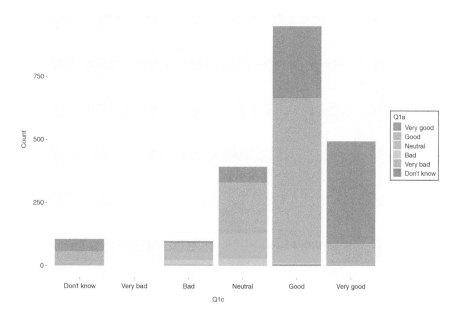

(c) Yes, Q1a and Q1c are related. Many people answered that Chattanooga as a place to work and live is good or very good.

3.49 Using area to compare categories instead of length, bar chart is a better option.

3.51 (b)

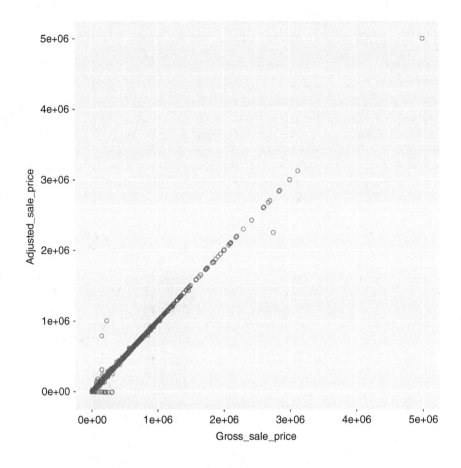

(c) What sticks out the most is negative adjusted sale prices. But there are also observations were both sale prices vary substantially.

3.53 It appears there is an exponential association between rental units affordable to average teacher, and owner units affordable to average teacher in Austin, Texas.

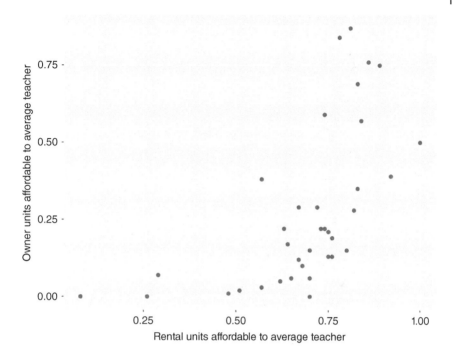

3.55 (a) 45.9. (b) 2.3. (c) 7781. (d) 11.

3.57 (a) No. (b) No. (c) Both years show no association between the two variables.

3.59 (a) 75 000. (b) 310 000.

3.61 (a) It is decreasing. (b) There's been a population increase.

3.63 The left chart shows the expected pattern, similar to the keys of a piano: number of days per month fluctuating between 28 and 31 days (except toward the end of the time period). In contrast, weather station 2 has a very incomplete rainfall record. For over 20 years, it does not have any data at all and the most recent data was sparse.

## Chapter 4

4.1 (a) $\overline{X} = 6.46$. (b) Median $= 7.21$.

4.3 (a) $\overline{X} = 5.50$. (b) Median $= 5.50$. (c) Mode $= 2$.

4.5   Median. Because when the distribution is highly skewed, it is questionable to use the mean to measure centrality.

4.7   (a) Right skewed. (b) Median. Because the data set is large and the distribution is skewed to the right.

4.11  It is not recommended. The yearly median income is not constant, thus the median over all years should perform very poorly.

4.13  No, number of tourists has been increasing steadily.

4.15  (a) $s^2 = 22.38$. (b) $s = 4.73$.

4.17  (a) $s^2 = 103.70$. (b) $s = 10.18$.

4.19  (a)

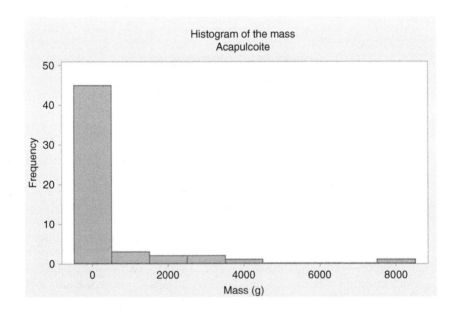

(b) Right-skewed distribution, mean greater than the median.
(c) Median.

(d)

Descriptive statistics: mass (g)

Statistics

| Variable | Mean | Median |
|---|---|---|
| Mass (g) | 490 | 51 |

(e)

Descriptive statistics: mass (g)

Statistics

| Variable | StDev |
|---|---|
| Mass (g) | 1279 |

4.21 (a) $\overline{X} = 23.89$. (b) Median $= 23.70$. (c) $s^2 = 7.20$. (d) $s = 2.68$.

4.23 75% of the exam scores were less than or equal to 82.

4.25 (a) 38. (b) Model A. (c) Model A. (d) Model A: $s = 1.98, IQR = 1.745$; Model B: $s = 0.975, IQR = 1$; Model A has greater variability in mpg.

4.27 $IQR = 18$.

4.29 $Q_1 = 95$ and $Q_3 = 170$.

4.31 $IQR = 134.75$. In contrast, $s = 1279.41$, likely influenced by the outliers in the data.

4.33 (b) $Q_1 = 13.95$. (c) Because, the instrument detected few <10 values, and $Q_1$ was above these values. (d) Median $= 16.12$. Same reasoning as in (c). (e) The mean is not reliable since it requires the specific value of every measurement.

4.35 (b) It is not reliable because when values are ordered from lowest to highest, $Q_1$ is not based on actually measured observations. The 25% quartile is within these unknown values. (c) Yes. It is reliable because when values are ordered from lowest to highest, the median is within actually measured observations.

4.37 (a) No quadrant dominates. $r$ is low. No apparent association among variables. (b) No quadrant dominates. $r$ is low. However, a nonlinear association among variables is clearly visible.

4.39 (a) The correlation is 0.87. There is a positive strong linear association between the variables. (b)

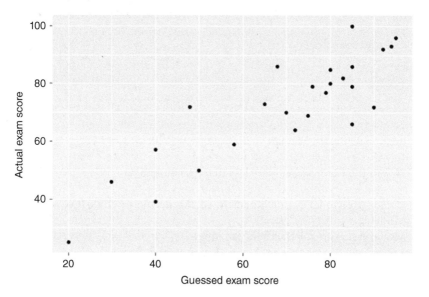

The graph indicates a moderate to strong linear association between the variables, as one variable increases the other variable also increases.

4.41 (a) $\overline{X} = 2310.3$. (b) Median= 2266.

4.43 Since some parking lots have only a few lots available, while others have many, the mean would be of little use. It would be more valuable to use the proportion of occupied lots.

4.45 (a) $\overline{X} = 3.95$. (b) $s = 0.56$. (c) Median= 13.

4.47 (a) $x_{min} = 999$, Median= 115 250, $\overline{x} = 160\ 698$ and $x_{max} = 1\ 000\ 001$. Minimum appears too low. It may be a code value for "not available." (b) It appears data is censored at 1 000 001. Thus, median house values of over 1 million appear as 1 000 001. (c) None, median does not depend on value of most observations. (d) Calculated sample mean is smaller than it should be.

4.49 (d) is the correct answer. Article is based on a measure of centrality of rents in cities throughout the United States. Thus, some modest two-bedroom apartments are cheaper, and some more expensive.

4.51 (a) $X_{min}$ = 3.17, $Q_1$ = 27, Median = 76.31, $Q_3$ = 87.22, $X_{max}$ = 611.88.
(b)

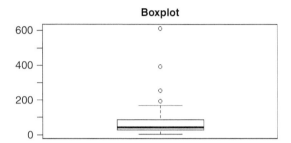

In the boxplot, it is observed that the data is skewed to the right and four observations are outliers. (c)

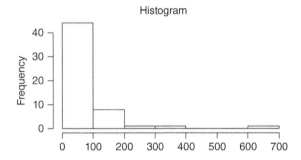

The distribution of the data is skewed to the right.

4.53 (a) Skewed to the right. (b) (i) In most cases, which are summarized by condition, the average length of stay is greater than the median. Therefore, the distribution for most of the summarized cases is skewed to the right. For example, the average duration for cerebral infarction is 20.3 and the median is 10. (ii) It makes sense because for most health conditions, people often do not stay long at the hospital, but there are some exceptions.

4.55  The least appropriate statement is item (a).

4.57  (a) True. (b) False. (c) False. (d) False. (e) True.

4.59  250 too, data has a lot of 250s.

4.61  (a) $\hat{x} = 58.61$. (b) The beach should not be closed.

4.63  (a) 0.739. (b) 0.727.

4.65  0.185.


## Chapter 5

5.1   (a) $S = \{RR, RB, RG, BB, BR, BG, GG, GB, GR\}$.
      (b) $\{RB, BR\}$.

5.3   The intersection is $\{4, 6\}$. The intersection of the two events is not empty.

5.5   (a) $B = \{4, 6\}$. (b) $A = \{\}$. (c) No.

5.7   Outcomes that are part of the intersection are ace of spades and ace of clubs.

5.9   Probabilities have long-term interpretations.

5.11  No, probabilities have long-term interpretations.

5.13  0.14.

5.15  (a) $P((B \text{ or } E)^c)$. (b) $P(B \text{ or } E)$. (c) $P(B \text{ or } E) - P(B \text{ and } E)$.

5.17  (a) 0.6. (b) 0.65. (c) 0.6.

5.19  (a) 0.8. (b) 0.75.

5.21  (a) 36. (b) $\frac{6}{36}$. (c) $\frac{2}{36}$.

5.23  (a) 0.09. (b) 0.65. (c) No, because $P(L \text{ and } U) \neq 0$. (d) Yes.

5.25  (a) 0.5/0.35. (b) 0.5/0.4. (c) $A$ and $B$ are not independent.

5.27 (a) *A* and *B* are not independent, $P(A)P(B) \neq 0.07$. (b) $0.07/0.35$. (c) $0.07/0.10$.

5.31 0.

5.33 $\frac{6}{90}$.

5.35 By Bayes Theorem.

5.37 12/13.

5.41 (a) 330/1296. (b) 149/1296. (c) 311/1178. (d) 145/1296. (e) 212/1296. (f) 0.

5.43 Yes, if events *A* and *B* are independent, $P(A$ *and* $B)$ will be "close" to $P(A)P(B)$. In fact, by the law of large numbers, the larger *n* is, the closer to $P(A$ *and* $B)$ gets to $P(A)P(B)$.

5.45 (a) 90/22 421. (b) 22 132/22 421. (c) 59/1864. (d) 120/5814. (e) 90/10 820. (f) No.

5.47 (a) 2243/12 741. (b) 3919/36 006.

5.49 24.

5.51 (a) Probability is higher now. (b) 0.0433.

5.53 No.

5.57 Tuesday.

5.61 (a) 1.0 (b) 0.22/0.47. (c) *A* to happen. (d) *B* to happen.

5.63 (a) 0.648. (b) 0.988.

5.65 (a) 0.26. (b) 0.52.

5.67 (a) 20/73. (b) 0. (c) 68/73.

5.69 (a) 4333/6831. (b) 1488/2398. (c) 2845/4433. (d) 1520/6831. (e) 5243/6831.

## Chapter 6

6.3 Cases A and C do not define an adequate probability function because the sum of the probabilities are not equal to 1. Also, in case C $P(X = 7)$ must be in one row, not two.

6.5

| $x$ | 1 | 2 | 3 | 4 |
|---|---|---|---|---|
| $P(X = x)$ | 0.488 | 0.249 | 0.127 | 0.134 . |

6.7 $P(X \leq 3) = 0.07$.

6.9 $E(X) = 0.5$.

6.11 (a)

| $x$ | 0 | 500 000 |
|---|---|---|
| $P(X = x)$ | 0.999 999 9 | 0.000 000 2 . |

(b) $0.1.

6.13 (a) $-875$. (b) 128,625. (c) 30,625. (d) 104,125.

6.15 $5.

6.17 Use precautionary alternative.

6.19 Lower, because the probability of success is high.

6.21 (a) 0.000 047 3. (b) 0.1611. (c) 0.9999. (d) 0.999 95. (e) 10. (f) 6. (g) 2.4494.

6.23 (a) Greater, because its $\pi$ is greater and the value of the random variable is the same. Also, the distribution of $X$ is skewed to the left while the distribution of $Y$ is skewed to the right. (b) $P(X = 4) = 0.1114$ and $P(Y = 4) = 0.0111$.

6.25 0.26.

6.27 (a) and. (d) are candidates to be modeled with a Poisson distribution.

6.29 (a) 1.5. (b) 0.1255.

6.31  0.0164.

6.33  (a) 0.40. (b) 0.60. (c) 0.075.

6.35  (a) 307.36. (b) 0.606. (c) 1.332.

6.39  (a) 0.229. (b) 3. (c) 1.597.

6.41  (a) 0.022. (b) The new strategy worked.

6.43  (a) 0.222. (b) 0.206. (c) 0.194. (d) 0.220.

6.45  (a) 1. (b) 0.056. (c) 0.040.

6.47  (a) 0.105. (b) 0.895.


## Chapter 7

7.1  (a) Yes, function is a rectangle with a total area of 1 and $f(x)$ is always nonnegative. (b) No, $f(x)$ can be negative.

7.5  (a) $\mu = 30$; $\sigma = 11.55$. (b) $\mu = 300$; $\sigma = 115.47$. (c) $\mu = 300$; $\sigma = 57.74$. (d) $\mu = 50$; $\sigma = 28.87$.

7.7  218.9.

7.9  (a) $\mu = 375$. (b) $\sigma = 43.30$. (d) $50/150 = 0.33$. (e) $125/150 = 0.83$. (f) 420.

7.11  $x$-Axis represents possible values of the normal random variable; $y$-axis represents values of the normal pdf.

7.13  (a) Lower than 0.5. Probability of an interval which is to the right of the mean. (b) Greater than 0.5. Probability of an interval which includes most of the area below the normal pdf. (c) Equal to 0.5. Mean is equal to the median for the standard normal distribution. (d) Not possible to tell just by common sense since interval has one limit to the left of the mean, the other to the right.

7.15  (a) Incorrect. Common sense indicates this probability should be <0.5. (b) Incorrect. It is 0.4562. (c) Correct. (d) Correct. (e) Correct.

7.17  (b) 0.9545. (c) 0.6827. (d) 0.5467.

7.19 (a) −1.96, 1.96. (b) 0.8416. (c) −0.4399. (d) −1.5, 1.15.

7.21 (a) Lower than 0.5. Probability of an interval which is to the right of the mean. (b) Greater than 0.5. Probability of an interval which includes most of the area below the normal pdf. (c) Lower than 0.5. Probability of an interval which is to the left of the mean.

7.23 (a) Not adequate. (b) Not adequate. (c) Adequate. (d) Adequate.

7.25 (a) 0. (b) 0.9987. (c) 0.

7.27 (a) 27.3780. (b) 52.1375.

7.29 First store.

7.31 29.64%.

7.33 (a) Greater than 0.5. Probability of an interval with lower limit to the left of the mean and goes to infinity. (b) 0.7665. (c) 68.74. (d) 96.09.

7.35 9.1906 tons.

7.37 0.

7.39 (b) 0.0868. (c) 0.0868 > 0.01, investment company's claim should not be rejected.

7.43 (a) 0.5655. (b) 0.0999.

7.45 (a) Continuous. (b) Continuous. (c) Discrete. (d) Continuous.

7.47 0.8.

7.49 (a) Less than 0.5. (b) Greater than 0.5. (c) Less than 0.5. (d) Can not be sure. (e) Greater than 0.5. (f) Less than 0.5.

7.51 (a) 0.0099. (b) 0.259.

7.53 Property of symmetry and they are at the same distance from the mean representing equivalent areas under the normal curve.

7.55 161.3437.

7.57 (a) 0.8465. (b) 0.0799. (c) 0.0736.

## Chapter 8

8.3   (a) 1.8974. (b) 0.9487. (c) 0.7746. (d) No.

8.5   (a) Greater than 0.5. Probability of an interval with upper limit to the right of the mean and goes to minus infinity. (b) For $n = 30$. (c) For $n = 20$.

8.7   (a) Not adequate. (b) Adequate.

8.9   (a) 0.1469. (b) 0.1469 > 0.1; the mean costs in 2015–2016 was no higher than 2013–2014.

8.11  (a) 0.6936. (b) 0.4124. (c) 0.2812. (d) 0.8286. (e) 0.

8.13  (a) Yes, normal population with known variance. (b) No, since $n < 30$ and variance is unknown.

8.15  For the first employee. The second employee's sample came from a highly skewed distribution. $n = 30$ may not be large enough to assume normality for the sample mean.

8.17  (a) Increase, probability is now 0.673. (b) Increase, probability is now 0.958.

8.19  (a) Mean is 5 inches, standard deviation is 0.0224. (b) 0. (c) 0. (d) Since $n > 30$, central limit theorem applies.

8.21  0.6869.

8.23  (a) Yes. (b) Yes. (c) No. (d) Yes.

8.25  (a) $n = 300$. (b) $n = 300$.

8.27  117.2611.

8.29  $\hat{p}$ does not have a domain of $0, \ldots, n$.

8.31  WRONG probability is smaller than the TRUE probability.

8.33  No, since highly asymmetric.

8.35  Standard error 0, because $\overline{X} = \mu$.

8.37  (a) $E(\hat{p}) = E\left(\dfrac{Y}{n}\right) = \dfrac{1}{n}(E(Y)) = \dfrac{1}{n}(n\pi) = \pi$.   (b)  $Var(\hat{p}) = \dfrac{1}{n^2}Var(Y) = \dfrac{\pi(1-\pi)}{n}$.

## Chapter 9

9.1 (a) (55.30, 61.69). (b) (10.94, 11.28). (c) (1017.97, 1047.73). (d) (93.03, 94.19).

9.3 (a) 100%. (b) 90%. (c) Confidence interval with 95%. (d) Confidence interval with $n = 25$.

9.5 (a) (24.39, 28.41). (b) We are 90% confident that the true mean MPG of the vehicle is between (24.39, 28.41). (c) No, level is chosen before seeing data.

9.7 (a) (3.48, 3.92). (b) We are 95% that the true mean price per pound of grass-fed beef is between ($3.48, $3.92).

9.9 (a) $n = 166$. (b) $n = 664$.

9.11 $n = 22$.

9.13 (a) Second. (b) Second. (c) Second, since $P(t > -2.4) = P(t < 2.4)$. (d) First.

9.15 (a) 4.032. (b) 1.665. (c) 2.499. (d) 1.972.

9.17 (14.50, 16.24).

9.19 (a) (496.27, 656.55). (b) We are 95% confident that the mean tax refund is between ($496.27, $656.55). (c) Tax refunds are normally distributed.

9.21 (a) (1216.89, 1555.11). (b) We are 95% confident that the mean repair costs is between ($1216.89, $1555.11). (c) Reasonable, by central limit theorem.

9.23 (26.15, 27.04) Assumption: fat percentage of cheese follows a normal distribution.

9.25 (a) Hold, since $34 > 15$ and $n(1 - \hat{p}) = 21 > 15$. (b) Hold, since $17 > 15$ and $n(1 - \hat{p}) = 132 > 15$. (c) Hold, since $214 > 15$ and $n(1 - \hat{p}) = 341 > 15$. (d) Does not hold, since $4 < 15$.

9.27 (a) (0.439, 0.574). (b) We are 90% confident that the proportion of school graduates who have job offers right after graduation is between (0.439, 0.574). (c) No, since 0.65 is not within the interval.

9.29 (0.0275, 0.0289). We are 95% confident that the proportion of children with levels of lead between 5 and 9 µg/dl is between (0.0275, 0.0289).

9.31 (a) 16%. (b) 5164.

9.33 (a) (70.98, 75.56). (b) We are 95% confident that the criminal domestic violence rate in country $X$ is between 70.98 and 75.56.

9.35 $n = 97$.

9.37 $n = 1$, because they chose a margin of error that is far bigger than the standard deviation of the population.

9.39 4.58.

9.41 (a) (17.27, 23.74). (b) We are 90% confident that the mean time it takes to do an oil change is between (17.27, 23.74). (c) Time it takes to do an oil change is normally distributed.

9.43 (5.57, 9.92).

9.45 (a) (45 630.53, 54 369.47). (b) Manufacturer's claim is not justified, because 45 000 is not within the interval (the mean trouble-free miles are, in fact, higher).

9.47 (a) (0.8112, 0.8413). (b) We are 80% confident that the proportion of Americans, either own a drone or know someone who does, is between (0.8112, 0.8413). (c) $n\hat{p} = 856 > 15$ and $n(1 - \hat{p}) = 180 > 15$.

## Chapter 10

10.1 (a) =. (b) ≥. (c) ≤. (d) <. (e) >. (f) ≠. (g) ≥.

10.3 (a) $H_o: \mu \leq 103.2$. (b) $H_o: \mu \geq 547$. (c) $H_o: \mu = 706.44$. (d) $H_o: \pi = 0.75$. (e) $H_o: \mu_1 - \mu_2 \leq 0$. (f) All coefficients are equal to zero.

10.5 There is a probability of 0.01 to Reject $H_o$, given that $H_o$ is true.

10.7 (a) Power $= 0.90$ and $1 - \alpha = 0.99$. (b) Power $= 0.80$ and $1 - \alpha = 0.95$. (c) Power $= 0.95$ and $1 - \alpha = 0.90$. (d) Power $= 0.50$ and $1 - \alpha = 0.99$.

10.9    There is a probability of 0.84 to reject that the expected battery life is at most 10.7 hours when it is greater than 10.7 hours.

10.13   Yes, $p$-value would be smaller than $\alpha$, since $\mu = 175$ is not in the confidence interval, at the same significance, $H_o$ would be rejected.

10.15   $H_o$: $\mu \geq 5.25$; $H_1$: $\mu < 5.25$.

10.17   $H_o$: $\mu = 425$; $H_1$: $\mu \neq 425$.

10.19   (a) Do not reject. (b) Do not reject. (c) Do not reject. (d) Reject. (e) Reject. (f) Do not reject.

10.21   (a) $p$-Value $= 0.0095 < 0.05$, Reject $H_o$. (b) $p$-Value $= 1 > 0.05$, Do not reject $H_o$. (c) $p$-Value $= 0.1773 > 0.05$, Do not reject $H_o$. (d) $p$-Value $= 0 < 0.05$, Reject $H_o$.

10.23   The procedure falls apart because the $p$-value cannot be found anymore. Specifically, the null does not provide a specific value for $\mu$ to compute a test statistic.

10.25   (d) $H_o$: $\mu = 40.5$; $H_1$: $\mu > 40.5$.

10.27   $p$-Value $= 0 < 0.05$, Reject $H_o$. Thus, at 5% significance, we conclude that there is evidence that mean voltage in his neighborhood is higher than 120 V.

10.29   (a) With 90% confidence, the true mean monthly cost of supplies is at most $714. (b) Lower tail.

10.31   (a) Do not reject. (b) There is no evidence of practical importance since null was not rejected. (c) Null is rejected, but practical importance is debatable since bound is so close to 5.25. (d) Reject, practical importance likely because upper bound is much lower than 5.25.

10.33   (a) 14. (b) 35. (c) 31.

10.35   (a) $0.02 < p$-value $< 0.05$, which is $<0.1$, Reject $H_o$. (b) $p$-Value $< 0.005$, which is $<0.05$, Reject $H_o$. (c) $p$-Value $< 0.005$, which is $<0.01$, Reject $H_o$. (d) $0.10 < p$-value $< 0.20$, which is $>0.05$, Do not reject $H_o$.

10.37   (a) $p$-Value $= 0.0323 < 0.05$, Reject $H_o$. (b) $p$-Value $= 1 > 0.05$, Do not reject $H_o$.

10.39 $p$-Value $= 0 < 0.01$, Reject $H_o$, Company's claim is not valid.

10.41 $p$-Value $= 0.2403 < 0.05$, Do not reject $H_o$, the company was unsuccessful in attaining the proposed mean repair time.

10.43 (a) $p$-Value $= 0.9388 < 0.01$, Do not reject $H_o$, no evidence against the news show claim. (b) (553.6902, 587.2934). (c) Monthly average miles are normally distributed.

10.45 (a) $H_o$: $\pi \geq 0.05$; $H_1$: $\pi < 0.05$. (b) $H_o$: $\pi \leq 0.15$; $H_1$: $\pi > 0.15$. (c) $H_o$: $\pi = 0.5$; $H_1$: $\pi \neq 0.5$.

10.47 (a) $p$-Value $= 0.4624 > 0.01$, Do not reject $H_o$. (b) $p$-Value $= 0.8898 > 0.01$, Do not reject $H_o$. (c) $p$-Value $= 0 < 0.01$, Reject $H_o$. (d) $p$-Value $= 0.0031 < 0.01$, Reject $H_o$.

10.49 (a) $p$-Value $= 0.0927 < 0.10$, Reject $H_o$. (b) $p$-Value $= 0.7367 > 0.10$, Do not reject $H_o$. (c) $p$-Value $= 0.9464 > 0.10$, Do not reject $H_o$. (d) $p$-Value $= 0 < 0.10$, Reject $H_o$.

10.51 (a) $H_o$: $\pi \leq 0.061$; $H_1$: $\pi > 0.061$. (b) $Z = 27.09$. (c) $p$-Value $= 0$. (d) $p$-Value $= 0 < 0.10$, Reject $H_o$. (e) At 10% significance, we conclude that the proportion of members of the LGBT community who seriously consider committing suicide is higher than that of the general population. (f) $n\pi_o < 15$, the condition is not satisfied. However, the same conclusion is reached when the exact method is used.

10.53 $p$-Value $= 0.7162 > 0.10$, Do not reject $H_o$, there is no support for the researcher's belief.

10.55 $p$-Value $= 0.9328 > 0.10$, Do not reject $H_o$, the political candidate claim's is right.

10.57 (c).

10.59 (a).

10.61 (b).

10.63 The probability of rejecting the null hypothesis, given that it is true, is 0.05.

10.65 Power $= 1$ (probability null rejected when false). Misleading because rejecting null may not imply practical importance.

10.67 $P(\text{rejecting } H_o | H_o \text{ is true}) = 0$ and thus, $P(\text{not rejecting } H_o | H_o \text{ is true}) = 1$. The only way the null can be rejected is if $p$-value $= 0$. Most importantly, $\beta$ will be higher. In summary, if the null is false, $\alpha = 0$ is too conservative and will make it very hard to reject the null.

10.69 If the null is rejected in a two-sided test, a one-sided test performed with the same data, at the same significance, would certainly reject the null too. On the other hand, rejecting the null with a one-sided test does not guarantee rejection of the null when performing a two-sided test since the $p$-value is always bigger. The two-sided test is a more conservative test than the one-sided test.

10.71 $p$-Value $= 0.0007 < 0.01$, Reject $H_o$. The training program worked.

10.73 (a) $H_o$: $\mu \le 59$; $H_1$: $\mu > 59$. (b) 3. (c) (ii). (d) Reject the null hypothesis. (e) At 1% significance, there is evidence that the mean amount spent in fragrance in her town is higher than $59. (f) Amount spent in fragrance follows a normal distribution.

10.75 (a) It will be underestimated. (b) More likely to reject the null when it is true.

10.77 (b) $p$-Value $= 0.3898 > 0.05$, Do not reject $H_o$.

10.79 Yes, normal assumption seems reasonable.

## Chapter 11

11.1 (a) 0.9405. (b) 0.9018.

11.3 (a) $(-5.003, -2.317)$. (b) With 90% confidence, it is concluded that the true difference in means is between $(-5.003, -2.317)$. Since this confidence interval does not include 0 and both limits are negative, there is evidence that mean of group 1 is less than group 2.

11.5 It would become wider since $Z_{\alpha/2}$ would be bigger.

11.7 (a) (0.94, 9.06). (b) With 95% confidence, it is concluded that the true difference in mean age of customers of two stores is between (0.94, 9.06).

Since this confidence interval does not include 0 and both limits are positive, there is evidence that mean age of costumers of store 1 is greater than store 2. (c) Difference in sample means of ages of costumers is normally distributed.

11.9 (a) With 90% confidence, the difference in mean monthly cost of store 1 and store 2 is \$714 or less. (b) Lower tail.

11.11 (b) and (c).

11.13 (a) $H_o: \mu_1 - \mu_2 \geq 0; H_1: \mu_1 - \mu_2 < 0$. (b) 5.1765. (c) $p$-Value $= 0.9999$. (d) Do not reject $H_0$. (e) At 5% signifcance, the mean miles per gallon for the tweaked motor is not better than the original motor.

11.15 $(-\infty, -214.5488)$.

11.17 (a) Men, because $(-0.22 + 0.0048)/2 = -0.1076$, a negative difference. (b) Interval becomes $-0.0048$ and $0.22$, but conclusion is the same. (c) Not statistically significant. (d) No! Empirical rule states that a normally distributed variable has most possible value within 3 sd of mean. But note how 3 sd would give negative ideal number of children for women. Central limit theorem still applies here, though some may argue that it is best to compare distributions, not means.

11.19 (a) Depends if data represents entire population. Some missing values but uncertain if these are closing schools. (b) $p$-Value $= 0.159 > 0.1$, Do not reject $H_o$. The mean number of males is equal to the mean number of female students for the 2014–2015 school year. (c) $(-1.73, 22.30)$. (d) Independent samples and equal variances. (e) Assumes 2013–2014 data was representative of 2014–2015 school year. Inference here does not account for school size. (f) $p$-Value $= 0.628 > 0.1$, Do not reject $H_o$. The mean number of females at Bridgeport school district is equal to the mean number of females at the Fairfield School District.

11.21 He's correct.

11.23 (a) 31. (b) $H_o$: mean difference in scores $= 0; H_1$: mean difference in scores $> 0$. (c) 3.81. (d) $p$-Value $= 0.047 > 0.01$, Do not reject $H_o$. (e) At 1% significance, there is no evidence against the mean guessed and actual scores being equal. (f) The guessed and actual scores are for same person.

11.25 (a) $H_o$: mean difference in scores $= 0$; $H_1$: mean difference in scores $\neq 0$. (b) 5.03. (c) *p*-Value $0.124 > 0.05$, Do not reject $H_o$. (d) At 5% significance, there is evidence that the mean scores of two cereals are same.

11.27 *p*-Value $= 0 < 0.10$, Reject $H_o$. At 10% significance, there is evidence that the proportion of males and females enrolled in schools are different.

11.29 $(-0.0129, 0.0169)$.

11.31 Assuming same data, limits would be strictly negative, but no impact on conclusion.

11.33 *p*-Value $= 0 < 0.01$, Reject $H_o$. At 1% significance, there is evidence that mean daily sales per marijuana extract for inhalation unit for August of 2016 was higher than for August of 2017.

11.35 Should use a paired *t*-test because superhero and villain violence was measured in each movie, samples are dependent.

11.37 (a) *p*-Value $= 0.018 < 0.05$, Reject $H_o$. At 5% significance, there is evidence that the mean number of art related to business for 2011 and 2012 was different. (b) No, histogram has left skewness.

11.39 Will have less power than confidence interval for difference.

11.41 (a) Lower bound is far above 64. (b) The confidence interval has a very large margin of error. Observe that the true deaths due to Hurricane Maria may be any value in the confidence interval: 4645 is not more likely than any other value in the range.

## Chapter 12

12.1 (a) Restaurants customers. (b) Restaurant score. (c) Terminal. (d) Terminals 1, 2, 3, and 4.

12.3 Observations of the response variable are independent, variances of each population are equal, and response variable observations come from a normal distribution.

12.5

| Source | DF | SS | MS | F | p-Value |
|---|---|---|---|---|---|
| Factor *A* | 4 | 113.12 | 28.28 | 1.63 | 0.188 |
| Error | 36 | 624.57 | 17.35 | | |
| Total | 40 | 737.69 | | | |

12.7 (a)

| Source | DF | SS | MS | F | p-Value |
|--------|-----|--------|-------|-------|---------|
| Factor $A$ | 3 | 2316 | 772 | 40.74 | 0.0038 |
| Error | 12 | 397.93 | 18.95 | | |
| Total | 15 | 2713.93 | | | |

(b) 4. (c) 180.93. (d) $H_o$: $\mu_1 = \mu_2 = \mu_3 = \mu_4$; $H_1$: at least two of the population means are not equal. (e) $p$-Value $= 0.0038 < 0.10$, Reject $H_o$. At 10% significance, there is evidence that at least two of the population means are not equal.

12.9 $p$-Value $= 0.017\ 329 < 0.05$, Reject $H_o$. At 5% significance, there is evidence that at least two of the population means on course grades in different programs are not equal.

12.11 (a) Confidence interval for the difference in mean score of terminals 3 and 4 includes zero. Thus, those restaurants statistically have the same mean score. (b) $H_o$: $\mu_3 - \mu_4 = 0$, $H_1$: $\mu_3 - \mu_4 \neq 0$. (c) No, since confidence interval for difference between $\mu_4$ and $\mu_3$ includes zero.

12.13 (a) $p$-Value $= 0 < 0.05$, Reject $H_o$. At 5% significance, there is evidence that at least two of the population means on incident response time in different time of day are not equal. (b) The mean incident response time on time of day $(12, 18]$ is statistically different than other times of day (greater). The means incident response time on time of days $[0, 6]$ and $(6, 12]$ are statistically equal. (c) Residuals do not follow a normal distribution.

12.15 (a) Yes, $p$-value $= 0.026 < 0.05$, Reject $H_o$. At 5% significance, there is an interaction effect among factor $A$ and factor $B$. (b) Yes, because interaction is significant. (c) Yes, because interaction is significant.

12.17 (a) Yes, $p$-value $= 0 < 0.05$, Reject $H_o$. At 5% significance, there is an interaction effect among marijuana product types and year. (b) Yes. (c) Yes. (d) No, interaction is significant.

12.19 (a) 3. (b) 2. (c) Yes. (d) Technically does not matter since interaction is signficant, but factor $A$ on its own does not appear to have an impact.

12.21 (a) $0.063 > 0.05$, Do not reject the null. (b) $0 < 0.05$, Reject the null. (c) $0 < 0.05$, Reject the null.

12.23 (a) Company customers; customer satisfaction scores; customer centers; four centers. (b) $\mu_i$ = mean customer satisfaction of center $i, i = 1, 2, 3, 4$; $H_o: \mu_1 = \mu_2 = \mu_3 = \mu_4$; $H_1$: at least two of the population means are not equal. (c) Mean score of center 3 has lower mean than rest. (d) No.

12.25 (a) $H_o: \mu_1 = \mu_3$; $H_1: \mu_1 \neq \mu_3$. Reject $H_o$. At 10% significance, the mean score of center 1 is different from center 3. (b) Center 3 has the lowest mean score.

12.27 Two-way ANOVA is more powerful, more likely to reject the null when it should than running two one-way ANOVA.

12.29 $0.068 > 0.05$. At 5% significance, it is concluded that there is no difference in mean delivery times.

12.31 (a) Yes. (b) No. (c) No.

## Chapter 13

13.1 Negative.

13.3 $E(Y)$ increases.

13.5 (a) It is estimated that every unit increase in $X$ increases the expected value of $Y$ by 11.6 units. (b) $\hat{Y} = 5.068$.

13.7 (a) Negative. (b) Positive. (c) Positive. (d) Close to zero. (e) Close to zero. (f) Negative.

13.9 (a) The asset's expected return is 44% higher than the market. (b) The asset's expected return is 71% smaller than the market (or 29% the expected market return). (c) The asset's expected return is 17%, and in the opposite direction as the expected market return. (d) Company C.

13.11 (a) Close to 1. (b) For every unit increase in model price, the expected value of selling price increases by 0.98 units. (c) Since it includes 1, the new model will have a good performance. (d) $255 102.09.

13.13 (a) $H_o: \beta_1 = 0$; $H_1: \beta_1 \neq 0$. (b) Do not reject $H_o$. (c) Do not reject $H_o$. (d) 4. (e) The average monthly transportation explains 0% of the variability in median home value. (f) 0.

**13.15** We are 99% confident that the slope coefficient of the linear regression model of affordability index of mortgages and rent is between (0.0907, 0.6473).

**13.17** (a) $r^2 = 0.0262$. The average monthly transportation cost explains 2.62% of the variability in median rent value. The remaining 97.38% is explained by the random variation component of the model. (b) $r = 0.1621$. There is a weak positive linear association between the average monthly transportation cost and median rent value.

**13.19** (b) Adjusted Sale Price explains 99.82% of the variation in Gross Sale Price.

**13.21** (a) 38.563. (b) We are 95% confident that the mean affordability index for mortgages (affordm15) is between (36.550, 40.576) when affordability index for rent (affordr15) is 53.8. (c) We are 95% confident that for a randomly chosen community, affordability index for mortgages will fall within (23.90, 53.23) when affordability index for rent is 53.8. (d) The model does not explain much of the variability in the response (pretty small $r^2 = 15.98\%$). It is a poor model for prediction.

**13.23** (b) $p$-Value $= 0 < 0.01$. Thus, the slope of drains 10 is statistically different to zero. (c) 99% CI: (84.95, 163.39), 99% PI: $(-102.45, 350.79)$.

**13.25** (b) $p$-Value $= 0 < 0.05$. Thus, the slope is statistically different to zero. (c) 95% CI: (23 216.70, 24 158.70), 95% PI: (18 996.60, 28 378.80).

**13.27** The example in this chapter is a two-sided test, $p$-value is twice as large.

**13.29** The normal probability plot looks good, thus no concern about the errors being normal. A visual inspection of histogram reaffirms normality. The standardized residuals versus fitted values does not present any major pattern, no strong evidence against the variance of errors being constant. Finally, the standardized residuals versus order indicates that errors are independent.

**13.31** Normality assumption appears questionable. The standardized residuals versus fitted values present funnel shape, thus the variance of errors may not be constant.

**13.35** $Y$ stands for the raw observation, while $\hat{Y}$ is the estimate of $E(Y)$ under the linear regression model.

13.37  5.7041.

13.39  (a) $H_o$: $\beta_1 = 0$; $H_1$: $\beta_1 \neq 0$. (b) $p$-Value $= 0 < 0.05$, Reject $H_o$. (c) Actual Score $= 18.63 + 0.7629*$Guessed Score. (d) 83.48. (e) The Guessed Score explains 75.94% of the variability in Actual Score. The remaining 24.06% is explained by the random variation component of the model. (f) $r = 0.8714$. There is strong positive linear association between the Guessed Score and Actual Score.

13.41  (a) $H_o$: $\beta_1 = 0$; $H_1$: $\beta_1 \neq 0$. (b) $p$-Value $= 0.227 > 0.01$, Do not reject $H_o$. (c) The number of absences explains 5.77% of the variability in the final grade. (d) $r = 0.2402$. There is a weak negative linear association between final grade and number of absences.

13.43  Religion or going to church does not necessarily cause obesity among young adults, since correlation does not imply causation.

13.45  The maximum hours playing video games in the data set was 19 hours. Estimating the starting salary when the predictor is 30 assumes the association between the variables continues to be approximately linear outside the range of predictor values used to construct the regression model.

13.49  (a) Score has NDA that must be removed. After removal, the scatterplot does not present any kind of association. (b) $p$-Value $= 0 < 0.05$, Reject $H_o$. The linear regression model is significant. (c) The teacher score explains 5.73% of the variability in the rate of misconduct. (d) $r = -0.2398$. There is weak negative linear association between rate of misconduct and teacher score. (e) Rate of misconduct per 100 students is 11.07 for a teacher score of 75. Prediction is not reliable because, despite the slope being significant, $r^2$ is low. (f) The histogram is right skewed, indicating the normality assumption does not hold. (g) Checks how closely estimates are to observations. Lots of variability in the scatterplot, and model estimates are not too reliable.

13.51  (a) Scatterplot does not show linear association, and points are spread out. (b) $p$-Value $= 0.131 > 0.05$, the slope is not significant. (c) The residuals follow a normal distribution, but unclear if error variance is constant.

13.53  The new model is statistically significant, the residuals look better, and normality and constant variance now hold, but $r^2 = 0.07$ is still low.

## Chapter 14

14.1 (a) $\hat{Y} = 6.79 + 3.97X_1 + 1.48X_2 - 2.85X_3$. (b) \$3970 is the estimated expected increase in revenue, associated to an increase of one-thousand dollars in expenditure in web advertising, while expenditure in print advertising and average surfboard volume are kept fixed. (c) \$2850 is the estimated expected decrease in revenue, associated to an increase of one liter on average surfboard volume for the week while expenditure in web and print advertising are kept fixed.

14.3 (b) Apparently there is a weak or moderate positive linear association between the two variables. (c) $\widehat{gas} = -1.24 + 0.380delivery + 0.108miles + 0.016cargoweight$. Expected gas consumptions increases by 0.0526 gallons when cargoweight increases one unit, while delivery and miles remain fixed. (d) $R^2 = 0.765$, the model explains 76.50% of the variability in gas consumption. (e) Model predicts that 9.31 gallons of gasoline will be consumed by the delivery truck.

14.5 (a) The model explains 15.8% of the variability in college freshmen GPA. (b) Not much better than using the response mean to predict values (the error using $\hat{y}$ to predict $y$ is only 15.8% smaller than the error using $\bar{y}$ to predict $y$).

14.7 No, coefficient estimates are a function of scale of measurements made.

14.9 0 and 1. $\hat{Y}$ can not perform worse than $\bar{y}$, so it falls between 0 and 1.

14.11 (a) 3. (b) $H_o$: $\beta_1 = \beta_2 = \beta_3 = 0$; $H_1$: at least one coefficient is not zero. (c) $p$-Value $= 0 < 0.05$, Reject $H_o$.

14.13 (a) $p$-Value $= 0 < 0.05$, Reject $H_o$. At 5% significance, we conclude that at least one of the coefficients differ from zero. (b) All $p$-values are less than 0.05, and the predictors are statistically significant, thus none should be removed.

14.15 $\beta_1$ and $\beta_2$.

14.17 (a) 307 04.01. According to the model, the predicted income per capita, when the percent below poverty is 11.6% and percent aged 25 without high school diploma is 19.3%, is \$30 704.01. (b) 10 370.8. According to the model, the predicted income per capita, when the percent below poverty is 20.5% and percent aged 25 without high school diploma is 41.6%, is \$10 370.8. (c) −3786.3. According to the model, the predicted income per capita, when the percent below poverty is 30.7% and percent aged 25 without high school diploma is 54.8%, is −$3786.3. Since income is positive, this prediction is unreliable.

14.19 (b) Data errors it seems. Also, *dai* are rounded. *ai* equation results are not rounded.

14.21 Because, if $n/k$ is too small, the power of the tests become low and $r^2$ becomes untrustworthy. The higher $n/k$, the better.

14.23 (a) $R^2 = 66.82$, $R_a^2 = 63.06$. (b) $R^2 = 63.39$, $R_a^2 = 62.48$. (c) The model with six predictors, since it has the highest $R_a^2$ of all the models.

14.25 (a) Predictors appear to be correlated among each other except carpool13.

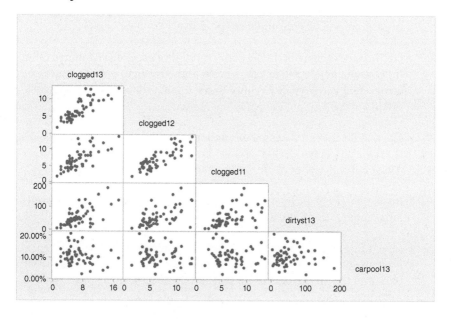

(b) carpool13 is insignificant. (c)

| Model | $R^2$ | $R_a^2$ |
|---|---|---|
| Without carpool13 | 72.51 | 70.92 |
| With carpool13 | 73.50 | 71.42 |

14.27 (a) With 95% confidence, the mean rate of clogged storm drain reports *clogged*13 is between (5.264, 6.361), and for a randomly chosen clogged storm drain, *clogged*13 will fall within (1.932, 9.694) when clogged12= 5.5, dirtyst13= 45. (b) With 95% confidence, the mean rate of clogged storm drain reports *clogged*13 is between (18.126, 26.025), and for a randomly chosen clogged storm drain, *clogged*13 will fall within (16.565, 27.585) when clogged12= 25, dirtyst13= 72.

14.31 Estimated intercept = 2.84 is value of the college freshmen GPA when the predictors high school GPA and combined math and critical reading SAT scores are equal to their mean values.

14.33 (a) Type of crime, and whether convict has children. (b) 8.

14.35 (a) Moisture percentage. (b) 1. (c) 5. (d) 2.

14.37 23.28 gallons.

14.39 (a) $H_o$: $\beta_1 = \beta_2 = 0$; $H_1$: at least one coefficient is not zero. (b) $p$-Value $= 0 < 0.01$, Reject $H_o$. (c) $R^2 = 0.2399$, the model explains 23.99% of the variability in the final course grade. (d) $R = 0.4898$. The correlation between final course grade and estimated final course grade is 0.4898. (e) $\widehat{grade} = 73.12 + 7.66sex$. (f) $\widehat{grade} = 80.78$.

14.41 (a) If the cheese type is hard, the moisture effect on cheese fat percentage is adjused by 0.67:
$\widehat{fat} = 37.86 - 0.23Moisture - 1.28Organic$.
(b) 31.62. (c) The model explains 57% of the variability in the fat percentage of cheese.

14.43 Mainly because slope estimates are dependent on scale of measurements.

14.47 (b) A linear association between *dirtyst*12 and the predictors *dirtyst*11, *clogged*12 is apparent. Also, these two predictors appear to be weakly or moderately correlated.

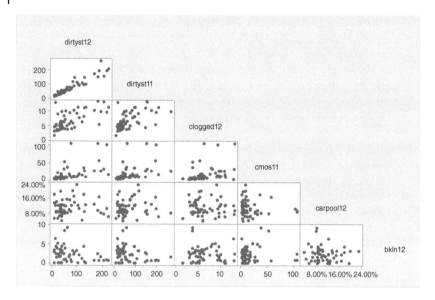

(c) $p$-Value $= 0 < 0.05$, at least one nonzero coefficient. (d) Based on $t$-tests, all variables are significant except *carpool*12. $\widehat{dirtyst}12 = 2.67 + 0.76dirtyst11 + 2.81clogged12 + 0.37cmos11 - 3.09bkln12$. (e) $R^2 = 0.928$, the model explains 92.8% of the variability in the rate of dirty streets *dirtyst*12. (f) 56.36. (g) Histogram indicates standardized residuals follow approximately a normal distribution. (h) Plot may have funnel form, but pattern may be due to sparseness of the data. (i) Association is approximately linear. Checks how well model does and outliers.

14.49 (a) There are some negative exponential associations between response and predictor variables, and also multicollinearity is present. (b) $H_o$:  $\beta_1 = \cdots = \beta_5 = 0$;  $H_1$:  at least one coefficient is not zero. (c) $p$-Value $= 0 < 0.05$, the regression model is statistically significant. (d) $\widehat{INC} = 76\,042 - 409UNEMP - 598NO\_HS - 896U18O64$. (e) This model has better $R_a^2$. We are using new and additional predictors. (f) $R^2 = 0.785$, the model explains 78.5% of the variability in the Per Capita Income.

14.51 (a) $p$-Value $= 0 < 0.05$ the regression model is statistically significant. (b) The model explains 99.82% of the variability in the Gross Sale Price.

14.53 (a) Many hypothesis tests would be required; making hypothesis testing errors likely. (b) $R^2$ stays the same or increases when more predictors are added to the model. Moreover, including unnecessary predictors in the model may cause overfitting.

## Chapter 15

15.1 (a) $H_o$: Ride-sharing use and gender are independent; $H_1$: Ride-sharing use and gender are dependent. (b) No, all expected cell counts are greater than 5. (c) Do not reject $H_o$, since $p$-value = 0.678 > 0.05. (d) At 5% significance, we conclude that ride-sharing use and gender are independent.

15.3 $p$-Value = 1 > 0.05, $H_o$ not rejected. At 5% significance, the quality rating is independent of the work shift at which it was produced.

15.5 $p$-Value = 0 < 0.05, Reject $H_o$. At 5% significance, the classification of alcohol related deaths is dependent of the gender of the person.

15.9 $p$-Value = 0 < 0.05, Reject $H_o$. At 5% significance, the vehicle type is dependent of whether vehicle was leased or not.

## Chapter 16

16.1 $p$-Value = 0.055 > 0.01, Do not reject the null. At 1% significance, we conclude that the Acapulcoite meteorites have a median mass of at most 50 g.

16.3 $p$-Value = 0 < 0.05, Reject $H_o$. At 5% significance, the median mass of Pallasite meteorites is different to 100 g. Wilcoxon Sign-Rank test assumes the distribution is symmetric. Mass of Pallasite meteorites appear to be asymmetric.

16.5 $p$-Value = 0.003 ≤ 0.10, Reject $H_o$. At 10% significance, we conclude that the ad campaign increased sales. The assumptions are the differences are independent and have a symmetric distribution.

16.7 (a) Yes, the null hypothesis is Rejected. (b) At 1% significance, the median answer from the younger group (44 or less) is higher than the median answer from the mature group. (c) Assuming samples are random and independent, and responses from younger people have the same shape as mature respondents.

16.9 (a) $H_o$: Median of Chattanoogan men rating of the city as a place to live is equal to Median of Chattanoogan women rating of the city as a place to live. $H_1$: Median of Chattanoogan men rating of the city as a place to live is greater than Median of Chattanoogan women rating of the city as a place to live. (b) 0.0823 > 0.05, the null hypothesis is not rejected. (c) At 5% significance, the median rating of Chattanoogan men of the city as a place to live is equal to the median rating of Chattanoogan women of the city as a place to live.

**16.13** Parametric: $Z$-test for $\mu$, $Z$-test for $\pi$, $t$-test for difference in $\mu_1$ and $\mu_2$. Nonparametric: Kruskal–Wallis test, Mann–Whitney test, and Wilcoxon Rank-Sum Test.

## Chapter 17

**17.1**  There is a decreasing trend, no seasonality.

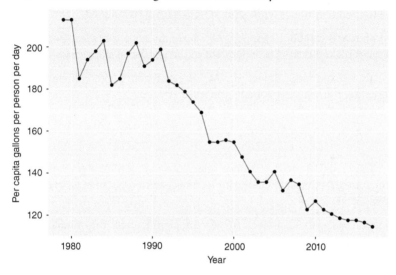

**17.3**  (a) No trend seen, possible seasonality.

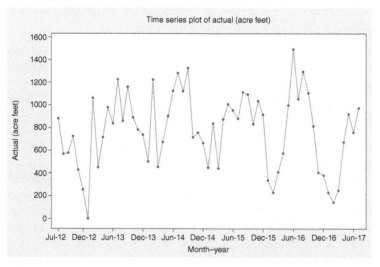

(b) Yes, but lots of variability, especially February.

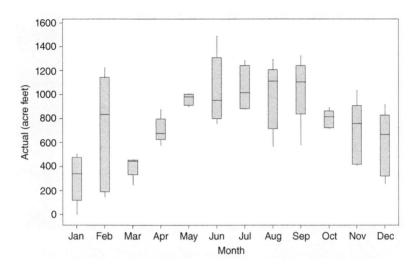

17.5

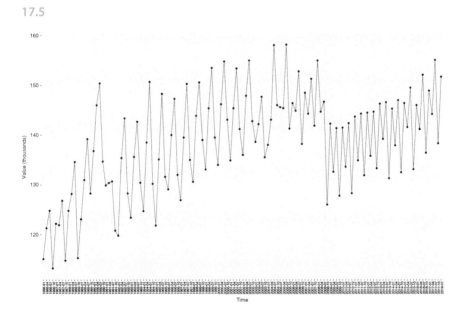

17.7 (a) $109 303. (b) $109 303. (c) Not good, since there is probably seasonal movements in snowboard sales.

**17.9** (a)

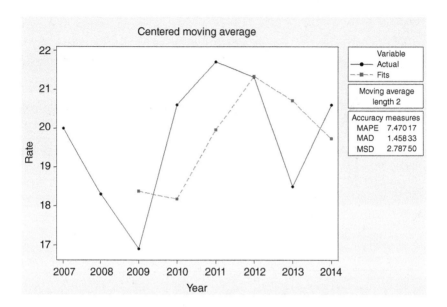

(b)

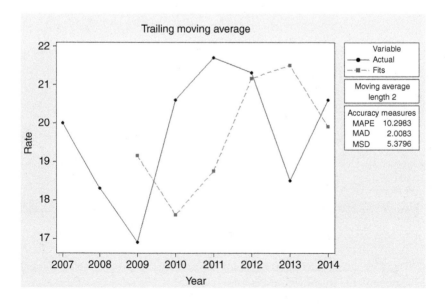

**17.11**

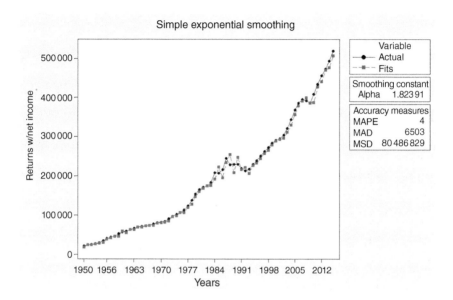

**17.13 (a)**

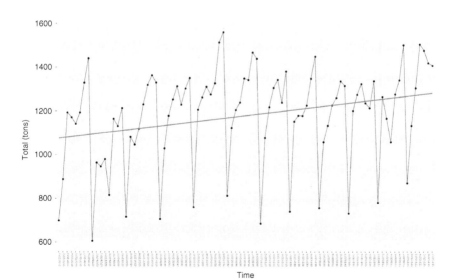

(b)

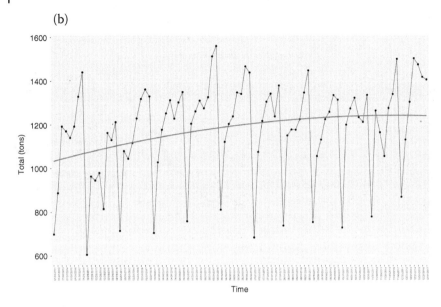

(c) Linear.

17.15  36 161.94, 36 311.97, 36 462.55.

17.17  2050.7 pounds.

17.19

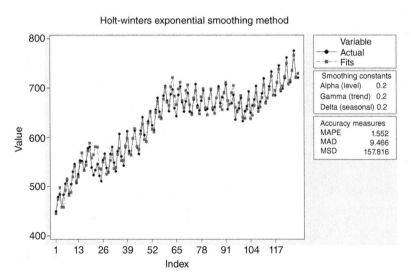

17.21 (a)

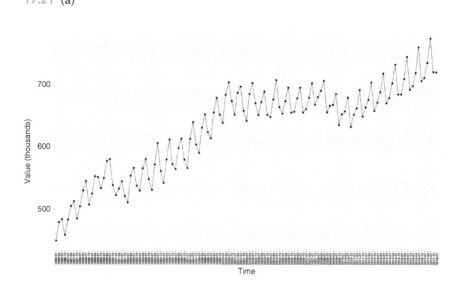

(b)

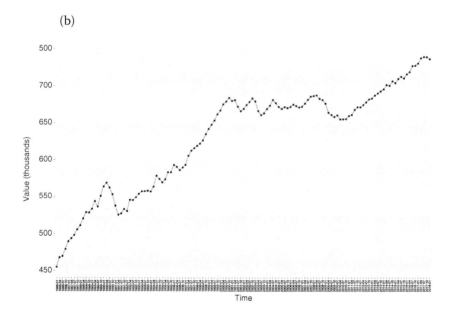

(c) Yes.

17.23 (a)

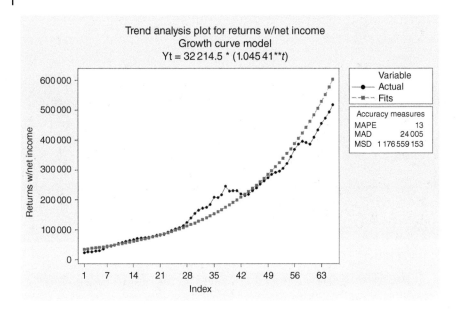

(b)

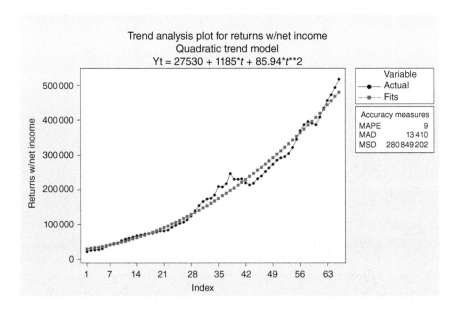

(c) Quadratic looks more random.

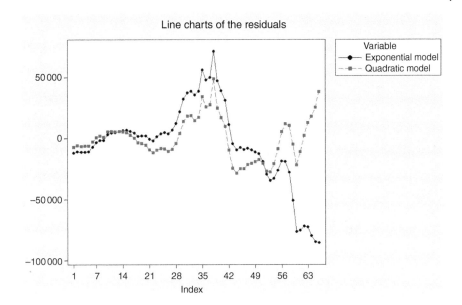

17.25 (a)

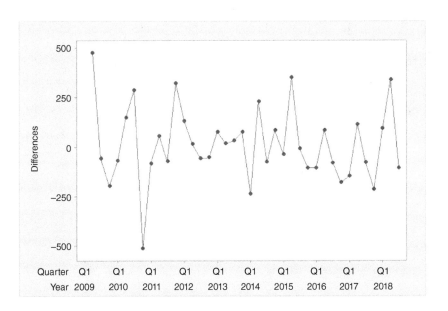

(b) No.

**17.27** (a)

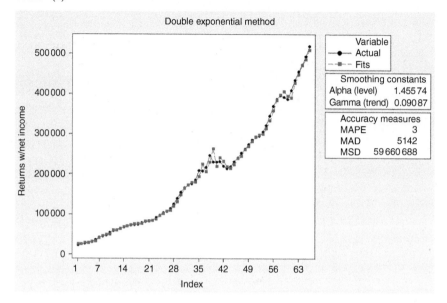

(b) Because data has trend.

**17.29** (a)

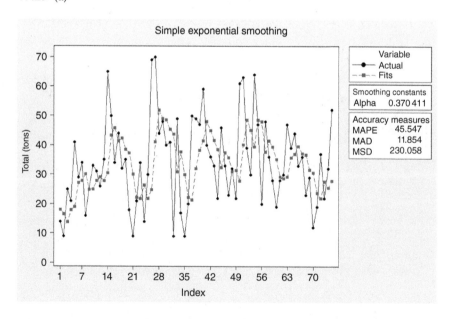

(b)

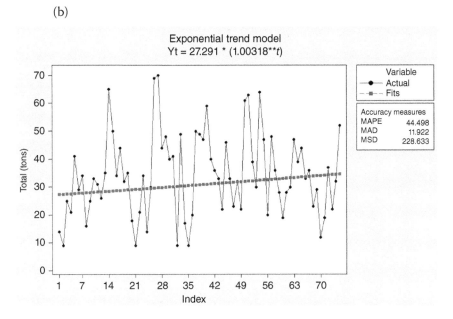

Exponential trend model
Yt = 27.291 * (1.00318**t)

(c) Simple exponential smoothing looks better.

(a)

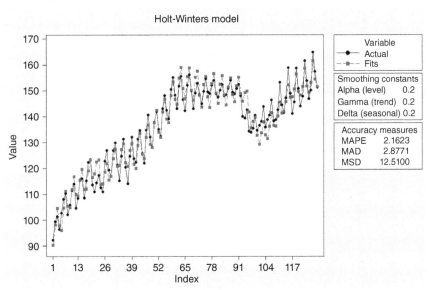

Holt-Winters model

(b)

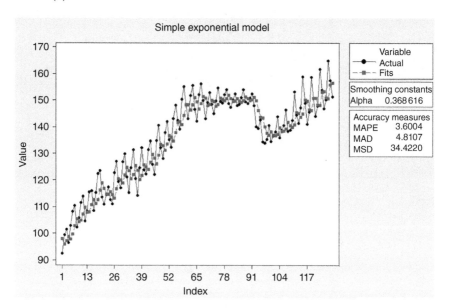

17.33 (a)

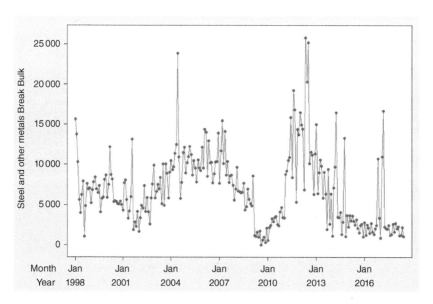

(b)

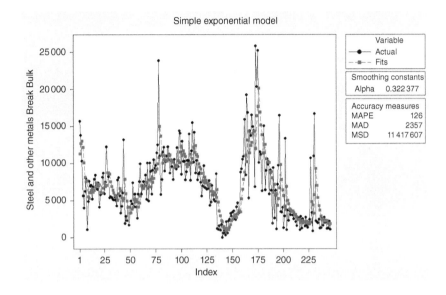

(c)

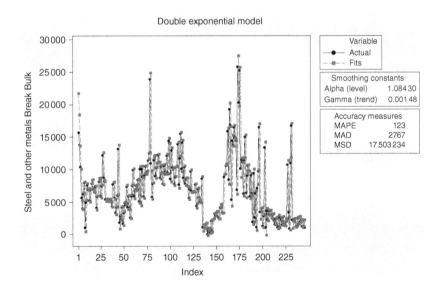

(d)

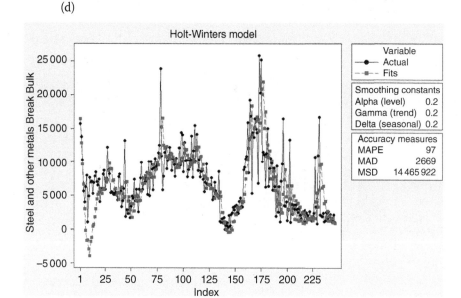

(e) Based on model errors, simple smooth method looks more homoscedastic and random.

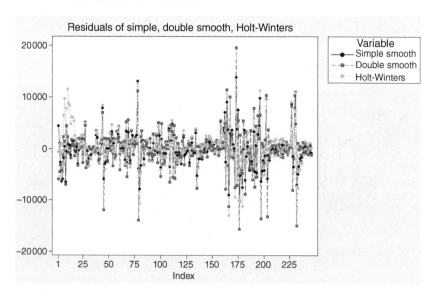

(f) Because in-sample errors may not estimate forecasting errors well. Better to use out of sample errors.

# Index

*Principles of Managerial Statistics and Data Science*, First Edition. Roberto Rivera.
© 2020 John Wiley & Sons, Inc. Published 2020 by John Wiley & Sons, Inc.
Companion website: www.wiley.com/go/principlesmanagerialstatisticsdatascience